EARLY DESERT FARMING AND IRRIGATION SETTLEMENTS: ARCHAEOLOGICAL INVESTIGATIONS IN THE PHOENIX SKY HARBOR CENTER VOLUME 2: DUTCH CANAL RUIN

Edited and compiled by

DAVID H. GREENWALD
M. ZYNIECKI
DAWN M. GREENWAL

With contributions by

RICHARD V. N. AHLSTROM
MARK L. CHENAULT
LINDA SCOTT CUMMINGS
DAVID H. GREENWALD
DAWN M. GREENWALD
SARAH L. HORTON
SCOTT KWIATKOWSKI
KATHLEEN S. MCQUESTION
THOMAS N. MOTSINGER
KIMBERLY SPURR
SUSAN K. STRATTON
KATHRYN L. WULLSTEIN
M. ZYNIECKI

SWCA, Inc.,
Environmental Consultants
Flagstaff and Tucson

SWCA Anthropological Research Paper Number 4

1994

General series editor - Robert C. Euler
Assistant editor - Richard V. N. Ahlstrom
Copy editor - Sally P. Bennett
Cover design - Christina Watkins

1994
SWCA, Inc., Environmental Consultants
Flagstaff and Tucson, Arizona

ISBN 1 – 931901 – 06 – 6

TABLE OF CONTENTS

APPENDIXES

List of Figures

List of Tables

PREFACE

This is the second of four volumes prepared for the City of Phoenix by SWCA, Inc., Environmental Consultants on the results of field investigations conducted at the Phoenix Sky Harbor Center. The research design and work plan for Dutch Canal Ruin are described in Volume 1, and synthetic studies and project summaries appear in Volume 4. Volume 2 presents the results of data recovery during May, August, September, and October of 1990 from the field house and farmstead loci and canal alignments constituting Dutch Canal Ruin.

Chapter 1 places Dutch Canal Ruin within the context of the Phoenix Sky Harbor Center and identifies applicable research objectives. Chapters 2 through 11 discuss individually each investigated area within the site; Chapters 12 through 17 present material culture analyses; and Chapters 18 and 19 discuss general conclusions at the site level.

Because Dutch Canal Ruin provided investigators an opportunity to examine extensive sections of canal networks, they focused on canal studies. During the data recovery phase, archaeologists sought to clarify the reasons for Hohokam occupation of the floodplain of the Salt River, a high-risk habitation area, and the relationship between this local network and the remainder of Canal System 2.

We would like to express again our gratitude to the City of Phoenix for affording us the opportunity to participate in the archaeological investigations at the Phoenix Sky Harbor Center. The preservation and understanding of the culture and accomplishments of those who preceded us in the valley of the Salt River depend on the continuing cooperative efforts of citizens, public officials, and scientists who appreciate their significance.

Steven W. Carothers
President, SWCA, Inc., Environmental Consultants

ACKNOWLEDGMENTS

SWCA successfully completed its field investigations, analysis, and report preparation for Dutch Canal Ruin because of the efforts of numerous individuals, and the contributions of each one deserve to be recognized. Dr. Richard V. N. Ahlstrom served as Principal Investigator, contributing helpful insights throughout the field and laboratory phases of the project as well as providing editorial assistance with the volume. The Project Manager was Mr. David H. Greenwald, who oversaw the day-to-day administration of the project and directed the analytical and reporting phases. Mr. Mark Zyniecki and Mr. Mark L. Chenault, Field Supervisors, assisted the Project Manager. During the field phase, Ms. Kathleen McQuestion and Ms. Sarah Horton served as Assistant Field Supervisors. The field crew members were Andy Arias, Lisa Beyer, Lee Black, Devora Block, Ken Boden, Eric Chambers, Rick Crandall, Joe Crary, Jim Harris, Paul Kelsey, Tom Meserli, Tom Motsinger, and David Purcell. The field lab was staffed by Paula McBrayer, Gina Gage, Molly O'Halloran, and Ginger White. Mr. Kirk C. Anderson served as Project Geomorphologist, assisting with canal studies and geomorphic interpretations. Mr. Charles Sternberg assisted with re-establishing the project grid and vertical datums. Royce Horton of Triple H Construction supplied heavy equipment, and Billy Ellenbarger operated the equipment with care and a delicate touch.

At various times, SWCA brought project consultants and specialists to the field to assist with interpretations and to have the opportunity to witness firsthand the relationships and associations of the cultural, natural, and disturbance processes. Mr. Fred Nials, who worked closely with Mr. Kirk Anderson, provided insights into the geomorphic processes associated with Dutch Canal Ruin and Pueblo Salado. Mr. Gary Huckleberry was also invited to participate in addressing geomorphic problems. Mr. Jerry Howard provided insights concerning the prehistoric canals, assisting with interpretations concerning formation processes and chronology. At various times, Mr. Todd Bostwick, City of Phoenix Archaeologist, visited the field, providing helpful suggestions and comparative information regarding the ephemeral nature of many of the features that were investigated by the crews.

Various individuals conducted the in-house analyses: Mr. Tom Motsinger and Ms. Sarah Horton (ceramic analysis); Ms. Dawn Greenwald, Ms. Kimberly Spurr, and Mr. Scott Kuhr (flaked stone analysis); Ms. Dawn Greenwald and Ms. Kathleen McQuestion (ground stone analysis); Ms. Kathryn Wullstein (human remains analysis); and Mr. Kirk Anderson (soil analysis). Others provided consulting services: Dr. Linda Scott Cummings (pollen analysis); Mr. Scott Kwiatkowski (macrobotanical analysis); Mr. Arthur Vokes (shell analysis); Ms. Susan Stratton (faunal analysis); Dr. Jeffrey Eighmy (archaeomagnetic analysis); Dr. Murry Tamers (radiocarbon analysis); and Ms. Lee Fratt (pigment analysis). All analytical needs and data management were coordinated by Ms. Kathleen McQuestion with assistance from Ms. Marianne Marek, Laboratory Director, and Ms. Gina Gage, Assistant Laboratory Director. Mr. Richard Henzler, Ms. Kimberly Spurr, Mr. Mark Cedarholm, Ms. Dawn Greenwald, and Mr. Tom Motsinger were instrumental in developing computer data management systems for the various files created for this project.

The dedication of several individuals made possible the production of the report. Ms. Nancy Mueller and Ms. Jean Ballagh provided their editorial skills. Ms. Margie Smith and Ms. Erica Berk produced the text and tables. Ms. Julie Hutchinson, Mr. Charles Sternberg, and Mr. Mike Stubing provided their drafting skills. The report was submitted to a peer review panel consisting of Drs. Patricia Crown, William Doelle, and Judy Brunson and was further revised and commented on by Mr. Todd Bostwick, Dr. Robert Euler, and Dr. Richard V. N. Ahlstrom. Mr. David Greenwald, Mr. Mark Zyniecki, and Ms. Dawn Greenwald then reviewed all comments, restructured portions of the volume, and thoroughly edited the report. Dr. Steven Carothers provided encouragement, support, and guidance throughout the project. Mr. Robert Wojtan of the Community and Economic Development Department of the City of Phoenix also provided encouragement and advice to project staff.

Undoubtedly, we have overlooked some of those who contributed. To them we apologize, and we hope they do not feel slighted. To all who have been a part of the Phoenix Sky Harbor Center Project, we gratefully acknowledge your contribution.

David H. Greenwald
Project Manager

ABSTRACT

This volume, which focuses on archaeological data recovery efforts at Dutch Canal Ruin, is the second of four prepared for the Phoenix Sky Harbor Center Development Project. Dutch Canal Ruin is located in the southern half of Section 10, Township 1 North, Range 3 East (Volume 1:1). Investigators identified 20 individual loci during the testing phase within the Phoenix Sky Harbor Center at Dutch Canal Ruin and excavated a sample of eight loci (Areas 1 through 8). During the monitoring of the remote parking facility in the eastern portion of the project area, SWCA discovered and excavated additional features and included them in the data recovery analysis (Areas 9 and 10). SWCA thus investigated half of the loci (Areas 1 through 10) in the project area during data recovery.

The majority of the remains that constituted Dutch Canal Ruin, including the canals, were associated with the pre-Classic periods. Investigators assigned 2 loci to the Classic and post-Classic periods and 10 to the pre-Classic periods but concluded that 6 could not be assigned and 1 contained both pre-Classic and Classic period components. Occupation at Dutch Canal Ruin began at seasonal field house loci during the Snaketown phase; at this time the first canals were constructed. Field house settlements and seasonal occupation continued through the Colonial period, with activities related primarily to agriculture. Although investigators identified Sedentary period remains in four loci, the early occupants of Dutch Canal Ruin apparently used the canals only until approximately A.D. 900 and then may have shifted their land-use strategies to the procurement of wild economic resources associated with the previous agricultural disturbance; they eventually allowed the area to return to its natural setting and environment.

The portion of the Phoenix Sky Harbor Center that contained Dutch Canal Ruin appeared to have been abandoned during the Soho phase, possibly as early as the Sacaton phase. The only features discovered in the project area that may have been associated with the Soho phase were four burials, including one from the Squaw Peak Parkway. Two intruded into pre-Classic period canals, one intruded into a pre-Classic period pit structure, and one was found adjacent to a canal. Burial goods from two of the interments suggested a Classic period association, which was supported by stratigraphic data in three cases. Interpreting the temporal placement of these burials was problematic; they may have represented limited use of the area during the Classic period in the form of resource procurement activities. Interment within the abandoned canals and the pit structure may have been a matter of convenience, in that the depressions offered by these features would have facilitated ease of burial.

The Hohokam reoccupied Dutch Canal Ruin during the Civano phase. These later inhabitants constructed more canals through the site area and occupied habitation areas, which remained small and simple, on a permanent basis. These settlements probably never exceeded one or two households, although the density of material culture items suggested that occupation had been intensive (permanent) and perhaps long in duration. At least one habitation locus, Area 8, appeared to have been inhabited during the Polvorón phase of the post-Classic period. The Classic and post-Classic period occupations at Dutch Canal Ruin shared similarities with Pueblo Salado, a Classic and post-Classic period site south of Dutch Canal Ruin that is discussed in Volume 3 of this series. Reoccupation of Dutch Canal Ruin may have resulted from an expansion of the Pueblo Salado settlement and use of the area for its agricultural potential.

In summary, Dutch Canal Ruin was used extensively as an agricultural zone during much of the Hohokam occupation of the Phoenix Basin. Some of the earliest canals within Canal System 2 extended through the site. Inhabitants of nearby village sites, such as La Ciudad, Los Solares, Pueblo Grande, and La Villa/Casa Chica, may have exploited the Dutch Canal Ruin area during the pre-Classic periods because of the deep alluvial soils. The area may have been abandoned as an agricultural zone after the Colonial period because of devastating floods. During the Sacaton and Soho phases, Dutch Canal Ruin may have been used casually for the exploitation of wild plant resources and hunting. The Hohokam established a permanent occupation during the Civano phase and extended their occupation into the post-Classic period. These later occupants relied on canal irrigation agriculture and wild resources for their subsistence. Dutch Canal Ruin was in an area considered high risk by today's standards, yet the Hohokam persisted in using this area, initially during the late Pioneer and Colonial periods and later during the Classic and post-Classic periods. The investigations at this site have provided information on the types of land-use patterns and settlement strategies used by the Hohokam during these periods on the geologic floodplain of the Salt River and have added to the growing data base concerning the Hohokam of the Phoenix Basin.

Although this volume focuses on descriptive statistics and data presentation for the 10 investigated areas of Dutch Canal Ruin, it also presents detailed analyses of the site's ceramics, flaked and ground stone, and botanical and faunal assemblages; a reconstruction of site chronology; and a discussion of the site within a regional perspective. The appendixes provide raw data from various analyses. For further interpretive and synthetic information, see Volume 4.

CHAPTER 1

DUTCH CANAL RUIN:
FIELD HOUSE AND FARMSTEAD IRRIGATION SETTLEMENTS

David H. Greenwald

INTRODUCTION

Development of the Phoenix Sky Harbor Center included the archaeological investigation of two sites: Dutch Canal Ruin (AZ T:12:62, ASM) and Pueblo Salado (AZ T:12:47, ASM) (Figure 1.1). SWCA, Inc., Environmental Consultants conducted data recovery at these two sites for the City of Phoenix during May, August, September, and October of 1990. This volume focuses on the investigative results at Dutch Canal Ruin, a scatter of field house and farmstead loci dispersed along a network of canal alignments (refer to Volume 1 for a discussion of the project and information on previous studies in the project area). Volume 3 concentrates on Pueblo Salado, a Classic period site that contained a compound and scattered hamlets.

The Phoenix Sky Harbor Center is located immediately west of Phoenix Sky Harbor International Airport, and the Dutch Canal Ruin site is in line with the northern runway (T1N, R3E, Sec. 10). After establishing the townsite of Phoenix in the 1860s, Euroamerican settlers farmed in the project area, which was southeast of the townsite. Accelerated growth followed World War II, and housing tracts slowly replaced agriculture as the principal use of the area. Commercial properties lined such main roadways as 16th, 20th, and 24th streets and Buckeye Road and were also adjacent to the Southern Pacific Railroad tracks that form the northern project boundary. Even with the agricultural activity in this portion of the project area, Hohokam remains were still visible on the surface in the 1930s at the time of the first archaeological recording (Midvale 1934, n.d.).

This study indicated that the Hohokam had used Dutch Canal Ruin as a farming locale during the late Pioneer and Colonial periods and again during the late Classic period (Chapter 18; Dean 1991:Table 3.3 and Figure 3.6). They constructed irrigation canals through the site as early as A.D. 600, the late Pioneer period, typically building field houses adjacent to the canals. They built and modified several canals during the pre-Classic periods, until approximately A.D. 900, the Santa Cruz phase, when they abandoned the first terrace as an agricultural zone. Before abandoning Dutch Canal Ruin, the farmers expanded their canal networks and farm systems onto the lower bajada or second terrace north of the Salt River. Evidence suggested that the Hohokam used Dutch Canal Ruin during the subsequent Sacaton and Soho phases on a limited basis, possibly as a resource exploitation zone accessed from settlements on the lower bajada. Farmers re-established canals at Dutch Canal Ruin during the Civano phase (A.D. 1300–1350) and built small, permanent settlements scattered along the routes of the canals. The post-Classic period (after A.D. 1350) was represented by a single settlement in the west-central portion of the site.

Dutch Canal Ruin, as recorded in archival documents, extended from northeast of the intersection of 24th Street and the Southern Pacific Railroad tracks southwest to approximately 14th Street north of Buckeye Road (Figure 1.1), an area of 2.1 km (slightly over 1 mile) east-west by 0.8 km (nearly 0.5 miles) north-south. The entire portion of the site within the Phoenix Sky Harbor Center boundaries was available for study from 24th Street west to 16th Street and from Buckeye Road north to the Southern Pacific Railroad tracks except for two small areas, one occupied by a church and the other by a commercial building, and the Squaw Peak Parkway corridor and its access into Phoenix Sky Harbor International Airport. Maximum dimensions of the site within the

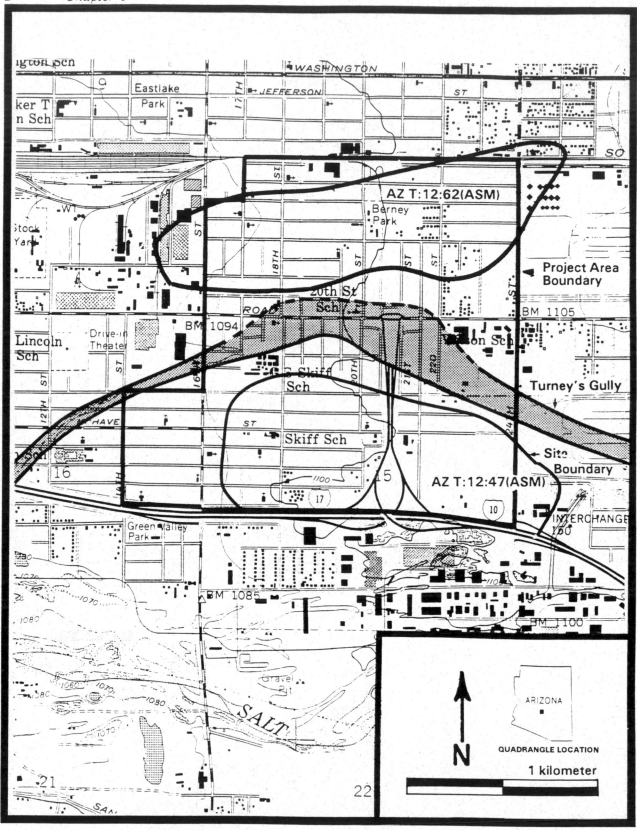

Figure 1.1. Phoenix Sky Harbor Center Project boundaries and site boundaries of Dutch Canal Ruin (AZ T:12:62, ASM) and Pueblo Salado (AZ T:12:47, ASM). Base map is Phoenix, Arizona, 7.5′ USGS topographic map, photo revised 1982.

project area were 1600 m east-west by 500 m north-south, or 800,000 m² (72 ha). Archaeologists excavated 7278 m² and monitored and excavated another 5980 m² during construction, amounting to 1.66% of the available site area. Investigators examined approximately half of all the cultural remains located during testing. The wide spatial distribution of the remains accounted for the apparently low percentage of the project area investigated. Table 1.1 summarizes the type of settlement and temporal affiliation for each of the investigated areas.

Table 1.1. Investigated Areas and Their Temporal Affiliations

Area Number	Type of Research Area	Temporal Affiliation
1	Field house locus	Colonial period, Santa Cruz phase
2	Field house locus	Colonial period, Gila Butte phase
3	Multiple field house/ farmstead loci	Colonial period
4	Limited activity/resource procurement and processing	Pre-Classic periods
5	Field house	Late Colonial/possibly early Sedentary period
6	Canals	Pioneer period, Snaketown phase/ Colonial period
7	Field house locus/canals	Pre-Classic periods: Colonial period, Santa Cruz phase Classic period, Civano phase
8	Hamlet	Post-Classic period, Polvorón phase
9	Field house/farmstead	Early Colonial period through late Classic period
10	Field house/farmstead	Colonial period/Classic period, Civano phase

Investigators identified five canal alignments at Dutch Canal Ruin (Figure 1.2; Table 1.2) and assigned them temporally based on the data recovery results, the findings from testing, and the results of earlier excavations within the Squaw Peak Parkway corridor. The South Main Canal (Alignment 8003) appeared to be the earliest canal to have been constructed, lasting from the late Pioneer period into the early Colonial period. Alignment 8005, a small channel originating in the South Main Canal and following a northwesterly trajectory, appeared to predate the North Main Canal (Alignment 8001), since the construction of the North Main Canal channels had destroyed this smaller channel. The North Main Canal alignment consisted of three channels; the earliest two

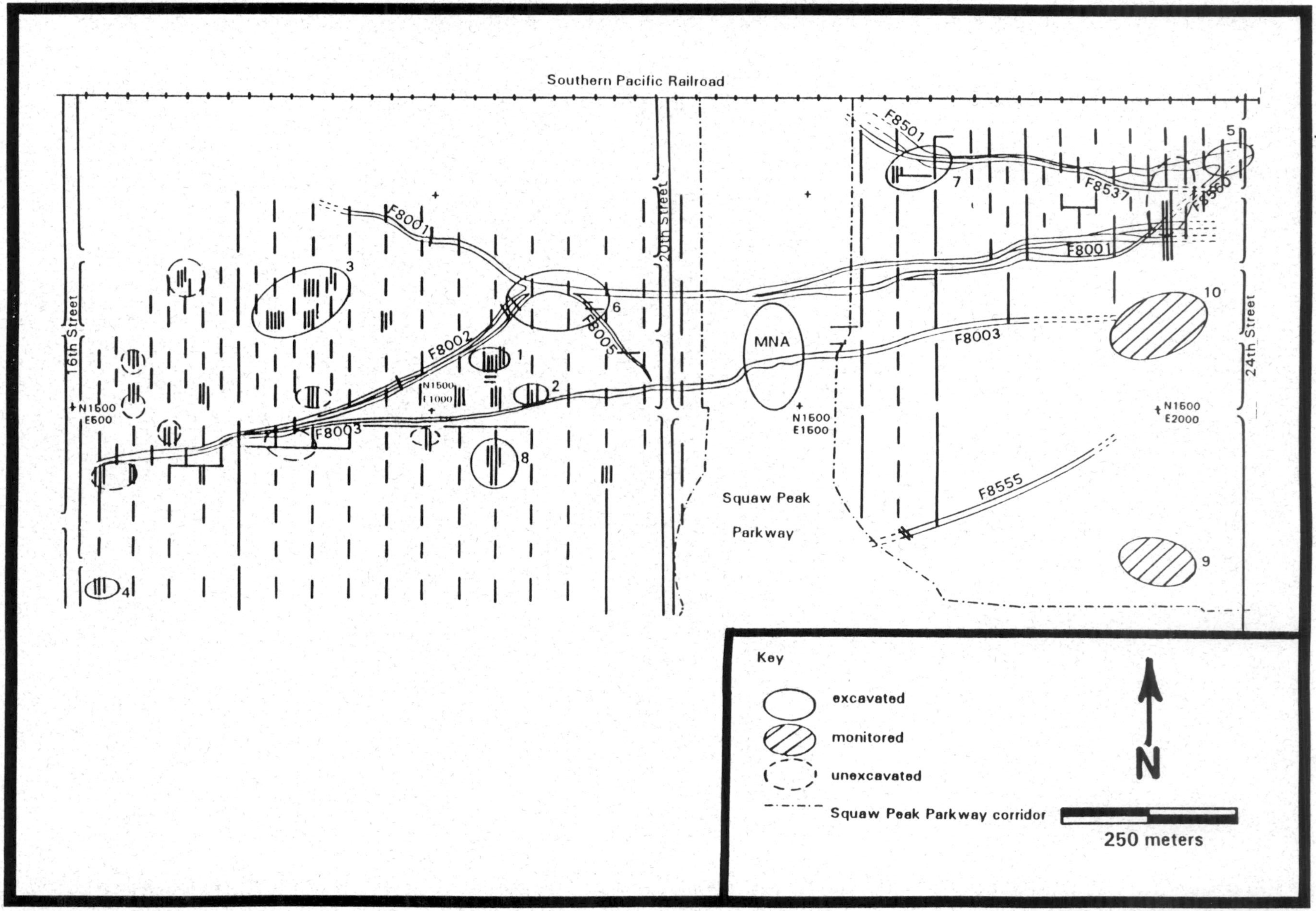

Figure 1.2. Dutch Canal Ruin as defined within the Phoenix Sky Harbor Center project area, showing canal alignments and feature locales. Investigated areas indicated by numbers (e.g., 8); the feature locale excavated by the Museum of Northern Arizona (MNA) is identified in the Squaw Peak Parkway corridor.

Table 1.2. Project Canals and Their Associated Alignment Numbers

Canal Name	Alignment Number
North Main Canal	8001
Southern Branch, North Main Canal	8002
South Main Canal	8003
Small Branch, South Main Canal	8005
Canal Viejo	8501
Canal Nuevo	8537
Canal Barranca	8555
Crosscut between 8001 and 8501 (North Main Canal)	8560

may have been built by the end of the Pioneer period and operated into the early Colonial period. North Main Canal 3, the latest in this group, dated to the Colonial period. North Main Canal 3 also connected the North and South Main Canal alignments in the western portion of the project area (Alignment 8002) and the North Main Canal alignment with Canal Viejo (Alignment 8501) in the eastern portion (Alignment 8560). North Main Canal 3 was constructed within the earlier channels, yet it maintained a more gradual hydraulic gradient. Canal Viejo probably dated to the early Colonial period on the basis of its stratigraphic position relative to that of North Main Canal 3. Canal Nuevo (Alignment 8537) was assigned to the late Classic period. The temporal placement of Canal Barranca (Alignment 8555) was somewhat problematic, in that field crews did not recover chronometric data from that channel. It may have dated to the Classic period.

Frank Midvale originally recorded Dutch Canal Ruin in the mid 1930s, although his work was neither complete nor official. Maps drawn by Midvale at that time, which are housed at the Mesa Southwest Museum and in Arizona State University's Anthropology Department, illustrate the locations of burned mounds, historic structures, streets and roadways as they existed at that time; canal alignments; and rough boundaries of the site. On one sketch map, several dots appear in two clusters: one at the east end and the other at the west end of the site. The implications of these dots had never been clear. Some researchers thought the dots represented trash mounds; they considered Dutch Canal Ruin to be a large habitation site. The results of this project suggested that Midvale's dots represented concentrations of artifacts, as their location corresponds with the field house loci documented by SWCA. The burned mound on Midvale's 1930 map (Volume 1:Figure 4.13) probably corresponds with Area 8, the only post-Classic settlement defined at Dutch Canal Ruin.

Dutch Canal Ruin was not assigned an official site number until 1986, when the Museum of Northern Arizona initiated test excavations and subsequent data recovery within the Squaw Peak Parkway (Greenwald and Ciolek-Torrello 1988a). The corridor of the Squaw Peak Parkway (Figure 1.2) bisects Dutch Canal Ruin and the Phoenix Sky Harbor Center into nearly equal east and west halves. The 1986 investigations revealed that the site contained four main canals (these extended through the freeway corridor from east to west) and associated field house structures and activity areas. During the testing and data recovery efforts, SWCA investigators repeatedly observed field house structures associated with canals. This chapter discusses these relationships and observations.

DUTCH CANAL RUIN AND THE PHOENIX SKY HARBOR CENTER

Environmental Setting

Dutch Canal Ruin occupied the northern extent of the first terrace of the geologic floodplain along the north side of the Salt River. Along with the deep alluvial soils on the first terrace, a number of other resources were available to Hohokam groups who used and occupied this portion of the Salt River Valley. Among those resources were cobbles and boulders for stone tools and construction; wood in the form of driftwood and species that populate riverine environments; desert hardwoods such as mesquite and possibly ironwood; faunal resources including small mammals, rodents, waterfowl, amphibians, and fish; and economically important wild plants used for construction, food, tools, and medicine. The location of Dutch Canal Ruin within this environmental zone enabled the Hohokam to exploit a wide variety of resources while practicing canal irrigation agriculture. The resources in the project area were widely used and have been documented as important to the Hohokam economy (Gasser and Kwiatkowski 1991). The Hohokam also utilized upland zones for various species of cactus, but investigators have found scant evidence of these resources in the botanical record from this site (Chapters 15 and 16; Cummings 1988; Ruppé 1988).

Physiography and Geology

The project area is located in the Phoenix Basin, which is within a physiographic region known as the Basin and Range province and characterized by a series of valleys divided by low mountain ranges. The Phoenix Basin lies at the confluence of the Salt and Gila rivers, two of the largest drainages in south-central Arizona. This is an area of broad, gently sloping alluvial valleys, interrupted by abrupt uplifts. The Salt River was a perennial stream prior to the construction of modern impounds along the Salt and Verde rivers. In the arid conditions of this desert region, the combination of arable soils and a permanent water source led to the application of canal irrigation to domesticated crops. The Salt River upstream from Tempe Butte is bordered by a series of terraces labeled the Lehi, Blue Point, Mesa, and Sawick, listed here from youngest to oldest (Volume 1:Table 3.1; Péwé 1978). Downstream from Tempe Butte, where the valley floor widens and gradually slopes toward the foothills that ring the Phoenix Basin, these terraces are indistinguishable; the gradual slope of the project area facilitated canal construction (Turney 1929). With the expansion of Canal System 2, between 20,000 (Howard 1991:5.25) and 30,000 acres (8,100–12,150 ha) became available for irrigation agriculture. Acreage estimates are based on the maximum extent of Canal System 2 (Figure 1.3) as defined by Turney (1929), including the geologic floodplain.

The well-drained soils, low in calcium carbonate, in the vicinity of Dutch Canal Ruin generally exhibit high agricultural potential. They are deep in most places, having been formed in alluvium as a result of periodic flooding. Geologically speaking, these are young soils. Descriptions of all identified soil types found at Dutch Canal Ruin are included in Volume 1 as Appendix B. The character of these soils changed as a direct result of human intervention, principally in the form of irrigation, which increased the clay content in the soils. Levels of organic matter probably also increased as a result of farming activities, and alterations in chemical composition may have occurred as well.

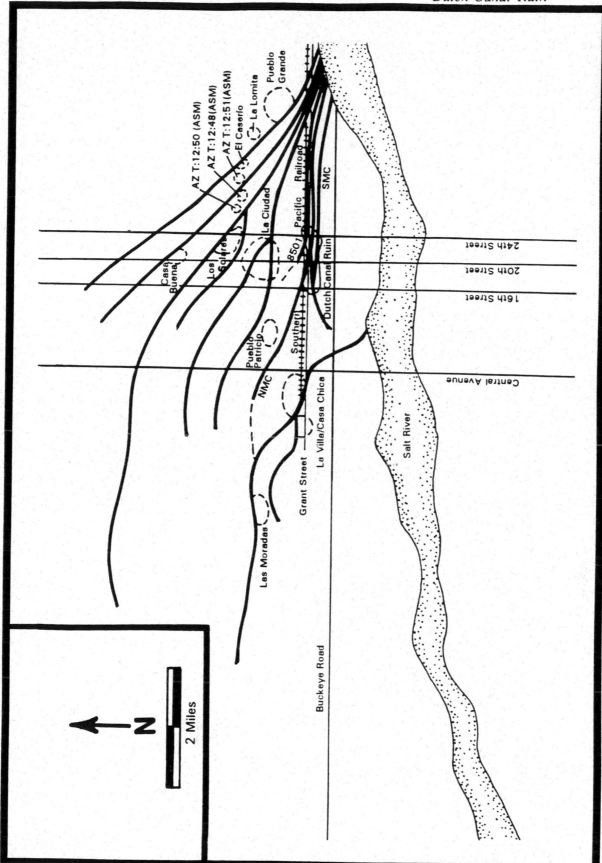

Figure 1.3. Pioneer and Colonial period canals and sites in Canal System 2 (compiled from Turney 1929; Howard 1991; and current investigations [T. Bostwick, personal communication 1992]).

Climate, Rainfall, and Temperature

Winter storms provide the greatest amounts of precipitation to the southern deserts, in the form of cyclonic frontal systems originating over the Pacific Ocean. The lower watershed of the Phoenix Basin often receives its moisture during the winter in the form of rain; snowfall at the higher elevations, which melts during the late winter months and early spring, contributes to elevated streamflows. Summer moisture comes from convection systems, often resulting in violent winds and rainstorms. Flash floods, wind, and hail can cause serious damage to crops and habitation areas, but the associated moisture often augments irrigation.

Streamflow is directly related to this bimodal precipitation pattern. Calculations of seasonal variability, as recorded for the Salt and Verde rivers between 1914 and 1979 (Graybill and Nials 1989), yielded peak flows during the late winter and early spring (February and March). This study determined that streamflow levels increased somewhat during the summer monsoon season and dropped again during the drier fall months.

Average rainfall in the Phoenix Basin is 88.9 mm (3.5 inches) for the summer months and 111.8 mm (4.4 inches) in winter, calculated from 1914 to 1979 (Graybill and Nials 1989:Table 2.3). With an average rainfall of just 200 mm (less than 8.0 inches) per year, successful farming would have required irrigation or the diversion of surface runoff in combination with moisture retention techniques.

Temperature means for the Phoenix Basin range from a summer high of 28°C (82°F) to a winter low of 14°C (58°F). Increases in summer temperatures greatly affect moisture evaporation levels. Soil moisture accumulates during the winter months, when evaporation and heating of the ground surface is low, and is rapidly lost in the high daytime temperatures of the summer months. Today, as in prehistoric times, summer monsoonal regimes do not provide enough moisture to support domestic crops without supplementation of moisture levels.

Vegetation

The Phoenix Basin is located in the Sonoran desertscrub biome, the largest and most arid subdivision within the Sonoran Desert (Turner and Brown 1982). The site area is located in a saltbush community with a transitional zone to a creosotebush-bursage community immediately north of Dutch Canal Ruin (Turner 1974). U.S. General Land Office maps produced during the 1860s indicate that extensive mesquite thickets grew on the first terrace west of the project area. Along drainages and river channels a mixture of riparian species can be found, including cottonwood, willow, mesquite, reeds, annual and perennial grasses, saltbush, and bursage. These types of vegetation may have affected the construction of canals and development of field systems at Dutch Canal Ruin. The Hohokam would generally have preferred open areas over mesquite bosques for two apparent reasons. First, development of fields in mesquite bosques would have required the clearing of trees to allow sufficient space for the fields and associated canals. Second, mesquite has long been recognized as a staple in the Hohokam diet, and the clearing of mesquite bosques would have reduced the availability of an important dietary resource. Saltbush communities also indicate the likelihood of available arable lands (Chapter 15).

Historic and Modern Use of the Dutch Canal Ruin Area

In 1868, early Phoenix settlers constructed their first irrigation canal, the Swilling Ditch, and built field systems and farms along the first terrace and lower bajada (Zarbin 1979, 1980). They finished the Dutch Ditch, a major lateral of the Swilling Ditch, by April 1869; it supplied irrigation water to areas shown on current maps to the south of the Southern Pacific Railroad tracks (Rodgers and Greenwald 1988:28–33). Demands for agricultural produce increased, and the Salt River Project Water Users' Association organized irrigation systems to meet the increased demands. Water users improved the irrigation systems and modified canal alignments to accommodate the increased number of acres under cultivation and the sophisticated irrigation techniques developed by the valley's farmers. That portion of the Dutch Canal Ruin within the project area was intensively farmed until the 1950s, although beginning in the late 1920s housing began to compete with farming for land as housing tracts were established to provide residential space for the small but growing city. Residential construction was slow at first, but following World War II, house construction began to flourish in the project area. By 1952, few lots were available for construction. The main house style was that of the Arizona bungalow. Outbuildings and privies in backyards were common, even after the city installed other sewer facilities.

Street patterns followed the grid developed elsewhere in Phoenix, using section lines for major corridors and quarter-section lines for secondary routes. Residential streets were typically set 300 feet on-center, providing each lot a maximum depth of 150 feet from the center of the street. Primary and secondary streets supported commercial properties, as did the area from 22nd Street east to 24th Street and from Buckeye Road north to Grant Street in the southeastern portion of Dutch Canal Ruin. Neighborhood churches were common landmarks, with schools and parks also present.

During the early 1930s, a series of aerial photographs was taken of the Phoenix Basin (USDA Soil Conservation Service 1934). In the photograph that included the Phoenix Sky Harbor Center area, nearly the entire project area either was under cultivation or contained newly established housing tracts (Volume 1:Figure 4.15). At this same time, Frank Midvale began mapping the site, despite the disturbance from historic farming and residential activities. The current investigations revealed that preservation of the prehistoric resources was good, and information gathered from the project area as a result of this undertaking will expand the understanding of Hohokam land-use strategies and settlement within what is now the metropolitan Phoenix area.

Phoenix Sky Harbor Center and Dutch Canal Ruin

Dutch Canal Ruin was a large site. Approximately 89 ha (220 acres, or, according to Midvale's maps [1934], 80–85% of the site) lay within the Phoenix Sky Harbor Center boundaries (Figure 1.1). The current channel of the Salt River is approximately 1.9 km (1.2 miles) to the south. During the Holocene and late Pleistocene epochs, periodic flooding of the Salt River deposited deep alluvium between the active channel and the second terrace. As the river meandered from north to south, it alternately built up and eroded away the alluvial terraces. The process apparently took thousands of years, although single flood events could have realigned the channel by several hundred feet or resulted in severe erosional downcutting such as occurred during the 1979 flood. Dutch Canal Ruin was susceptible to the effects of flooding, and the Hohokam certainly understood the risks of establishing canals and habitation loci upon the first terrace. This perhaps explains why they established their larger settlements not at Dutch Canal Ruin but a short distance away at an elevation less prone to inundation by the Salt River.

Dutch Canal Ruin was located on the first terrace near its interface with the second terrace. The first terrace represents the geologic floodplain, whereas the second terrace represents the lower bajada slope. The individual river terraces defined by Péwé (1978) upstream from Tempe Butte cannot be identified along the north side of the Salt River in this portion of the Phoenix Basin. Nevertheless, the geology of the project area may include poorly developed remnants of the Blue Point and Lehi terraces.

Data recovery efforts in eight areas at Dutch Canal Ruin emphasized excavation, with monitoring and excavation at two additional loci. Although field crews found cultural remains in other areas of Dutch Canal Ruin, investigators determined that excavation of all feature clusters would have produced redundant information and was not warranted. They therefore focused on a representative sample of all such clusters, including information from the earlier excavations sponsored by the Arizona Department of Transportation within the Squaw Peak Parkway (Greenwald and Ciolek-Torrello 1988a). Figure 1.2 illustrates the location of each investigated area, including the two areas defined through monitoring. Other feature clusters have been indicated on Figure 1.2 to show the distribution of physical remains in the project area.

Because of the extensive historic and modern modifications to the project area, archaeologists could gain only limited information about the physical features that may have existed during the Hohokam occupation at Dutch Canal Ruin. Historic maps and geomorphic studies (Volume 1:Chapter 3 and Figure 3.2) conducted during the testing phase provided information on two floodplain channels that bisected the project area. The northernmost channel had been filled by natural processes prior to historic use of the project area. Archaeological features dating as early as A.D. 600 existed in its upper levels (Chapter 4). Although investigators could not determine the age of this channel, geomorphic evidence indicated that it predated the Hohokam occupation. The second channel was faintly visible on the surface in the western portion of the Phoenix Sky Harbor Center near Buckeye Road and 16th Street, appearing as a topographic change with relief of as much as 1 m. This channel was a recognizable and perhaps prominent surface feature during the historic settlement and subsequent growth of Phoenix. It is indicated on Turney's maps (1924, 1929) as a gully representing a channel of the braided Salt River (Volume 1:Figure 3.2). Referred to as Turney's Gully, this channel effectively formed the southernmost extent of Dutch Canal Ruin.

Although investigations within the Phoenix Sky Harbor Center encompassed an area in excess of 2.6 km^2(1 mi^2), only the southern boundary of Dutch Canal Ruin could confidently be defined. This was due to the linear nature of the site as it followed the canal alignments and because Midvale's records (1934, n.d.) indicated that the site may have extended beyond the eastern and western project boundaries. Canal alignments exited the project area in three locations (Volume 1:Figures 10.1 and 10.2). Alignment 8001 (the North Main Canal) exited at the northwest corner of the project area, near 16th Street and the Southern Pacific Railroad tracks (Buchanan Street). The combined routes of Alignments 8002 and 8003 (the South Main Canal) exited at 16th Street and Sherman Street. The combined routes of Alignments 8501 and 8537 (Canals Viejo and Nuevo) exited the project area at the intersection of the Southern Pacific Railroad tracks and the Squaw Peak Parkway. In each instance, investigators identified habitation features in relation to the canal alignments where the canals exited the project area, as well as along the extreme eastern project boundary. Assuming that Midvale's records were generally accurate and basing their interpretation on the results of the present undertaking, the investigating team concluded that the site extended beyond the project area boundaries. Midvale's original site boundaries could not be confirmed, however, without additional work in those areas. Because field houses were associated with the canals (Figure 1.2) within the project area, habitation features were likely to be dispersed along the canals between large habitation sites, and thus archaeologists may not be able to determine specific boundaries without the presence of natural features such as Turney's Gully.

DUTCH CANAL RUIN AND ITS RELATIONSHIP
TO CANAL SYSTEM 2 AND THE HOHOKAM REGION

Dutch Canal Ruin was a series of field house and farmstead loci scattered along an extensive network of canals (Volume 1:Chapter 1; for discussions refer to Cable and Doyel 1984; Crown 1985a; Gregory 1991:160–163; Masse 1991:198–199). These areas, defined as irrigation settlements (or habitation locales associated with irrigation facilities), were seldom occupied on a permanent basis. Only two such settlements at Dutch Canal Ruin, Areas 8 and 10, appeared to have been permanent. Area 8 dated to the Polvorón phase, and Area 10 dated primarily to the Civano phase.

Irrigation settlements can range from field house loci to villages, with each individual habitation locus considered an irrigation settlement. In the case of Dutch Canal Ruin, SWCA's investigators used the general spatial association of multiple activity loci, or settlements, to define the distribution and variability of the site's components. An irrigation settlement should not be confused with an irrigation community, defined as a set of interrelated sites sharing a common irrigation network (Howard 1987:211). An irrigation settlement may be one component of an irrigation community. Similarly, two or more irrigation communities form community networks, the larger political and economic units of the irrigation agriculture complex (Masse 1991:202). Canal System 2, with its multiple settlements and canal alignments, was a community network that evolved from a simple system located on the geologic floodplain (Figure 1.3) to a complex, multichanneled network of canals, field systems, and habitation locales that included portions of the lower bajada (Cable 1991; Howard 1990).

Few researchers question that the canals associated with Dutch Canal Ruin were some of the earliest canals in Turney's Canal System 2. Residents constructed, operated, and maintained a small main canal, Alignment 8003, during the late Pioneer period. They continued to operate and expand the canal networks at the site into the Colonial period and established additional irrigation settlements, including other field house locales and farmsteads. The first terrace presented the fewest engineering problems and obstacles to the early canal builders (Turney 1929), although steeper grades contributed to erosional problems (Volume 1:Chapter 8; D. H. Greenwald 1988a:84–85). Some settlements within the area covered by Canal System 2 (Figure 1.3) predated canal construction. Among these settlements were structures that dated to the pre-Vahki or Red Mountain phase (Cable and Doyel 1987), the earliest recognized phase of the Hohokam culture (Chapter 18; Dean 1991:Table 3.3 and Figure 3.6).

During the pre-Classic periods, Dutch Canal Ruin consisted of a series of field house loci and farmstead sites, none of which exhibited evidence of year-round use. Field house loci were areas of seasonal activity that incorporated the use of some type of structural feature (Crown 1983; see also Cable and Doyel 1984:259–279 for discussions of functions of field houses and farmsteads). Although the term *field house* has been applied to types of sites other than those related strictly to farming (Ward 1978), the main activities associated with pre-Classic period structures at Dutch Canal Ruin were related to agriculture. Farmsteads have been defined elsewhere as areas of habitation consisting of one or two houses adjacent to fields and occupied by a single family or an extended family (Masse 1991:199). Such may have been the case in Area 3 at Dutch Canal Ruin, where investigators established that multiple structures had been grouped together. They found no evidence to support the idea of permanent occupation of any of the pre-Classic period habitation areas at Dutch Canal Ruin. The informal nature of the structures and the lack of storage pits indicated that the Hohokam had seasonally occupied all the pre-Classic period areas of the site.

If this general interpretation of the pre-Classic period occupation of Dutch Canal Ruin is accepted, one question becomes immediately apparent: Where did the occupants of Dutch Canal

Ruin build their homes? At this time investigators can only speculate on the answer. Researchers have examined land-use patterns of the Hohokam through analogy with historic and modern groups and through reconstruction of past behavior based on physical remains. Moreover, archaeologists have developed models and applied them to questions concerning prehistoric behavior. Because some prehistoric groups relied on a diverse subsistence base, they often used areas extending 5 km or more as catchment zones to supply needed resources (Flannery 1976:105–111). The Hohokam probably farmed locations no more than 5 km away from the village (Cable and Doyel 1984:260; Chisholm 1962; Crown 1985a:90) and established their major villages containing platform mounds at intervals of approximately 5 km (Gregory and Nials 1985). The first terrace, with its extensive canals and deep alluvial soils, constituted part of the resource zone available for exploitation by permanent settlements located on the second terrace. In addition to being used for farming, the first terrace offered a variety of wild plant and animal resources that could have been exploited by members of permanent settlements.

RESEARCH AREAS DEFINED

The testing phase at Dutch Canal Ruin identified 18 areas containing prehistoric features. Monitors discovered 2 additional areas, Research Areas 9 and 10, during construction of the remote parking lot between 22nd and 24th streets north of Buckeye Road (Figure 1.2). They recorded, excavated, and sampled the features in these two areas during the construction phase. From among the other 18 areas, they selected 8 for further study. The selection process is discussed individually for each research area.

Research Area 1

Testing identified two pit structures, an outdoor activity area, and an inhumation within an area that measured 20 × 20 m (Figure 1.4). Using ceramic association, archaeologists assigned Area 1 to the Colonial period. Because of the lack of morphological characteristics such as floor and wall plastering, clay-lined hearths, and so forth, investigators determined that the architectural remains were field houses. They could not determine the relationship of the inhumation to the other features. Although field crews found the burial partially below one of the structures, analysts later considered it intrusive, probably postdating the pre-Classic period occupation of Area 1.

The investigating team selected Area 1 for excavation because of its potential to answer questions about Colonial period settlement composition at the site, that is, spatial configuration, specific functions and functional relations of features, and the nature of activities conducted there. Also, the City of Phoenix deemed it necessary to remove the burial to protect it from construction damage and because of the potential for scientific information. Archaeologists removed the remains and handled them in a professional and respectful manner.

Research Area 2

Located immediately north of the South Main Canal (Alignment 8003), this area measured 14 × 14 m (Figure 1.4). Excavators identified a pit structure and a possible structure approximately 5 m apart during testing and dated them to the late Pioneer or Colonial periods. They considered these features to have functioned as either a field house or a small farmstead locale because of their informal morphology and limited use. Investigators assigned these remains to the Colonial period although few data clarified their temporal association.

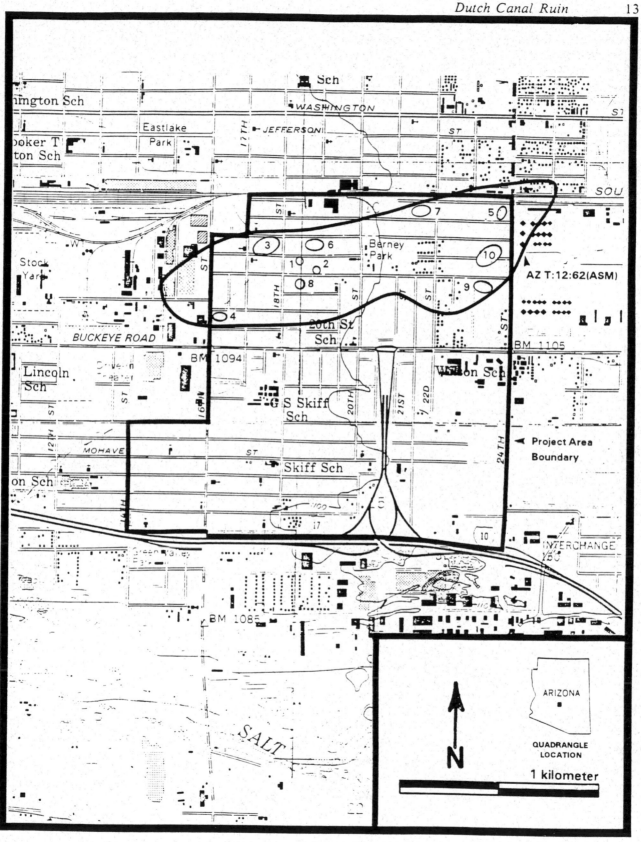

Figure 1.4. Location of research areas in Dutch Canal Ruin in relation to project boundaries and the Phoenix Sky Harbor Center. Base map is Phoenix, Arizona, 7.5′ USGS topographic map, photo revised 1982.

Area 2 was selected for further study to address questions relating to the spatial and functional characteristics of field house settlements and to examine the relationship between this particular settlement and the canal. Although SWCA's investigators could identify only one feature as a pit structure, they found other activity areas during the data recovery phase. By closely examining the stratigraphic position of the habitation features relative to that of the canal, archaeologists expected to define a temporal relationship between them. Examination of a similar situation in the Squaw Peak Parkway indicated that the canal had been abandoned prior to the establishment of the farmstead settlement in that location (Greenwald and Ciolek-Torrello 1988a). Major issues for Area 2 focused on settlement function, contemporaneity of habitation and canal features, and land-use strategies.

Research Area 3

The testing phase revealed that this area, which contained seven structures, a possible structure, three pits, and a stain, was the largest defined concentration of habitation features at Dutch Canal Ruin (Figure 1.4), measuring approximately 100 × 120 m. Investigators identified two goals: (1) to define the spatial relationships and variability among feature types; and (2) to establish a well-defined chronology for the features in this area to approach the issue of periodic reuse of the area versus occupation and settlement by a larger group. Analysts also recognized this area for its potential for dating flood deposits, as at least one structure contained laminated fill thought to be fluvial. The investigating team selected Area 3 for examination because of the size of the settlement area, its potential for producing chronometric data for interpreting the use of the first terrace, and its potential for dating flood deposits.

Field crews uncovered 7 pit structures, 11 pits, and a rock cluster in four spatially discrete loci within Area 3. Chronometric results suggested an occupation primarily associated with the Santa Cruz phase of the Colonial period. The occupation appeared to have been short term, seasonal, recurring, and associated with agriculture and wild resource exploitation.

Research Area 4

Area 4 was located along 16th Street south of Tonto Street in the extreme southwestern corner of the portion of Dutch Canal Ruin within the project area (Figure 1.4). The area was small, measuring approximately 10 × 12 m. Investigators identified two possible pit structures in this location during the testing phase. The location of a field house settlement in such close proximity to Turney's Gully (Figure 1.1) seemed aberrant when compared to other feature concentrations at the site. The area was also spatially removed from the canal network, although it was immediately south and downslope of an area thought to be a prehistoric field. The investigating team selected Area 4 for further study because of its location, with the goal of defining other economic strategies that the Hohokam may have employed at Dutch Canal Ruin. Turney's Gully was a channel of the braided Salt River and may have served as a natural area into which fields drained or where irrigation run-off was intentionally channeled. As a result of increased moisture, Turney's Gully may have become an economically important microenvironment supporting a wide variety of both plant and animal resources (Volume 4:Chapter 7). Processing settlements that closely resembled field house loci may have been strategically positioned along the margins of this gully, affording residents easy access to both the field systems and this microenvironment. Analysts conducted further studies in Area 4 to examine the variability of land-use strategies employed at Dutch Canal Ruin. The poorly preserved features in this area provided limited data. Investigators found evidence indicating activities associated with food processing but could not define specific resources.

Excavation revealed that no structures were present in Area 4. One of the features thought during testing to be a pit structure was determined to be a large, basin-shaped pit; the other was a scatter of fire-cracked rock and charcoal. These features supported the implication that Area 4 was associated with resource processing and had been used on a limited basis during the pre-Classic periods.

Research Area 5

Area 5, in the extreme eastern portion of the project area (Figure 1.2), consisted of a pit structure positioned between the banks of two canals within an area that measured 8 × 10 m. The structure was stratigraphically higher than the canals and postdated their use. Area 5 was selected because of its potential for refining canal chronology and providing information about subsistence-related activities: investigators thought they could perhaps elaborate on the existing chronology for the canals by further investigating the relationship of the pit structure to the canals and dating the structure through archaeomagnetic and radiocarbon techniques. Excavation of the structure and associated features was expected to help to address questions about land-use strategies and temporal components at Dutch Canal Ruin.

Further investigation of the canal channels determined that the channels were separate canal alignments. Both canals dated to the pre-Classic periods, with one confidently dated to the Colonial period. The limited-occupation pit structure was archaeomagnetically dated to the late Colonial period.

Research Area 6

Turney's maps (1924, 1929) of prehistoric canal systems show a bifurcation in the vicinity of 20th Street north of Grant Street (Figure 1.2; see also Volume 1:Figures 4.11, 4.14, and 4.17). The testing phase of this project located such a feature farther to the west. The research team selected this area for further investigation because of its potential for yielding data on canal mechanics and morphology. However, they found no habitation features at this location.

Studies in this area centered on the bifurcation and on how the various canal alignments related to one another. A portion of the area that measured approximately 70 × 150 m (including the canal features) was investigated as Area 6; in particular, field crews attempted to find intrachannel features such as headgates, weirs, and plunge pools to study their morphology and function and to collect additional chronometric data concerning the site's canals and the sequence of the channels. Investigators expected that canal studies at this location would provide information on the operation of early systems located on the floodplain. The results of the investigations indicated that all of the canals represented sequential channels associated with the late Pioneer or Colonial periods. Although field crews found a split in the canal alignments, none of the evidence supported the idea of a bifurcated system.

Research Area 7

The investigating team also selected Area 7 for study because of its potential to advance canal studies. During the testing phase they recognized this area as geomorphically and culturally complex, and several questions remained unresolved following testing. Investigators had two primary research goals: to examine the unusual sedimentation in this location and to pose questions concerning channel erosion. Earlier researchers had observed erosion of canal banks and basal channel units in the Squaw Peak Parkway corridor (Greenwald and Ciolek-Torrello 1988a), and Area 7 appeared to represent a similar event. Also, analysts had not adequately dated either of the two parallel channels in this location (Figure 1.2) during the testing phase. In addition, field crews excavated a burned pit structure in this area because of its preservation and potential for contributing to chronological data. Investigators decided to excavate this structure after weeks of effort in other areas from which they recovered few datable samples. Area 7 measured approximately 48 × 60 m.

Excavations at Area 7 revealed that the Hohokam had constructed erosion control devices in the canal channels, forming "baffles" to force the water away from the downslope bank when the directions of the canals were changed. The baffle feature was constructed of stone, similar to rip-rap, and incorporated a wooden structure. Investigators noted wide variations in sediment accumulation between the two canals, indicating variations in flow efficiency and possibly greater variability in flow cycles. As a result of these studies, analysts determined that one channel was a Classic period canal and the other a pre-Classic period canal. They attributed many of the differences between the two canals to changes in canal engineering efficiency and cyclic use. Although erosion was a problem during the early occupation of the site, by the late Colonial period the Hohokam had developed more efficient engineering designs that reduced erosion and had used these designs at Dutch Canal Ruin. The pit structure was a field house that probably dated to the pre-Classic periods.

Research Area 8

Investigators selected Area 8 for further study because it seemed to be a Classic period component consisting of five, or possibly six, pits, a trash deposit, and possible structure surfaces and walls. Classic period use of the project area appeared to compare closely with the pre-Classic period occupation in terms of land-use strategies. Small, scattered settlements—which may have been permanently settled hamlets rather than the earlier field house and farmstead settlements—were introduced on the first terrace at this time. Classic period components on the geologic floodplain reflected a change in settlement strategy. Investigations focused on questions relating to the use and reoccupation of the geologic floodplain and Classic period settlement, subsistence, and land-use strategies. Excavation revealed Area 8 to be a Polvorón phase hamlet that covered an area of 22 × 34 m.

Twenty-four habitation features were found in Area 8 during data recovery, consisting of 2 pit structures, 2 cremations, 19 pit features, and an adobe wall segment. The bulk of the materials recovered from Dutch Canal Ruin came from Area 8, supporting the interpretation that it had been a permanently occupied settlement. Although Area 8 may have been established initially during the Civano phase, the high frequency of Roosevelt Red Ware (including Gila and Tonto Polychrome) sherds and the results of radiocarbon assays supported the interpretation of an occupation during the Polvorón phase.

Areas Not Selected for Investigation

In each of the areas not selected for further investigation, archaeologists identified at least one pit structure or possible pit structure during testing. Two of the areas contained three structures; however, one of these areas, in the extreme eastern portion of the project area adjacent to Area

5 (Figure 1.2), contained widely dispersed rather than clustered structural features. Because they did not produce temporally diagnostic artifacts, five of the nine areas could not be assigned to a specific period or phase. Examination of the stratigraphic positioning of the features indicated, however, that they existed within the same stratigraphic units as other pre-Classic features at the site.

Investigators recognized two patterns in the distribution of features at Dutch Canal Ruin (Figure 1.2). First, the concentration of habitation features was greater in the western portion of the project area; the features appeared to be more closely clustered here than at any other area of Dutch Canal Ruin. Second, a greater concentration of habitation features lay in proximity to the South Main Canal (Alignment 8003) than to any of the other canal alignments, perhaps because the South Main Canal alignment had been used during two different periods. It was the first canal constructed through the project area, between A.D. 600 and 700 during the Pioneer period. Portions of it (in the extreme western extent of the project area) were then reused between A.D. 800 and 900 during the Colonial period. The vicinity of this reused portion of the canal contained the greater concentration of habitation loci.

The four areas adjacent to the South Main Canal that could be assigned to specific periods appeared to be Colonial in age. The two areas in the vicinity of N1500, E800 (Figure 1.2) were not considered one habitation locus because they lay on opposite sides of the canal. Though they may have existed simultaneously with the operation of the canal during the Colonial period, these two settlements may not have been absolutely contemporaneous with each other, or they may have been associated with separate field systems on opposite sides of the canal. The grouping of other defined areas may be possible, especially those in the extreme western portion of the project area. Because of the absence of temporally diagnostic ceramics and the spatial separation of each cluster of features, analysts did not attempt this sort of classification. The shortest distance between any two defined areas was slightly more than 50 m; more often, the areas were at least 100 m apart.

The spatial patterning of the habitation areas coincided with the distribution of features mapped by Midvale in the 1930s. An undated map by Midvale (Volume 1:Figure 4.12; Midvale n.d.) shows the extent of surface remains that were apparent in the 1930s. When Midvale's map is compared with the concentration of habitation features and general canal alignments found during investigations within the Phoenix Sky Harbor Center, the distribution pattern of the two data sources is very similar (Figure 1.5). Figure 1.5 shows two concentrations. The western concentration extended from approximately 18th Street to the west across 16th Street and perhaps as far west as the projected location of 15th Street. The eastern concentration was divided by 24th Street; the remains west of 24th Street were south of the Southern Pacific Railroad tracks, while those east of 24th Street were north of the railroad tracks. Midvale indicated no surface remains in the vicinity of 20th Street and the present Squaw Peak Parkway corridor; however, extensive trenching by the Museum of Northern Arizona (Greenwald and Ciolek-Torrello 1988a) located subsurface remains. Aerial photographs (USDA Soil Conservation Service 1934) show that the freeway corridor and adjacent areas were intensively farmed, and thus surface expressions may not have been recognizable at the time Midvale began mapping Dutch Canal Ruin.

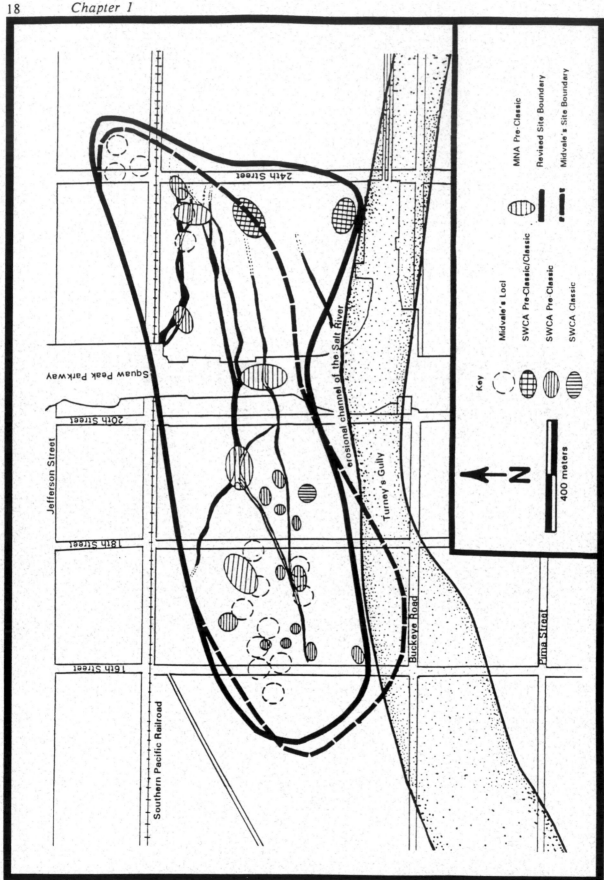

Figure 1.5. Distribution of feature concentrations mapped by Midvale (1934) and re-identified during the Phoenix Sky Harbor Center investigations.

RESEARCH OBJECTIVES FOR DUTCH CANAL RUIN

Investigators identified five problem domains for the data recovery phase within the Phoenix Sky Harbor Center (Volume 1:Chapter 10) and addressed each one at Dutch Canal Ruin.

Problem Domain I

The study of site structure, settlement composition, and settlement typology contributed to investigating problems associated with Hohokam social organization and social change. Dutch Canal Ruin throughout its occupational history was consistently an area of specific use focused on agriculture. Field house, farmstead, and small hamlet settlements were scattered along the various canals. These limited-use settlements were generally occupied seasonally by small groups that were never larger than extended families. Residents participated in a variety of activities associated with agricultural production, including canal construction, maintenance, and operation; field tending and cultivating; and processing and exploitation of economically important wild food sources. The temporary or seasonal occupation pattern of the pre-Classic period remains supported the notion that the first terrace had been used as an agricultural zone by members of larger, probably permanent settlements located on the second terrace. Given the linear pattern of the canal network in the Phoenix Sky Harbor Center, residential settlements associated with project field house and farmstead settlements may have been located at sites such as Pueblo Grande, La Ciudad/Los Solares, Pueblo Patricio, and La Villa/Casa Chica. The re-establishment of canals and habitation loci during the Classic period suggested that changes within the Canal System 2 sociopolitical system, population pressures on existing productive capabilities, environmental effects, or a combination of these factors had encouraged expansion into the Dutch Canal Ruin agricultural zone. The Hohokam re-established farming on the first terrace late in the Classic period, beginning in the Civano phase and extending into the Polvorón phase of the post-Classic period.

Problem Domain II

After defining the spatial and temporal parameters of the remains found in the Phoenix Sky Harbor Center and reconstructing the function of the settlements, investigators sought to demonstrate the variable use pattern of the Hohokam at Dutch Canal Ruin in the area defined as Canal System 2. During the Classic period a complex sociopolitical structure developed throughout the Hohokam region, a hierarchical system evolved from the expansion and control of canal systems (Gregory 1991:173, 181). In general, the hierarchical system established within irrigation communities is that of village, hamlet, farmstead, and field house, presented here in descending order of relative size. Problem Domain II focused on the relationship of the project resources to the largest recognized sociopolitical unit in the Hohokam regional system, the irrigation community (Gregory 1991). By defining the function of Dutch Canal Ruin resources and establishing a chronological sequence for each component, investigators realized the objectives of Problem Domain II and demonstrated the role of Dutch Canal Ruin in Canal System 2 (Chapter 18).

Problem Domain III

Because they recognized the first terrace as an area of high risk, Hohokam farmers moved their field systems and canals to higher ground to reduce the damaging effects of Salt River floods while increasing the amount of land under cultivation (Howard 1991:5.14–5.15). Canals and field systems on the geologic floodplain were more susceptible to flooding than those on the lower bajada.

Expansion of canal and field systems during the Colonial period onto the lower bajada would have reduced the impact of flood damage on the agricultural system as a whole. Investigators considering Problem Domain III hoped to define the events, both cultural and environmental, that affected Hohokam land-use strategies in the project area. In this regard, analysts found cultural events and their influence on the use of the geologic floodplain easier to reconstruct than environmental events. Researchers could posit geologic factors, but specific events posed formidable interpretation problems. The archaeological record has shown that habitation areas consisted of settlements requiring low energy input. Initially, the labor invested in canal construction and field development was high, and a single high-magnitude flood could have rendered the canals and fields useless. A preference for locating habitation sites and canal and field systems at higher elevations was demonstrated by the overall abandonment of the first terrace during the Sedentary period. Archaeologists have not defined specific factors that led to the reuse of the first terrace by Hohokam groups, but they may be linked to environmental factors.

Problem Domain IV

Many projects conducted in the Phoenix Basin have contributed to understanding Hohokam canal irrigation technology. The canal networks were a main focus of investigation at Dutch Canal Ruin, where extensive sections of canals were available for study. Investigators concentrated on the temporal affiliation, morphologies, and mechanics of the canals and associated features. Although Problem Domain IV also included the study of irrigated fields, the results of these efforts at Dutch Canal Ruin depended on investigations conducted at Pueblo Salado. Investigators had little doubt that the soils throughout much of Dutch Canal Ruin had been modified by prehistoric irrigation, even though they did not positively identify specific field areas.

Problem Domain V

Establishing chronological relationships and refining chronology are integral to every archaeological project. Analysts for this project expected to establish a project chronology, revise or refine existing chronometric information for the Dutch Canal Ruin features, and contribute to the refinement of Hohokam chronology in general. These efforts generally yielded good results. Radiocarbon dates for pre-Classic features, however, were consistently poor, perhaps due to long-term contamination of the radiocarbon samples from frequent saturation. Also, ceramics, especially temporally diagnostic types, were often poorly represented in the assemblage. Nevertheless, the ceramic assemblages were consistent with the limited duration of occupation for each area and were therefore considered reliable. Investigators emphasized stratigraphic associations, especially in developing chronologies of construction and use events for canals and for habitation features that were directly associated with canals. Although the archaeological remains and their potential to produce chronometric data were limited, analysts attained the general goals of Problem Domain V.

FIELD METHODS

Specific field methods used during the data recovery phase are discussed in Volume 1, Chapter 11, and briefly reviewed here. After re-establishing the grid system and the vertical controls at the site from various reference points established during the testing phase, field crews opened backhoe trenches that contained features selected for further investigation. Because of surface modifications in the project area following the testing phase, the filled backhoe trenches were not visible on the surface. Investigators sometimes had to place a trench perpendicular to the original trench to help locate it. After re-excavating the trenches, field crews cleaned the trench faces to relocate features exposed in profile. Once the features were located, investigators defined the areas to be mechanically stripped.

Mechanical stripping was accomplished with a backhoe equipped with a specially built seven-foot-wide bucket with a smooth rather than a toothed blade to produce a clean surface. An archaeologist monitored the depth of stripping to ensure that only the historic disturbance zone (30–50 cm in thickness) was removed, facilitating hand excavation. Field crews also removed fill from features that were deeply buried by alluvial materials to expedite their excavation. Investigators proposed to strip an area at least 5 m wide on all sides of structures to study extramural areas associated with them; however, they could not always do so because of existing streets and utilities or historic disturbance.

Field crews excavated by hand according to metric control units (1 × 1 m, 1 × 2 m, 2 × 2 m, etc.) to define features in horizontal exposure. Once they had defined a feature, they confined their excavations to the feature itself. They divided large features such as structures into halves or quarters, depending on size, and excavated the features according to either natural or arbitrary levels. In structures, if the fill lying above the floor could be distinguished from later deposits by a roof fall zone, investigators designated the deposits above the roof fall as structure fill and removed them but considered those below the roof fall to be floor fill. Field crews recorded the exact provenience of all floor contact artifacts and collected the artifacts separately from the floor fill zone. If they could not define a roof fall zone, they designated an arbitrary 10-cm level as the floor fill zone. They collected all artifacts, faunal remains, and raw materials, as well as construction daub that contained impressions.

Field crews also routinely collected pollen and flotation samples from each floor unit, from floor features, and from under floor artifacts. They collected macrobotanical samples—usually corn kernels, cob fragments, and wood charcoal—when they observed macrobotanical remains. From among the inventory of samples, investigators chose a select number for analysis. They also sampled extramural features according to fill sequence and depth.

When possible, researchers collected chronometric samples (which included all charcoal samples that might be large enough to yield results) and archaeomagnetic samples from features that contained sufficient burned material to produce results. Due to the paucity of charcoal from project features, investigators frequently collected composite samples in an attempt to recover enough for dating purposes. They also collected archaeomagnetic samples as detrital samples from clays in canals.

Field crews used screens of 1/4- and 1/8-inch mesh to facilitate the recovery of artifacts. Initially, they screened only floor fill and pit fill units. Later, as they determined that few materials were associated with project features, they screened all fill units. Overburden was not screened, as it was usually removed through mechanical stripping.

22

CHAPTER 2

DUTCH CANAL RUIN, AREA 1

M. Zyniecki

Research Area 1 included two pit structures, an inhumation, two prehistoric pits, and one historic or modern pit (Figure 2.1). Along with the rest of Dutch Canal Ruin, Area 1 was situated within the northernmost channel of the Salt River floodplain. Though difficult to detect, this channel was identified within the Phoenix Sky Harbor Center project area as a broad, shallow feature with an east-west orientation. Relative and absolute dates confirmed the conclusions from testing that the area had been occupied during the Colonial period. The evidence indicated that Area 1 was a farmstead, a locale with more than one field house with accompanying extramural features. The term applies both to contemporaneous structures and to structures for which contemporaneity is a possibility but cannot be established.

Area 1 was adjacent to the south side of the recently constructed Sky Harbor Circle, within the project grid at N1550-1570, E1070-1090 (Figure 1.2) and in T1N, R3E, Sec. 10, NW¼SW¼NE¼SW¼ on the Phoenix, Arizona, 7.5 minute USGS topographic map (Figure 1.4). Areas 2 and 8 and two canal features were nearby; Area 1 was 40 m northwest of Area 2, 90 m north of Area 8, 50 m north of the South Main Canal, and 20 m southeast of Alignment 8002, a crosscut canal situated between the North Main Canal and South Main Canal alignments. During the testing phase investigators determined that Area 2 was an isolated pre-Classic period habitation area that may have functioned as a field house or small farmstead locale (Chapter 3). Area 8 was thought to be a small Classic period farmstead that was more complex and occupied for a longer period than the pre-Classic period components of Dutch Canal Ruin (Chapter 9).

RESULTS OF TESTING

The backhoe operator excavated nine trenches, totaling 190 m in length, in Area 1 during testing. Six of the trenches, each 25 m in length, were spaced at 5-m intervals to provide a more intensive investigation of this area. Investigators identified two pit structures, an outdoor activity area, and an inhumation in the trench faces and assigned the pit structures to the Colonial period (Volume 1:Chapter 9). The inhumation was located below one of the structures, but analysts found no evidence to indicate that the burial intruded on the structure. Testing results indicated, based on a more formal pit structure revealed in one of the structure profiles (Feature 1-1), that Area 1 was a small farmstead locale.

RESEARCH OBJECTIVES

In general, investigators had three research objectives for Area 1: to determine the nature and extent of features in this area, to characterize the Hohokam occupation and its chronology, and to determine the relationship between this and other areas of Dutch Canal Ruin and Canal System 2. The research questions pertaining to investigations in Area 1 involved site structure; settlement composition and typology; spatial, functional, and temporal variation of settlements in Canal System 2; settlement and land use in the geologic floodplain; and general refinement of the Hohokam chronology (Volume 1:Chapter 10).

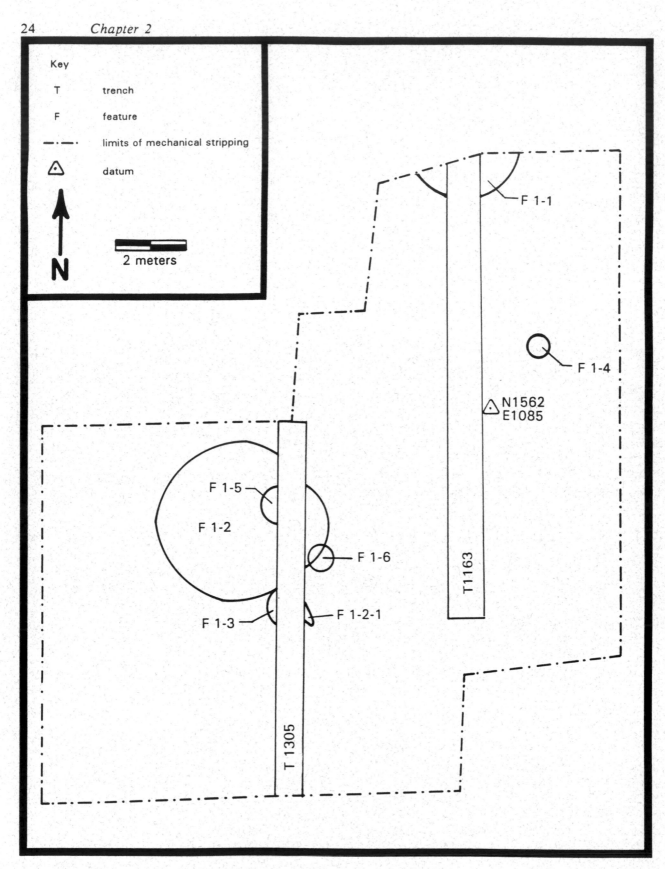

Figure 2.1. Area 1, Dutch Canal Ruin.

FIELD METHODS AND STRATEGIES

Backhoe Trenching

Because of the intensive nature of trenching in Area 1 during testing, investigators dug no new trenches during data recovery but reopened approximately 13 m of Trench 1163 and 10.5 m of Trench 1305 (Figure 2.1). At this point, they found that a portion of Feature 1-1, a pit structure, was within the right-of-way for Sky Harbor Circle and thus was not accessible for excavation.

Stripping

Field crews stripped approximately 189 m² mechanically or by hand in Area 1 (Figure 2.1). Stripping removed overburden to the level of detection for the three features discovered during testing and revealed an additional three pits: two prehistoric and one historic or modern.

Investigators hoped to strip an area extending outward 5 m around the two known structures and the inhumation but could not because of the proximity of Sky Harbor Circle and its accompanying landscaping to the north of Area 1. To offset these circumstances, crews stripped a greater area to the south of the known features. They uncovered no additional features during this process; moreover, investigators doubted that stripping the area according to the original plan would have exposed any other features.

Hand Excavations

Crews used the standard hand excavation methods described in Chapter 1 for all features in Area 1. The only exception, Feature 1-1, was excavated as a whole unit instead of in halves or quadrants because approximately half of the structure was inaccessible.

STRATIGRAPHY AND ASSOCIATION

The natural stratigraphy throughout Dutch Canal Ruin exhibited similar types of strata, although the thickness of strata varied. Investigators identified three strata in the trenches of Area 1. The uppermost, Stratum 1, was a zone of historic and modern disturbance that extended 25 cm below the present ground surface (bps). This stratum was generally loose with sandy lenses, pebbles, and historic trash. In places it was very compact, possibly due to the presence of former alleyways and streets (Kirk Anderson, personal communication 1990). The boundary between Stratum 1 and Stratum 2 varied between gradual and abrupt. Stratum 1 corresponded to the recently disturbed zone described by Nials and Anderson (Volume 1:Chapter 3).

Texturally, Stratum 2 was clay loam characterized by dark, organic-rich clays and sandy clays that probably resulted from overbank deposits of the Salt River, combined with prehistoric and historic irrigation practices. This middle stratum began at a depth of 25 cm and extended to 50 cm bps. Stratum 2 was one of the Holocene soils reported during testing (Volume 1:Chapter 3) and was probably a Gilman loam (Volume 1:Appendix B).

The lowest stratum exposed in Area 1, Stratum 3, consisted of prehistoric floodplain silts and fine sand. The features in Area 1 were situated in this stratum. The soil was hard, compact, slightly sticky, and slightly plastic when wet. The textural definition of this soil was silty loam. Stratum 3 was equivalent to the clayey silts reported by Nials and Anderson (Volume 1:Chapter 3).

CULTURAL FEATURES

Investigators discovered two structures and an inhumation during testing. Mechanical and hand stripping of the areas around these features revealed two additional prehistoric pits and a historic or modern pit.

Structures

The excavation results supported the initial assessment that the two structures in Area 1 dated to the Colonial period. During testing, investigators thought Feature 1-1 was a formal pit structure because it was well defined in profile; excavation revealed that it was not as formal as originally thought. Feature 1-2 was originally defined as a pit structure, and this definition was also supported by the excavations.

Feature 1-1

Testing Feature: 1009
Trench Number: 1163
Horizontal Provenience: N 1568.08–1569.24; E 1083.91–1086.88
Vertical Provenience: 4.04–4.21 meters below datum (mbd)
Feature Type: Pit Structure

Investigators identified Feature 1-1, an oval pit structure, during testing. Subsequent construction of Sky Harbor Circle prevented complete excavation because the sidewalk and landscaping next to the road sat atop the feature. One-third to one-half of the structure was excavated (Figure 2.2), an area of 1.11 × 2.90 m with a maximum depth of 0.16 m. The testing profile was 3.80 m in length (Volume 1:Appendix A), resulting in a projected plan of 2.90 × 3.80 m. During testing field crews recorded the top of the feature as 3.99 mbd, as opposed to the 4.04 mbd recorded during data recovery (Volume 1:Appendix A). The height of the walls was 10–16 cm, with a slope ranging from nearly vertical in places to approximately 45°. The pit walls were unprepared and detectable only from a slight change in soil color. The floor of this structure, though leveled and compacted from use, was also unprepared. The description of the plan as oval was based on the portion of the structure that was excavated.

No entryway or other interior features were discernible. Given current knowledge about standard Hohokam construction techniques and what appeared to be the remnants of burned beams, investigators concluded that the roof and upper walls had probably been constructed of posts or beams with sticks, brush, or reeds woven between or attached to them.

Field crews recovered two sherds, a polishing stone, and a grinding slab stained with hematite and a possible black pigment from this structure. The small number of artifacts suggested purposeful abandonment of the structure. The burned organic material and beams on the floor and in the fill suggested that the structure had burned after abandonment.

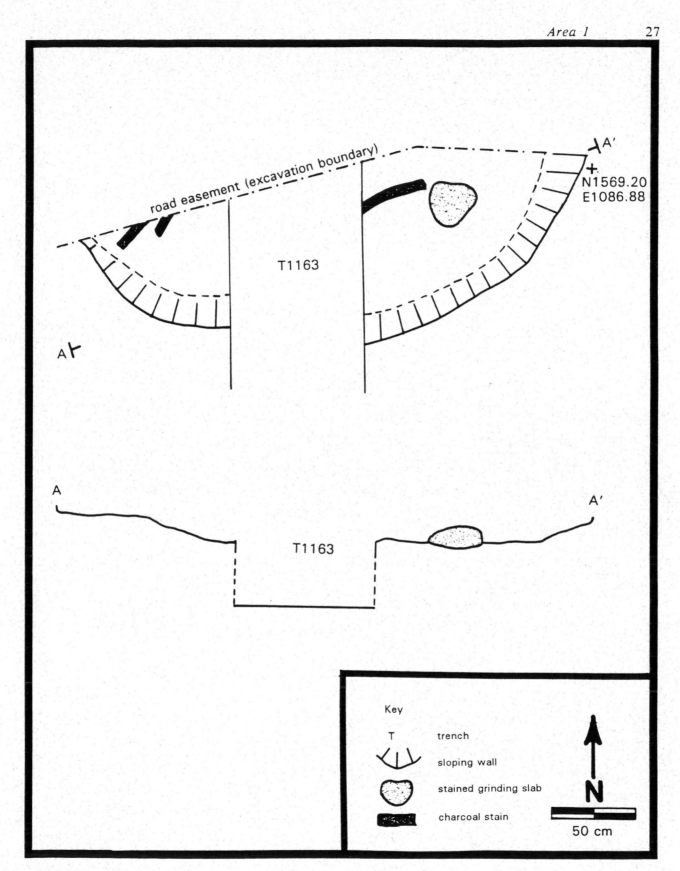

Figure 2.2. Area 1, Feature 1-1, plan and cross section.

Feature 1-2

Testing Feature: 1054
Trench Number: 1305
Horizontal Provenience: N 1556.67–1560.88; E 1076.70–1081.55
Vertical Provenience: 4.20–4.44 mbd
Feature Type: Pit Structure

Feature 1-2 was a roughly circular pit structure with an appended entryway (Figure 2.3), Subfeature 1-2-1, opening to the south-southeast. The structure measured 4.48 m northeast-southwest by 4.28 m northwest-southeast. The walls and floor of the structure were not formally prepared. The floor was uniform, compacted through use, slightly basin shaped, and approximately 15.10 m^2 in area. The edges of the floor were 5 cm higher than the center. The walls were almost vertical in the eastern portion of the structure and had a slope of between 45° and 60° in the western portion. Field crews recorded the top of the walls at 3.88 mbd during testing (Volume 1:Appendix A). A historic or modern pit, Feature 1-6, intruded into the southeast corner of the structure, and Feature 1-5, a pit with evidence of burning, intruded into the fill of the structure. Field crews recovered fewer than 10 artifacts from the floor fill and a single piece of ground stone from floor contact.

The entryway extended approximately 1.2 m beyond the structure. A backhoe trench obliterated most of the entrance and an inhumation, Feature 1-3, that was directly below the entrance. However, field crews identified a threshold or step within the structure near the entrance. Although no hearth was present, crews observed two informal fire areas, Subfeatures 1-2-11 and 1-2-12, on the floor (Figure 2.3). These areas consisted of oxidized portions of the floor surface. Subfeature 1-2-11 was 40 cm in diameter, and Subfeature 1-2-12 was 22 × 38 cm. Nine postholes were excavated: investigators determined that three were primary postholes because of their diameters (11–13 cm) and depths (7–13 cm) and that six were secondary postholes having diameters between 5 cm and 7 cm and depths ranging between 2 cm and 5 cm. Seven of the postholes made a roughly circular pattern 20–90 cm inside the outline of the wall.

After abandonment, Feature 1-2 was inundated and a layer of coarse sands was deposited on the floor. Subsequently, the structure burned and collapsed, and alluvial deposits ultimately filled the remaining depression. Sometime during this sequence of events, Feature 1-3 may have intruded into the entrance to Feature 1-2. Investigators based their interpretation on the fact that inhumation was a preferred burial practice during the Classic period, not the pre-Classic, and that the inhumation lay directly below the entrance.

Burial

Feature 1-3

Testing Feature: 1053
Trench Number: 1305
Horizontal Provenience: N 1556.00–1556.90; E 1079.94–1080.25
Vertical Provenience: 4.36–4.68 mbd
Feature Type: Primary Inhumation

Originally identified during testing, Feature 1-3 was a primary inhumation located directly below the entrance to Feature 1-2. Although the backhoe removed most of the burial during

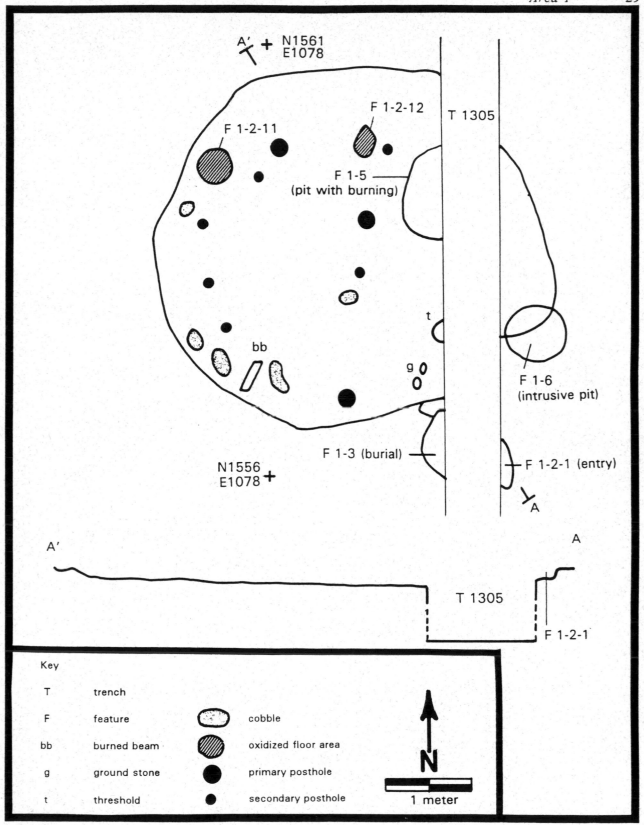

Figure 2.3. Area 1, Feature 1-2, plan and cross section.

testing, a portion of the right side of the skeleton remained in situ. No grave goods were associated with the interred. The profile drawn of the burial during testing showed a pit bottom below the remains, with possible pit walls to the sides; excavation did not reveal a pit.

Extramural Features

Feature 1-4

Horizontal Provenience: N 1563.59–1564.08; E 1087.08–1087.66
Vertical Provenience: 3.91–4.20 mbd
Feature Type: Pit with Burning

Feature 1-4 was a pit with evidence of burning in the fill. The pit was 51 × 61 cm in plan with a conical shape in cross section and a depth of 29 cm. Most of the feature fill consisted of a slightly oxidized soil mixed with charcoal. The base of the pit was covered by a thin layer of ash, but field crews found no evidence of oxidation on the sides or bottom of the pit. This feature was discovered during extramural stripping operations between the pit structures, Features 1-1 and 1-2 (Figure 2.1). The top of the pit was 13 cm above the top of the walls of Feature 1-1 and 29 cm above the top of the walls of Feature 1-2.

Feature 1-5

Trench Number: 1305
Horizontal Provenience: N 1559.93–1560.29; E 1079.62–1080.20
Vertical Provenience: 4.23–4.34 mbd
Feature Type: Pit with Burning

Feature 1-5 was a pit with evidence of burning that intruded into the fill of Feature 1-2, a pit structure. The bottom of the pit ended 10 cm above the floor of the structure. The pit was 50 × 60 cm in plan, an irregular oval in shape, and 12 cm in depth. The cross section showed a nearly vertical south wall, a sloping north wall, and a shallow, basin-shaped bottom. Field crews recovered no artifacts from the pit. The fill contained charcoal flecks, oxidized soil, and burned areas. Lack of oxidation of the pit edges combined with oxidized areas within the fill indicated that either Feature 1-5 contained the refuse from a fire pit or else minimal burning or burning of very short duration had occurred in the pit. The location of Feature 1-5 in the fill of Feature 1-2 indicated the pit was constructed after the pit structure had burned, collapsed, and filled.

Feature 1-6

Horizontal Provenience: N 1557.38–1558.09; E 1081.99–1082.67
Vertical Provenience: 4.12–4.14 mbd
Feature Type: Historic or Modern Pit

Feature 1-6 was a nearly circular pit that intruded into the southeast corner of Feature 1-2, a pit structure. The pit measured 71 × 68 cm and was 2 cm in depth. Field crews found no artifacts in the pit. The fill consisted of homogeneous silty clays with some charcoal flecks. This pit appeared to be historic, probably machine excavated, because of the almost perfectly circular plan, the near-vertical sides, and the level bottom, and probably was associated with a public utility installation or construction activity.

MATERIAL CULTURE

The small artifact assemblage recovered from Area 1 consisted of ceramics and flaked and ground stone. The best represented of these three categories, flaked stone, contained only 19 artifacts. The small number of artifacts probably reflected low-intensity use of this area.

Ceramics

Investigators recovered only 18 sherds from Area 1 (Table 2.1), from three of the six identified features. The small quantity of ceramics pointed to the limited and short duration of use in this portion of Dutch Canal Ruin. Of these sherds, 13 (72.2%) were Gila Plain, Salt Variety, and the remaining 5 (27.8%) were Santa Cruz Red-on-buff. Four of the Gila Plain sherds were associated with the pit structures (2 with Feature 1-1 and 2 with Feature 1-2). The additional 9 Gila Plain sherds and the 5 Santa Cruz Red-on-buff sherds were present in the fill of Feature 1-4, a pit with evidence of burning.

Table 2.1. Area 1, Ceramic Frequencies by Feature

Ceramic Ware/Type	Features			Total
	1-1 Pit Structure	1-2 Pit Structure	1-4 Pit with Burning	
Hohokam Plain Ware				
Gila Plain, Salt Variety	2	2	9	13
Hohokam Decorated Ware				
Santa Cruz Red-on-buff			5	5
Total	2	2	14	18

Analysts garnered little information about the temporal associations of Area 1 from such a small ceramic assemblage. Feature 1-4 contained the only diagnostic sherds; it appeared to have been utilized during the Santa Cruz phase of the Colonial period, which agreed with the conclusions drawn from testing about the general period of occupation of the area.

Flaked Stone

Of the 19 flaked stone artifacts collected from Area 1 (Table 2.2), 17 were debitage, one was a core, and one was a large primary flake. Analysts observed no use wear on any of the artifacts. The most common material category was basalt, accounting for 73.7% of the artifacts. Ten of the flaked stone artifacts were associated with features.

Table 2.2. Area 1, Flaked Stone Morphological Type by Material Type

Morphological Type	Material Type				Total
	Basalt	Igneous Other	Quartzite	Metasediment	
Angular Debris		1			1
Flake Fragment	6			1	7
PRB[1] Flake	1	1			2
Whole Flake	6		1		7
Core		1			1
Large Primary Flake	1				1
Total	14	3	1	1	19

[1]Platform-remnant-bearing

Feature 1-2 contained 9 items of debitage (8 of basalt and 1 of igneous other material). Several other artifacts probably were associated with Feature 1-2. Field crews recovered a large primary flake and a whole flake, both of basalt, from a 2 × 2-m stripping unit above the entrance, immediately south of the structure. Another 2 × 4-m stripping unit to the south of the structure contained several flaked stone artifacts, including a core of igneous other material. This was the same material and texture as the platform-remnant-bearing (PRB) flake of igneous other material recovered within Feature 1-2. A single metasediment flake fragment was present in Feature 1-6. Since Feature 1-6 was a historic or modern pit that intruded into Feature 1-2, this flake fragment either had come from Feature 1-2 or had been introduced from overburden when the pit was dug.

Investigators recovered no formal flaked tools from Area 1, and none of the recovered artifacts exhibited use wear, making it difficult to discuss the function and implications of the flaked stone assemblage in this area. The fact that the core and one of the PRB flakes were composed of the same material demonstrated that the Hohokam had produced flaked stone artifacts during the occupation of Feature 1-2. As with the other material culture categories, the low number of flaked stone artifacts in Area 1 emphasized the short duration and minimal intensity of occupation and use of the area. This conclusion was underscored by the lack of use wear on any of the artifacts.

Ground Stone

As with the other artifact categories, the amount of ground stone in Area 1 was small, consisting of one polishing stone, two grinding slabs, two pieces of possible ground stone, and an unmodified manuport of greenstone. The polishing stone, made of rhyolite, and a grinding slab of igneous material were from Feature 1-1. The ground stone from Feature 1-2 consisted of a greenstone grinding slab; of the two possible ground stone pieces, one was of granite, and the other was greenstone. Both of the grinding slabs had fine-textured surfaces. All of the ground stone and the manuport were cobbles, presumably from the Salt River, with no evidence of production modification.

The two grinding slabs were the only examples of this artifact type recovered from Dutch Canal Ruin. Generally smaller than metates, their function may have been different from or may have included more than simply food processing (Chapter 14) or other subsistence-related activities. The utilized surface of both grinding slabs was fine in texture, and both had pigment adhering to the grinding surface. The pigment (hematite and possibly another type) was probably used for decorative purposes. The presence of the polishing stone in Feature 1-1 and the pigment on the grinding slab suggested at least the possibility of decorated ceramic production in Area 1, although investigators found no other evidence for ceramic production. Alternatively, the pigment could have been used for body adornment, such as facial painting.

Subsistence Remains

The few subsistence remains recovered from Area 1 were botanical. None of the macrobotanical (flotation) samples collected was productive. Pollen analysis from the microbotanical samples indicated that exploitation of some wild species may have taken place, and the presence of pollen from plant types known to grow in disturbed soils provided tenuous evidence for agriculture in the immediate area.

Analysts detected no carbonized plant remains in the three flotation samples from Area 1 (Chapter 15 and Appendix C). The samples came from the floor fill level in the southern half of Feature 1-1, the floor fill level in the southeast quadrant of Feature 1-2, and the fill from the primary inhumation, Feature 1-3.

The three pollen samples submitted for analysis were more productive (Chapter 16:Figure 1). One sample, from the floor under a burned roof beam in Feature 1-1, contained a slightly elevated frequency of Gramineae pollen, indicating that grass seeds may have been processed in the structure (Chapter 16) or that grass mats may have been present (Linda Scott Cummings, personal communication 1991). A sample collected from beneath a rock on the floor of Feature 1-2 suggested that *Prosopis* sp. (mesquite), Cheno-ams (*Chenopodium* sp. and *Amaranthus* sp.), and one of the Solanaceae (potato/tomato family) could have been processed in the structure (Chapter 16). The Solanaceae family includes, for example, *Lycium* sp. (wolfberry), *Solanum* sp. (wild potato), and *Physalis* sp. (ground cherry), as well as what are commonly called chili peppers. The third pollen sample from Area 1 came from the fill immediately above the primary inhumation in Feature 1-3. This sample, like the sample from Feature 1-1, contained an elevated frequency of Gramineae and a small aggregate of the pollen. Given the context of the sample and the presence of the aggregate, analysts considered it possible that a grass mat had been placed with the interred in Feature 1-3 (Chapter 16).

The pollen samples may indicate some sort of economic activity at Area 1. Although the analyzed macro- and microbotanical samples yielded no direct evidence for agricultural activities, archaeologists inferred the possibility of agriculture from the presence of the Cheno-am pollens. Cheno-ams are known to grow in areas of disturbed soil such as might be found in and near agricultural fields. The lack of charred Cheno-ams in the flotation samples and the only slightly elevated frequencies of the pollen in the sample from Feature 1-2 may indicate that these plants were restricted to minor growth as a result of agricultural activities (Chapter 15).

Human Remains

The one inhumation, Feature 1-3, contained the remains of a male 25–35 years old whose age was determined by dental evidence (Volume 4:Chapter 10). Except for a first molar from the left maxilla, only portions of the right side of the body were recovered, because a backhoe had removed the remainder. The right cranial bones and portions of the right clavicle, right humerus, right mandible, right maxilla, and four ribs were recovered. The right mandible contained all three molars, and the right maxilla had three molars, the first premolar, and both incisors. The right maxillary first molar exhibited extreme buccal-lingual wear. The condition of the bone indicated that the missing teeth were lost after death. The only evidence of pathology on the remains was porotic scarring on the frontal bone, indicating porotic hyperostosis, an anemic condition caused by iron deficiency. Analysts could not collect metric data because of the incompleteness of the skeleton.

CHRONOLOGY

Determining the temporal relationships between the features in Area 1, between Area 1 and the other areas of Dutch Canal Ruin, and between Dutch Canal Ruin and other sites along Canal System 2 (Turney 1929) was a major objective of the research orientation for Dutch Canal Ruin. Investigators expected that by coming to understand these relationships they could then address the research questions pertaining to the temporal assignment of cultural manifestations and changes in these manifestations through time. Chronological analyses included relative dating employing stratigraphy and diagnostic ceramics and absolute dating using radiocarbon and archaeomagnetic methods.

Area 1 appeared to have been used repeatedly over a relatively long period of time. The one absolute date from the area indicated that Feature 1-2 had been utilized sometime during the middle of the Colonial period. This agreed with ceramic data that indicated utilization of a pit during the Santa Cruz phase. Stratigraphic evidence proved that one or possibly two features postdated Feature 1-2. The evidence as a whole suggested that the occupants had used Area 1 most intensively during the Colonial period, specifically, during the Santa Cruz phase.

Stratigraphy

As noted above, all features in Area 1 were located in the third stratum from the surface, which consisted of prehistoric flood deposits from the Salt River. Some superpositioning and intrusion appeared between the features. The inhumation, Feature 1-3, may have intruded into the entrance of Feature 1-2, a pit structure. Feature 1-3 was directly below the entrance of Feature 1-2, but the burial pit outline was not discerned. If the burial was temporally affiliated with the Classic period, Feature 1-3 must have intruded into the entrance of Feature 1-2 because of the position of the human remains. If, however, the burial was associated with the pre-Classic periods, the inhumation may have predated the structure.

Two pits, Features 1-5 and 1-6, postdated the pit structure designated Feature 1-2. Feature 1-5, a pit with evidence of burning, was in and superposed on the feature fill of the pit structure. A modern or historic pit, Feature 1-6, intruded into and was superposed on the south wall of Feature 1-2. These stratigraphic associations indicated that the two pit structures (Features 1-1 and 1-2) and a third pit (Feature 1-4) may have been roughly contemporaneous. The top elevations recorded for each feature were within 11 cm of each other and indicated a slope toward the river. Investigators could not, however, determine exact contemporaneity.

The stratigraphic relationships of the features in Area 1 indicated that the area had been occupied, abandoned, and then utilized at least once more. The most intensive period of use appeared to have coincided with the pit structures (Features 1-1 and 1-2) and the pit with evidence of burning (Feature 1-4). Despite the lack of direct evidence, investigators assumed (as the most likely possibility) that the inhumation, Feature 1-3, intruded into Feature 1-2. The final period of use in Area 1 was represented by the historic or modern pit, Feature 1-6.

Ceramics

Of the 18 sherds recovered from Area 1, 5 were diagnostic in terms of temporal association. Field crews identified these sherds as Santa Cruz Red-on-buff and removed them from the fill of Feature 1-4, a pit with evidence of burning. The association of Feature 1-4 with the Santa Cruz phase agreed with the archaeomagnetic date obtained from Feature 1-2.

Absolute Dates

As can be expected for field house contexts on the floodplain of the Salt River, samples suitable for radiocarbon dating were not abundant. The only radiocarbon sample collected in Area 1 came from Feature 1-2. Because of the small size of the sample after processing by Beta Analytic, Inc., and the likelihood of contamination, this sample was not analyzed.

Only one archaeomagnetic sample was available from Area 1, from an oxidized area on the floor of the pit structure, Feature 1-2. Given the location of the oxidation (about 18 cm from the wall) and its circular shape, investigators thought it represented an informal fire area within the structure. The visual date for the sample contained two intervals, A.D. 675-725 and A.D. 850-940 (Appendix A); the visual date was used rather than the 95% residual date range, as suggested by Eighmy and Baker (1991:9).

In a recent essay, Dean (1991) examined the independent chronometric dates for the Hohokam. The results of his investigation indicated that the maximum range for the Colonial period was A.D. 700-1100, with a probable range for the Gila Butte phase of ca. A.D. 775-850/900 and a probable range for the Santa Cruz phase of A.D. 850/900-950/1000 (Dean 1991:Table 3.3). These results suggested that the earlier of the two visual archaeomagnetic date intervals, A.D. 675-725, was too early for the pit structure. The remaining date range, A.D. 850-940, indicated that the burned floor dated to sometime between late in the Gila Butte phase and the middle of the Santa Cruz phase. This date matched the Colonial period assignment of Area 1 based on testing results (Volume 1:Chapter 9) and concurred, in part, with the Santa Cruz phase association of the ceramics in Feature 1-4. Features 1-2 and 1-4 were thought to be roughly contemporaneous because of their relative proximity, about 7 m apart, and the small difference in top elevations, 3 cm. Therefore, analysts proposed that the oxidized area in Feature 1-2 had been fired in the interval A.D. 850–940.

SUMMARY

Three features—the two pit structures (Features 1-1 and 1-2) and a pit (Feature 1-4)—seemed to be roughly contemporaneous during this phase. A second pit (Feature 1-5), constructed in the fill of Feature 1-2, may or may not have been from the Colonial period. Temporally, diagnostic sherds from Feature 1-4 and an archaeomagnetic sample from Feature 1-2 suggested that the inhabitants had occupied Area 1 during the early to middle portion of the Santa Cruz phase. Although investigators could not date the primary inhumation or define its stratigraphic relationship, the presence of the burial suggested the possibility of a Classic period use of Area 1.

The features in Area 1 may have been related to nearby agricultural fields and activities associated with the fields, canal maintenance, and temporary habitation. The morphology and attributes of the two pit structures indicated that these structures were probably seasonally occupied field houses. In addition, the ground stone assemblage within Feature 1-1 suggested that the inhabitants had both produced ceramics and processed botanical resources in the area. The macrobotanical samples from features in Area 1 were unproductive. Nevertheless, the pollen samples provided evidence for the possible exploitation of wild resources and weedy species associated with soil-disturbing, possibly agricultural, activities. The botanical analyses provided no direct evidence for agriculture or related activities within the area.

The architecture within Area 1 was consistent with the view that the area had been seasonally occupied. The architectural details of the pit structures (e.g., lack of floor and wall plaster, formal entrances, etc.) indicated that their construction had not involved the investment of time or effort required for more formal, permanently occupied pit houses. The floor area of the one complete structure was approximately 15.1 m^2. The lack of internal features, other than postholes, and the meager artifact assemblage indicated the limited use of these structures. An archaeomagnetic assay from an informal fire area on the floor of one of the structures pointed to use of the structure between the late Gila Butte and the mid Santa Cruz phases. This agreed with the conclusion drawn from the testing evidence that Area 1 had been occupied during the Colonial period.

The two prehistoric extramural features in the area were pits with evidence of burning. One of these pits contained Santa Cruz Red-on-buff sherds, the only temporally diagnostic artifacts recovered from Area 1. An inhumation provided evidence for the possible use of the area during the Classic period.

Botanical remains provided no direct evidence for agriculture. However, the presence of weedy plant remains commonly associated with soil-disturbing activities suggested that the occupants had undertaken agricultural activities in the vicinity. In addition, pollen samples indicated that they may also have exploited wild plant resources.

In summary, the archaeological evidence from Area 1 indicated that the area was a farmstead that had been occupied during the Santa Cruz phase. Evidence for ceramic production indicated that the Hohokam had also pursued activities other than those related to agriculture. They occupied the area on a seasonal basis, probably during the growing season and most intensely during the harvest. Microbotanical remains suggested that the inhabitants had also exploited Cheno-ams and other wild resources.

CHAPTER 3

DUTCH CANAL RUIN, AREA 2

Kathleen S. McQuestion
David H. Greenwald

Research Area 2 was in the western portion of Dutch Canal Ruin immediately north of South Main Canal 1 (Alignment 8003) between 18th and 20th streets (Figure 1.2). Area 2 consisted of a pit structure with an associated extramural activity area or possible second structure located to its southwest (Figure 3.1). Investigators found artifacts, representing a trash deposit, in the upper fill of the canal, indicating that residents had disposed of refuse in the canal channel during its latter periods of use and after it was no longer functional. Area 2 was thus defined on the basis of the spatial proximity of these three features and their apparent temporal associations.

Area 2 was located within the project grid at N1505–1521, E1127–1140, immediately north of Sherman Street, and in T1N, R3E, Sec. 10, SE¼NE¼SW¼ on the Phoenix, Arizona, 7.5 minute USGS topographic map (Figure 1.4). This location had formerly been a residential area; however, field crews noted few impacts to the prehistoric remains. The sidewalk along the north side of Sherman Street had little effect on the cultural resources. Adjacent utility lines—probably a 4-inch water line and a gas line—extended into the upper deposits of the South Main Canal, but their impact was also minimal because they were stratigraphically shallow.

The backhoe operator excavated three trenches spaced at 5-m intervals in Area 2 during the testing phase. Each trench was 25 m in length, beginning at the sidewalk and extending northward. Trench 1173 was the originally excavated systematic trench. As standard procedure during the testing phase, additional trenches were dug 5 m to each side of trenches that contained cultural features, and Trenches 1312 and 1313 were placed to either side of Trench 1173 after the South Main Canal and Feature 2-2, a pit structure, were discovered.

Overall, Area 2 appeared similar to Area 1 in relative size, composition, function, and temporal association. However, its geographic location and its spatial relationship to a main canal feature offered additional research opportunities. During the testing phase, excavators noted that the pit structure and activity area originated at approximately the same elevation as the canal (Figure 3.2). This relationship, plus the potential for addressing problem domains relative to functional and temporal questions, led to the selection of Area 2 for inclusion in the Phoenix Sky Harbor Center data recovery program.

RESULTS OF TESTING

During the testing phase, investigators identified Feature 1039, believed to be a pit structure, in Trench 1173. To locate other nearby features, they added judgmentally placed backhoe trenches 5 m to each side of Trench 1173: Trench 1312 to the west and Trench 1313 to the east. Field crews found Feature 1055, identified as a possible structure from its profile morphology, in the west face of Trench 1312. No additional features appeared in Trench 1313. All trenches exposed the South Main Canal at their southern extremes. Trenches could not be dug farther to the south than the N1506 grid line because Sherman Street and its associated sidewalks extended east-west through this portion of the project area. During data recovery, the South Main Canal (Alignment 8003), Feature

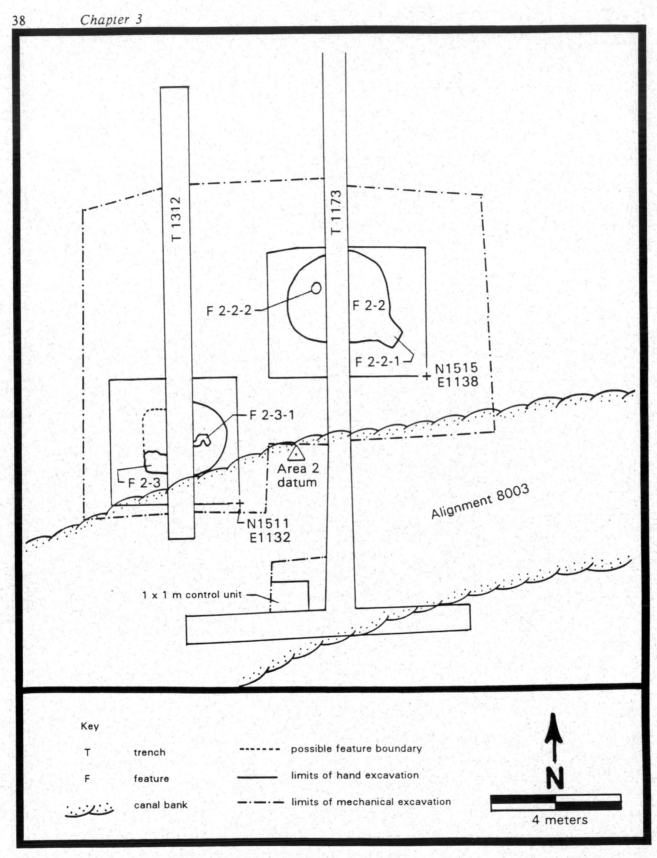

Figure 3.1. Area 2, Dutch Canal Ruin.

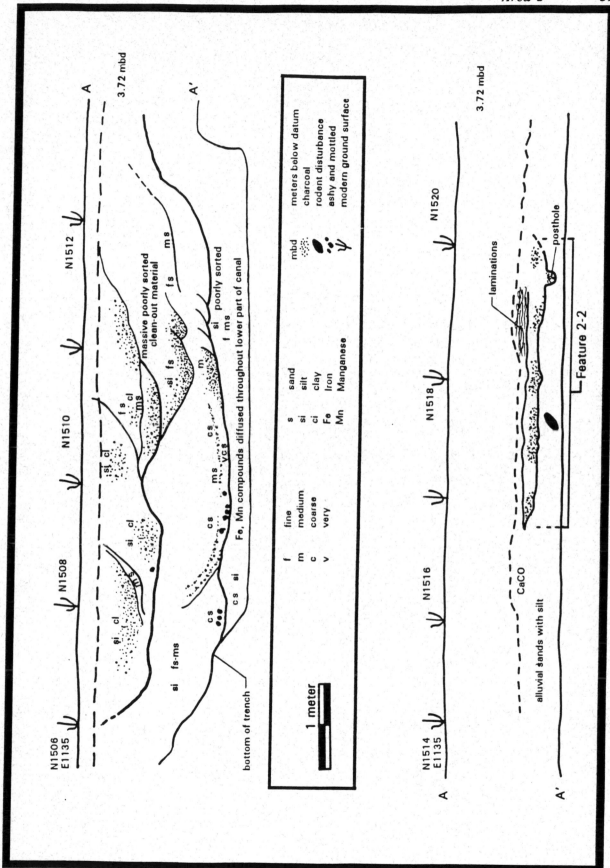

Figure 3.2. Area 2, profile of Backhoe Trench 1173 showing the relationship of the South Main Canal (Feature 2-1) to the pit structure (Feature 2-2).

1039 (the pit structure), and Feature 1055 (the activity area) were designated Features 2-1, 2-2, and 2-3, respectively (Figure 3.1).

The presence of a Gila Butte sherd and archaeomagnetic dating of basal oxides in the South Main Canal (Ciolek-Torrello and Greenwald 1988:Table 24) indicated that the Hohokam may have occupied Area 2 during the Colonial period. Investigators easily recognized the spatial relationship of the pit structure and its associated activity area to the canal alignment. However, they did not define the specific occupational sequence of the area during the testing phase, and they designed data recovery efforts specifically to address this issue.

RESEARCH OBJECTIVES

Because excavators found at least two pit structures along the north side of the South Main Canal alignment during the testing phase, they defined Area 2 as an isolated habitation area, perhaps a field house or a small farmstead locale. Through study of this area, investigators intended to provide a basis for defining feature clusters in mixed or concentrated units elsewhere at the site (such as Area 3) by developing a better understanding of the composition and function of a field house or farmstead locale.

The investigating team also expected to determine the number, kind, and distribution of surface and subsurface remains. For example, no "trash mounds" are indicated on any of Midvale's maps (1934) of the Dutch Canal Ruin area in the vicinity of Area 2. Similarly, no surface artifacts were located in Area 2 during BRW's pedestrian survey of the project area (1986:Figure 4).

Another goal was to determine the relationship of cultural remains to the natural stratigraphy of the project area. Because possible habitation features lay adjacent to a main canal feature in Area 2, investigators hoped that examination of the area's stratigraphy would provide information about the site's occupational sequence, which they considered important to understanding land-use strategies.

Systematic collection during testing recovered only ceramics. Of 25 total sherds, 17 were from Trench 1313, 7 were from Trench 1312, and 1 was from Trench 1173. Analysts identified 9 different types, predominantly nondiagnostic Hohokam wares. The temporally or culturally affiliated wares were one Gila Butte Red-on-buff sherd and one Lower Colorado Buff Ware ceramic. Nondiagnostic wares were Gila Plain, Salt Variety (12), Wingfield Plain (5), Gila Plain, Gila Variety (2), and single examples of an unidentified plainware, a plain redware, an unidentified smudged type, and an undecorated Hohokam Buff Ware.

Through further investigation of Area 2, archaeologists attempted to gain information on land-use strategies, canal and field maintenance activities, and site selection processes. They focused on the association between structures and canals and the developmental processes of fields and canal systems within larger habitation zones, such as the La Ciudad Occupation Zone.

SWCA's investigating team expected to apply information gained during excavations in Area 2 to research questions pertaining to the significance of field house sites, the development of the canal system, and the occupation and use of the geologic floodplain. They also sought archaeological evidence to support the notion of an occupational hiatus on the first terrace during the Sedentary period of Hohokam occupation (Volume 1:Chapter 6; see also Nials and Gregory 1989; Nials, Gregory, and Graybill 1989).

FIELD METHODS AND STRATEGIES

Backhoe Trenching

During the data recovery phase field crews reopened two of the three trenches excavated during the testing phase (totaling 40.7 m) to relocate Features 2-1, 2-2, and 2-3. Despite refacing and re-examining the trenches, they failed to identify any additional features. An additional east-west trench, 1173A, excavated perpendicular to Trench 1173 assisted in relocating it and in exposing an area of Feature 2-1, the South Main Canal.

Mechanical Stripping

Mechanical stripping of overburden deposits and peripheral areas surrounding possible structural features enabled investigators to identify the limits of those features (Features 2-2 and 2-3) and search for any additional extramural features. To allow for controlled excavation and sampling of the South Main Canal (Feature 2-1), field crews mechanically stripped an area 1.50 m north-south by 1.65 m east-west from ground surface to a depth of 1.07 m, exposing deposits at the approximate center of the most recent channel of the South Main Canal. They discontinued stripping within the fill of Feature 2-1 at a level containing a concentration of sherds and lithics. This level appeared to be the contact between the last flow episodes and the postabandonment deposits.

Crews also mechanically stripped a rectangular area around Features 2-2 and 2-3 (N 1510.80-1521.30, E 1127.20-1140.00 [134.4 m^2]) to locate additional features. The majority of the exposed area was west of Feature 2-2 and north of Feature 2-3. Overburden deposits were stripped from ground surface, at 3.20 mbd, to a depth of approximately 4.06 mbd. Investigators revealed no additional features and collected only a few flaked stone, ground stone, and ceramic items from the stripped area and its backdirt.

Hand Excavations

Archaeologists conducted hand excavations using a 1 × 1-m control excavation unit at N1570.5, E1133.0, within the canal fill of Feature 2-1, from 3.99 mbd to 4.29 mbd. They extracted an archaeomagnetic sample from 4.28 mbd, at the base of the excavation unit, in a sediment stratum of fine, dark brown, silty clay, approximately 88 cm above the base of the canal. These deposits resulted from either very slow moving water or standing water that permitted a rich botanical environment to develop. Analysts believe these sediments represented either late flow deposits or postabandonment deposits.

When the perimeters of Features 2-2 and 2-3 could not be defined in plan view by backhoe stripping, hand stripping became necessary. Hand stripping of the remaining overburden above Feature 2-2 encompassed an area 4.0 m north-south by 5.0 m east-west at N1515-1519, E1133-1138. Stripping continued from 3.83 mbd to 4.08 mbd, where the feature could be outlined in plan view.

The hand stripping unit around Feature 2-3 encompassed an area 4.0 m north-south by 4.0 m east-west at N1511-1515, E1128-1132. Hand stripping of the feature began at 3.82 mbd and continued until a controlled excavation unit was established at the level of feature detection, 3.85

mbd. Hand excavation continued within the features after their outlines had been defined according to standard methods.

STRATIGRAPHY AND ASSOCIATION

Natural stratigraphy in Area 2 consisted of three identifiable strata. The upper stratum extended approximately 25 cm below the present ground surface. This was the recently disturbed zone described by Nials and Anderson (Volume 1:Chapter 3). The deposit was loosely compacted and contained sandy lenses with pebbles. This stratum contained historic trash and pits and the historic Dutch Ditch. The boundary with the underlying blocky clays was gradual to abrupt.

The second stratum, texturally described as a clay loam, consisted of blocky clays that were the result of historic and prehistoric farming of Salt River floodplain deposits. The clay loam was high in organic matter, and its high clay content allowed for disturbance by natural pedogenic and bioturbation processes. This stratum consisted of Holocene soils with high agricultural potential (Volume 1:Chapter 3 and Appendix B).

The third stratum in Area 2 was a silty loam. This stratum consisted of massive prehistoric floodplain silts and fine sands that were hard, compact, and slightly sticky and slightly plastic when wet, corresponding to the poorly sorted sands reported by Nials and Anderson (Volume 1:Chapter 3). Both Features 2-2 and 2-3 lay within this prehistoric silty loam stratum. Feature 2-1, the South Main Canal, had been excavated into that stratum and was situated stratigraphically at the soil horizon boundary between the silty loam and the overlying blocky clay stratum.

CULTURAL FEATURES

Investigators relocated the three cultural features—the South Main Canal (Feature 2-1), a pit structure (Feature 2-2), and a possible structure or activity area (Feature 2-3)—during investigation of Area 2. They uncovered no extramural features. However, they did define three subfeatures during data recovery phase operations: an entry (Subfeature 2-2-1) and a fire pit (Subfeature 2-2-2) identified as part of the pit structure and a concentration of fire-cracked rocks (Subfeature 2-3-1) associated with the possible structure or activity area (although it appeared to have been deposited somewhat later and was located at a slightly higher elevation).

Structures

Trench 1173 bisected Feature 2-2, a pit structure located approximately 3 m north of the canal (Figure 3.1). A possible second pit structure, Feature 2-3, was approximately 5 m southwest of Feature 2-2, immediately adjacent to the north bank of the canal in Trench 1312. Because Feature 2-2 was oriented with its entryway to the southeast, investigators expected to find no other contemporaneous structures east or southeast of the feature, given its proximity to the canal.

Feature 2-2

Testing Feature: 1039
Trench Number: 1173
Horizontal Provenience: N 1516.00–1518.85; E 1133.76–1136.86

Vertical Provenience: 3.99–4.24 meters below datum (mbd)
Feature Type: Pit Structure

Feature 2-2 was a relatively circular pit structure, measuring 2.85 m north-south by 3.10 m east-west, with an entryway oriented to the southeast (Figure 3.3). The rectangular ramped entryway measured 60 cm in length and 85 cm in width. The entryway sloped up gradually (approximately 18 cm) from the floor of the structure to the prehistoric ground surface. The walls were highest at the interior opening of the entryway, measuring 25 cm in height at this location.

The earthen pit walls were for the most part relatively vertical, with only a slight curvature to the floor. In the northwestern and southwestern portions of the structure (Figure 3.3) a gently sloping curvature joined the floor and walls. Wall heights for the structure ranged from 11 cm to 25 cm.

Neither the floor of the structure nor the entryway had been formally prepared; both had been compacted through use. The floor was fairly level, varying only 4 cm in elevation at various locations across the surface. Its state of preservation was poor; rodent disturbance and bioturbation were apparent in the mottled floor deposits. The floor lay 97 cm below the present ground surface and 28 cm below the prehistoric surface, as determined by the entryway's exterior opening. The floor was covered with a very thin, medium- to coarse-grained sand deposit, consisting of a heterogenous composition of angular sand grains. Their angularity implied that the sands had not been transported far from their origin, whether by canal or by alluvial processes. Six artifacts were on the floor: 2 flakes, 1 mano, 1 possible ground stone item, and 2 ceramics, including 1 worked sherd.

The only interior feature was the fire pit (Subfeature 2-2-2), located opposite the entryway, 45 cm from the wall in the northwest quadrant of the structure. Fire pits are commonly located near the walls of small structures such as Feature 2-2 (Crown 1985a:76). The fire pit measured approximately 30 cm in diameter and 8 cm in depth and showed signs of disturbance on its eastern edge. The fill consisted of charcoal flecks and slightly ashy soil, but no sample for radiocarbon dating was available. A heavily oxidized area, which may have burned as drafts came through the entryway, extended 14 cm northwest of the feature.

Another oxidized area noted in the eastern half of the structure was not designated as a feature. Located approximately 75 cm northwest of the entryway, the burned area measured 25 cm north-south by 30 cm east-west. The intensity of burning in this area was low, as indicated by the pink to orange color of the oxidized soil, much lighter than the red appearance of the fire pit oxidation. Immediately northwest of the oxidized area, adjacent to Trench 1173, was a dense concentration of charcoal-rich soil. This deposit had probably resulted from burning in the lightly oxidized area, or the charcoal could have been related to a hearth or another fire pit that was more centrally located within the structure before being removed by the backhoe. Alternatively, the oxidized and charcoal-stained soil areas could have resulted from catastrophic burning of the structure. Another possible explanation is that the slightly oxidized area was a less formal burning area. The more formal fire pit, Subfeature 2-2-2, may have been evidence of reoccupation of the pit structure during another season or a longer visit to the structure; however, Crown (1985a:83) reported that historically, Piman huts used overnight had hearths, a practice also documented ethnographically for the Navajo (Russell 1978:39). Ash accumulations have been associated with Navajo field houses when a structure was occupied for more than a few days or was reoccupied at a later date.

Two circular organic stains, approximately 3 cm in diameter, were exposed on the floor in the eastern half of the structure. They appeared darker than the floor surface and were thought to be

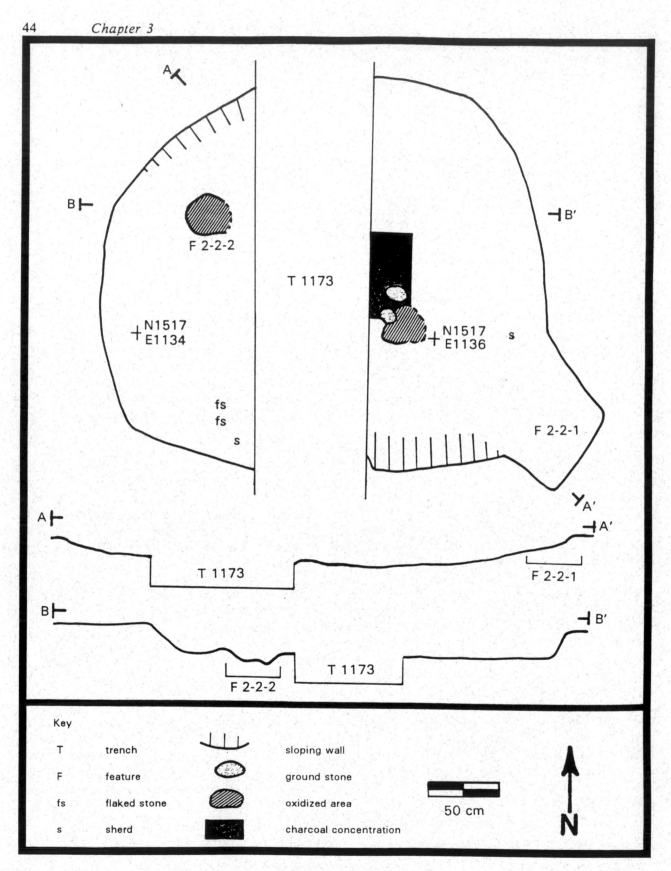

Figure 3.3. Area 2, Feature 2-2, plan and cross sections.

possible postholes. After excavation, investigators dismissed these stains as rodent holes. The organic-stained appearance probably resulted from rodents disturbing charcoal-rich floor deposits. The posthole profiled on the western half of the structure during testing may have been a similar manifestation, but field crews could not relocate it during this phase of the investigation.

Although field crews found no posthole features, they did uncover other evidence for the construction of the roof and upper wall. They observed burned daub roofing material in some areas of the structure, just above the floor. The roof had probably been wattle and daub, but due to the ephemeral nature of such superstructures, excavators found no evidence of any support posts in association with this structure.

Apparently the pit structure burned, the roof collapsed onto the floor, and the feature was subsequently covered by alluvial deposits. Analysts could not determine whether the coarse angular sands that covered the floor had been deposited by natural or cultural means. In all likelihood, however, because the sands were beneath the charcoal layer, they probably existed in this location naturally as a component of the poorly sorted sand unit into which cultural features were occasionally dug (Chapter 18; Volume 1:Chapter 3). However, they could have been used in the construction of the superstructure and deposited across the floor when the wattle-and-daub covering collapsed. Alternatively, these sands could have been washed into the depression of the structure as a result of a flood episode. It seemed unlikely that residents of this structure would have intentionally covered the floor with the sands, as this is not recorded as a common practice among Hohokam groups.

Feature 2-3

Testing Feature: 1055
Trench Number: 1312
Horizontal Provenience: N 1512.05–N 1512.74; E 1129.07–E 1129.72
Vertical Provenience: 3.94–4.15 mbd
Feature Type: Possible Pit Structure

Feature 2-3 was a possible pit structure or an extramural activity area (Figure 3.4). Preservation was poor, and the feature was difficult to define, probably because of limited use. Controlled excavation units did not reveal the upper perimeter of the feature, which was profiled during testing as 45 cm in depth.

At the base of the south end of Feature 2-3 there appeared to be a curved wall-floor juncture, which may have represented a pit associated with an extramural surface or a pit structure. It was similar to the fairly level ash lens originally profiled. Field crews followed a relatively vertical earthen pit wall, ranging from 12 cm to 18 cm in height, to a point approximately 60 cm west of Trench 1312. They identified an indefinite boundary, possibly the remainder of the west and north walls, only by a difference in compaction. Silts and sands within the feature were slightly more compact than the loose, fine, sandy silts surrounding it. An interior surface was not located in that area. Field crews identified no walls east of the backhoe trench and could not determine whether the feature extended beyond the trench; in fact, the backhoe may have removed the eastern half of the feature.

Investigators identified a horizontal surface on the basis of compaction and the presence of a scattered deposit of charcoal flecks that contained some ash. Sands with coarse- to very coarse-grained calcium carbonate accumulations lay immediately below the feature, indicating culturally

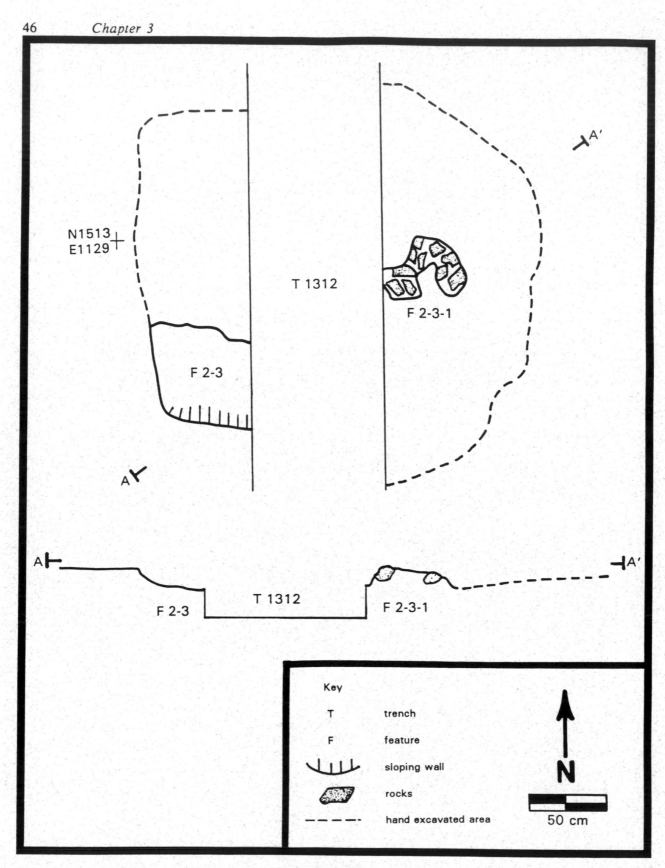

Figure 3.4. Area 2, Feature 2-3, plan and cross section.

sterile deposits. The surface had been compacted through use, but no preparation was apparent. The area measured a maximum of 69 cm north-south by 65 cm east-west and was distinguishable as a surface adjacent to the pit wall. Located 86 cm below the modern surface, this surface was 23–33 cm below the level at which the pit originated. The floor and pit wall remnants indicated that the plan of the pit structure may have been rectangular. The ephemeral nature of the feature represented extremely limited use, possibly even a single event, which would have resulted in little surface compaction. No artifacts lay on the remnant surface.

The fire-cracked rock concentration located east of Trench 1312 was designated as Subfeature 2-3-1 because of its proximity to Feature 2-3. However, investigators established no direct association between them, and the rock concentration may have been an extramural feature of Feature 2-2. Situated at 3.96–4.06 mbd, the rocks in Subfeature 2-3-1 were nearly level with the surface on which Feature 2-2 (the pit structure) originated but were 13 cm above the compacted surface identified for Feature 2-3 (4.19 mbd). The burned rock shown in the testing profile at 4.10 mbd was probably associated at one time with the fire-cracked rocks of Feature 2-3-1.

Subfeature 2-3-1 consisted of nine stones in a horseshoe-shaped arrangement with its opening to the south. The stones were fractured river cobbles, ranging from 9 cm to 12 cm in diameter. Two of the burned rocks may have been used for grinding activities. Very few charcoal flecks, no more than in the surrounding soil, appeared in the associated fill. The function of the feature was unknown. Due to the unnatural arrangement of the rocks, the concentration did not appear to be material associated in any way with the canal, such as spoils from canal maintenance activities. Archaeologists often find ground stone items used secondarily in cooking features, and possibly the rocks were dumped there after having been used elsewhere.

Field crews found no daub or adobe material that might indicate a structure more permanent than a lean-to or a windbreak during the investigations of Feature 2-3 or Subfeature 2-3-1. A windbreak or shade feature constructed along the north bank of the canal might have left such ephemeral remains, and runoff from fields to the north or water escaping the banks of the canal might have accumulated in the depression, concentrating the charcoal. The presence of charcoal and fire-cracked rock also could have resulted from trash having been dumped into the depression.

Canals and Trash Deposition

The South Main Canal, Alignment 8003, was dated to the late Pioneer and early Colonial periods by stratigraphic association, archaeomagnetic dating techniques, and ceramic associations (Ciolek-Torrello and Greenwald 1988). During the earlier study, an archaeomagnetic sample collected from the basal oxide deposits of the canal yielded a date of A.D. 700–775, the preferred interval. Since then, the Southwest VGP curve has been revised; when Dr. Jeffrey Eighmy reanalyzed that sample, it yielded a revised date of A.D. 650–725. Stratigraphically, the sample collected during the current project should have dated later than the sample from 1988. In fact, it produced two date ranges, the earliest visually determined to A.D. 700–800 and the latest to A.D. 875–1250. The earlier date fit well with the interpretations of the South Main Canal and supported the earlier results. The ceramic assemblage recovered from Area 2 in this canal, however, appeared to be mixed, containing both pre-Classic and Classic period types. This mixing of types was probably a result of the proximity of Area 8 to Area 2 and the fact that the majority of sherds recovered from the canal came from the upper or later channels. Late channels would have been open longer and would have been more likely to accumulate late ceramic types.

The South Main Canal, as its name implies, was used as a main distributional channel. Although its greatest width at this point was nearly 7.0 m, the widest reconstructible channel was only approximately 3.6 m wide with a depth of 67 cm (Figure 3.2). The second-earliest channel, however, appeared to have been over 90 cm in depth. Investigations in Area 6 (Chapter 7) and the South Main Canal's association with Alignment 8005 indicated that this canal was the earliest one constructed at Dutch Canal Ruin. It had the steepest gradient (1.92–2.78 m/km) recorded for any of the canals in the project area. Typical cross sections exhibited basal channel and bank erosion, with inner-channel migration, usually southward, commonly represented in the sediment history. Sediments were coarse, especially in the basal sections, and oxide deposits were usually well developed. These characteristics supported the contentions that the South Main Canal had been used extensively, water velocity had been high during early channel use, erosion had been a constant problem, and flow events had been of such duration that oxides had freely developed due to long periods of saturation (D. H. Greenwald 1988a).

In the location of Area 2, the South Main Canal could not have functioned later than A.D. 850 because the southern branch of the North Main Canal (Alignment 8002, also referred to as North Main Canal 3) reused that portion of the South Main Canal in the extreme western portion of the project area (Figure 1.2). North Main Canal 3 was placed in use by approximately A.D. 850/900. Given the stratigraphic evidence, investigators had no doubt that North Main Canal 3 postdated the use of the South Main Canal alignment, and therefore the Classic period ceramics in the upper fill of the South Main Canal could only have been deposited there through postabandonment processes.

The spatial relationship of the South Main Canal to the habitation features in Area 2 is shown in Figure 3.1. Investigators considered that determining contemporaneity for these features and defining their function within this spatial association were important research topics. Establishing the temporal relationship between the habitation features and the canal was one means of assessing the chronological interpretations for the canal. As shown in Figure 3.2, the prehistoric ground surface of the habitation feature and the level of origination of the canal compared very closely and suggested contemporaneity between the canal and the features. This notion was supported by the temporal association of the sherds recovered from the canal's fill.

Trash was deposited only in the canal in Area 2. No sheet trash had accumulated, as would be expected if occupants had used the habitation features on a year-round basis. Trash disposal in the canal may have occurred at any time during the occupation of the area, but debris may not have accumulated due to the high velocity of flow. Trash had accumulated, however, in the upper channel, at the interface between flow episode deposits and abandonment deposits. This trash accumulation indicated that Area 2 had been occupied at least periodically until the abandonment of the South Main Canal. The archaeomagnetic sample interval and the stratigraphic associations indicated an occupation during the Gila Butte phase. If the later of the two preferred date ranges from the archaeomagnetic samples is used, Area 2 and the South Main Canal may have been abandoned at approximately A.D. 700–800, in the latter part of the Snaketown phase or during the Gila Butte phase.

MATERIAL CULTURE

During the data recovery phase, investigators collected and analyzed 249 artifacts from Area 2. Of that total, 154 artifacts (61.8%) were ceramic, 86 (34.5%) were flaked stone, 8 (3.2%) were ground stone, and 1 (0.4%) was shell. Only 1 sherd came from a nonfeature context; however, 3 ground stone and 2 flaked stone artifacts (2.0% of the assemblage) lay in overburden deposits that were not directly associated with a feature. Absent from the Area 2 assemblage were human and faunal remains.

The control excavation unit in Feature 2-1, the South Main Canal, produced 1 ground stone and 4 flaked stone items, in addition to 118 ceramics (49.4% of the total artifact assemblage for Area 2). The pit structure, Feature 2-2, yielded 2 ground stone artifacts, 17 ceramics, and 67 flaked stone items, 34.5% of the Area 2 assemblage. The most ephemeral feature, Feature 2-3, the possible pit structure, contained 13 chipped stone artifacts, 1 piece of shell, and 18 ceramics, or 12.8% of the Area 2 artifact assemblage. Finally, the fire-cracked rock concentration, Subfeature 2-3-1, contained 2 ground stone artifacts, 0.8% of the assemblage.

Almost 50% of all the artifacts recovered from Area 2 came from a 1 × 1-m control excavation unit in the canal that was only 29 cm in depth. This fact showed not only the density of accumulated artifacts in the canal, but also the paucity of cultural remains recovered from the habitation features, a pattern typically associated with isolated, short-term habitation areas (Crown 1985a:Table 5).

Ceramics

Field crews recovered three types of Hohokam Plain Ware (136 sherds, 88.3% of the Area 2 assemblage), two Hohokam Decorated Ware types (16 sherds, 10.4%), and one Hohokam Red Ware type (2 sherds, 1.3%) from Area 2 (Table 3.1). In order of descending abundance, Hohokam Plain Ware included Gila Plain, Salt Variety; Squaw Peak Plain; and Gila Plain, Gila Variety. Hohokam Indeterminate Buff and Hohokam Indeterminate Red-on-buff were the only Hohokam Buff Ware types present. Analysts identified the Hohokam Red Ware sherds as Salt Red (Table 12.5).

As shown in Table 3.1, 76.6% of the ceramic assemblage was from Feature 2-1, the canal. Only one sherd came from a nonfeature provenience in Area 2. The remainder of the collection (22.7%) was equally divided between Feature 2-2, the pit structure, and Feature 2-3, the possible pit structure or activity area.

Eight sherds were from jars and eight were from bowls. Analysts could not identify the vessel form of the majority of the sherds (89.6%). From the control excavation unit in the South Main Canal, Feature 2-1, at least three Hohokam Plain Ware bowls were represented by five sherds. A Hohokam Red Ware bowl and a Hohokam Indeterminate Buff Ware bowl were also represented. In addition, at least four Hohokam Plain Ware jars were represented by five sherds, as was one Indeterminate Hohokam Red-on-buff jar.

Less can be said about the vessel forms represented at the other two features in Area 2. Feature fill deposits in the pit structure, Feature 2-2, contained one sherd from an incised Hohokam Indeterminate Red-on-buff bowl. During excavation, in-field identification of the incised red-on-buff sherd classified it as Snaketown Red-on-buff. This sherd was destroyed during thin-section petrographic analysis. One of the plainware sherds recovered from the floor contact in Feature 2-2 was a worked sherd ground on two edges. One Hohokam Plain Ware jar and a Hohokam Indeterminate Buff Ware jar were represented by two sherds found in Feature 2-3. Analysts had difficulty in establishing temporal associations for Area 2 using the results of the ware/type ceramic analysis from data recovery work at Dutch Canal Ruin. Because of the temporal range associated with Gila Plain, Salt Variety (ca. A.D. 300-1450+) and Gila Plain, Gila Variety (ca. A.D. 400-1350+) ceramics, these ceramics provided little aid in the relative dating of noncanal features in Area 2. Few temporally diagnostic ceramics were present in the assemblage, and none

Table 3.1. Area 2, Ceramic Frequencies by Feature

Feature Number	Ware Type						Total
	Gila Plain, Salt	Gila Plain, Gila	Squaw Peak, Plain	Salt Red	Indeterminate Red	Indeterminate Buff	
NF[1]	1 *100.0						1 **0.6
2-1	80 67.8	5 4.2	23 19.5	2 1.7	2 1.7	6 5.1	118 76.6
2-2	10 58.8	3 17.6	1 5.9		1 5.9	2 11.8	17 11.0
2-3	9 50.0	4 22.2			4 22.2	1 5.6	18 11.7
Total	100 *64.9	12 7.8	24 15.6	2 1.3	7 4.5	9 5.8	154 100.0

*Row %
**Column %
[1]Nonfeature context

came from well-preserved contexts such as floor fill or floor contact proveniences (not a surprising result, since field houses typically yield few diagnostic ceramic types). The presence of one incised red-on-buff sherd in the fill of Feature 2-2 suggested a Snaketown or Gila Butte phase affiliation; researchers have detailed that the technique of incising ceramics died out by the end of the Gila Butte phase (Haury 1976:228). Possibly the structure was abandoned before the end of the Gila Butte phase, but one diagnostic sherd is insufficient evidence for establishing the temporal affiliation for the structure.

One of the 20 Lower Colorado Buff Ware sherds recovered during testing of Dutch Canal Ruin came from Area 2. Field crews found the intrusive ceramic in a 5 × 5-m backdirt collection unit from Trench 1173, at N1520, E1135, just north of the pit structure, Feature 2-2. The sherd possibly represented limited use of the area and temporal affiliation with the middle to late Sedentary or early Classic period occupation, as documented at Las Colinas (Beckwith 1988:220), a major village site west of Dutch Canal Ruin in Canal System 2. The sherd might also have been temporally associated with the occupation of Area 8 at Dutch Canal Ruin, approximately 60 m southwest of Area 2. Area 8 yielded 45 Lower Colorado Buff Ware ceramics during the testing (10 sherds) and data recovery (35 sherds) phases.

The single Gila Butte Red-on-buff ceramic found during testing was recovered from the 5 × 5-m backdirt collection unit in Trench 1313 at N1505, E1140. This sherd probably came from South Main Canal deposits, corresponding temporally with the period of use proposed for the South Main Canal.

Ceramics provided evidence that the uppermost channel of the South Main Canal had not been actively flowing at the end of the Colonial period. One Hohokam Plain Ware sherd of indeterminate form, located in the canal fill, exhibited a shoulder, a trait often associated with the Sedentary period. Several of the ceramics in the canal deposits were heavily coated with calcium carbonates, indicative of a depositional environment produced by standing or slow-moving water. Although the canal was abandoned during the Colonial period, it continued to serve as a moisture catchment area, evidenced by the heavy accumulations of calcium carbonates on the artifacts. Most of the ceramics excavated from the fill had sharp edges, although a few were well rounded, as if transported by water. Two Salt Red ceramics, manufactured between A.D. 1300 and 1450, were also recovered from the uppermost channel of the South Main Canal, indicating that infilling of the channel had occurred during the Classic period of occupation. The quantities of artifacts found in the upper levels of the canal indicated intentional deposition of trash, probably associated with the later occupation of Area 2. The existence of later ceramic types probably resulted from use of the general area by residents of Area 8.

Flaked Stone

Investigators collected and analyzed 86 flaked stone artifacts from Area 2, just over 19.7% of the flaked stone assemblage recovered from pre-Classic areas at Dutch Canal Ruin. Analysts identified seven morphological types, 93.0% of which were debitage, and five raw material types (Table 3.2). Well over half (61.6%) of the material present was basalt. When combined with other types of igneous material, artifacts of volcanic origin constituted 88.4% of the assemblage.

The three features at Area 2 yielded 84 flaked stone artifacts. Of these, 67 (77.9% of the total flaked stone assemblage from Area 2) came from the pit structure, Feature 2-2. The remainder were distributed between the possible pit structure (13, or 15.1%) and the South Main Canal (4, or 4.7%).

Two of the four artifacts from the South Main Canal were core/hammerstones of basalt. The others were a whole flake, also of basalt, and a battered hammerstone of some other igneous material. Given the rather large size of the cores and the hammerstone and the fact that none of the four artifacts exhibited random battering as would be expected had the artifacts been transported by high velocity flows, investigators concluded that the items had probably been dumped in the canal either when water was moving through it slowly or after its abandonment. Because of the limited amount of reduction, these tools exhibited potential for further use and were, therefore, designated as trash based on context.

Field crews recovered 67 flaked stone artifacts from the pit structure, Feature 2-2: 17 from feature fill, 48 from floor fill (Table 3.3), and 2 from floor contact. The feature fill yielded a relatively small number of artifacts because crews did not systematically screen these deposits but collected only observed artifacts.

Of the 17 pieces of debitage from the pit structure feature fill, basalt and igneous other material were evenly represented, with 8 items each. One quartz flake fragment was also present in the

Table 3.2. Area 2, Flaked Stone Morphological Type by Material Type

Morphological Type	Material Type					Total
	Quartz	Chert	Basalt	Igneous Other	Quartzite	
Angular Debris			7	4		11
			*63.6	36.4		**12.8
			**13.2	17.4		
Flake Fragment	1	1	20	7	3	32
	3.1	3.1	62.5	21.9	9.4	37.2
	100.0	100.0	37.7	30.4	37.5	
PRB[1] Flake			4	3		7
			57.1	42.9		8.1
			7.5	13.0		
Whole Flake			17	8	5	30
			56.7	26.7	16.7	34.9
			32.1	34.8	62.5	
Used Flake			1			1
			100.0			1.2
			1.9			
Hammerstone/ Core			3			3
			100.0			3.5
			5.7			
Hammerstone			1	1		2
			50.0	50.0		2.3
			1.9	4.3		
Total	1	1	53	23	8	86
	*1.2	1.2	61.6	26.7	9.3	100.0

*Row %
**Column %
[1]Platform-remnant-bearing

Table 3.3. Feature 2-2, Floor Fill, Flaked Stone Morphological Type by Material Type

Morphological Type		Material Type			Total
	Chert	Basalt	Igneous Other	Quartzite	
Angular Debris		7	2		9
		*77.8	22.2		**18.8
		**24.1	18.2		
Flake Fragment	1	13	5	2	21
	4.8	61.9	23.8	9.5	43.8
	100.0	44.8	45.5	28.6	
PRB[1] Flake		1	1		2
		50.0	50.0		4.2
		3.5	9.1		
Whole Flake		6	3	5	14
		42.9	21.4	35.7	29.2
		20.7	27.3	71.4	
Hammerstone/ Core		1			1
		100.0			2.1
		3.5			
Hammerstone		1			1
		100.0			2.1
		3.5			
Total	1	29	11	7	48
	*2.1	60.4	22.9	14.6	100.0

*Row %
**Column %
[1] Platform-remnant-bearing

feature fill. The 48 flaked stone artifacts found in the Feature 2-2 floor fill deposits included debitage and two tool types: a core/hammerstone and a hammerstone (Table 3.3).

The Feature 2-2 flaked stone assemblage (a core/hammerstone, a hammerstone, and debitage) suggested that occupants of the structure had reduced cores there to produce core tools or flakes suitable either for immediate use or for further reduction into tools. The two types of hammerstone suggested that occupants had also conducted both tool production and maintenance and food processing activities in the structure. Core/hammerstones have been associated with shaping and roughening the surfaces of grinding tools (Dodd 1979:237–239). Hammerstones were used for food processing, as indicated by their blunt surfaces and use-wear patterns, and were also used as percussors during core reduction.

Field crews recovered 13 pieces of debitage (including 6 flake fragments, 4 platform-remnant-bearing flakes, and 3 whole flakes) from Feature 2-3, the possible pit structure. The fill from three control excavation units contained 7 pieces of debitage. Four items lay in the 19 cm of floor fill deposits overlying the definable contact surface in Feature 2-3. No flaked stone artifacts were in direct contact with the surface.

Ground Stone

Investigators recovered eight ground stone items from Area 2. The 1 × 1-m control excavation unit in the South Main Canal (Feature 2-1) yielded one piece of ground stone. Two ground stone artifacts came from the pit structure, Feature 2-2, and two came from the fire-cracked rock concentration, Subfeature 2-3-1. Three artifacts were present in overburden above the east half of Feature 2-2.

A fragment of a metate not further specified (NFS) was recovered from the control excavation unit dug in Feature 2-1. The fragment, made of vesicular basalt, had a coarse surface texture, and it was probably part of a slab or trough metate.

Field crews recovered two burned ground stone artifacts from the floor of Feature 2-2, the pit structure. These artifacts were northwest of an area in which carbonized material was concentrated in the soil, parallel with the entryway, near the center of the structure. A complete round/oval mano of coarse-textured diorite had been utilized on two of its waterworn flat surfaces; however, the irregularly shaped cobble showed no evidence of production investment. A well-rounded quartzite cobble, three-quarters complete, lay 5 cm away. This waterworn cobble had possibly been used for grinding purposes, but its slightly polished appearance may also have been created by burning.

Two burned items that may once have been used for grinding activities were present in Subfeature 2-3-1, the fire-cracked rock concentration. One item was a waterworn sandstone cobble fragment that was not shaped. The other fragment had two flat surfaces, one with possible pecking wear, and was of granitic material. Use wear was often difficult to distinguish on other ground stone artifacts of similar material in the project collection. Analysts classified both samples as possible ground stone.

Field crews recovered two rectangular/loaf manos and a trough metate fragment from the overburden removed from above Feature 2-2 during mechanical stripping. All were composed of materials with a coarse surface texture. One mano and the metate fragment were of quartz-bearing basalt; the other mano was of vesicular basalt. Both manos had been completely shaped. One had two use surfaces, and the other was ground on only one surface. These manos had been used with a trough metate. In fact, on one of them the only well-polished area that remained was the trough wear on the ends of the tool. Both manos were broken late in the process of pecking to rejuvenate use surfaces. Only a fragment of the trough metate was present, although analysts determined that the tool had been an open-ended type.

The metate fragment found in the canal had undoubtedly been discarded, because in its recovered condition its function would have been very limited. Investigators could not determine when it had been discarded. Both of the ground stone artifacts on the floor of the pit structure, Feature 2-2, had probably been there since before the structure burned, unless their burned condition was the result of some earlier event. The informal round/oval mano was not shaped; replacement of the mano may have involved only a trip to the closest river cobble source. Whether the little-used tool had been curated with the intent of reuse or had been expediently used and discarded could not be determined.

The two ground stone items in the rock concentration did not occur there naturally. The Hohokam on occasion used broken pieces of ground stone for heating rocks in cooking activities. Because investigators did not recover other pieces of these fragmented items during the excavation, they could not determine how the fragments had come to be associated with the feature. Evidence of cultural modification for grinding activities was not obvious or extensive, and analysts could not assess the functional implication of these items within the ground stone assemblage.

The two floor artifacts from Feature 2-2 and the fire-cracked rocks from Subfeature 2-3-1 were associated with the Area 2 occupation. Of these, production investment could be measured on only one of the artifacts. No effort had been invested in producing the round/oval mano, implying it was an expediently used tool. In addition, the cobble had probably originated from a local source, such as the closest streambed gravel outcrop. Procurement strategies were influenced by the availability of raw materials, as the Salt River and its braided streambeds offered reliable and convenient sources.

Analysts could not conclusively determine whether the formal ground stone artifacts recovered from the overburden were in fact associated with the occupation of Feature 2-2. Quite possibly, their presence was the result of postabandonment discard activities. This notion was supported by the fragmentary nature of the remains and the fact that none of the materials showed evidence of the extent of burning that would be expected had they been within or on top of the structure when it burned.

Shell

Field crews collected one shell item from Feature 2-3, the possible structure. The artifact was located between 3.82 mbd and 4.03 mbd, in the excavated portion of the 2 × 2-m grid unit at N1513, E1328, north of the definable contact surface. The shell came from fill deposits of this feature, which would have overlain floor fill materials. The artifact was a worked fragment of *Laevicardium* sp. The form of the item could not be determined.

Subsistence Remains

Feature 2-2, the pit structure, yielded two pollen samples and one flotation sample for analysis. Field crews collected one pollen sample and the flotation sample from the floor of the structure and collected the second pollen sample from the fire pit.

The pollen floor sample selected was from the west half of the structure. Analysis supported the notion of both plant processing and possible natural transport (Chapter 16). An aggregate of *Zea mays* pollen demonstrated that corn had been processed in Feature 2-2. Elevated Gramineae pollen counts and the presence of Cruciferae pollen indicated that members of the grass and mustard family also had probably been processed in the structure. Economic genera of the Gramineae family would primarily include *Sorghum, Sporobolus* (dropseed), *Panicum* (panic grass), and *Oryzopsis* (rice grass) (Kearney and Peebles 1960:112–115, 119, 142). A small quantity of *Prosopis* (mesquite) pollen was also present. Cummings (Chapter 16) suggests that the mesquite pollen might represent food preparation, including the processing of pods, or that the pollen could have been deposited naturally by wind or rain. One could expect more than a small quantity of pollen if mesquite or mesquite pods were being stored in the structure.

The fire pit, Subfeature 2-2-2, produced a small quantity of Liliaceae (lily family) pollen. Also, small quantities of Liguliflorae-type Compositae (sunflower family) pollen were present in the sample. Cummings suggests that this might reflect contamination, possibly by rodent activity, or that some member of the Compositae group was present as a prehistoric weed. Cummings notes specifically the lack of *Zea mays* pollen in the sample from the fire pit (Chapter 16).

Pollen types identified in Area 2 implied occupation during the summer and fall. Mesquite pollen resulting from mesquite pod harvesting would indicate summer occupation. Corn potentially can be harvested in late June and October, grasses typically in summer and fall.

In comparison to the other seven pre-Classic period pit structures that were analyzed for pollen at Dutch Canal Ruin, Feature 2-2 was one of two that contained pollen from the mustard family, and it was the only pit structure to have elevated pollen from the grass family. The presence of mesquite and corn pollen, however, was common in these types of structures.

Of the five pre-Classic period pits that underwent pollen analysis at Dutch Canal Ruin, Subfeature 2-2-2, the fire pit, was the only pit that contained Liliaceae pollen. This pollen type did not appear in the pit structure samples (Chapter 16).

Flotation analysis was conducted on a sample extracted from the lower 5 cm (4.14–4.19 mbd) of floor fill in the western half of Feature 2-2, the pit structure. Kwiatkowski (Chapter 15) classified the 2.75-liter sample as unproductive in terms of carbonized plant taxa present. Wood charcoal made up less than 0.5 ml of the 7.5 ml light fraction. Analysts identified only two miscellaneous round fibers and one white, styloid crystal.

CHRONOLOGY

Stratigraphy

The entryway of Feature 2-2, the pit structure, terminated at a prehistoric surface at approximately 3.93 mbd. As originally profiled, the northern boundary of Feature 2-3 originated at 3.95 mbd. The southern boundary of the pit was not used to illustrate this association, as the corresponding extramural surface may have been affected by its proximity to the canal and possible canal construction spoils material. Both features lay within the silty loam stratum that consisted of prehistoric floodplain silts and sands. The features originated within 7 cm of the surface at which the initial channel of the South Main Canal was dug (4.02 mbd). The uppermost channel and the two channels immediately below it were truncated by modern disturbance.

Feature 2-2 was buried by alluvial silts and sands. An indefinite, irregular contact separated those poorly sorted sands from the Holocene soils above, indicating an erosional period prior to the formation of the blocky clays. Calcium carbonate accumulation at that contact implied that an environment with standing water had existed before the blocky clays associated with prehistoric and historic irrigation filled Feature 2-2. The presence of the calcium carbonates in a horizontal pattern indicated that a surface had once existed at this level or slightly above it. The canal banks would have served to retard natural runoff, and standing water could have occurred frequently along the north side of the canals. The location of these carbonates illustrated the approximate level of the prehistoric surface associated with both the canal and Feature 2-2.

Ceramics

Two Gila Butte ceramics were present in Area 2 of Dutch Canal Ruin: one possibly from the fill of the South Main Canal and one, an incised sherd, from the fill of the pit structure, Feature 2-2. Two sherds alone would not substantiate a Gila Butte phase occupation, but combined with other available stratigraphic and absolute dating information, they supported a plausible Colonial period designation.

Wingfield Plain and Lower Colorado Buff Ware sherds indicated a later period of use, also suggested by the existence of the shouldered vessel form and the two Salt Red sherds found in the South Main Canal deposits. Although investigators confidently dated the abandonment of the canal to the Colonial period, the presence of these types indicated that the channel had not been completely filled by the Classic period. Thick accumulations of calcium carbonate on the ceramics from this provenience indicated that the sherds had been inundated by standing or slow-moving water after deposition.

Absolute Dates

Field crews collected one datable radiocarbon sample from Area 2, a composite sample extracted from the control excavation unit in the South Main Canal. Investigators did not submit the sample for dating but curated it in the event that archaeologists increase the reliability of radiocarbon results from canal features.

Analysis of an archaeomagnetic sediment sample (CSUSED-80) taken from the postabandonment clays in the basal sediments of the uppermost channel of the South Main Canal resulted in a visual date of A.D. 700–800 and A.D. 875–1250, with a declination date of A.D. 925–1675 (Appendix A). The visual date ranges corresponded to dates proposed for the Snaketown and early Gila Butte phases, at the boundary between the Pioneer and Colonial periods, and the late Colonial through early Classic periods. The declination date range represented a broader time span, from the Santa Cruz phase of the Colonial period, through the Sedentary period, to the Soho phase of the Classic period (Dean 1991).

The Museum of Northern Arizona sampled basal oxides from South Main Canal 1 in the Squaw Peak Parkway corridor (Ciolek-Torrello and Greenwald 1988:159). That analysis produced date intervals of A.D. pre-700–720 and 915–940 when initially plotted. When replotted for this study on an updated curve, SWCV590, the date ranges were refined (Appendix A). Visual dates were A.D. 650–725, 900–975, and 1550–1650, and declination dates were A.D. 650–1025 and 1325–1725. The refined visual dates were associated with the Snaketown phase, the Santa Cruz phase, or the post-Classic period of the Hohokam chronology (Dean 1991). Eighmy and Howard (1991) have discussed the reliability of dating canal sediments.

The construction of the South Main Canal in Area 2 of Dutch Canal Ruin dated to at least the Gila Butte phase based on stratigraphic association with Feature 2-2. The results of the two archaeomagnetic samples, ceramic associations, and stratigraphic associations indicated that the construction of the South Main Canal can be dated more confidently to the Snaketown phase.

SUMMARY

Feature 2-2, a pit structure, and Feature 2-3, an ephemeral feature that may possibly be the remnants of another pit structure, constituted the limited remains associated with a major irrigation canal found in Area 2. Investigators had difficulty in conclusively establishing the structure of Area 2 because of the indefinite nature of Feature 2-3. The structural composition of this locality, therefore, must be discussed by relying primarily on evidence associated with the pit structure.

Usually, before isolated structures are classified as field houses, their association to agricultural fields or features must be established. Gregory (1991:163) defined field houses as "individual structures and facilities established solely for the purpose of tending agricultural fields . . . inferred to have been occupied during periods of planting, growing, and/or harvesting of crops." The pit structure, located approximately 3 m from the north bank of the South Main Canal, was associated with this feature both spatially and temporally. Proximity to an irrigation canal provides justification for a field house designation even if agricultural fields cannot be identified. Distance from a major village site can also be used to infer a field house classification, or at least a limited-activity structure.

It is safe to assume that the amount of labor expended to construct a habitation unit or any intramural feature is directly related to the amount of time that will be spent at the locality. The pit structure in Area 2 was more formal than a windbreak, lean-to, or ramada feature, implying either repeated use of the same locality or a semipermanent occupation of seasonal duration. The presence of only one Colonial habitation feature, few artifacts, and a small amount of accumulated trash indicated occupancy of short duration by a relatively small number of people, possibly a subnuclear family (Cable and Doyel 1984:275). Seasonal or short-term occupation was implied by the absence of such features as storage space, midden areas, and burials.

The floor area of Feature 2-2 (7.7 m^2) was within the range established by Crown (1985a) for small structures. Overall, structure size was close to the mean structure size established for Snaketown and Gila Butte phase field houses (Cable and Doyel 1985:282). Russell (1978:36) described field houses used by the Navajo for intermittent summer occupation that were probably similar to those built by the Hohokam. These circular brush shades were usually 2.0 m to 4.0 m in diameter and commonly accommodated six to ten people overnight. For Feature 2-2, limited use was implied by the limited number of interior and extramural features; seasonality was suggested by the lack of a formal central hearth or fire pit to provide warmth during cooler months. The relative lack of extramural features did not indicate that the occupants had conducted no specialized activities at this location but that their activities (such as planting, weeding, and harvesting fields or maintaining or controlling canal flow) had left no archaeological evidence.

The lack of diversity and the limited number of artifacts in the Area 2 assemblage implied that the Hohokam had undertaken few activities within the area. The evidence of flaked and ground stone manufacturing and maintenance activities and food processing was limited as well. Ground stone items located at floor contact in the pit structure provided evidence for limited grinding activities. Formal grinding tools may have been carried back and forth from more permanent occupation sites or cached off-site, resulting in a general absence of the tools at field house sites (Russell 1978:39). The predominance of plainware ceramics indicated that inhabitants had conducted utilitarian activities such as cooking in Area 2. Another attribute of field houses is that they contain extremely small amounts of trade goods (Cable and Mitchell 1988:Table 11.1). Only one shell fragment was recovered from Area 2 during excavations. Similarly, archaeologists usually do not find items associated with rituals at field house sites, and investigators found none in Area 2. However, the Hohokam could have participated in ritual activities at the site without leaving evidence for such activity. Ellis (1978:65) reported that Pueblo people living away from their primary residence still observed small rituals, such as tossing corn meal to the sun each morning.

Subsistence remains supported the designation of Feature 2-2 as a field house. Corn pollen aggregates implied that portions of the plant, probably ears or cobs, had been present within the pit structure. Lack of faunal bone indicated that occupants had not discarded animal resources in this structure. Subsistence remains found in the fire pit of Feature 2-2, including Liliaceae pollen, may have indicated on-site consumption and exploitation of wild food resources, possibly ones that thrived near the canal during the growing season (Volume 4:Chapter 7).

Field crews found no surface indications of the intact subsurface features that existed in Area 2. Investigations at this locus established the relationship between the natural stratigraphy and the cultural remains. The habitation and canal construction events were dated, as expected, within the Hohokam occupational sequence. The data recovery efforts fully realized the significance and information potential of Area 2.

CHAPTER 4

DUTCH CANAL RUIN, AREA 3

M. Zyniecki

Research Area 3 of Dutch Canal Ruin was the largest and contained the most features of any of the pre-Classic period areas investigated for the Phoenix Sky Harbor Center Archaeological Project. A major research objective for Area 3 was to determine whether the features resulted from repeated use of the area by small groups or a contemporaneous occupation by a large group. Investigators excavated seven pit structures, 11 pits, and a rock cluster within an area measuring 66 × 100 m during data recovery efforts in Area 3 (Figure 4.1). Two of the structures recorded during testing, Features 1044 and 1045 (Volume 1:Appendix A), lay underneath the reconstructed portion of Grant Street and therefore were not investigated during data recovery.

Area 3 was located immediately north of Grant Street (Figure 1.4) and was within the project grid at N1614–1680, E770–870 (Figure 1.2). The area was approximately 200 m west-northwest of Area 1 and about 100 m north of Alignment 8002, the southern branch of the North Main Canal (Chapter 7). The North Main Canal, Alignment 8001, was 73 m north-northeast of Area 3. Area 3 was in T1N, R3E, Sec. 10, SE¼NW¼NW¼SW¼ on the Phoenix, Arizona, 7.5 minute USGS topographic map (Figure 1.4).

Temporally diagnostic ceramics and chronometric samples supported the testing conclusions that the Hohokam had occupied Area 3 during the Colonial period and suggested that the most intensive occupations had developed during the Santa Cruz phase. Chronological data suggested two or three separate occupations of Area 3. These same data also indicated five groups of contemporaneous features. Botanical analyses provided convincing evidence that the Hohokam inhabitants of the area had cultivated corn and had exploited cholla cactus and other nondomesticated plants. The area functioned as a farmstead locale occupied repeatedly over time. The occupations were temporary, presumably seasonal, and related to agricultural and subsistence resource exploitation. The proximity of Area 3 to the canals indicated that the prehistoric inhabitants of the area had probably also engaged in canal maintenance and water regulation activities.

RESULTS OF TESTING

Field crews excavated 20 trenches, with a total length of 500 m, in Area 3 during the testing phase. All trenches were 25 m in length and oriented north-south. Because of the number of features in the southwestern part of the area, more intensive trenching took place in this portion of Area 3, consisting of five trenches spaced at 5-m intervals.

During the testing phase, crews recorded seven structures, one possible structure, three pits, and one stain. Although investigators did not assign pits to a more specific type in the testing results, descriptions indicated that two of the pits contained evidence of burning, while the third pit type was unspecified. The testing results also suggested the possibility that the stain was a poorly defined structure (Volume 1:Appendix A).

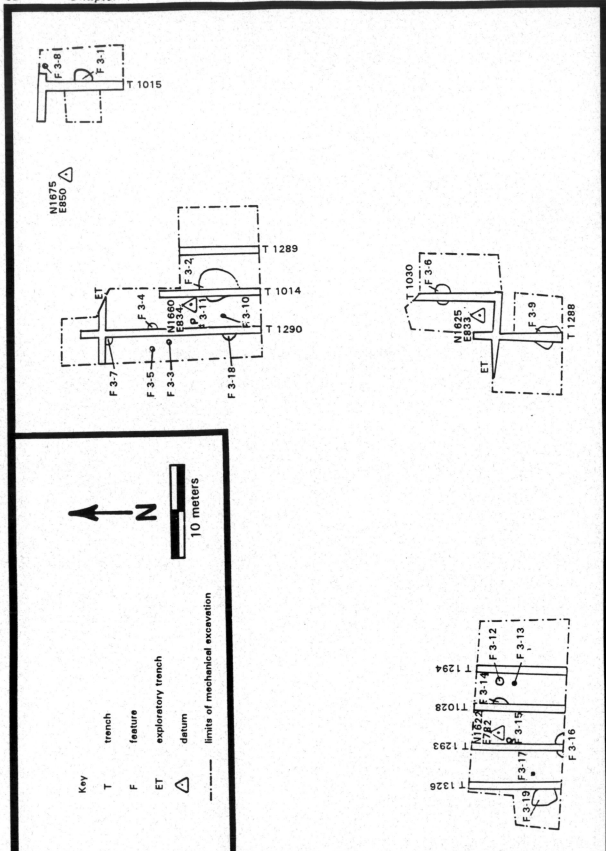

Key

T   trench

F   feature

ET   exploratory trench

△   datum

—···—   limits of mechanical excavation

N

10 meters

Figure 4.1. Area 3, Dutch Canal Ruin.

RESEARCH OBJECTIVES

Through excavations in Area 3 archaeologists intended to determine whether any spatial or temporal patterning existed in the distribution of habitation features and whether the cultural remains could be attributed to reuse of the area over time or to a large contemporaneous occupation. As with the other research areas of Dutch Canal Ruin, the investigating team also designed the research objectives to explicate this area's role in Dutch Canal Ruin as a whole, within Canal System 2, and in relation to other sites within the region (Problem Domain II).

FIELD METHODS AND STRATEGIES

Backhoe Trenching

Field crews reopened portions of 10 trenches, a total of 116 m, during data recovery excavations in Area 3 (Figure 4.1). In addition, they excavated three exploratory trenches to facilitate the relocation of testing trenches. The exploratory trenches were oriented east-west (perpendicular to the testing trenches) and totaled just over 24 m in length.

Mechanical Stripping

Investigators intended to strip an area extending 5 m around each of the structures and did so for all features except those at the south end of Area 3 adjacent to Grant Street. About 500 m² were stripped in four basic blocks centered around the features identified during testing (Figure 4.1). This stripping process revealed nine additional features, all pits of one type or another.

Hand Excavations

Because of the low number of artifacts recovered from the pre-Classic period structures in Area 1, investigators increased the amount of fill screened from structures in Area 3. In addition to the floor fill (the 10 cm of fill directly above the floor of a structure), field crews also screened the fill above floor fill.

STRATIGRAPHY AND ASSOCIATION

The trench profiles in Area 3 revealed three natural strata. Stratum 1, the uppermost, was the historic disturbance and fill zone consisting of grading material for homes, roads, and other construction and corresponded to the recently disturbed zone defined during testing (Volume 1:Chapter 3). This stratum was relatively loose, with sand lenses, pebbles, and historic and modern trash. Stratum 1 extended from the surface to a depth of roughly 30 cm below the present surface (bps).

Stratum 2, the middle stratum, was a clay loam consisting of irrigation clays and silts that had settled on prehistoric and historic fields. The large amount of clay within the soil produced a high degree of shrink/swell action, and the amount of bioturbation was also extremely high. Stratum 2 was massive, with grain sizes from fine sand to silt to clay. The depth of this stratum was approximately 25-50 cm bps. Stratum 2 corresponded to the Holocene soils described by Nials and

Anderson (Volume 1:Chapter 3) for the testing phase of this project and was probably a Gilman loam (Volume 1:Appendix B).

Stratum 3, the lowest, was a silty loam resulting from prehistoric flooding of the Salt River. This stratum contained all the prehistoric features in Area 3. The top of some of the features must have been near the interface of Strata 2 and 3, because a few of the features were filled with a matrix similar to Stratum 2. Stratum 3 extended from 50 cm bps to the bottom of the trenches, roughly 1.5 m bps. This stratum was equivalent to the clayey silts reported for the testing phase (Volume 1:Chapter 3).

CULTURAL FEATURES

Structures

All structures excavated in Area 3 were temporally associated with the Colonial period, which was consistent with the interpretations of testing results. However, excavations revealed that some structures varied from the testing phase descriptions, no doubt due to the limited amount of information available during testing. Feature 3-4, Testing Feature 1046, was originally described as a possible structure (Volume 1:Appendix A), but excavation revealed a large pit with evidence of burning. Also, the features identified during testing as a stain, Feature 3-19 (Testing Feature 1058; see Volume 1:Appendix A), and a pit, Feature 3-6 (Testing Feature 1006; see Volume 1:Appendix A), were discovered during excavation to be pit structures. In sum, investigators excavated seven pit structures in Area 3.

The structures in Area 3 were distributed throughout the area (Figure 4.1). Feature 3-1 was in the northeast corner of the area, Feature 3-2 was toward the center, Features 3-6 and 3-9 were in the southeast, and Features 3-14, 3-16, and 3-19 were in the southwest.

Investigators attempted to recognize spatial patterning among features in Area 3. This led to the suggestion that some of the features belonged to contemporaneous groups. However, analysts could not define the relationships between the groups, and either the structures within the groups had no discernible entrances or the entrances were not oriented toward each other.

Feature 3-1

Testing Feature: 1004
Trench Number: 1015
Horizontal Provenience: N 1671.95–1674.27; E 860.19–862.95
Vertical Provenience: 4.84–5.04 meters below datum (mbd)
Feature Type: Pit Structure

Feature 3-1 was an oval pit structure with dimensions of 2.32 m north-south by 2.76 m east-west and was oriented with the possible entrance toward the west (Figure 4.2). Trenching during the testing phase removed almost the entire west wall of the structure. Field crews excavated a small section of wall, 57 × 12 cm, on the west side of the trench. Given the symmetrical outline of the rest of the structure, investigators thought that this portion of wall may have been a remnant of the entrance. Ranging between 11 cm and 18 cm in height, the walls in this pit structure were not prepared, and portions were oxidized. A historic or modern pit filled with rocks intruded into

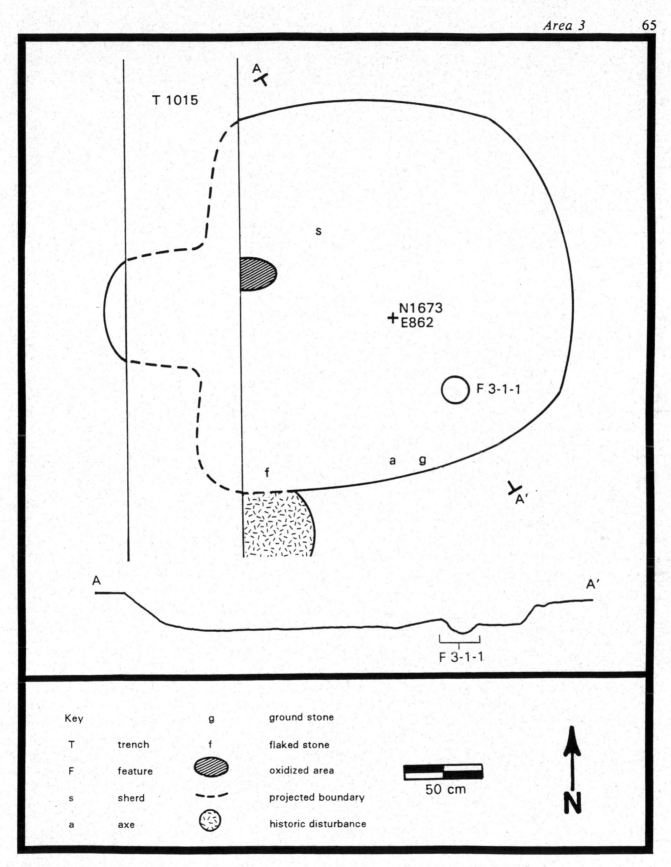

T 1015

N1673
E862

F 3-1-1

s

f

a   g

A   A'

F 3-1-1

Key

T	trench	g	ground stone
F	feature	f	flaked stone
s	sherd		oxidized area
a	axe		projected boundary
			historic disturbance

50 cm

N

Figure 4.2. Area 3, Feature 3-1, plan and cross section.

the south wall near the east face of the trench. The floor was not prepared; it had been compacted through use and had probably not been a level surface when occupied. A charcoal deposit across the floor made it easy to define during excavation.

A fire pit, Subfeature 3-1-1, was in the southeastern portion of the structure. The fire pit was 15.0 cm in diameter and 4.5 cm in depth. Its unlined edges and bottom were oxidized. On the east side of the trench, investigators observed an informal fire area consisting of an oxidized portion of the floor, similar to those in Feature 1-2 (Chapter 2). This area was about 20 cm in diameter. Field crews located no other floor features. Although crews observed no postholes, the layer of charcoal-laden fill above the floor and the presence of oxidized, burned soil and burned daub on the floor argued for a post-and-stick or brush superstructure covered with mud or adobe.

The pit structure had apparently burned and collapsed. Of the 23 sherds collected in Feature 3-1, field crews retrieved 21 more than 10 cm above the floor. The same was true for 25 of the 54 flaked stone artifacts. Feature 3-1 had probably burned before it was abandoned and was then used for a small amount of trash disposal. Alternatively, or in combination with the trash disposal, the depression may have trapped sheetwash. Feature 3-1 was subsequently filled with blocky clays, probably the result of prehistoric and historic irrigation farming.

Feature 3-2

Testing Feature: 1005
Trench Number: 1014   .
Horizontal Provenience: N 1654.52–1658.88; E 833.94–838.35
Vertical Provenience: 4.94–5.14 mbd
Feature Type: Pit Structure

Feature 3-2 was a roughly square pit structure covering an area measuring 4.0 × 4.2 m (Figure 4.3). The remnants of an entrance and step, Subfeature 3-2-3, were located along the east face of Trench 1014 oriented to the north-northwest. Most of the entrance was removed by the trench, making it difficult to determine its exact shape. The step extended about 40 cm into the structure and was about 6 cm in height. A 20 cm portion of the vertical east wall of the entrance remained, 12–14 cm in height.

As is typical of pit structures, the walls were unprepared. They were 9–18 cm in height, averaging about 12 cm. The slope of the walls varied from 90° to approximately 50° from the horizontal. The floor also was unprepared but had been compacted through use; it inclined slightly, with the northern portion 2–3 cm higher than the southern. Walls and floor were very easily defined in the west half of the structure because laminated strata of dark clays and tan silts had filled that portion. None of these laminates appeared in the east half of the feature. The walls in the east half were defined by a slight difference in color, and the floor by a small difference in compaction.

A fire pit, Subfeature 3-2-1, was situated near the center of the structure. The fire pit measured 31 × 34 cm in plan, with a depth of 15 cm. The edges of the pit were highly oxidized and burned. The upper portion of the fill contained the same dark clays as the western half of the structure, and the bottom was filled with a layer of ash. An overhanging lip of oxidized soil on the east side of Feature 3-2-1 gave the cross section of that portion the appearance of a bell-shaped pit. This lip could have been the result of the original pit construction or of some disturbance caused by the deposition of the dark clays. Another floor feature observed in this pit structure, Subfeature 3-2-2,

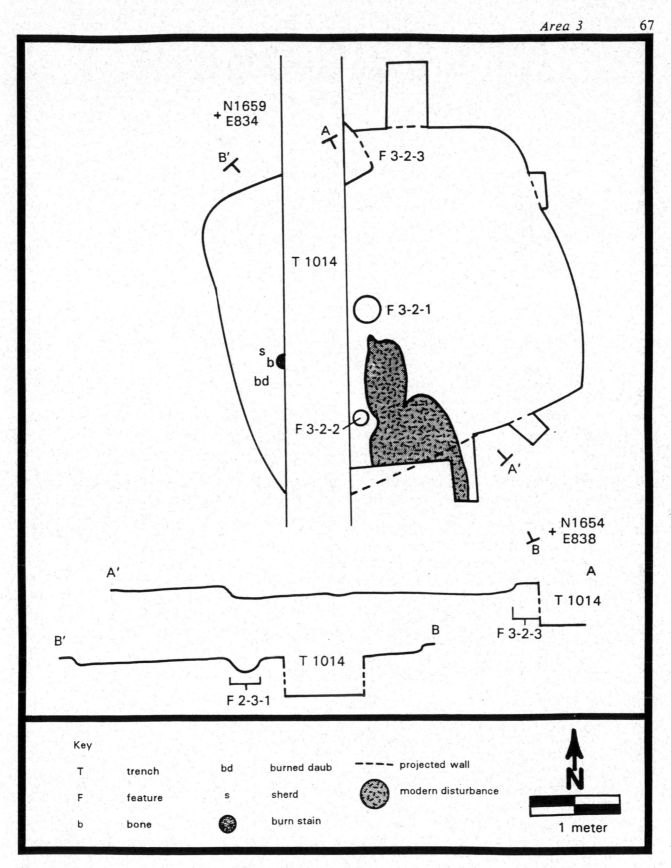

Figure 4.3. Area 3, Feature 3-2, plan and cross sections.

a shallow depression with a diameter of 14 cm and a depth of 5 cm, probably had functioned as a posthole or post support. This feature was about 60 cm from the south wall of the structure. Given its location about 1 m from the wall, investigators surmised that several more posts had probably supported a superstructure covered with mud or adobe. Additionally, field crews noted a small, charcoal-stained area, 12 cm in diameter, along the west edge of the trench; it may have represented an informal fire area or the remnants of a burned portion of the superstructure or roof.

Feature 3-2 may have burned and collapsed, an interpretation supported by the presence of a small piece of burned daub and the burned stain on the floor. However, this structure had been catastrophically flooded, as shown by the laminated clays and silts in the west half, and this episode could also account for the presence of the burned daub. After the structure had been abandoned, it probably collapsed in a way that covered and thereby protected the east half, leaving the west half exposed to subsequent inundation and infilling with flood deposits.

Feature 3-6

Testing Feature: 1006
Trench Number: 1030
Horizontal Provenience: N 1627.41–1629.37; E 834.40–836.52
Vertical Provenience: 4.72–4.85 mbd
Feature Type: Pit Structure

Feature 3-6 was a roughly circular structure originally identified as a pit from the testing phase profile. The structure ranged from 1.96 m to 2.08 m in diameter (Figure 4.4). Field crews did not locate an entryway. The pit walls, not formally prepared, were between 8 cm and 12 cm in height, and sections were oxidized. The south and east walls had a slope of less than 45°, while the north and west walls were more nearly vertical. The floor of the pit structure was compacted through use and was almost level, with a rise of 2 cm from west to east. Several discontinuous charcoal lenses within the fill may have represented either different occupational episodes or occupation and burning of the structure after abandonment. Field crews observed no interior features.

Burned daub in the feature indicated that the superstructure of Feature 3-6 had probably been composed of posts covered with poles or brush that was then layered with mud or adobe. The presence of daub, the charcoal lenses, and the oxidized wall sections indicated that the structure had burned before being filled with eolian and alluvial deposits.

Feature 3-9

Testing Feature: 1040
Trench Number: 1288
Horizontal Provenience: N 1613.78–1616.82; E 828.67–831.61
Vertical Provenience: 4.90–5.27 mbd
Feature Type: Pit Structure

Feature 3-9 was an irregularly shaped pit structure measuring 2.77 × 2.95 m (Figure 4.5). The structure's rectangular, appended entrance, Subfeature 3-9-1, was most likely a ramp and was located in the southeast corner. The interior opening of the entrance was 1 cm above the structure floor, while the exterior opening was 6 cm above the floor. Trenching during testing had removed a portion of the entrance, and the remainder measured 92 cm by approximately 80–85 cm.

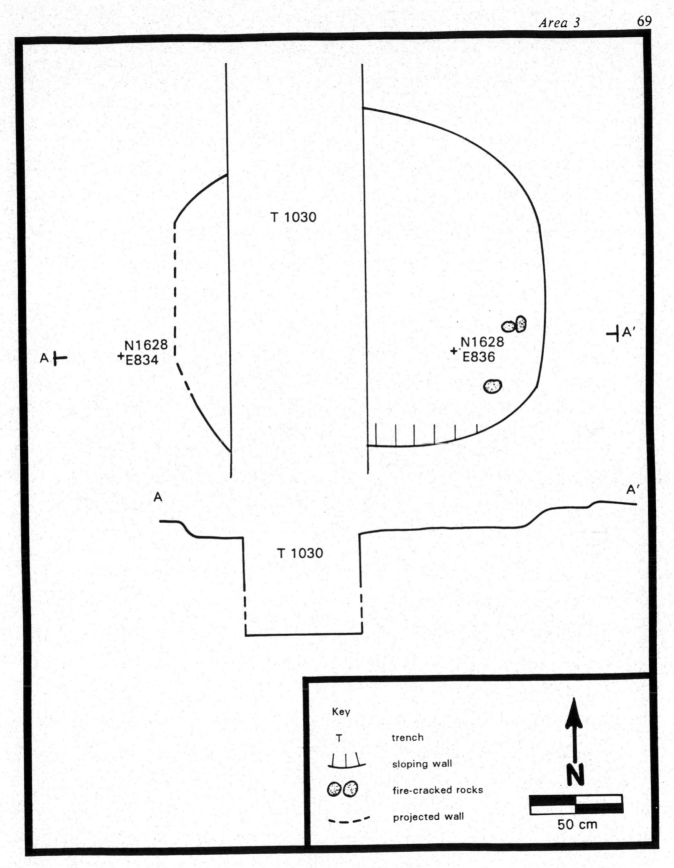

Figure 4.4. Area 3, Feature 3-6, plan and cross section.

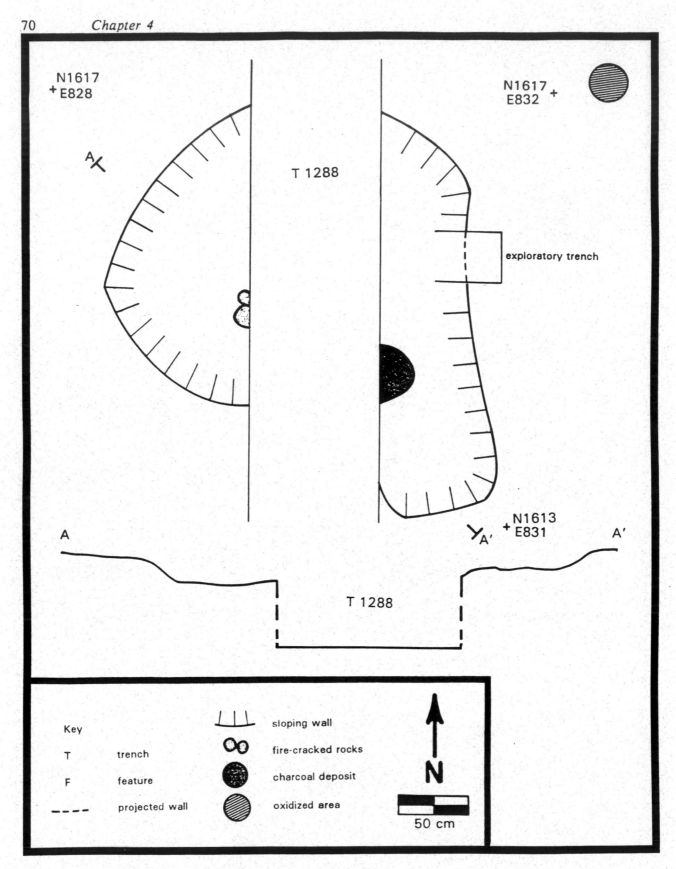

Figure 4.5. Area 3, Feature 3-9, plan and cross section.

The walls and floor of the entrance, like the walls and floor of the structure, were unprepared, and their juncture was curved. The curve began 12–15 cm from the wall. Field crews noted a difference of 20 cm between the floor of Subfeature 3-9-1 at the exterior opening and the prehistoric occupation surface. This difference in elevation appeared as a slope across a horizontal distance of 13–15 cm. The walls of the pit for the structure were 9–23 cm in height, with the tallest walls on the east side. The juncture of the walls and floor within the pit was curved, beginning 12–15 cm from the wall. The floor declined from south to north with a difference of 3 cm in elevation. Although in generally good condition, the floor was difficult to discern toward the north.

An oxidized lens thought to be the edge of a fire pit or hearth was located roughly in the center of the floor. A rainstorm destroyed the lens before it could be excavated. The oxidation measured 35 cm across with a depth of 6 cm. The high degree of oxidization in the soil profile indicated that the entire possible fire pit or hearth was probably just as oxidized. On the floor between the fire pit and the entrance was a roughly oval concentration of charcoal. It was partially destroyed by the same rainstorm that destroyed the fire pit. The remaining portion of the concentration measured 40 × 23 cm. The charcoal concentration may have represented the refuse from cleaning the fire pit. Neither postholes nor daub were observed in the structure. The superstructure had probably been of post-and-brush or stick construction.

Field crews observed an oxidized, ashy area with charcoal flecks on the prehistoric occupation surface outside the structure. This area was 28 cm in diameter and was located 1.67 m northeast of the structure. Most likely, the area was an informal fire area.

A relatively large number of fire-cracked rocks lay in the fill of the west half of the structure in a postabandonment context. Burned charcoal twigs were uncovered along the northeast and south walls. Investigators recorded no oxidation of the walls and floor and found no burned roof fall or daub in the fill. The evidence suggested that Feature 3-9 had collapsed after abandonment and had been used as a trash area.

Feature 3-14

Testing Feature: 1001
Trench Number: 1028
Horizontal Provenience: N 1621.35–1622.93; E 785.45–786.15
Vertical Provenience: 4.91–4.95 mbd
Feature Type: Pit Structure

Feature 3-14 was a remnant of a pit structure (Figure 4.6). Interior dimensions of the remaining portion of the feature, including the remaining floor segment, were 1.58 × 0.70 m; exterior dimensions, including the floor and a posthole, were 1.58 × 0.98 m. The upper portion and the west side of the structure were inadvertently removed during mechanical stripping activities. Only the floor on the east side of the structure remained. A trash-filled historic or modern pit intruded into the south end of the floor. The floor remnant was unprepared and declined from east to west, with a difference of 4 cm between the two sides. Field crews observed no entrance. This pit structure, identified during testing, was at a depth of 15 cm below the plow zone (Volume 1:Appendix A).

A fire pit or hearth, measuring 28 cm across and 8 cm in depth, was observed in profile during the testing phase but was destroyed while reopening the trench for the data recovery phase. An additional fire pit, Subfeature 3-14-1, was approximately 5 cm from the hearth or fire pit that had been destroyed. The backhoe trench removed a small part of the west side of Subfeature 3-14-1.

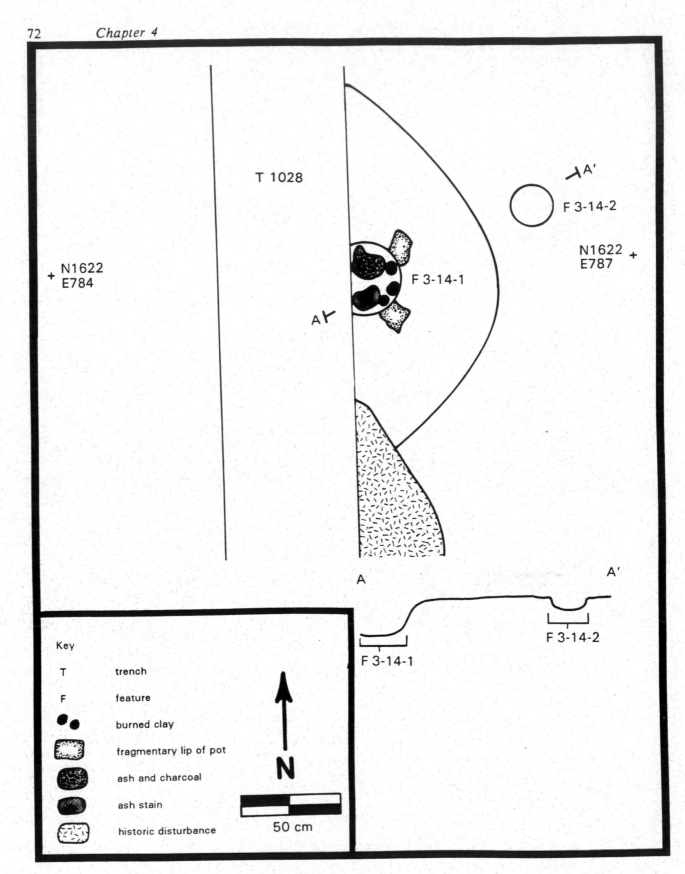

Figure 4.6. Area 3, Feature 3-14, plan and cross section.

The remaining portion of the fire pit was 30 cm in diameter and 18 cm in depth. Ash was present throughout the fill of the pit, with the greatest amount of ash at the bottom. The soil at the base of the pit was oxidized. A primary posthole, Subfeature 3-14-2, was situated 15 cm east-northeast of the floor remnant. The posthole was 18 × 17 cm in plan and had a depth of 5 cm. The fill of the posthole contained charcoal flecks, suggesting that the post, and possibly the structure, had burned.

The stratigraphic position of Feature 3-14 may have been similar to that of the two pits, Features 3-12 and 3-13, located northeast of the pit structure. The excavated portion of Feature 3-14 was in Stratum 3. A layer of clays similar to the matrix in Stratum 2 covered part of the structure floor; these clays were naturally deposited and did not represent a floor preparation. As with Features 3-12 and 3-13, the top of the pit structure was probably either in Stratum 2 or near the interface between Strata 2 and 3. The superstructure of the feature had probably been constructed of posts, with a brush covering that may have had an external layer of mud or adobe.

Feature 3-16

Testing Feature: 1050
Trench Number: 1293
Horizontal Provenience: N 1613.85–1615.18; E 778.70–781.75
Vertical Provenience: 4.98–5.26 mbd
Feature Type: Pit Structure

Feature 3-16 was a basin-shaped pit structure that was probably oval in plan, with a north-facing entrance, Subfeature 3-16-1 (Figure 4.7). The exact shape of the structure was difficult to determine, because its southern half was under one of the newly constructed streets of the Phoenix Sky Harbor Center. The excavated portion of the structure measured 3.05 m east-west by 1.33 m north-south with a depth of 28 cm. During testing, the north-south profile of the structure measured 2.03 m in length, suggesting that the maximum dimensions for the structure were 3.05 × 2.03 m. The west half of the entrance was destroyed by the backhoe during testing; the remaining portion was 23 × 23 cm. The shape of the entrance may have been oval, but the fragmentary nature of the remnant made the estimation of shape questionable. However, the entrance did seem to be ramped, with a difference in elevation of 3 cm between the floor at the exterior opening and the floor at the interior opening. Walls as such did not exist in Feature 3-16. Investigators noted a gentle slope from the center of the floor to the top of the pit. The difference in elevation between the floor center and the top of the pit structure was 28 cm. The floor and walls were not prepared.

No evidence for internal features of any kind was present in the excavated portion of Feature 3-16. An area of charcoal staining recorded along the floor from the entrance toward the center of the feature may have been associated with a fire pit, hearth, or informal fire feature that was removed by backhoe trenching. Field crews observed no roof fall material. The superstructure may have been of brush, without a mud or adobe covering.

The fill of the structure contained some charcoal flecks and more than 50 rocks and fire-cracked rocks; the soil was not stained to any great degree. This evidence indicated that the depression of the abandoned and collapsed structure had been used as a trash area for discarding the refuse from a roasting pit or horno. The roasting feature was probably to the south of Feature 3-16, under the landscaping and new road, since investigators could not locate it in the large, mechanically stripped area to the north of the feature.

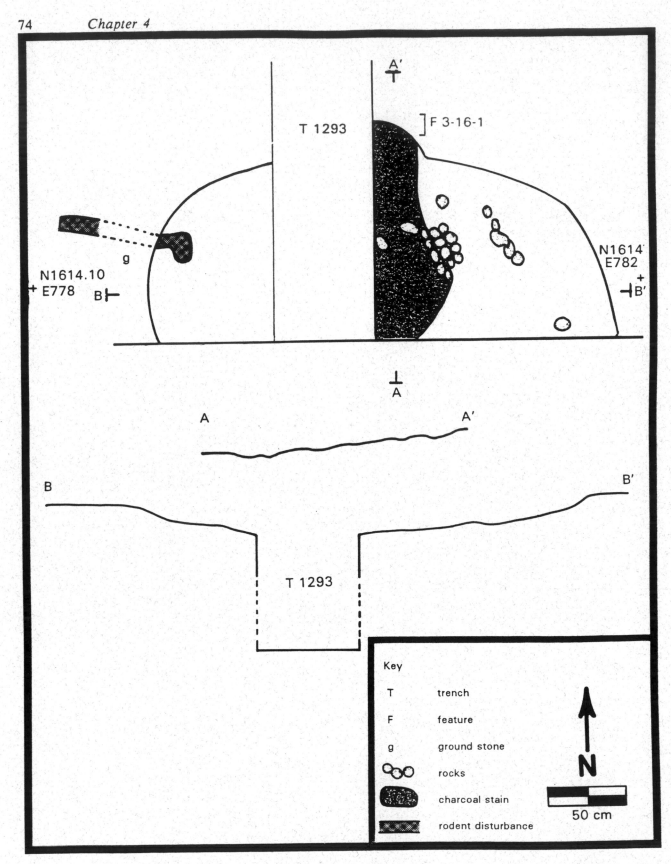

Figure 4.7. Area 3, Feature 3-16, plan and cross sections.

Feature 3-19

Testing Feature: 1058
Trench Number: 1326
Horizontal Provenience: N 1614.62–1617.87; E 772.44–774.43
Vertical Provenience: 4.97–5.23 mbd
Feature Type: Pit Structure

Feature 3-19 was a roughly oval pit structure with a shallow basin cross section and an entrance in line with the long axis of the oval (Figure 4.8). The east edge of the structure was removed by Trench 1326. The structure measured 3.3 × 2.1 m in plan, with an estimated original width of 2.4 m. The depth of the structure from the highest point of detection to the lowest point on the floor was 26 cm. The entrance, Subfeature 3-19-1, opened to the north-northwest and was irregular in shape, measuring 80 × 30 cm, with the long axis parallel to the opening. The walls of the entrance were 5–9 cm high on the east side and 19–23 cm high on the west side. The floor and walls of the entrance were not formally prepared. Subfeature 3-19-1 was not a ramp-type entrance, but the floor sloped gently to walls that were inclined approximately 45°. The tops of the walls seemed to gradually flatten to the exterior surface. The height of the walls and slope of the floor indicated that the exterior opening of the entrance had probably been along its eastern side.

The floor of the structure, defined by a layer of burned roof material, had been compacted through use but was not completely level. The walls were simply the gently sloping sides of the shallow pit that had been excavated for the structure. Neither the floor nor the walls had been formally prepared. Wall height ranged between 21 cm and 26 cm, with an estimated mean of 23 cm.

Field crews observed no internal features in Feature 3-19 but did record a highly oxidized area along the long axis of the structure running through the entrance. This area of oxidation was 30 × 40 cm and irregular in shape. It most likely represented an informal fire area within the structure. A layer of what appeared to be burned roof material indicated that the superstructure had probably been of pole or post construction with brush covering, though crews uncovered no evidence for posts or postholes. An area of organic and charcoal staining, measuring 100 × 90 cm, appeared on the prehistoric occupation surface approximately 1 m west of the structure. This area probably represented a single episode of dumping because of its ephemeral nature and lack of depth.

The layer of burned roof material indicated that Feature 3-19 had burned. Scattered areas of coarse sand along the floor suggested that this burning had occurred during use or shortly after abandonment. If the structure had burned after a prolonged period of abandonment, a more distinct layer of deposited materials would have been expected between the floor and the burned roof material. After the burning of the structure, the remaining depression was filled with eluvium and alluvium.

Extramural Features

Investigators excavated 12 extramural features during the data recovery phase in Area 3. As noted above, 3 of the features were reported during testing, and 9 more were discovered during excavations. Of the total, 4 were fire pits, 6 were pits with evidence of burning, 1 was a pit with

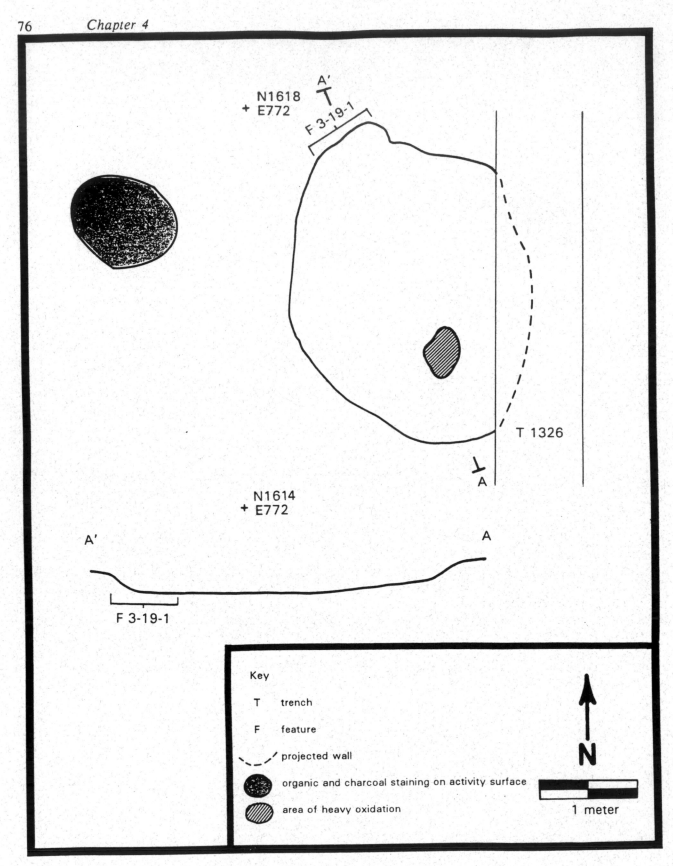

Figure 4.8. Area 3, Feature 3-19, plan and cross section.

rock, 1 was an unspecified pit, and 1 was a rock cluster. An extramural activity surface was identified during excavation of Feature 3-10, a fire pit. The activity surface was not given a separate feature number and is included in the description of Feature 3-10.

Feature 3-3

Horizontal Provenience: N 1662.55–1662.94; E 829.25–829.53
Vertical Provenience: 4.90–5.01 mbd
Feature Type: Pit with Burning

Feature 3-3 was an unlined, oval pit with evidence of burning that measured 38 × 43 cm in plan with a depth of 11 cm. The sides of the pit sloped, and the bottom was level. The fill of the pit contained oxidized soil and burned organic material, although the walls did not exhibit signs of either. Several fragments of burned corn were recovered from the fill.

Feature 3-4

Testing Feature: 1046
Trench Number: 1290
Horizontal Provenience: N 1664.00–1664.40; E 830.70–831.90
Vertical Provenience: 5.00–5.20 mbd
Feature Type: Pit with Burning

Feature 3-4 was a pit with evidence of burning. The irregularly shaped pit was unlined and measured 1.20 × 1.35 m in plan and 20 cm in depth. The fill contained a large amount of charcoal flecks and burned adobe, but the pit walls showed no signs of burning.

Feature 3-5

Horizontal Provenience: N 1664.59–1665.05; E 828.29–828.59
Vertical Provenience: 4.87–4.92 mbd
Feature Type: Pit with Rock

Feature 3-5 was a shallow, oval pit with five fire-cracked rocks. The pit was 46 × 29 cm in plan and had a depth of 5 cm. The matrix surrounding the pit contained noticeably more charcoal flecks than the fill of the pit. No oxidation of either the fill or the pit walls had occurred.

Feature 3-7

Testing Feature: 1047
Trench Number: 1290
Horizontal Provenience: N 1669.55–1670.55; E 829.11–830.10
Vertical Provenience: 5.70–6.05 mbd
Feature Type: Pit with Burning

Feature 3-7 was a large, roughly circular pit. Trench 1290 and a perpendicular exploratory trench, Trench 1290A, removed almost three-quarters of the feature. The remainder of the pit was

1.00 × 1.01 m in plan and 35 cm in depth. Fill in the feature contained a high density of charcoal flecks and some burned daub or adobe, as well as a noticeably greater amount of clay than the surrounding matrix. The greater amount of clay in the fill was probably the result of ponding after abandonment of the pit. Field crews observed no evidence of burning along the pit walls.

Feature 3-8

Horizontal Provenience: N 1677.31–1677.52; E 863.06–863.30
Vertical Provenience: 4.72–4.77 mbd
Feature Type: Fire Pit

Feature 3-8 was a shallow, slightly oval pit measuring 20 × 24 cm in plan and 5 cm in depth. The pit was discovered during mechanical stripping in the vicinity of Feature 3-1. The fill of the pit was ashy, with charcoal scattered throughout. The bottom and sides of the pit were oxidized. An area of oxidation and charcoal flecks contiguous with the southern edge of the pit indicated where the prehistoric surface must have been. Shovel scraping between Features 3-8 and 3-1 revealed flat-lying artifacts and a prehistoric occupation surface between these two features.

Feature 3-10

Horizontal Provenience: N 1656.67–1656.83; E 832.15–832.32
Vertical Provenience: 5.01–5.06 mbd
Feature Type: Fire Pit

Feature 3-10 was a circular fire pit with a diameter of 17 cm and a depth of 5 cm. The pit was in an activity surface that measured approximately 2.36 × 1.75 m. The fill of the fire pit contained ash, charcoal flecks, and oxidized soils. The sides and base of the fire pit also were oxidized.

Feature 3-11

Horizontal Provenience: N 1659.58–1659.98; E 832.18–832.55
Vertical Provenience: 4.90–4.98 mbd
Feature Type: Fire Pit

Feature 3-11 was a roughly circular exterior fire pit that measured 40 × 37 cm in plan. The pit was basin shaped with a depth of 8 cm. Fill in the fire pit consisted of silts mixed with clumps of ash, charcoal flecks, and oxidized soils. Oxidation was present on portions of the pit walls.

Feature 3-12

Horizontal Provenience: N 1621.57–1622.58; E 787.67–788.78
Vertical Provenience: 4.87–4.96 mbd
Feature Type: Pit with Burning

Feature 3-12 was an oval pit with evidence of burning that measured 1.11 × 1.01 m in plan with a depth of 0.09 m. The top of this feature was removed during mechanical stripping operations. The matrix in the pit was mostly sandy clays, but charcoal flecks, organic inclusions, and pieces of burned daub also were present. The stratum containing the pit was composed of silty loams, while that above the pit was of sandy clays. The similarity of the pit fill to the upper stratum suggested that the top of Feature 3-12 had originally been in the upper, sandy clay stratum or near the interface of the two strata.

Feature 3-13

Horizontal Provenience: N 1620.20–1649.49; E 787.61–787.61
Vertical Provenience: 4.84–4.95 mbd
Feature Type: Pit with Burning

Feature 3-13 was a circular, basin-shaped pit with evidence of burning. The diameter of the pit was 28 cm, and the depth was 11 cm. Although the fill of the feature contained charcoal staining, no evidence indicated that the pit walls or bottom had burned. The top of Feature 3-13 was removed during mechanical stripping operations. As with Feature 3-12, the remaining portion of Feature 3-13 was filled with a matrix similar to the stratum above the pit. Likewise, the top of Feature 3-13 probably lay within the upper stratum or near the interface of the two strata.

Feature 3-15

Testing Feature: 1049
Trench Number: 1293
Horizontal Provenience: N 1620.65–1621.05; E 780.37–780.77
Vertical Provenience: 5.11–5.18 mbd
Feature Type: Pit with Burning

Feature 3-15 was an approximately circular pit with evidence of burning. The pit was 40 cm in diameter with a depth of 7 cm. The top of the pit was accidentally removed during mechanical stripping operations. During the testing phase, investigators measured the profile of this feature as 27 cm in depth. Small pieces and flecks of charcoal appeared throughout the fill of the pit. The base of the pit had charcoal staining but was not oxidized. The fill contained a higher concentration of clay than the surrounding matrix. Originally, the top of the pit had probably been in Stratum 2 or near the interface of Strata 2 and 3.

Feature 3-17

Horizontal Provenience: N 1617.63–1617.87; E 776.55–776.74
Vertical Provenience: 4.98–5.04 mbd
Feature Type: Rock Cluster

Feature 3-17 consisted of a cluster of seven rocks. The cluster measured 24 × 19 cm in plan and was 6 cm in height. The matrix surrounding the stones contained a very low density of charcoal flecks. The feature could have been an informal fire area or a pile of discarded rocks from another thermal feature.

Feature 3-18

Trench Number: 1290
Horizontal Provenience: N 1654.80–1655.93; E 829.40–830.00
Vertical Provenience: 4.88–5.04 mbd
Feature Type: Fire Pit

Feature 3-18 was a fire pit in what was most likely an extramural activity area. The feature measured 113 × 60 cm in plan with a depth of 15 cm. At least half of the feature was removed by the backhoe trench. Based on the remaining portion of the feature, the pit was probably irregularly shaped, although basin shaped in cross section. The feature had not been formally prepared. The fill of the fire pit consisted of dark clays, similar to those found in Feature 3-2, with ash concentrations toward the bottom of the pit. The fill contained pieces of burned daub and small pieces of charcoal, and portions of the fill were oxidized. A portion of the bottom of the pit was oxidized, and field crews recovered a fragment of burned corn from this area. Feature 3-18 was about 4 m west of Feature 3-2 (Figure 4.1), with the top 6 cm higher than the highest detectable level of Feature 3-2. The similarity of fill, however, indicated that the two features were roughly contemporaneous. The prehistoric occupation surface that contained Feature 3-18 was probably an activity area extramural to Feature 3-2.

MATERIAL CULTURE

As can be expected in field house and farmstead loci, archaeologists recovered relatively little evidence for material culture. However, compared to other pre-Classic period areas of Dutch Canal Ruin investigated for the Phoenix Sky Harbor Center Project, Area 3 had the largest artifact assemblage for the simple reason that it contained the most features. Field crews collected 143 sherds, 148 flaked stone items, and 31 ground stone artifacts.

Ceramics

All of the 143 sherds found in Area 3 were Hohokam wares (Table 4.1). The 110 undecorated sherds represented 76.9% of the total assemblage for Area 3; the other 33 (23.1%) were either red-on-buff or buff types. Of the total ceramic assemblage, 81 sherds (56.6%) were associated with features (Table 4.2). Two features, 3-1 and 3-2, accounted for over one-third (37.8%) of the sherds from Area 3 and two-thirds (66.7%) of the sherds recovered from features (n=54).

As is usual in pre-Classic period field house contexts, temporally diagnostic sherds were few. Seventeen sherds (11.9%) in the Area 3 ceramic assemblage were temporally diagnostic; of the sherds recovered from features, seven (8.4%) were temporally diagnostic. One Gila Butte Red-on-buff sherd came from Feature 3-2, a pit structure, and six Santa Cruz Red-on-buff sherds came from Feature 3-4, a pit with evidence of burning.

The recovery of the single Gila Butte Red-on-buff sherd from Feature 3-2 did not provide convincing evidence that this feature had been occupied during the early Colonial period, because it was present in feature fill partially composed of stratified flood deposits. This sherd had probably been secondarily deposited in the feature. In contrast, Feature 3-4 contained six Santa Cruz Red-on-buff sherds in fill that was consistent with a primary deposition context, indicating that this pit may have been used during that phase. The ceramic evidence, therefore, seemed to suggest that

Table 4.1. Area 3, Ceramic Frequencies

Ceramic Ware/Type	Absolute Frequency	Percent Ware	Percent Assemblage
Hohokam Plain Ware	109		**76.2**
Gila Plain, Salt Variety	62	56.9	43.3
Gila Plain, Gila Variety	11	10.1	7.7
Squaw Peak Plain	33	30.3	23.1
Wingfield Plain	3	2.8	2.1
Hohokam Red Ware	1		**0.7**
Squaw Peak Red	1	100.0	0.7
Hohokam Decorated Ware	33		**23.1**
Gila Butte Red-on-buff	1	3.0	0.7
Santa Cruz Red-on-buff	16	48.5	11.2
Indeterminate Red-on-buff	6	18.2	4.2
Indeterminate Buff	10	30.3	7.0
Total	143		**100.0**

Table 4.2. Area 3, Ceramic Frequencies by Feature

Ceramic Ware/Type	Feature								Total
	3-1	3-2	3-3	3-4	3-9	3-11	3-12	3-19	
Hohokam Plain Ware									*68
Gila Plain, Salt Variety	18	10	1	4	3				36
Gila Plain, Gila Variety		3			3			1	7
Squaw Peak Plain	5	14					6		25
Hohokam Red Ware									*1
Squaw Peak Red		1							1
Hohokam Decorated Ware									*12
Gila Butte, Red-on-buff	1								1
Santa Cruz, Red-on-buff				6					6
Indeterminate Red-on-buff	1								1
Indeterminate Buff	1					2	1		4
Total	23	31	1	10	6	2	7	1	81

*Total ware

Area 3 had been at least a locus of limited activities, if not of occupation, during the Santa Cruz phase of the Colonial period.

Flaked Stone

Of the 148 flaked stone artifacts collected from Area 3, 127 (85.8%) were recovered from features. In addition, two pit structures, Features 3-1 and 3-9, accounted for 92 artifacts, 62.2% of the total assemblage and 72.4% of the artifacts from features. The largest morphological categories in Area 3 were whole flakes and flake fragments, which included 70.9% of the total flaked stone assemblage (Table 4.3).

Table 4.3. Area 3, Flaked Stone Morphological Type by Material Type

Morphological Type	Material Type						Total
	Quartz	Chert	Basalt	Igneous Other	Quartzite	Metamorphic Other	
Angular Debris			4	4		1	9
Flake Fragment		1	31	24			56
PRB[1] Flake	1		13	7	3		24
Whole Flake		3	29	16	1		49
Core			3				3
Core/Hammerstone			2				2
Hammerstone			1				1
Cobble Uniface				1			1
Other Biface			1				1
Large Primary Flake				1			1
Indeterminate Tool			1				1
Total	**1**	**4**	**85**	**53**	**4**	**1**	**148**

[1]Platform-remnant-bearing

Only 10 (6.8%) artifacts were not debitage. Cores, core tools, and core/hammerstones made up 3.4% of the assemblage. Additionally, the categories of cobble uniface, other biface, large primary flake, and indeterminate tool contained 1 (0.7%) artifact each. Various raw material categories were represented, including basalt (57.4%), igneous other material (35.8%), quartzite (2.7%), chert (2.7%), quartz (0.7%), and metamorphic other material (0.7%).

Field crews recovered flaked stone artifacts from 8 of the 19 features in Area 3 (Table 4.4). Of these features, 5 were pit structures, and 3 were pits with evidence of burning. The pit structures contained 89.0% of the artifacts associated with features, and the pits with evidence of burning contained 11.0%. All cores and tools in Area 3 except for the single other biface came from features.

Table 4.4. Area 3, Flaked Stone Morphological Type by Feature Type

Morphological Type		Feature Type		Total
	Pit Structure	Pit with Burning	Nonfeature Context	
Angular Debris	7	2		9
Flake Fragment	41	4	11	56
PRB[1] Flake	18		6	24
Whole Flake	41	5	3	49
Core	2	1		3
Core Tool	1			1
Core/Hammerstone	1			1
Hammerstone		1		1
Cobble Uniface	1			1
Other Biface			1	1
Large Primary Flake		1		1
Indeterminate Tool	1			1
Total	**113**	**14**	**21**	**148**

[1]Platform-remnant-bearing

As noted, the two pit structures, Features 3-1 and 3-9, contained most of the artifacts. Feature 3-1 had 53 pieces of debitage. It also contained 1 core/hammerstone of basalt. Debitage from Feature 3-1 comprised igneous other material (27), basalt (23), chert (2), and metamorphic other material (1). Field crews recovered 36 pieces of debitage, a basalt core, and a cobble uniface of igneous other material with scraping use wear from Feature 3-9. Debitage from Feature 3-9 was composed of basalt (25), igneous other material (9), quartz (1), and quartzite (1). The context of the artifacts in both features indicated that these features had been used as refuse areas after the structures had collapsed.

The other three pit structures with flaked stone artifacts were Features 3-2, 3-16, and 3-19. Feature 3-2 yielded 7 pieces of debitage, Feature 3-16 contained 5, and 6 came from Feature 3-19. Feature 3-16 contained artifacts of basalt and igneous other material, and material types found in the other two structures were chert, basalt, igneous other material, and quartzite. Tools found in Feature 3-2 were a core/hammerstone and an indeterminate tool with use wear, both of basalt; Feature 3-16 contained a basalt core.

Pits with evidence of burning that contained flaked stone artifacts were Features 3-4, 3-7, and 3-12. Feature 3-4 contained 4 pieces of debitage and a basalt hammerstone. Feature 3-7 contained 2 items of debitage. Feature 3-12 had 5 pieces of debitage, a basalt core, and a large primary flake of igneous other material with no use wear.

Twenty pieces of debitage and 1 basalt biface from Area 3 were not associated with features. Of the debitage, 12 items were composed of basalt, 7 were of igneous other material, and 1 was of quartzite.

The presence of debitage and cores in the flaked stone assemblage indicated that the area's inhabitants had reduced material there. As expected for the "expedient" industry of the Hohokam, however, tool makers had apparently produced very few tools. The ratio of debitage to tools in Area 3 was 19.7:1, higher than the ratio of 11.5:1 for Dutch Canal Ruin as a whole. The Area 3 ratio was also much higher than the 6.1:1 ratio obtained from noncanal features in the Dutch Canal Ruin excavations for the Squaw Peak Parkway (D. H. Greenwald 1988c:139). The Area 3 assemblage was small, and analysts must be cautious in making inferences. Nevertheless, the total lack of utilized flakes and minimal use wear on most tools suggested either that the occupants had practiced only minor tool use in Area 3 or that they had removed their tools when they abandoned the area.

Ground Stone

As with the other artifact categories, the number of ground stone artifacts (n=31) was higher in Area 3 than in the other pre-Classic areas of Dutch Canal Ruin. Half of the ground stone artifacts, 16 pieces, came from the west half of Feature 3-9, a pit structure, which had probably been used as a refuse area after the structure had been abandoned and had collapsed.

Investigators identified 6 artifacts (19.4%) in the ground stone assemblage from Area 3 as stone tools: 2 manos, 2 metates, a handstone, and a maul (Table 4.5). The remainder of the assemblage was composed either of pieces of possible ground stone or of ground stone not further specified (NFS), with each category containing about 40% of the assemblage. The entire ground stone assemblage was produced from eight types of material, with quartzite and sandstone the two most frequent, each accounting for about one-third of the assemblage.

Twenty-three (74.2%) pieces of ground stone or possible ground stone were associated with features in Area 3. As noted, field crews recovered 16 lithics from Feature 3-9; of these, 7 were ground stone NFS, and 6 were possible ground stone. Two metates NFS and a mano NFS, all of granite, were also present. Four other features accounted for 7 ground stone items. Feature 3-6 contained a ground stone NFS item made from sandstone and 2 items classified as possible ground stone. A complete greenstone maul and a vesicular basalt ground stone NFS came from Feature 3-1. Finally, Feature 3-16 contained one ground stone NFS of granite, and Feature 3-17 had a possible ground stone item.

Table 4.5. Area 3, Ground Stone Morphological Type by Material Type

Morphological Type	Material Type								Total
	Quartzite	Sand-stone	Vesicular Basalt	Green-stone	Gneiss	Granite	Meta-sediment	Igneous NFS	
Ground Stone NFS	2	7	1			1	1		12
Possible Ground Stone	6	3			1	2		1	13
Maul				1					1
Mano NFS						1			1
Metate NFS						2			2
Handstone	1								1
Round/Oval Mano	1								1
Total	10	10	1	1	1	6	1	1	31

NFS = not further specified

Of the 8 pieces of ground stone retrieved from nonfeature contexts in Area 3, 6 (19.4% of the assemblage) were associated with Feature 3-10, either from the prehistoric occupation surface or from the fill within 10 cm of the surface. Five of these artifacts were of quartzite: analysts classified 2 as possible ground stone, one as ground stone NFS, one as a handstone with one use surface, and one as a round/oval mano. In addition, a possible ground stone item of sandstone was also associated with the prehistoric occupation surface.

The only ground stone artifact recovered in Area 3 that had been purposely manufactured was the maul collected from the floor of Feature 3-1. Modification through use was the norm for the other tools in the assemblage. River cobbles were the predominant source of material. Cobbles of appropriate size, shape, and material probably were selected for a task and put directly to use with little or no alteration or preparation.

Ground stone artifacts in Area 3 were mostly fragmentary and too small to classify as "active" or "passive" tool types and were in many cases too small to classify definitely as ground stone. Analysts identified five complete and partial milling stone tools, including two manos, a handstone, and two metates. A mano and one of the metates were present in the floor of Feature 3-9, and the other metate came from in the feature fill. Although the metates were not identifiable as to type because of their fragmentary nature, they were definitely not trough metates. These artifacts may have constituted a slab grinding system used for the processing of wild plant resources (Greenwald 1991). The remaining mano and handstone were associated with the activity surface at Feature 3-10. These could also represent part of a slab grinding system. The maul from Feature 3-1 indicated energy-intensive use, which could have been related to activities such as field clearing and maintenance, canal construction and maintenance, construction of structures, or any combination of these.

Twenty (64.5%) of the 31 artifacts in the ground stone assemblage had been burned, including 14 of 16 pieces (87.5%) in Feature 3-9, 1 of 2 from Feature 3-1, all 3 pieces from Feature 3-6, and 2 of 6 (33.3%) associated with the Feature 3-10 activity surface. Most of these artifacts were possible ground stone or ground stone NFS. The only burned tools were the two metate fragments and the mano fragment from Feature 3-9. None of the burned artifacts had been utilized to any great extent. It seemed likely that appropriate cobbles had been chosen for grinding tasks and then, before extensive use, had been used as heating stones, which caused them to break.

Shell

Field crews recovered a single piece of freshwater shell in Area 3. It was an unworked *Anodonta* sp. fragment from Feature 3-2, a pit structure.

Subsistence Remains

The subsistence remains from Area 3 were consistent with activities and functions of pre-Classic period Hohokam field house contexts. The botanical evidence suggested that the pre-Classic period features in this area of Dutch Canal Ruin had been in or near fields of corn (Chapter 15).

Botanical Remains

Analysts investigated eight flotation samples, one vegetation sample, and seven pollen samples from Area 3. Samples from five of the seven structures in the area, Features 3-1, 3-2, 3-9, 3-16, and 3-19, were included in the analysis, along with samples from two pits, Features 3-3 and 3-18. The vegetation sample came from Feature 3-3.

Using the criterion that a productive sample should contain charred nonwoody botanical remains identifiable to at least the family level (Chapter 15), analysts considered that four of the flotation samples were productive. These samples came from Features 3-1, 3-2, 3-3, and 3-9. The productive portions of the samples from Features 3-2, 3-3, and 3-9, as well as the vegetation sample, contained only *Zea mays* fragments and parts. Analysts identified a seed from a Cheno-am and three sprangletop-type grass grains in the sample from Feature 3-1.

Investigators ascribed plant-related activities to these features based on the macrobotanical remains (Chapter 15). The large number of charred corn remains in Feature 3-3, a pit with evidence of burning, suggested that corn might have been used as fuel, and the absence of oxidation of the pit base and walls indicated that the corn might have been in a context of secondary refuse (Schiffer 1987:18). The charred corn remains, which came from the floor fill levels in pit structures Features 3-2 and 3-9, also indicated that corn had been used as fuel in these features. The presence of the burned corn remains in the structures could have resulted from the processing of corn prior to storage, the consumption and discard of corn in the area, or a combination of these activities (Chapter 15).

The sample from Feature 3-2 also contained wood charcoal with characteristics of mesquite. Analysts could determine no special activity from this sample because it could have been the result of using wood as fuel or the burning of the wood superstructure (Chapter 15). In Feature 3-1, the presence of the burned grass grains suggested at least the possibility of a grass-covered superstructure that had burned, but this premise seemed unlikely given the lack of other burned grass parts.

Six of the pollen samples contained pollen indicative of subsistence activities (Chapter 16). The pollen samples, like the flotation samples, contained a large amount of *Zea mays* remains. Subfeatures 3-1-1 (a fire pit within Feature 3-1) and Features 3-2, 3-3, and 3-19 all contained corn pollen. All these samples indicated subsistence activities related to corn agriculture. The fire pit contained burned corn pollen, further suggesting that corn had been cooked there or had been used as fuel (Chapter 16).

In addition to the corn pollen, the sample from Feature 3-2 included Cruciferae (mustard family) pollen, and the sample from Feature 3-3 had Liguliflorae pollen. The greens and seeds of Cruciferae could have been used for food (Chapter 16). Similarly, the greens and roots of dandelion and chicory, two of the plants in the Liguliflorae family, are ethnographically known food sources (Chapter 16). As suggested by the pollen samples taken from Features 3-2 and 3-3, both domesticates and wild plants may have been exploited at this area of Dutch Canal Ruin.

The pollen sample from Feature 3-16, a pit structure, provided further evidence that the inhabitants of Area 3 had exploited native plants. After the collapse of the structure, the remaining depression served as a discard area for a relatively large amount of fire-cracked rock from a roasting pit or horno. Even though the pollen sample came from the floor of the structure, it might still represent roasting activity rather than other functions within the structure. The pollen included a relatively large quantity (20%) of Liguliflorae-type Compositae pollen and a small amount of *Cylindropuntia* sp. (cholla cactus) pollen. Cholla buds can be collected and roasted, and the fruit, or *tuna*, is also collected and eaten. Cholla buds are known as one of the more predominant wild food resources in the project area (Castetter and Bell 1942:63), and wild resources were undoubtedly utilized in Area 3.

In summary, the botanical evidence in Area 3 indicated that the inhabitants of this area had exploited both agricultural and wild resources. Corn was present in both macro- and microbotanical samples. The wild plants that had been exploited included plants from the mustard family, possibly dandelion or chicory, and cholla cactus.

Faunal Remains

As was the case with the artifact categories, Area 3 had more faunal remains than the other pre-Classic areas of Dutch Canal Ruin and accounted for 16.0% of the total faunal assemblage from the site (Chapter 17). Six features contained 60 whole or partial bones (Appendix D). Feature 3-12, a pit with evidence of burning, accounted for 31 (51.7%) of the faunal specimens from Area 3. Bones and fragments from the pit represented *Thomomys bottae* (Valley pocket gopher), *Sylvilagus* sp. (cottontail rabbit), *Lepus* sp. (jackrabbit), a medium-sized mammal that could not be further specified, and unidentified fragments, which accounted for 25 (80.6%) of the 31 specimens. Of the bones from Feature 3-12, 28 (87.1%) were burned. Feature 3-2, a pit structure, yielded 13 unburned faunal fragments: *Dipodomys* sp. (kangaroo rat) (n=1), *Lepus* sp. (n=2), an unidentified medium-sized mammal (n=2), and 8 unidentified fragments. Subfeature 3-14-1, the fire pit in Feature 3-14, contained 12 specimens: an ulna fragment from a *Sylvilagus* sp. and 11 unidentified burned fragments from small mammals. The final 4 specimens in Area 3 were from an unspecified medium-sized mammal in Feature 3-18, a burned fragment from a *Lepus* sp. in Feature 3-11, and 2 unidentified fragments from Feature 3-9.

The large quantity of burned bone and charcoal flecks in the fill of Feature 3-12, a pit with evidence of burning but without oxidation of the pit walls, suggested that the final use of this pit was for refuse rather than as a fire or roasting pit. However, the burned bone in the fire pit indicated that small mammals had been cooked in this feature. The field crews collected faunal specimens (all unburned) mostly from the flood deposits on the west half of Feature 3-2, the structure, which indicated that these specimens appeared in a secondary context. The same was true for the single unburned bone retrieved from the flood deposits in Feature 3-18. The single burned jackrabbit fragment collected from Feature 3-11 indicated that this fire pit had been used for cooking.

The faunal assemblage from Area 3 represented small and medium-sized mammals, which might be a pattern for pre-Classic period field house and farmstead contexts (Chapter 17). The quantity of identifiable bone was too small to further specify patterns in the types of animals exploited. Some evidence indicated that the animals had been cooked both within structures and in extramural features.

CHRONOLOGY

One of the major research goals in Area 3 was to determine whether the features were the result of repeated use of the area through time or of contemporaneous use of a large area equivalent to a hamlet or large farmstead. The evidence suggested that the Hohokam had repeatedly occupied Area 3 during the Colonial period.

Ceramic artifacts were associated with the Gila Butte and Santa Cruz phases, and archaeomagnetic samples from four fire pits were associated with the late Gila Butte through the early Sacaton phases. As sometimes happens with radiocarbon samples from field house contexts on the Salt River floodplain, the dates from these samples were much earlier than expected. One of the radiocarbon dates predated the Hohokam sequence, and the other was from the Pioneer period, both unlikely temporal affiliations for Area 3.

Some features in Area 3 may have been exactly contemporaneous. For example, Features 3-1 and 3-8, a pit structure and a fire pit, may have been used in association with one another. The same was possible for Features 3-2 and 3-18, a pit structure and fire pit. Features 3-12, 3-13, 3-14, and 3-15, consisting of a pit structure with three associated pits, were probably all contemporaneous. Finally, Features 3-10 and 3-11 might represent contemporaneous fire pits in an activity area. The grouping of contemporaneous features indicated that Area 3 had been repeatedly occupied, with similar activities taking place each time.

Stratigraphy

All features in Area 3 were contained within the same stratigraphic level, Stratum 3. This level was a silty loam, the result of prehistoric flooding of the Salt River. Because all features lay within the same geologic unit with no superpositioning or intrusion of features, investigators determined nothing certain about the relative chronological relationships between features. They did, however, draw some tentative conclusions concerning stratigraphic relationships.

As noted earlier, Area 3 was excavated in four basic blocks (Figure 4.1). Three of these blocks appeared to contain five groups of "roughly contemporaneous" features, that is, features used during a time interval equivalent to two or three generations. The idea of rough contemporaneity is based on the hypothesis that features in relative proximity with similar top elevations should be similar in age when those features are in a context with minimal topographic relief. In other words, roughly contemporaneous features will be closely grouped and located on or very near the same prehistoric occupation surface. With its location on the geologic floodplain of the Salt River, Area 3 certainly meets the criterion for minimal relief. Within five roughly contemporaneous groups of features, none were further than 15 m apart and most less than 9 m. The maximum difference between the top elevations of contemporaneous features was 13 cm, with most less than 10 cm. Top elevations for the features in each group are listed in Table 4.6, and the locations of features within four of the groups are shown in Figure 4.9. The fifth group, Testing Features 1044 and 1045, is not illustrated because these features were inaccessible during data recovery. The combination of spatial proximity, similarity of elevations, and minimal relief suggested that features in a group may have been constructed on the same prehistoric surface, making them roughly contemporaneous.

Table 4.6. Area 3, Groups of Roughly Contemporaneous Features

Group	Feature Number	Type	Top Elevation*	Maximum Difference
1	3-1	Pit structure	4.67**	5 cm
	3-8	Fire pit	4.72	
2	3-2	Pit structure	4.79**	9 cm
	3-4	Pit with burning	4.81**	
	3-5	Pit with rock	4.87	
	3-18	Fire pit	4.88	
3	3-3	Pit with burning	4.90	11 cm
	3-10	Fire pit	5.01	
	3-11	Fire pit	4.90	
4	3-12	Pit with burning	4.90	13 cm
	3-13	Pit with burning	4.84	
	3-14	Pit structure	4.86**	
	3-15	Pit with burning	4.84**	
	3-16	Pit structure	4.85**	
	3-17	Rock cluster	4.98	
	3-19	Pit structure	4.97**	
5	1044	Pit structure	4.52**	1 cm
	1045	Pit structure	4.53**	

*meters below datum
**from testing

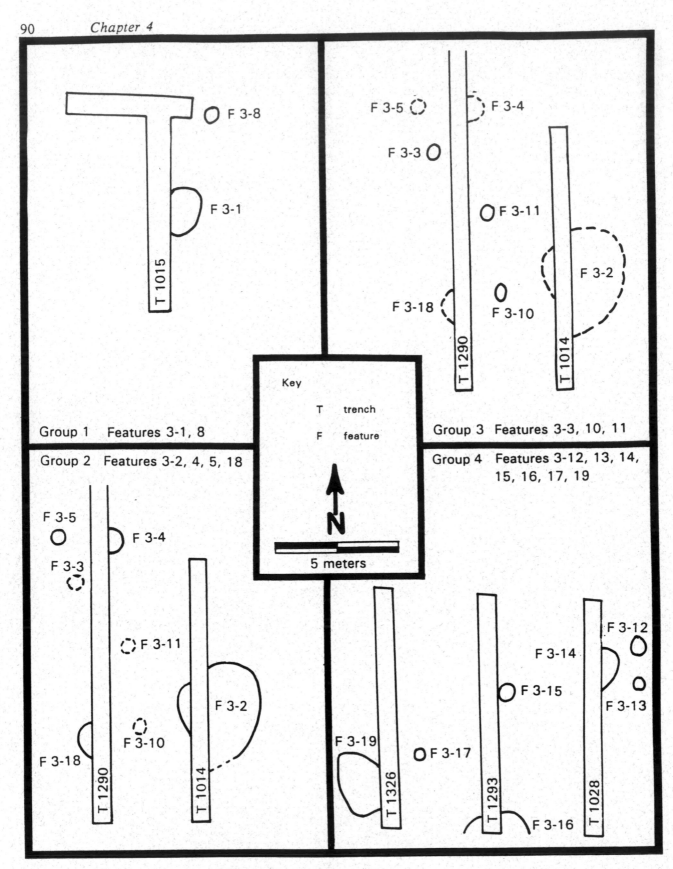

Group 1   Features 3-1, 8

Group 2   Features 3-2, 4, 5, 18

Group 3   Features 3-3, 10, 11

Group 4   Features 3-12, 13, 14,
15, 16, 17, 19

Key

T   trench

F   feature

N

5 meters

Figure 4.9. Area 3, roughly contemporaneous features.

The first group of roughly contemporaneous features includes Features 3-1 and 3-8, a pit structure and fire pit. These two features were less than 4 m apart and had a difference of 5 cm between their top elevations. Shovel scraping between the two features uncovered a number of flat-lying artifacts and a prehistoric occupation surface common to both features. This evidence supported the hypothesis for rough contemporaneity and indicated a strong probability for precise contemporaneity between the features.

The second group of features that may have been roughly contemporaneous consisted of Feature 3-2 (a pit structure), Feature 3-4 (a pit with burning), Feature 3-5 (a pit with rock), and Feature 3-18 (a fire pit). The distance between Feature 3-2 and Feature 3-18 was 4 m, and the distance between Feature 3-4 and Feature 3-5 was 2.5 m. The intervening distance between the two pairs was 7 m. The maximum difference in the upper elevations was 9 cm. In addition to the similarity in top elevations and the spatial proximity, the inference of rough contemporaneity between Feature 3-2 and Feature 3-18 was further strengthened by the fact that both features were filled with the same flood-deposited clays and silts. These two features may represent a structure and an associated extramural fire pit.

The third roughly contemporaneous group of features consisted of Features 3-10 and 3-11, both fire pits, and Feature 3-3, a pit with evidence of burning. The maximum difference between the tops of the features was 11 cm, and Features 3-10 and 3-11 had the same top elevation. Horizontally, the features were within 7 m of each other. The features in this group were situated in the middle of the features in the second group. Despite their physical proximity, the two groups were not considered contemporaneous because of differences in top elevations and in the dates from archaeomagnetic samples. The preferred date range for the archaeomagnetic assay from the fire pit in Feature 3-2 was A.D. 940–1015, while the preferred date ranges for assays from Features 3-10 and 3-11 were A.D. 830–925 and 850–940, respectively. Although these data confirmed the lack of contemporaneity between the two groups, they supported the idea of rough contemporaneity between features in the third group and suggested the possibility that Features 3-11 and 3-10 (with the associated activity surface) had been utilized at the same time.

The fourth group of features included Features 3-14, 3-16, and 3-19 (pit structures); Features 3-12, 3-13, and 3-15 (pits with evidence of burning); and Feature 3-17 (a rock cluster). All of these features were within 14 m of each other. The maximum difference in elevation of the feature tops was 13 cm. A subset of these (Features 3-12, 3-13, 3-14, and 3-15) may have been precisely contemporaneous, since these features were within a diameter of 7 m, with none of the features more than 4.5 m from one another. Elevations of the tops of the features in the subset were all within 6 cm. The fill in Features 3-12, 3-13, 3-14, and 3-15 provided evidence for the possibility of precise contemporaneity of for these four features. All were within Stratum 3 and filled with a matrix similar to that of Stratum 2, the next higher stratum. The tops of the four features must have been near the interface of the two strata. At the very least, Features 3-12, 3-13, 3-14, and 3-15 were utilized within a short period of time in geologic terms, making them roughly contemporaneous. Given the spatial proximity, the relative similarity of the top elevations, and the feature types, investigators posited that these features represented a structure and associated extramural features, all contemporaneous.

The fifth group consisted of two structures (Features 1044 and 1045) identified during testing but not excavated, as noted above. Although these two features were within approximately 7 m of the fourth group, their upper elevations were over 30 cm higher than those of the features in the fourth group, making it unlikely that the two groups of structures were even roughly contemporaneous. However, the structures in the fifth group were probably roughly, if not precisely, contemporaneous to each other, because their upper elevations were the same (within 1 cm), and they were within 30 cm of each other in horizontal distance.

The designation of these features as roughly or precisely contemporaneous is tenuous at best when based solely on elevation data and spatial proximity. Much greater spans of time than three generations could separate the periods of use of any or all of these features. Corroborative evidence for these groups of features, however, made this possibility unlikely. In the first group, a prehistoric occupation surface common to both features was identified. The same flood deposits appeared in two features of the second group. Archaeomagnetic data supported the rough contemporaneity of the features in the third group. Stratigraphic evidence strengthened the argument for precise contemporaneity for a subset of the fourth group and for rough contemporaneity of the group as a whole. Although features may be considered roughly or precisely contemporaneous based on elevations, archaeologists defined no implied or inferred relationship between groups based solely on elevations because of the distance between most groups.

Ceramics

Of the 143 sherds collected from Area 3, 17 (11.9%) were temporally diagnostic. One of the 17 sherds was Gila Butte Red-on-buff, and the rest were Santa Cruz Red-on-buff. The Gila Butte sherd and 6 of the Santa Cruz sherds were associated with features. Although the sherds suggested the possibility of Area 3 occupation during the entire Colonial period, the evidence was not conclusive. The Gila Butte sherd was present in stratified flood deposits within Feature 3-2, a pit structure. This one sherd could easily have washed in and thus may be completely unrelated to the feature, or it could have been associated with the structure and mixed with the flood deposits during the infilling process. The Santa Cruz Red-on-buff sherds associated with a feature were all collected from Feature 3-4, a pit with burning. Because of the context of the sherds, investigators thought it probable that this pit had been used during the Santa Cruz phase.

Absolute Dates

Investigators recovered two radiocarbon and four archaeomagnetic samples from Area 3. As was the case for the earlier excavation at Dutch Canal Ruin (Ciolek-Torrello and Greenwald 1988:165), the radiocarbon dates were much earlier than expected. Analysts calibrated the dates using the University of Washington Quaternary Isotope Laboratory Radiocarbon Calibration Program, Revision 2.0, a computer program developed for calibration of radiocarbon dates based on the method described by Stuiver and Becker (1986) (Appendix B). One sample, thought to be pieces of the burned superstructure of Feature 3-19, had a calibrated date of 900–390 B.C., which would place the structure in the Red Mountain phase (Dean 1991:Table 3.2), more than 1100 years before the beginning of the Colonial period. The other sample, a composite of charcoal from the superstructure of Feature 3-9, had a calibrated date of A.D. 60–381, which placed the ramada in the Red Mountain or the Vahki phase.

Four archaeomagnetic samples came from fire pits: Subfeatures 3-1-1 and 3-2-1 and Features 3-10 and 3-11 (Appendix A). Investigators used the visual dates for the samples instead of the 95% residual date range (Eighmy and Baker 1991:9). The most probable phase affiliations for these features, based on the ceramic data from testing and data recovery, were Gila Butte to Sacaton, circa A.D. 775–1000 (Dean 1991:Table 3.3). Analysts based the preferred date ranges for each sample on the inference for Colonial period affiliation from the ceramic data and the date range for the Colonial period developed by Dean. Features 3-10 and 3-11 both dated from the late Gila Butte to the middle Santa Cruz phase, A.D. 830–925 and 850–940, respectively. Subfeatures 3-1-1 and 3-2-1 were both temporally affiliated with the middle Santa Cruz to the early Sacaton phase, A.D. 900–975 and 940–1015, respectively.

Given the rough contemporaneity of Features 3-10 and 3-11, discussed above, and the results of the archaeomagnetic dating, investigators suggested that these two features were exactly contemporaneous. Therefore, the two fire pits and the surface associated with Feature 3-10 appeared to be a limited activity area probably related to the processing of subsistence resources.

The radiocarbon dates were unacceptably early. Excavations at Dutch Canal Ruin in the Squaw Peak Parkway corridor yielded two other radiocarbon dates from noncanal features (Ciolek-Torrello and Greenwald 1988:Table 24). One of the features, a Snaketown phase pit structure, was dated 100–400 years too early, a problem ascribed to the use of old wood (Ciolek-Torrello and Greenwald 1988:165). Likewise, the initial tendency of investigators during this study was to attribute the early dates from Area 3 to the old wood problem (Schiffer 1982), an unsatisfactory solution for which there was no justification other than the anomalous nature of the dates.

Cable (1985) cogently, but at length, argued that factors other than old wood could account for radiocarbon dates that were earlier than expected. He thought it more likely that they resulted from the use of older live trees and other mitigating factors, including ring bias in the sample (inner vs. outer rings), the dispersion of the sample (grab vs. composite samples), and fragmentation (size of charcoal pieces in sample). He provided tables giving the average age disparity between the dated event and the target date of a sample for a variety of tree ages at the time of use and for a variety of conditions based on the mitigating factors (Cable 1985:Tables 12 and 13).

Even under extremely biased conditions, such as a sample of inner wood from a 300-year-old tree, the two radiocarbon samples from Area 3 were too old. Using the average age disparity for these conditions, the samples would be 250 years older than the target date (Cable 1985:Table 12). This would adjust the calibrated date of the sample for Feature 3-9 to A.D. 310–631 and for Feature 3-19 to 650–140 B.C., dates that were still too early given the other evidence from excavation and testing.

Alternatively, Schiffer (1987:312) stated that, given the dry climatic conditions of the Sonoran desert and the prevalent hardwood tree species, wood used for fuel by the Hohokam was likely to be affected by the old wood problem (especially for Pioneer and Colonial period sites). River driftwood collected for fuel would also have been more likely to be old wood (Schiffer 1987:311). In construction wood, the effect of old wood was unknown because of the lack of data about the use of old versus freshly cut wood in Hohokam structures. Researchers proposed that both of the radiocarbon samples analyzed for Area 3 were pieces of burned superstructure.

Neither Cable's nor Schiffer's argument satisfactorily explained the earlier-than-expected radiocarbon dates. However, given the location of Area 3 on the floodplain, investigators thought that the dates were more likely to have been affected by the old wood problem than by average age disparity in the samples.

DISCUSSION

Two, possibly three, distinct periods of occupation in Area 3 were discernible in the data. The earliest and latest occupations contained field houses with associated extramural features, while the middle occupation was represented by fire pits and an activity area. The morphology of the structures and the nature and amount of the material culture provided evidence for temporary, seasonal occupation. The spatial, stratigraphic, and temporal associations among features indicated that the function of Area 3 had not changed drastically over time and that it was related to temporary seasonal occupation, agriculture, and subsistence resource exploitation.

Dates from four archaeomagnetic samples provided evidence that the first occupation of Area 3 had occurred in the northeastern portion of the area. The samples came from two extramural fire pits, Features 3-10 and 3-11, and the fire pits in two pit structures, Subfeatures 3-1-1 and 3-2-1. The ranges of accepted dates overlapped except for the dates for Feature 3-10 and Subfeature 3-2-1 (Appendix B). A number of other features were adjacent to these four dated features, including two extramural fire pits, Features 3-8 and 3-18; two pits with evidence of burning, Features 3-3 and 3-4; and a pit with rock, Feature 3-5. The features in this portion of Area 3 were field houses with associated features. However, stratigraphic and chronometric data suggested that these features could be divided into earlier and later groups.

The archaeomagnetic dates indicated that Features 3-10 and 3-11, the fire pits, may have been utilized before the structures, Features 3-1 and 3-2, were constructed. The date ranges for the two fire pits were earlier than those for the structures, and, as noted, Feature 3-10 predated Subfeature 3-2-1. In addition, Feature 3-11 and Subfeature 3-2-1 had only the year A.D. 940 in common, so Feature 3-11 probably also predated Subfeature 3-2-1. From stratigraphic information and the proximity of features, investigators determined that Features 3-10, 3-11, and 3-3 may have predated Features 3-1, 3-2, 3-4, 3-5, 3-8, and 3-18. The earlier group would have belonged to a limited activity area and the later group to a farmstead locale.

The chronometric data indicated that the earlier features had been utilized sometime between the late Gila Butte and the middle Santa Cruz phases and the later features between the middle Santa Cruz and the early Sacaton phases. Diagnostic ceramics placed the later features in the Santa Cruz phase. Therefore, the earlier features had probably been in use during or immediately after the transition between the Gila Butte and Santa Cruz phases.

Investigators based their determination of the final occupation on stratigraphic evidence. Features 3-12, 3-13, and 3-15, pits with evidence of burning, and Feature 3-14, a pit structure, represented the final occupation. These features, in the southwest corner of Area 3, lay in Stratum 3 but were filled with a matrix similar to the next higher stratum. Presumably, the tops of the features were near the interface of the two strata. This occupation probably represented another farmstead locus.

Features 3-12, 3-13, 3-14, and 3-15 were thought to be roughly contemporaneous with three other features. However, Features 3-12, 3-13, 3-14, and 3-15 postdated the other features, as their stratigraphic positions implied. These same stratigraphic data indicated that these features postdated all the features in Area 3 and could be temporally affiliated with the last part of the Colonial period or later.

The evidence suggested that the Hohokam had repeatedly used Area 3 as a farmstead locale. The features did not result from either a single large occupation or a smaller one of longer duration. Investigators did not determine the reasons for the continued and repeated use of Area 3. However, the area may have been located on a slight rise (Volume 1:Chapter 9), which would have made this a more advantageous area for a farmstead locale within the Salt River floodplain.

Macro- and microbotanical evidence indicated that the inhabitants of Area 3 had used corn more than nondomesticated plants. However, pollen samples from Features 3-2, 3-3, and 3-16 demonstrated that the inhabitants had collected and used wild resource plants, both weedy types and cactus. Faunal evidence indicated that the inhabitants had also exploited small and medium-sized mammals. Flaked stone reduction occurred within Area 3, although tools were scarce. The flaked stone analysis for the Squaw Peak Parkway excavations (D. H. Greenwald 1988c:139) indicated that tools were probably used near the canals and associated features. The ground stone assemblage provided evidence for possible decorated ceramic production; possible field, structure, or canal maintenance; and the roasting of foodstuffs. Ceramic data concerning function were limited. Four sherds of the entire assemblage could definitely be assigned a vessel form, and all were bowls. The structures in Area 3 were informal, that is, in most cases lacking prepared floors, walls, clay-lined hearths, and formal entrances. Combined, these data demonstrated that the occupants of Dutch Canal Ruin had conducted activities related to agriculture, exploitation of wild resources, and possible maintenance of canals but, for the most part, outside of Area 3.

SUMMARY

Area 3 was a locus of temporary and, presumably, seasonal use comprising field houses and associated features related to exploitation of subsistence resources and possibly canal maintenance. Most likely, occupation had occurred during the Santa Cruz phase but possibly also during the end of the Gila Butte and the beginning of the Sacaton phases.

According to acceptable chronometric data, the area was occupied during the interval between the late Gila Butte and the early Sacaton phases. Ceramic data suggested that the primary period of occupation was during the Santa Cruz phase. At least two separate periods of occupation were discernible in the data. The first was dated to the early Gila Butte through early Sacaton phases, based on archaeomagnetic assays. This occupation consisted of at least two structures, five associated extramural features, and an activity surface. This occupation may have been further divided. The earlier subgroup occupation would have included two fire pits and the activity surface; the later subgroup occupation would have included the two structures and three other pit features. In this case, the earlier occupation dated to the early Gila Butte through middle Santa Cruz phases, and the later dated to the middle Santa Cruz through early Sacaton phases.

The date of the second discernible occupation was based on stratigraphic evidence and was, therefore, a relative date. Neither chronometric samples nor temporally diagnostic artifacts were available for this second occupation. A pit structure and three extramural pits constituted the evidence for the second occupation of Area 3. After comparing the stratigraphy and absolute and relative dates for the first occupation, investigators determined that the second occupation would date no earlier than the middle Santa Cruz through early Sacaton phases.

Informal small structures and the paucity of artifacts pointed to temporary episodes of occupation. Because of the evidence for agriculture in the botanical samples, these temporary episodes had probably been seasonal, generally at planting and harvesting times. Investigators found evidence not only for corn agriculture, but also for the exploitation of cholla cactus and the plant family containing chicory and dandelion.

The flaked stone assemblage demonstrated that Area 3 was a locus not of tool use but of production. A ground stone maul with battering wear suggested activities related to field or canal maintenance, structure construction, fuel procurement, or any combination of these activities.

Area 3, in general, was a field house locus occupied between the late Gila Butte and early Sacaton phases. Occupations were temporary and tended to be seasonal. Activities in the area were related to the exploitation of corn and other subsistence resources.

CHAPTER 5

DUTCH CANAL RUIN, AREA 4

Mark L. Chenault

Unlike other research areas at Dutch Canal Ruin, Area 4 was a locus of resource procurement and processing. Investigators found no evidence of architecture and therefore concluded that this area was probably not a residential locus. Evidence of heating or food processing appeared in the form of a possible mano and thermally altered rock.

Area 4 was in the southwest corner of Dutch Canal Ruin, just east of 16th Street and between the former locations of Tonto Street and Maricopa Street (Figure 1.2) and at N1330–1375 and E520–550 within the project grid system. Its legal coordinates were T1N, R3E, Sec. 10, SW¼SW¼SW¼ on the Phoenix, Arizona, 7.5 minute USGS topographic map (Figure 1.4).

Investigators identified two possible pit structures in Area 4 during testing (Testing Features 1043 and 1066; see Volume 1:Appendix A). However, excavation revealed that no structures were present. One of the features found during testing was a large, basin-shaped pit that lacked such characteristics of a structure as a flat compacted floor. During the data recovery phase this pit was recorded as Feature 4-1, a pit not further specified (NFS). The second feature was an ill-defined area of more than 30 pieces of fire-cracked rock and charcoal without definite boundaries or morphology. Damage from modern construction and perhaps the subsequent razing of modern structures prior to the latest phase of development had destroyed most of the integrity of this area of Dutch Canal Ruin.

FIELD METHODS AND STRATEGIES

Field crews reopened two backhoe trenches (1222 and 1328) excavated during the testing phase and examined them for cultural features (Figure 5.1). They relocated and excavated one feature (Feature 4-1) identified during testing. They could not mechanically strip the area west of Feature 4-1 due to large buried pieces of concrete that were too heavy for removal by the backhoe; therefore, they concentrated the stripping east of Feature 4-1. They located no other cultural features during this stripping process, although they did find scattered ash, charcoal flecks, and fire-cracked rock in the immediate area.

CULTURAL FEATURES

Feature Number: 4-1
Horizontal Provenience: N 1252.0–1253.5; E 528.0–530.0
Vertical Provenience: 5.72–5.89 mbd
Feature Type: Pit NFS

This large, basin-shaped pit was oval to subrectangular in plan view (Figure 5.2). The pit measured 1.92 × 1.25 m and 17 cm in depth; it had been dug into the sterile silty substratum. The sides of the pit were neither oxidized nor lined. The fill of this feature consisted of a sandy silt matrix with a large amount of ash and charcoal. A thin lens of ash and charcoal, denser than the rest of the fill, lay across the base of the feature at the contact between the sterile substratum and

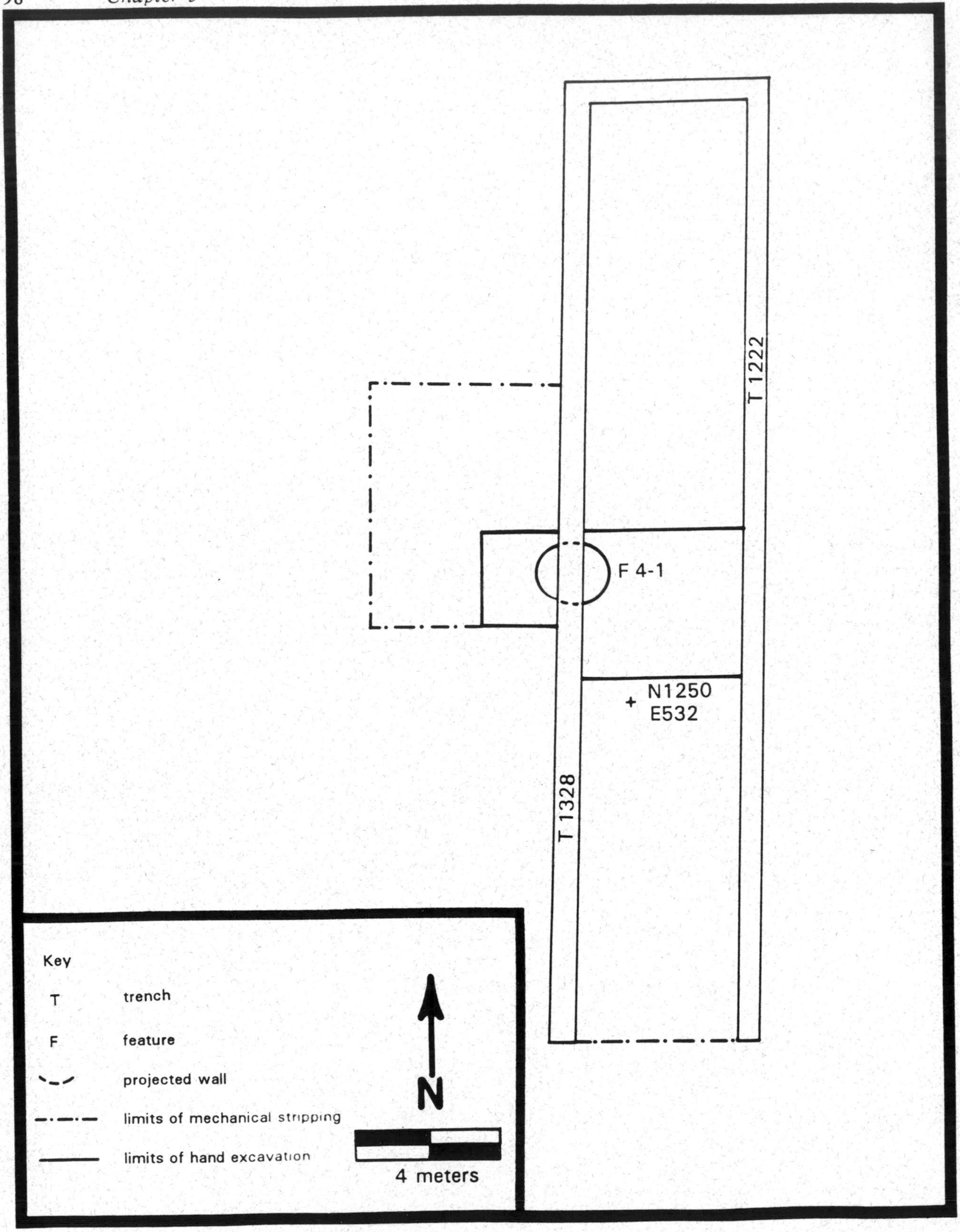

Key

T   trench

F   feature

⌒   projected wall

—·—·—   limits of mechanical stripping

——   limits of hand excavation

N

4 meters

T 1222

T 1328

F 4-1

+ N1250
  E532

Figure 5.1. Area 4, Dutch Canal Ruin.

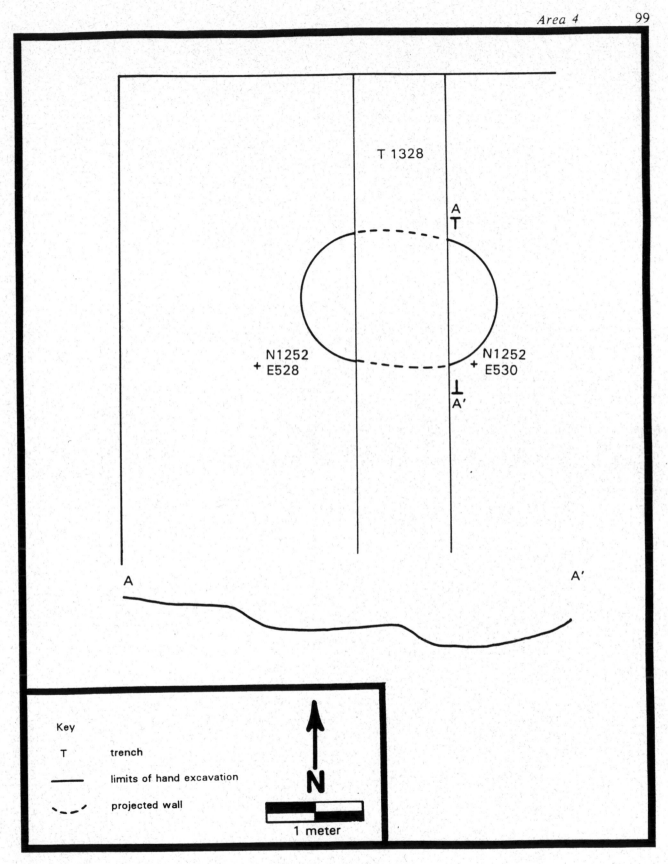

Figure 5.2. Area 4, Feature 4-1, plan and cross section.

the feature fill. Because field crews found no evidence of in situ burning, the ash and charcoal may have been secondary deposits. The fill, sides, and base of the feature were heavily disturbed by rodent activity. Artifacts recovered from the feature included ceramics, flaked stone, ground stone, and shell. Investigators collected but did not analyze a radiocarbon sample and two flotation samples. They could not determine the function of this pit feature given the existing evidence.

MATERIAL CULTURE

Although field crews recovered ceramics, flaked stone, ground stone, and shell artifacts, the artifact assemblage was very small and limited in diversity of types. Compared to other pre-Classic areas from Dutch Canal Ruin, however, the Area 4 assemblage was generally larger.

Ceramics

The ceramic assemblage from Area 4 consisted of 95 plainware sherds. These included 85 Gila Plain, Salt Variety; 8 Gila Plain, Gila Variety; and 2 Squaw Peak Plain sherds (Chapter 12). The low frequency and lack of diversity of ceramic types was indicative of limited activities, probably related to subsistence. In this regard, Area 4 compared with temporarily occupied field house areas at Dutch Canal Ruin that produced similar ceramic assemblages.

Flaked Stone

The flaked stone assemblage from Area 4 also was limited (Table 5.1). Out of 71 flaked lithic items, 66 (93.0%) were debitage, 2 (2.8%) were cores, and 3 (4.2%) were retouched items. None of the retouched tools displayed use wear, and the projectile point was the only tool with a high degree of energy invested in its manufacture. The debitage was composed of a relatively large diversity of material types, but cores and tools were confined to three types: basalt, igneous other material, and quartz.

Feature 4-1 contained 52.1% of the flaked stone artifacts from Area 4: 33 items of debitage, 2 cores, 1 projectile point, and 1 indeterminate tool. The basalt projectile point was stemmed, with serrated blades (Chapter 13:Figure 13.1b). Field crews recovered one flake uniface and 33 items of debitage from the rest of the area.

Though the sample itself was small, the relative frequency of retouched artifacts was high compared to assemblages recovered from other areas at Dutch Canal Ruin, particularly those from pre-Classic assemblages. This statistic, coupled with the low diversity of forms, suggested a limited-activity use for the area. This interpretation supported previous expectations (Volume 1:Chapter 10) that Area 4, being located some distance from the canals, might have been associated with the exploitation of wild resources.

Ground Stone

Field crews recovered three fragments of ground stone NFS from the fill of Feature 4-1. Two, sandstone fragments that were pecked on both surfaces and an edge, probably represented a mano. The third was a metasediment cobble of unknown function. The probable mano was medium textured, and its presence suggested that the inhabitants had processed food near Feature 4-1.

Table 5.1. Area 4, Flaked Stone Morphological Type by Material Type

Morphological Type	Material Type						Total
	Quartz	Chert	Basalt	Igneous Other	Quartzite	Meta[1] Other	
Angular Debris			2	1	1	1	5
			*40.0	20.0	20.0	20.0	**7.0
			**4.8	5.3	25.0	100.0	
Flake Fragment	1		15	4	3		23
	4.3		65.2	17.4	13.0		32.4
	33.3		35.7	21.1	75.0		
PRB[2] Flake			2				2
			100.0				2.8
			4.8				
Whole Flake	1	2	20	13			36
	2.8	5.6	55.6	36.1			50.7
	33.3	100.0	47.6	68.4			
Core			1	1			2
			50.0	50.0			2.8
			2.4	5.3			
Flake Uniface			1				1
			100.0				1.4
			2.4				
Projectile Point			1				1
			100.0				1.4
			2.4				
Indeterminate Tool	1						1
	100.0						1.4
	33.3						
Total	**3**	**2**	**42**	**19**	**4**	**1**	**71**
	*4.2	2.8	59.2	26.8	5.6	1.4	100.0

*Row %
**Column %
[1]Metamorphic
[2]Platform-remnant-bearing

Shell

Field crews recovered 11 pieces of *Anodonta* shell, representing at least 2 individual specimens, from Area 4. *Anodonta* is a freshwater bivalve that requires clean, flowing water to survive. The appearance of these shells in Area 4 might have been evidence of exploitation as a food resource, as none exhibited evidence of modification.

SUMMARY

Area 4 appeared to have been used for limited activities related to resource procurement and processing. The results of the analyses of material culture indicated that Area 4 was probably a nonagricultural, limited activity locus. Although investigations did not produce any chronometric samples and the artifact assemblage did not contain temporally diagnostic items, the occupation in the area appeared to date to the pre-Classic periods. This assignment was supported by the deep stratigraphic position of Feature 4-1.

Modern construction of buildings along the east side of 16th Street had apparently obliterated at least some of the cultural features in the area (Greenwald and Chenault 1991). Quantities of fire-cracked rock suggested the presence of a roasting pit or horno in the area, but investigators found no other evidence of such a feature. The only cultural feature defined during excavation was a pit NFS, Feature 4-1. The ash and charcoal in Feature 4-1 did not appear to be associated with the feature and probably represented secondary deposition, possibly originating from a nearby thermal feature such as an horno. The presence of fire-cracked and heat-altered rocks supported the notion that an horno had been present and that the inhabitants had processed economic resources, either cultigens or wild resources, in the area. In general, Area 4 appeared to deviate from the other investigated areas in that the focus of activity had apparently been plant procurement and processing.

CHAPTER 6

DUTCH CANAL RUIN, AREA 5

Sarah L. Horton
David H. Greenwald

Area 5, defined as a small habitation area, was stratigraphically positioned above two canal channels, Alignments 8560 and 8501 (Figure 1.2). The habitation area, as exposed during the testing phase, consisted of a single small, subterranean structure with a fire pit (Figure 6.1). The position of the canals in the Trench 2037 profile suggested that they might represent a bifurcated alignment.

Area 5 was located in the extreme eastern portion of Dutch Canal Ruin (Figure 1.2). This area also contained surface artifacts recorded during the pedestrian survey of the project area by BRW (1986), indicating the potential for intact buried remains. Area 5 was located in T1N, R3E, Sec. 10, NE¼NE¼SE¼ on the Phoenix, Arizona, 7.5 minute USGS topographic map (Figure 1.4) and originally lay between Harrison and Buchanan streets, within 50 m of 24th Street. Harrison and Buchanan streets no longer exist in the project area in the vicinity of Area 5 but can be found on various recent maps of Phoenix. Area 5 lay within the project grid at N1800-1860, E2050-2100.

Investigators expected that the superpositioning of the structure with its datable fire pit could provide corroborative chronometric data on the abandonment of the canals, as well as the function and age of the structure and possibly of associated habitation features. Field crews also examined the temporal and spatial relationships of the canals, evaluated their functions, and collected information regarding canal morphology and mechanics. Other habitation features and canal alignments lay in the immediate vicinity of Area 5, but the investigating team did not observe in other locations the specific spatial relationship of the features described above and believed that the data potential to be gained through further study of this location would probably not be matched in these other areas. Although the dating of canal sediments has progressed significantly in recent years (Ciolek-Torrello and Greenwald 1988; Eighmy and Howard 1991; Henderson 1989; Waters, Howard, and Greenwald 1988), archaeologists have found few associations in which canals intruded into habitation features, providing independent dates. Therefore, investigators selected Area 5 for excavation to address a number of issues relating to canal chronology in the project area.

RESULTS OF TESTING

The numerous trenches excavated within and surrounding Area 5 provided a fairly complete record of the distribution of the prehistoric irrigation canals and habitation features. Feature 5-1, the pit structure in Trench 2037, was the easternmost structure identified at Dutch Canal Ruin. During testing, field crews also recorded three other structures dispersed across the eastern portion of the site and several pit features. The ceramic assemblage recovered from trenches in this general area during the testing phase consisted of late Colonial and Sedentary period types. Ceramics from the surface were primarily redware types, supporting a Classic period affiliation. The cultural features and canal alignments in Area 5 and the adjacent areas were not well dated during the testing efforts. After the testing phase investigators concluded that the two canal channels located

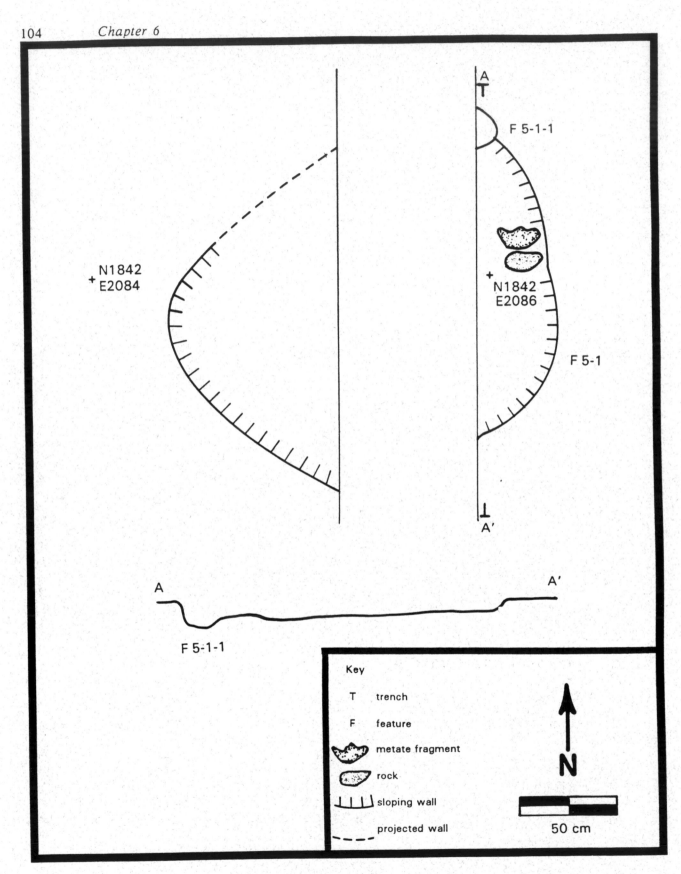

Figure 6.1. Area 5, Feature 5-1, plan and cross section.

beneath the pit structure dated to the pre-Classic periods, as did all canals at Dutch Canal Ruin. The southern channel was thought to be associated either with Alignment 8560, representing the latest channel in that alignment, or with North Main Canal 3 (Ciolek-Torrello and Greenwald 1988:153–154; D. H. Greenwald 1988a:55–65). The former interpretation was supported by the data recovery efforts.

RESEARCH OBJECTIVES

A main focus of investigation in Area 5 was to define the association between the two canal channels and to determine the presence or absence of a bifurcation within the canal alignments at this particular location. Alignment 8501 formed part of the northernmost canal alignment within the project area. Given the ceramic assemblage and the small settlements associated with this canal, investigators assigned the canal to the Colonial period. SWCA's investigators at first thought the southern channel had served as a crosscut canal that joined Alignment 8501 with Alignment 8001, but it was, in fact, the original alignment of North Main Canal 3 (Alignment 8560), which dated to the late Colonial period.

Archaeologists had to establish a chronological sequence for Area 5, as for all areas of Dutch Canal Ruin, to reconstruct prehistoric events at this location in association with habitation activities and canal use. Investigators recognized the superposition of the pit structure over the canals as one means of establishing an ending date for the use of the canals, if the age of the structure could be determined. They addressed the building and refining of the chronology for Area 5 through archaeomagnetic dating, radiocarbon dating, ceramic seriation, and stratigraphic association.

Canal studies were a major research focus because of the extent of the canal systems in the project area and their apparent early age. Because of the spatial association of the two canals in Area 5, investigators anticipated that they could address morphological and functional questions about Hohokam canal irrigation technology. The possible existence of a bifurcation within this canal system implied various measures to effectively channel water. Researchers expected that locating and defining types of intracanal features required for such a system to function, especially those features related to erosion and water diversion such as plunge pools, headgates, baffles, and dams, would provide significant information on canal morphology and feature function. Prehistoric re-engineering of the channels might have accounted for the spatial association of the two channels. If so, some definition of the temporal relationship of the two canals would be required to establish the spatial relationship.

Other research issues such as settlement patterns and settlement composition and typology also applied to Area 5. Investigators hoped that answers to these research questions would clarify the relation of Area 5 to Dutch Canal Ruin in terms of its settlement composition and type and how it compared to the overall settlement pattern of the site. Data recovery efforts successfully contributed to addressing all these questions.

FIELD METHODS AND STRATEGIES

Backhoe Trenching

Although field crews excavated several trenches during testing in Area 5, they reopened only one (Trench 2037) for data recovery, because it exposed all the features to be investigated in this area: a pit structure with a fire pit and two canal channels. Excavators tentatively defined the floor of the structure on both the east and west faces of the trench, based on the elevation of the hearth and on somewhat indistinct soil differences.

Hand Excavations

The normal procedure for excavating a structure during the Phoenix Sky Harbor Center data recovery phase was to divide it into four equal quadrants that field crews excavated and sampled as individual units. However, in this case distinguishing the structure outlines from the surrounding matrix was difficult, and the small size of the structure precluded the use of quadrants. Therefore, investigators used 2 × 2-m excavation units to begin excavation, and after defining the structure's limits, they divided it into east and west halves, based on the location of the backhoe trench.

Field crews used trowels and small hand picks to remove the hard-packed clay fill in 10-cm levels. Because the feature remained indistinct after stripping, they initiated hand excavation of the 2 × 2-m units at the trench edges where they saw some indication of structure morphology (i.e., the floor and the fire pit) in profile. The excavation was expanded to both the east and the west away from the trench edges to locate the walls or boundaries of the feature. After excavating the first 10-cm level in some excavation units, investigators determined that all or most of these units were outside the feature, indicating that the structure was extremely small.

STRATIGRAPHY AND ASSOCIATION

Field investigators recorded three strata in Area 5. The upper level consisted of the recently disturbed zone (0–54 cm), comprising a plow zone and residential disturbance. Although this level varied in depth and content over most of the project area, in Area 5 it contained large cobbles, asphalt and cement chunks, and other forms of construction trash within a sandy matrix, indicating that historically the area had been "filled" or elevated artificially. Field crews designated the disturbance zone Stratum 1 and mechanically stripped it away to access undisturbed deposits.

The interface between the first and second strata was obvious. Stratum 2 (54–82 cm) consisted of dark brown chunky clays that formed a regular horizontal stratum over most of the eastern half of Dutch Canal Ruin. The structure, Feature 5-1, was within this stratum. Given the location of Area 5 relative to the geologic floodplain, investigators speculated that this organically rich clay layer may have resulted from flood episodes that had periodically inundated Dutch Canal Ruin. A second source of these clays may have been irrigation activities, conducted during both prehistoric and historic use of the project area. The fact that Feature 5-1 not only was dug into the organic clays but also was filled and buried by them indicated that this depositional sequence had been an ongoing process before, during, and after the prehistoric occupation of the project area.

Stratum 3 (82–185 cm) consisted of a sterile, fine-grained, fairly well sorted, light brown silty clay loam. This stratum equated to the C horizon of the Avondale clay loam (Volume 1:Appendix B). Stratum 3 predated the prehistoric use of Area 5 and was the soil layer into which most cultural features, including canals, had been dug. Field crews did not reach the base of Stratum 3 and a well-developed calcium carbonate level in any of the trenches dug in this portion of Dutch Canal Ruin, a consequence of the deep alluvial deposits associated with the geologic floodplain.

CULTURAL FEATURES

Area 5 contained cultural features representing prehistoric habitation activities and canals. The habitation was a single pit structure located within the upper fill of the two canal channels. The northern channel had not been dated previously and represented the northernmost canal alignment in the project area. The southern channel proved to be the later of the two canal alignments in Area 5, previously identified as a channel in the North Main Canal alignment (Greenwald and Ciolek-Torrello 1988a).

Feature 5-1

Horizontal Provenience: N 1840.94–N 1841.72; E 2084.38–E 2086.35
Vertical Provenience: 1.53–1.73 mbd
Feature Type: Pit Structure

As defined during excavation, this structure was small, measuring 1.78 m north-south by 1.97 m east-west. From its level of detection to the floor, the depth varied from 5 cm to 9 cm. Its overall morphology was informal and asymmetrical in plan, with the fire pit positioned along the wall in the northeast corner (Figure 6.1). Trench 2037 bisected the feature into nearly equal halves, removing over one-third of its floor area.

The floor of the structure was not prepared and was poorly preserved. The floor contact zone, identified from the trench profile, was further defined by flat-lying artifacts, a slight soil color change, a change in soil consistency to less sandy, a decrease in charcoal density, and the horizontal separation of overlying fill at floor contact. Similarly, the low wall remnants exhibited no evidence of preparation and were poorly preserved but detectable as an abrupt upward sloping of the surface, forming a rounded wall-floor juncture. From the center of the structure to the walls, a rather gentle or gradual upward slope resulted in a basin-shaped profile. The walls had been created as a result of the pit having been dug into the clay-rich substrate.

In plan, the structure was trapezoidal with rounded corners. Determination of the boundaries or lower walls was possible from observed differences between the structure fill and surrounding clays. Although the fill was slightly lighter in color than the clays, the primary difference was in the consistency. The fill contained a higher percentage of sand than the clays, and the density of charcoal flecks was greater within the structure boundaries. Field crews discovered no floor features indicative of roof construction; as with most ephemeral structures, a simple brush covering probably had been erected over the pit. Because the structure had been seasonally occupied and used during summer months, shade from the sun and good ventilation were probably the principal design considerations. The presence of a fire pit, however, suggested that food may have been processed or heat provided in the structure, lending support to the idea that the structure had also been used during cooler months of the year.

The only floor feature within the structure was a fire pit, Subfeature 5-1-1, that measured 25 cm in diameter and 17 cm in depth. Although the fire pit was bisected by the backhoe trench, it appeared to be oval in plan. It was located along the northeast edge of the structure (Figure 6.1) and had been dug into the dark brown clays that served as the structure's floor and lower walls. No lining was present in the fire pit, but the pit edges were highly oxidized from an open fire. Investigators collected an archaeomagnetic sample from the pit edges to help date the structure and provide chronometric data for the canals beneath it.

The informal architectural characteristics of the pit structure, its small size, and the location of the fire pit adjacent to the wall suggested that this structure was a locus of limited use, probably inhabited by only one or two individuals. The only activities that archaeologists inferred had taken place within the structure were related to use of the fire pit. Floor artifacts were few, although field crews found a trough metate fragment and another possible piece of ground stone along the east wall. Investigators could not determine the function of these two items because of their recovered condition. They may have been used as some type of support for an interior post or as a rest for a basket or pot. In all likelihood, Feature 5-1 was a field house associated with seasonal occupation for agriculture or wild plant procurement.

Canals

Figure 6.2 shows the canal alignments identified in the immediate vicinity of Area 5 during the testing phase. Alignment 8501 (Canal Viejo) is the northernmost canal, and Alignment 8560 angles from northeast to southwest connecting Alignment 8501 with Alignment 8001 (Figure 1.2). During testing, investigators determined that Alignment 8560, designated North Main Canal 3, was the youngest of the three channels that constituted the North Main Canal alignment (D. H. Greenwald 1988a:55–65). As a result, existing chronometric and sediment information from the Squaw Peak Parkway investigations and from the Sky Harbor testing phase was available for comparison. However, Alignment 8501 was a newly discovered canal that required further study regarding its chronological placement and use history.

Canal Viejo (Alignment 8501) varied in size, with a minimum width of 2.20 m and a depth range of 0.75–1.13 m, as recorded in various trenches at Dutch Canal Ruin. The sediments of this canal consisted of coarse basal sands grading to medium- and fine-grained sand lenses intermixed with clayey silt lenses throughout the upper two-thirds of the channel. The general trend was a fining-upward sequence within the canal fill, indicating a decrease in water velocity and, perhaps, decreasing efficiency (Volume 1:Chapter 3).

In contrast, Alignment 8001 was represented by fine sandy silts intermixed with light-colored silty clay bands in the basal portion. Wider bands of clay deposits, with occasional sand lenses, appeared in the midsection with varying colored, stratified clay bands. The upper one-third of the channel was filled with chocolate-brown clays that gradually graded into the overlying dark clay soils found throughout this portion of the project area. The sediment history of this channel indicated that the water velocity had consistently been slower than that of Canal Viejo and that the upper fill may have been deposited by standing water rich in plant life. Because of a very distinct and rather thin parabolic clay lens recognized in it, Alignment 8001 could be traced from one trench to another, so that investigators could recognize the orientation of Alignment 8501 across the site because of its distinctive sediments. This canal had experienced considerable erosion during its use life, resulting in segments as wide as 10.0 m and in multiple channels. However, Trench 2037 in Area 5 and Trench 2022 in Area 7 (Chapter 8:Figure 8.3) showed well-preserved parabolic channels.

Alignment 8560 was originally identified as a crosscut canal connecting Alignment 8501 with Alignment 8001. After investigators determined that this channel belonged with North Main Canal 3, they re-evaluated its function based on its trajectory (which extended across the project area) and method of construction. North Main Canal 3, the youngest channel in Alignment 8001, consistently followed earlier canals while maintaining a west-southwest heading. At times, North Main Canal 3 was excavated into the earlier canals or paralleled them until its orientation shifted

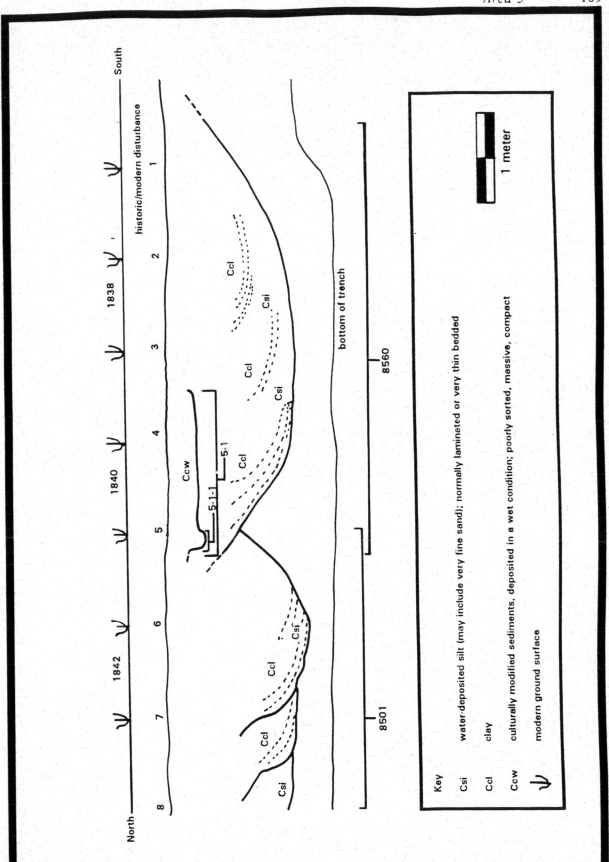

Figure 6.2. Area 5, profile of Trench 2037 showing the relationship of Alignment 8501 to Alignment 8560, with Feature 5-1 superimposed.

farther south. Field investigators first recognized this utilization of an earlier channel in Alignment 8003 near the western extent of the project area (Volume 1:Figure 8.2). Other consistent aspects of North Main Canal 3 were its parabolic shape and fine dark organic deposits. In Area 5, Alignment 8560 was less than 2.0 m in width and was filled with the dark, chocolate-brown clays.

Because of the variations in the sediments of these canals, investigators determined that Alignment 8560 represented a later period of use than Alignment 8501. No evidence of alternating flows between the two channels was present, which would have supported the premise of a controlled bifurcated system. In all likelihood, Alignment 8560 was either a re-engineering of Alignment 8501 or a completely separate canal that followed the earlier Alignment 8501, diverging from that channel at this location and then following a southwesterly trajectory.

Neither alignment contained high amounts of ferrous or manganese oxides when compared to other project canals such as Alignments 8537 and 8003 and North Main Canals 1 and 2 in Alignment 8001. In the profile exposures of Trench 2037, field crews observed no oxides in association with either canal. Researchers have generally attributed oxide development in canals or at their bases to increased periods of saturation and mineral leaching as a result of saturation and water stagnation (Volume 1:Chapter 8). The amount of mineral accumulation varied, being directly affected by soil composition, amounts of mineral present in the soils, and saturation levels. In general, however, the lower amounts of basal oxides in this trench suggested that saturation levels in both canals were lower than in other canals and that these two canals had been used less frequently than others. For Alignment 8560 (North Main Canal 3), investigators postulated an episodic flow regime, based on previous studies of this channel in the Squaw Peak Parkway (D. H. Greenwald 1988a:63–65).

MATERIAL CULTURE

The overall material culture assemblage from Area 5 was very small, a direct correlate of the ephemeral nature of the occupation and the sparsity of cultural features. The artifacts recovered from this area included ceramic, flaked stone, and ground stone items. Although the artifact assemblage was limited, the variability of subsistence remains ranged widely, including weed species, cultigens, mesquite wood, and agave fibers.

Ceramics

Field crews recovered 21 sherds from excavations in Area 5, all Hohokam undecorated wares. By analyzing temper, analysts classified 20 sherds (95.2%) as Gila Plain, Salt Variety and 1 (4.8%) as Salt Red. All of the ceramics lacked temporally diagnostic attributes.

Crews also recovered 4 Gila Plain, Salt Variety sherds from the fill of Feature 5-1, the pit structure. They collected the remaining 17 sherds from the various 2 × 2-m excavation units that they dug beyond Feature 5-1 to define its extent. Given the spatial association of the excavation units to the structure, investigators concluded that most of these sherds were probably related temporally to the structure's use.

Flaked Stone

The flaked stone assemblage consisted of 45 items of debitage. No tool forms were present, and field crews recovered only one item, a quartzite whole flake, from Feature 5-1. Material types were dominated by igneous other material and basalt, with quartzite, obsidian, and metamorphic materials minimally represented (Table 6.1). Although the flaked stone assemblage provided few data, investigators posited that the inhabitants had produced flakes in the immediate vicinity of Feature 5-1. None of the items exhibited use wear, suggesting that after suitable flakes were produced they were removed from this location and used elsewhere. The lack of evidence for flake utilization supported the premise of an ephemeral occupation of Area 5, and the limited number of items recovered from Feature 5-1 implied that tool makers had conducted few activities associated with flake production and use within this structure.

Table 6.1. Area 5, Flaked Stone Morphological Type by Material Type

Morphological Type		Material Type				Total
	Obsidian	Basalt	Igneous Other	Quartzite	Metamorphic Other	
Angular Debris			3 *100.0 **12.0			3 **6.7
Flake Fragment		11 40.7 78.6	13 48.1 52.0	2 7.4 50.0	1 3.7 100.0	27 60.0
Whole Flake	1 6.7 100.0	3 20.0 21.4	9 60.0 36.0	2 13.3 50.0		15 33.3
Total	1 *2.2	14 31.1	25 55.6	4 8.9	1 2.2	45 100.0

*Row %
**Column %

Ground Stone

The ground stone collection also was limited in number. Field crews recovered one trough metate fragment and one item identified as possible ground stone from the floor of the structure. The metate fragment was of vesicular basalt and completely shaped. The possible ground stone item was a metasedimentary material that may have had a small amount of grinding on its surface. Its presence on the floor of the structure implied that the occupants had used it in some way; however, since analysts could not determine whether it was culturally modified, they could not assess its function.

The presence of the metate suggested that the occupants of Area 5 had processed food there. Previous studies concerning the function of ground stone have demonstrated that items with coarse surface textures, such as this trough metate, were used for the grinding of large seeds such as corn, whereas fine-textured items were used for smaller seed resources, such as Cheno-ams (Ackerly 1979:326; Halbirt 1985:28, 1989:204). The location of both items within Feature 5-1 and the incomplete form of the trough metate suggested that the last use of these items had not been associated with food processing. Their position within the structure may have been related either to a support for a pot or basket or to an interior support post for the superstructure.

Subsistence Remains

Analysts examined the flotation and pollen samples to address whether the subsistence strategies in Area 5 were economically related to or associated with the construction of the pit structure. Investigators submitted for analysis two pollen samples collected from the floor of the structure: one from beneath the metate fragment and the other from the western half of the structure. The results indicated probable subsistence processing within the structure. Both samples contained large quantities of Cheno-am pollen, including pollen aggregates. A small amount of corn (*Zea mays*) pollen also was present in the samples, indicating some association of the use of the structure with corn agriculture. Because corn pollen is not transported by wind, its presence in the structure would have to have been cultural. The two pollen samples were very similar in content, as would be expected from a small structure, even though they were collected from different contexts, one from beneath the metate fragment and the other from the open floor.

One flotation sample from the floor of Feature 5-1 produced 6 pieces of mesquite wood charcoal, 4 agave fibers, 2 Cheno-am seed fragments, 6 indeterminate and 2 miscellaneous seed fragments, 6 unknown round fibers, 1 corn cupule, 2 cupule fragments, and 1 corn kernel fragment. Several activities associated with these plant taxa may be inferred. Because the wood charcoal was collected from an unburned structure, the mesquite charcoal may have represented fuel. The agave fibers may have been used for a number of purposes, such as cordage, matting, waste from the consumption of agave hearts or leaves, or clothing.

A few of the Cheno-am seeds collected from Dutch Canal Ruin loci were uncarbonized and therefore appeared to be recent contaminants. However, the two seed fragments from Area 5 were charred and thus seemed more likely to be part of the prehistoric botanical assemblage. The presence of the Cheno-am seeds may have been related to subsistence activities, including processing, consumption, or storage.

The corn remains may have resulted from a number of activities. Either the burning of cobs as fuel (thus disposing of a by-product of corn harvesting) or the overcooking of roasting ears could have created the charred remains (Chapter 15). Corn probably had been processed in this location prior to storage or had been eaten here.

Although investigators could not determine exact subsistence activities, the presence of these taxa indicated that the site inhabitants had probably processed plants and consumed food at Area 5. The two forms of botanical analysis provided similar results, both suggesting that the occupants had practiced corn agriculture or exploited and consumed weedy species, specifically, Cheno-ams. Although agave appeared in the macrobotanical assemblage, archaeologists would not expect to find agave pollen in conjunction with other by-products of the plant, because agave was probably not harvested until after it flowered. It may have been reduced in mass at the site of harvest to minimize its bulk during transport (Fish et al. 1985). Agave fibers could have been deposited in Feature 5-1 as food, clothing, or raw materials. The presence of mesquite charcoal is not unusual. Mesquite was probably available in many nearby areas and was probably readily exploited for fuel and construction materials. Previous studies have not supported the idea that mesquite was a food resource during the pre-Classic period occupation of Dutch Canal Ruin (Cummings 1988; Ruppé 1988), although researchers recognize the beans and green pods as an important food resource. Two explanations are posited for this: either mesquite procurement was not a focus of the subsistence strategies during this time at Dutch Canal Ruin, or mesquite was not abundantly available in the immediate area and was therefore not exploited when the Dutch Canal Ruin field houses were used. Because agriculture was the focus of the pre-Classic period settlement of Dutch Canal Ruin, the occupants of the area probably harvested mesquite in other locations, near mesquite bosques. Historic records indicate the presence of a mesquite bosque approximately one mile west of Dutch Canal Ruin (Scott Kwiatkowski, personal communication 1991), and if this bosque had existed prehistorically, village groups along the north side of the Salt River would undoubtedly have exploited it.

CHRONOLOGY

One of the research goals for Area 5 was to determine the temporal relationship of the pit structure to the canals. Analysts used several methods to investigate this chronological association, including relative stratigraphic relationships, ceramics, and archaeomagnetic and radiocarbon dating methods. They have defined the stratigraphic relationships, resulting in an understanding of the sequence of events in Area 5. Ceramic analysis, however, provided virtually no information on the range of possible use dates for the structure. Both the archaeomagnetic sample recovered from the fire pit and the radiocarbon sample retrieved from the fill of the pit structure yielded dates.

Defining the stratigraphic associations between the pit structure and the two canal channels in Area 5 revealed the sequence of events in this location. In the testing phase, investigators defined the relationship of the pit structure to the canals: the pit structure clearly postdated the canals, as it was built upon the alluvial deposits that filled them. Investigators expected to more clearly understand the abandonment of the canals by dating the period of use of the structure. Researchers had previously analyzed few samples, and those results supported neither archaeomagnetic nor radiocarbon dates of the canals at Dutch Canal Ruin. If an "ending" date could be determined, whether by dating a fire pit archaeomagnetically or by dating the pit structure through radiocarbon assay, previous results and interpretations could be evaluated. The location of Feature 5-1 over the canals carried the potential to evaluate both dating methods as they had previously been applied to canal chronology.

Investigators did not define the relationship of Alignment 8560 to Alignment 8501 during the testing phase. Further investigation in Area 5 determined that these two channels were not contemporaneous and did not represent a bifurcated system. Instead, Alignment 8501, Canal Viejo, was an earlier channel that was intruded on by North Main Canal 3. Previously, researchers had dated North Main Canal 3 to A.D. 850–935 (Ciolek-Torrello and Greenwald 1988:159), with a revised date, based on the latest Southwestern VGP curve, of A.D. 830–940 (Appendix A). These results were based on samples recovered from the upper levels of North Main Canal 3, which should be considered postabandonment deposits. As a result, the date range for North Main Canal 3 should reflect a terminal date. In all likelihood, North Main Canal 3 had been abandoned by A.D. 900. This inference was supported by the recent streamflow reconstruction for the Salt River (Graybill

1989:33–38) that indicated an extremely high moisture regime during A.D. 899, which had probably resulted in a severe flooding event that could have left the entire Hohokam canal network along the Salt River inoperable (Nials, Gregory, and Graybill 1989:65–66). The A.D. 899 flow was unmatched by any other flow during the streamflow reconstruction period of A.D. 740–1370. Researchers calculated a recurrence interval of 632 years for the magnitude of this flow (Nials, Gregory, and Graybill 1989). With these data, investigators could not project the latest use of either canal beyond A.D. 935 or 940, and with the high probability of the devastating flow during A.D. 899, the latest canal, North Main Canal 3, had most likely been abandoned by A.D. 900.

The ceramics offered no relevant chronological information for the area. Gila Plain ceramics spanned practically the entire Hohokam occupation in the Phoenix Valley. The presence of the redware sherd was anomalous, however. Salt Red is normally associated with the late Classic period, but all the other evidence collected from this locale indicated that the occupation in Area 5 dated to the pre-Classic periods. Investigators identified a Classic period component nearby (Chapter 11), and this single redware sherd appeared in fill that postdated both the structure and the canals; these data indicated that the sherd was present because of postdepositional processes associated with later use of the project area.

The archaeomagnetic sample (Sample AZ T:12:62 [ASM]-6) collected from the fire pit in the pit structure produced a visual date range of A.D. 925–975. The reliability of this date interval was well supported by the low alpha 95 reading of 1.67 at 15 steps of demagnetization. A second interval, A.D. 1550–1650, was sufficiently late to be discounted as a possibility. The corresponding declination date for this sample was A.D. 905–1020. Both dates would be acceptable, because they indicate a period of use that postdated the estimated abandonment date of the canals. Accepting the period of use of the pit structure as during the tenth century, and thereby accepting the date intervals for the archaeomagnetic sample as A.D. 925–975, supported the date derived from canal sediments for North Main Canal 3. As a result, investigators achieved, at least in part, the research goal of evaluating the temporal assignments of the canals; their research supported the earlier dates assigned to North Main Canal 3 and defined the stratigraphic association between Alignment 8501 (Canal Viejo) and Alignment 8560 (North Main Canal 3).

A radiocarbon sample of wood charcoal recovered from the fill of the pit structure and submitted for dating as a second means of evaluating occupational history and canal use in Area 5 was extremely small and required additional processing through use of an accelerator. Using Stuiver and Becker's (1986) formulas for radiocarbon calibration, analysts calculated a date range of A.D. 1218–1297. Investigators considered this date range extremely late given other lines of evidence, including the results of the archaeomagnetic sample. However, radiocarbon dating in general at Dutch Canal Ruin resulted in extremely poor results, with many samples producing intervals earlier than 1 B.C., some as early as 500 B.C. Analysts attributed these aberrant dates, all several years too early, to organic contamination from periods of saturation when the first terrace was flooded or was irrigated prehistorically and historically. By contrast, the interval for this particular sample appeared to be too late. The charcoal recovered from the fill of Feature 5-1 could have been intrusive from the Classic period occupation associated with nearby Area 10. If, in fact, the charcoal came from the Area 10 occupation, the date interval would fit the expected range for that area, including the estimated degree of contamination. As a result, the radiocarbon assay did not clarify the temporal assignments given to either the habitation episode or the canal use sequence.

SUMMARY

The Area 5 structure was one of the simplest in terms of organizational patterns (Gregory 1987:184). A social group that may have consisted of a nuclear family or smaller corporate unit probably occupied the single pit structure. The occupants used this habitation area on a seasonal basis and (as indicated by the small size and informal design of the structure and the extremely limited artifact assemblage) probably on an episodic basis, to maintain canals, regulate irrigation water, and plant, cultivate, and harvest fields. They probably used the structure only during the late spring and summer months, when they prepared fields for planting and then tended crops as needed until the harvest season. The date of use for the pit structure indicated one of the latest occupations at Dutch Canal Ruin during the pre-Classic periods. Only Area 9, south of Area 5, exhibited evidence of occupation during the Sedentary period (Chapter 10). Substantial support has been generated by streamflow studies for the premise that a devastating flood occurred during A.D. 899 and would have rendered Canal System 2 useless until the Hohokam rebuilt the canal heads and re-excavated the canals. Located on the first terrace, Area 5 and Dutch Canal Ruin would have been extremely vulnerable to the effects of such a flood. Recognizing the risks associated with farming in this location, residents may have relocated fields farther north on the bajada slopes of the second terrace. Evidence for use of the project area subsequent to A.D. 899 and until the late Classic period was minimal. The pit structure, Feature 5-1, in Area 5 may represent a brief attempt at exploiting this high-risk location after the abandonment of the pre-Classic period canals.

Area 5 functioned as a field house locus during the tenth century A.D. Specific activities other than those previously defined for field houses associated with canal irrigation (Cable and Doyel 1985) were not readily apparent at this location. Previously inferred functions of field houses included canal maintenance, water regulation, and planting, cultivating, and harvesting. Within Feature 5-1, activities relating to food processing and consumption were suggested by the presence of the fire pit, a trough metate fragment, and the charred botanical remains. Other related activities may have included the storage of tools during periods of nonuse and temporary storage of produce prior to transport to the permanent settlement (Wilcox 1978:26).

Given the archaeomagnetic results and the stratigraphic relationships of the pit structure to the canals, investigators concluded that the occupation of Area 5 had occurred toward the end of the pre-Classic period settlement of the site. Analysts have not defined pre-Classic period canals that postdated North Main Canal either within the project area or among the identified canals at Dutch Canal Ruin; only Alignment 8555, located south of Area 5, has not been positively dated. It is difficult to conceive that Hohokam farmers practiced corn agriculture in the lower desert without methods to increase soil moisture over that available from direct precipitation. Two explanations may serve: first, a later pre-Classic period canal existed that archaeologists have not as yet defined; second, the Hohokam living at La Ciudad after the canal system had been abandoned following the flood of A.D. 899 practiced other forms of farming in the vicinity of Dutch Canal Ruin in areas of greater available soil moisture.

Although not previously documented, corn agriculture could have been practiced and other economic plants exploited in areas that retained elevated moisture levels due to the physical alteration of the landscape or indirect application of water to a given area (Volume 4:Chapter 7). Crosswhite (1981) described similar instances of economically rich areas that border fields, with retention of moisture by restricted drainage. Various nonirrigation farming techniques are considered in Volume 4, Chapter 7, including microenvironments that the Hohokam may have used at Dutch Canal Ruin.

In summary, while field house settlements in the Phoenix Basin often have been equated with agriculture-related activities, various lines of evidence support their use in conjunction with other economic activities as well (Cable and Doyel 1985). In Area 5 at Dutch Canal Ruin, the botanical record implied activities associated with both corn agriculture and the exploitation of weedy species. Area 5 could have been used in direct association with canal irrigation or after the abandonment of the canals when other resources were readily available on the first terrace.

CHAPTER 7

DUTCH CANAL RUIN, AREA 6

David H. Greenwald

Research Area 6 contained four separate canals that either merged or split at this point (Figure 7.1). A small canal that followed a northwestern trajectory, Alignment 8005, "merged" in this location with a set of three larger canals, collectively designated Alignment 8001. Researchers had previously called this latter collection of three canals the North Main Canal alignment of Dutch Canal Ruin (Greenwald and Ciolek-Torrello 1988a).

Area 6 was in the northern portion of the project area, north of Grant Street and east of 18th Street (Figure 1.2). Field crews noted a light scatter of surface artifacts throughout the area. Investigations were confined to an area that measured approximately 70 × 150 m in T1N, R3E, Sec. 10, SW¼NE¼SW¼ on the Phoenix, Arizona, 7.5 minute USGS topographic map (Figure 1.4).

Given the orientation of Alignment 8001, investigators believed that these three canals corresponded to Canal Patricio, the name given by Frank Midvale (1968) to a canal alignment in this location and named for H. R. Patrick, who was one of the first to map Hohokam canals in the Phoenix Basin. Although surface evidence of the canals in this portion of the project area was visible as late as the mid 1930s, archaeologists had not confirmed their existence until excavations in the Squaw Peak Parkway corridor in 1986 (Greenwald and Ciolek-Torrello 1988a). During the testing phase of the Phoenix Sky Harbor Center project, field investigations conducted in areas adjacent to the Squaw Peak Parkway by SWCA identified the North Main Canal alignment and traced it from one end of the project area to the other, an extent of one linear mile from 24th Street on the east to 16th Street on the west (Figure 1.2; Volume 1:Chapters 6 and 8). Testing located seven canals and associated habitation features in the northern portion of the Phoenix Sky Harbor Center and provided an opportunity to study Hohokam canals across an area 1.6 km (1.0 mile) in length by 0.81 km (0.5 mile) in width. Among the canal features identified was the bifurcation mapped by Turney (1924; 1929) that apparently split the North Main Canal alignment into northern and southern branches. Investigators selected Area 6 for further study because of its information potential regarding the complex of canals and the bifurcation of the canal system.

The testing phase revealed that Area 6 at Dutch Canal Ruin contained three of the main canals (North Main Canals 1, 2, and 3) identified during the investigations of the Squaw Peak Parkway (D. H. Greenwald 1988a:55–65) and a smaller channel (Alignment 8005) that originated in South Main Canal 1, located a short distance (110 m) to the south (Volume 1:Chapter 8). Turney indicated on his 1924 and 1929 maps of prehistoric canals in the Phoenix Basin a bifurcation of the canal alignment approximately 200 m east of Area 6. Earlier investigations in that area (Greenwald and Ciolek-Torrello 1988a) found no evidence of a bifurcation. During testing of the Phoenix Sky Harbor Center, SWCA located the apparent bifurcation by digging a series of backhoe trenches and examining the exposed canal profiles. Alignment 8001, the North Main Canal alignment, continued in a westerly orientation from the Squaw Peak Parkway until reaching the former vicinity of 18th Street, where the canal alignment appeared to separate. At this point, two canals angled to the northwest, while two others angled to the southwest, supporting the inference that this split was Turney's bifurcation.

At the beginning of this study, because it was what excavators expected to find, a canal bifurcation was defined as a division or split in a canal producing two channels that functioned

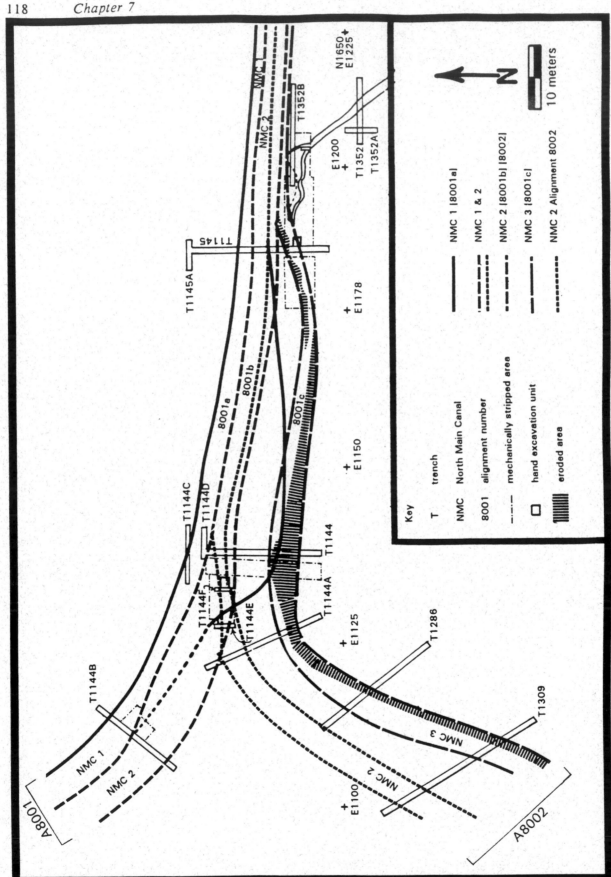

Figure 7.1. Area 6, Dutch Canal Ruin.

simultaneously. In other words, both channels must have been functional and capable of carrying water at the same time or through some means of alternating flows. However, based on both vertical and horizontal channel configurations, no two channels within the project area exhibited the characteristics defined above, and the observed splits in canals resulting from realignments over time are referred to in this discussion as bifurcations.

RESULTS OF TESTING

Testing in the area north of Grant Street and east of 18th Street located three canal alignments, which became the focus of this study (Figure 7.1). The first, the North Main Canal alignment (8001), extended westward through the project area from 24th Street to Area 6, where it angled to the northwest until it exited the project area; this was the earliest main canal identified at this location. The second, another alignment of the North Main Canal (Alignment 8002), departed from Alignment 8001 at the bifurcation or fork and angled to the southwest until leaving the project area at 16th Street; this served as the southern extension of this canal alignment and postdated the northern alignment. The third, a small channel (Alignment 8005), investigators traced from its intersection with the North Main Canal alignment to its origin, the South Main Canal alignment (8003), where the latter canal crossed 20th Street (Figure 1.2). Analysts could not temporally relate this canal to the other canals from the results of the testing phase.

RESEARCH OBJECTIVES

Investigators selected Area 6 for further investigation because of its potential for gathering information on the age, function, morphology, and spatial relationship of the canals to the geologic floodplain of the Salt River and on the habitation features that were scattered along the canals. Archaeologists considered the temporal and functional aspects of these canals important to understanding land-use strategies associated with the geologic floodplain. If the diverging canals in Area 6 indeed represented a functional bifurcation of the canal alignment, further information regarding the morphology of the canals and the function of the bifurcation would provide substantial insights into early pre-Classic period Hohokam canal technology. The construction, use, and abandonment of the canals may have been directly affected by flood events of the Salt River. Their location on the geologic floodplain increased their susceptibility to discharges from the river, and the abandonment of these canals may have been linked directly to flooding events. Review of the streamflow retrodiction studies completed for the Salt River (Graybill 1989; Nials, Gregory, and Graybill 1989) and comparison of those data with chronometric information recovered from investigations of the canals at Area 6 were selected as appropriate methods for addressing the impacts of Salt River flooding on Hohokam land-use strategies. For this reason, analysts regarded temporal information about the canals as fundamental for reconstructing the abandonment of those channels located on the geologic floodplain within the project area and Canal System 2.

Investigators structured the research orientation for Area 6 to evaluate the temporal associations, mechanical relationships, and functions of these features. They examined canal engineering and morphological attributes to provide a better understanding of (1) land-use strategies on the geologic floodplain; (2) the model of land use proposed by Greenwald and Ciolek-Torrello (1988a), including the support given to that model by the recent streamflow retrodiction studies completed by the Arizona State Museum (Graybill 1989; Nials, Gregory, and Graybill 1989); and (3) the mechanics and spatial relationships of the system.

Temporal Associations

Although existing data permitted analysts to date the canals in Area 6 according to both calendric dates and the current Hohokam chronology (Dean 1991), many of the available chronometric data relied on ceramic and stratigraphic association or the use of a promising but relatively untested method, detrital magnetism (Eighmy and Howard 1991). The latter method is a recognized geologic dating tool, and its application to events in the last 10,000 years has tremendous potential. Evaluation continues as more dates are produced. Dutch Canal Ruin was one of the first sites to be dated by detrital magnetism, as part of the investigations sponsored by the Arizona Department of Transportation in the Squaw Peak Parkway corridor (Greenwald and Ciolek-Torrello 1988a). At that time, analysts considered the results highly reliable. The Southwestern master VGP curve was later updated through continued collection and analysis of archaeomagnetic samples. The project goals, as they related to the chronology of the canals, were to refine the existing chronometric data for the canals. To this end, data recovery included collection of additional archaeomagnetic samples and attempts to gather other datable samples such as ceramics to aid in the evaluation process.

Canal Function

Although investigators understood the functions of certain alignments prior to this project, they assigned functions to some channels during the testing phase on the basis of limited information available from trench profiles. They were concerned with the presence and function of headgates, Turney's bifurcation, crosscut canals, and turnouts. To address function, they used horizontal excavation units to expose features and canal channels in three dimensions. Additional trenching initially helped to delimit canals and larger features, such as the erosional area associated with the bifurcation.

Canal Morphology

Morphological attributes of canals are useful in interpreting the function of canals, how they were built, the effects of use and disuse, and the need for maintenance and repair. The study of the morphology of associated features was also important for addressing questions of how canals worked, how water was regulated, maintenance requirements, and, perhaps, non-canal related activities, such as processing areas, storage ponds, or dipping pools. Archaeologists have directed much effort toward canal studies of late, but few projects have compared to the Phoenix Sky Harbor Center Project in terms of the large area available for studying pre-Classic period canal systems.

FIELD METHODS AND STRATEGIES

Backhoe Trenching

Backhoe trenches provided an efficient method of examining the morphology of canals and reconstructing their stratigraphic histories through exposure of canal profiles. However, trenches provided only a two-dimensional view of the canals and limited information to what was exposed in each profile. Mechanical stripping and hand excavations provided a three-dimensional exposure of the canal channels and revealed additional data.

During the testing phase, field crews excavated two systematic trenches and two judgmental trenches within Area 6. The placement of the two systematic trenches, 1144 and 1145, coincided with the location of Alignment 8001. Investigators observed a distinct widening of the channels in Trench 1144. Trench 1143, the next systematic trench to the west, revealed no channels, and investigators could not determine the cause of the widening from the profiles in Trench 1144 alone. Other systematic trenches to the northwest and southwest of Area 6 revealed that the canal alignments had changed direction. Field crews dug two judgmental trenches, 1286 and 1309, to verify this apparent change in orientation. Both trenches located a canal that investigators then thought was the southern alignment of the bifurcation indicated by Turney (1924, 1929).

For the data recovery phase field crews reopened Trenches 1144 and 1145 and dug other trenches to provide further information and support for the previous observations. They dug eight additional trenches (Figure 7.1), totaling 70 m. Placement of trenches depended on the projected orientation of the canal channels. Once the orientations were defined, investigators placed the trenches as much as possible at right angles to the canals to achieve perpendicular cross sections. One tentative explanation for the widening effect of the channels in Trench 1144 was that trenches had cut the channels at oblique angles, resulting in a distorted profile exposure. Further investigation, however, determined that this was an area of extensive lateral channel erosion.

Mechanical Stripping

Field crews excavated five mechanical stripping units, amounting to 232 m², adjacent to backhoe trenches (Figure 7.1). Investigators expected that horizontal exposure of the channels would help resolve the sequence of canal use (and in particular clarify the relationship of Alignment 8005 and Alignment 8001), define the morphology of the bifurcation of the channels, and further interpret the relationship of Alignments 8001 and 8002. Once they had clearly defined the upper limits of the channels, field crews halted the mechanical stripping and initiated hand excavations for finer control.

Hand Excavations

Field crews placed hand excavation units in four locations (Figure 7.1). Excavations concentrated at the intersection of Alignments 8001 and 8005, east of Trench 1145, clearly indicated that Alignment 8005 had been cut by the later 8001c channel (North Main Canal 3). An attempt to locate Alignment 8005 along the north side of the 8001c channel was unsuccessful; later construction of Alignment 8001 may have destroyed the earlier 8005 channel.

Crews also excavated by hand at the anticipated location of the Alignment 8001 bifurcation. Efforts to identify this bifurcation failed due to rechanneling of the canals and to historic or modern disturbance. Investigators observed remnants of the channels but did not recover definitive evidence to support the premise of an intentional bifurcation.

Other hand excavation units were concentrated at Alignment 8001c adjacent to Trench 1145 and at Alignment 8001a, North Main Canal 1, adjacent to Trench 1144b. Investigators placed both units to examine canal morphology, retrieve chronometric samples, and recover ceramics for dating purposes.

RESULTS OF INVESTIGATIONS

Detailed descriptions have been reported of previous studies associated with the canals that extend through Area 6 (D. H. Greenwald 1988a:55–65; Volume 1:Chapter 8), and the reader is directed to those discussions to compare interpretations and approaches. The focus of the present discussion, however, is on the three research issues identified above: temporal associations, canal function, and canal morphology.

Temporal Associations

The North Main Canal alignment consisted of three canals east of the split and four canals to the west. Investigators found it difficult to distinguish North Main Canal 1 from North Main Canal 2 in those portions of the project area where the two were not superimposed; by contrast, North Main Canal 3 was distinct in that it was consistently shallower than the other two channels and its sediment composition differed from that of the other channels. The relative ages of North Main Canals 1 and 2 appeared to be very similar, but North Main Canal 1 was obviously the earlier of the two. In fact, the northern branch of North Main Canal 2 may have been a major remodeling of North Main Canal 1 rather than a separate canal. It is, however, treated as a separate canal here because it was readily identifiable across the entire project area.

The area expected to provide the best opportunity for evaluation of the supposed bifurcation was extensively disturbed by historic activities, which limited examination of the premise that the two channels of North Main Canal 2 had carried water simultaneously. Nevertheless, the stratigraphic positioning of the divided channels of North Main Canal 2 (Alignments 8002 and 8001b) argued against a functioning bifurcation; the division instead appeared to represent sequential events.

Canal Function

The archaeologic literature identifies four basic functions of Hohokam canals: primary distribution, secondary distribution, delivery to fields, and delivery between main canals. The largest canals, or those that exhibit the potential for carrying the largest volume of water, have been identified as main canals or primary distribution canals (Masse 1987:17). The next level is secondary distribution canals, those that were designed to transport water from main canals to field areas (Masse 1987:17). Canals that delivered water directly to fields are referred to as field ditches or field lateral canals (Masse 1987:17) and were usually small compared to main and secondary distribution canals. The sediment history of these ditches often reflects having been deposited by slow-moving water, evidenced by accumulated fine-grained deposits high in organic content (D. H. Greenwald 1988a:72). Crosscut canals transported water from one main canal to another. A modern-day example of such a canal in Phoenix is the Arizona Cross-Cut Canal that connects the Arizona Canal with the Grand Canal in the vicinity of Papago Park. With the exception of Alignment 8005, only main canals were identified in Area 6. The function of Alignment 8005 was equivocal because of its size; investigators designated it as a crosscut canal during the testing phase (Volume 1:Chapter 8).

Archaeologists equated North Main Canals 1 and 2 in Alignment 8001 with Canal Patricio (Turney 1924) on the basis of their estimated age and orientation (D. H. Greenwald 1988a, 1988b:4; Midvale 1945, 1968). These canals would have transferred water from their headgates along the Salt River to various locations on the geologic floodplain and onto the second terrace and therefore were considered main canals. Canal Patricio extended at least to Pueblo Patricio, approximately 8.9 km (5.5 miles) downstream of its canal head in the vicinity of the Park of Four Waters (Howard 1991:5.13).

Prior to the current study, investigators documented only one pre-Classic period canal alignment following the orientation of Canal Patricio through Dutch Canal Ruin. As a result of the work conducted in the Squaw Peak Parkway corridor, this alignment (designated Alignment 8001) was inferred to be Canal Patricio. The current study documented a second canal (Alignment 8501; see Figure 1.2) that could also represent Canal Patricio because of its orientation within the project area, its projected alignment beyond the project area, and its relative age. This second canal was approximately 125 m north of Alignment 8001. Alignment 8501 paralleled Alignment 8001, curving northward east of the Squaw Peak Parkway corridor where it exited the project area. The orientation of Alignments 8001 and 8501 suggested that they would both extend through Pueblo Patricio approximately 2 km from the center of Dutch Canal Ruin (Chapter 8). The bifurcation recorded by Turney (1924), however, appeared to coincide with the bifurcation in Area 6. Investigators documented no such feature east of the Squaw Peak Parkway (Alignments 8501 and 8560 were not contemporaneous); therefore, Alignment 8001 was more likely the alignment identified as Canal Patricio.

Alignment 8002 equated principally with North Main Canal 3, although a later channel of North Main Canal 2 also was represented in Alignment 8002. Archaeologists confidently dated North Main Canal 3 to the Colonial period (Ciolek-Torrello and Greenwald 1988:159, 163), and this temporal assignment was further supported by the current project results and a reassessment of the earlier archaeomagnetic dates from the Squaw Peak Parkway (Appendix A). Both channels of Alignment 8002 functioned as main canals. After exiting the project area at 16th Street, Alignment 8002 extended for an unknown distance to the west. It may have been built to supply field systems located as far west as present-day Central or 7th avenues, in the vicinity of La Villa and Casa Chica (Howard 1991:5.13).

Two aspects of Alignment 8002 were noteworthy. First, it indicated that Hohokam groups had attempted to utilize the geologic floodplain until approximately A.D. 900 or 950 by following earlier alignments, despite steep gradients and erosional problems. Second, although an active flood channel was (in places) less than 225 m to the south of the alignment within the project area, the Hohokam located farms and temporary settlements in this area of high risk for flooding even after they had established canals and farms on the lower bajada in an area of lower risk. Farming along Alignment 8002 continued on the geologic floodplain in the project area until the end of the Colonial period.

The function of Alignment 8005 was problematic. Although originally identified as a crosscut canal, this small channel may have functioned as the original canal that served the area between Dutch Canal Ruin and Pueblo Patricio. Because of its small size, however, it would have carried only a fraction of the volume of water of the larger North Main Canals 1 and 2. Temporally, this canal was one of the earliest in the project area. Its association with Alignment 8003 has been demonstrated, and since it was later superimposed on and removed by Alignment 8001, it predated Alignment 8001. Its size may be directly related to its early age and to the area that it served.

Canal Morphology

This section describes the morphological attributes of each canal segment within Area 6 and presents information concerning the depositional history and composition of each channel. The discussion is organized by alignment, and when multiple channels were present in an alignment, as in the case of Alignment 8001, the individual canals are discussed in chronological order, from the earliest to the latest.

Alignment 8001

Alignment 8001 consisted of North Main Canals 1, 2, and 3 (Alignments 8001a, b, and c, respectively) until the canal alignments separated. At that point, Alignment 8001 continued in a northwesterly direction and comprised North Main Canals 1 and 2. Alignment 8002, consisting of a later channel of North Main Canal 2 and North Main Canal 3, originated at the separation and followed a southwesterly trajectory (Figure 7.1).

Upstream from the canal separation, North Main Canals 1, 2, and 3 exhibited general parabolic shapes. The uniformity of canal shape was also consistent downstream from the split in North Main Canal 1 and in both branches of North Main Canal 2, although the profiles exhibited frequent rechanneling, a product either of natural scouring or of intentional re-excavation. As a result, these two canals had remnants of multiple channels and at first appeared to be deep, wide canals (Figures 7.2 and 7.3). Unlike the other two canals, North Main Canal 3 exhibited a more uniform shape, indicating greater stability (Figure 7.2). This later canal retained its symmetrical, parabolic shape throughout its use, although profile exposures throughout the project area exhibited some internal rechanneling. The profile exposure of Trench 1144 revealed an exception in a wide cross section that showed what appeared to be an overbank flow and an erosional area extending to the south (Figures 7.1 and 7.3).

Alignment 8001a—North Main Canal 1

North Main Canal 2 frequently superimposed on North Main Canal 1 within the project area. Although in Area 6 the two channels were somewhat distinct, investigators had difficulty distinguishing them in Trench 1144 because of their superpositioning (Figure 7.2). Where the upper portions of the canal walls were preserved in North Main Canal 1, they appeared as a broad parabolic U, sloping gently at approximately a 45° angle. Minor southward channel migration commonly appeared among the canal sediments. None of the profile exposures in Area 6 showed complete separation of the two channels, which supported the original argument that North Main Canal 1 had been abandoned for only a short period before the earlier channel of North Main Canal 2 was built. North Main Canal 2, at least in its earliest use, was probably part of a major re-engineering of North Main Canal 1 rather than a completely new alignment that simply paralleled North Main Canal 1 through the Squaw Peak Parkway corridor (D. H. Greenwald 1988a:86).

Given the remnants of North Main Canal 1, investigators did not see radical channel migration in trench exposures east and west of the canal split (Figures 7.2 and 7.4) and, hence, thought that channel migration had not been a problem. In the vicinity of Trench 1144, the location of the split, North Main Canal 1 became a broad, eroded area approximately 14.7 m in width that was expressed in profile as a shallow depression containing coarse and medium-coarse sands in its basal units. Later erosional areas, created after the depression had filled with sediments, were present in the profile of Trench 1144.

The broad eroded area attributed to North Main Canal 1 at this location was due to a change in orientation of the canal to the northwest. This change lowered the velocity of the water, particularly toward the north bank, causing the coarser materials to drop from suspension. At the same time, the channel widened as the increased water velocity counteracted the change in

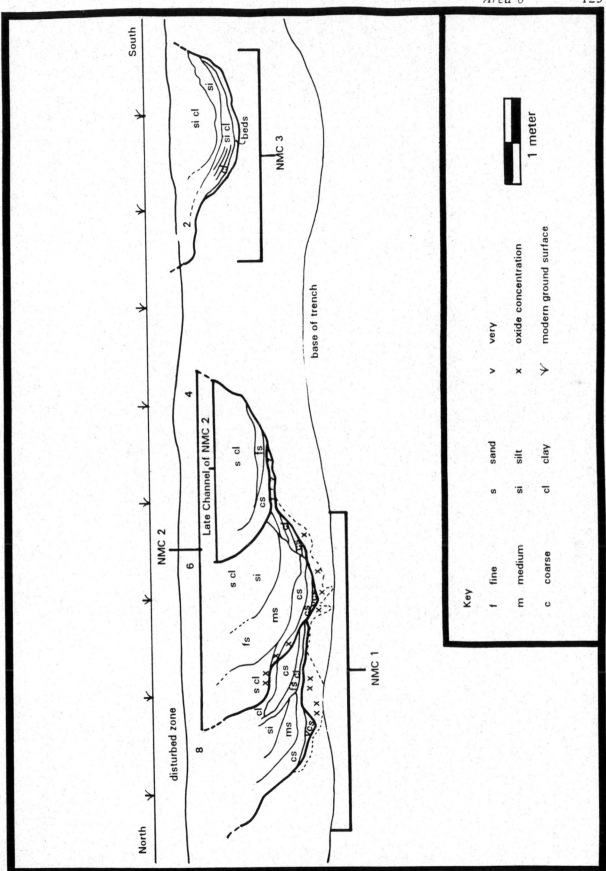

Figure 7.2. Area 6, east profile of Trench 1145 showing Alignment 8001, the North Main Canal.

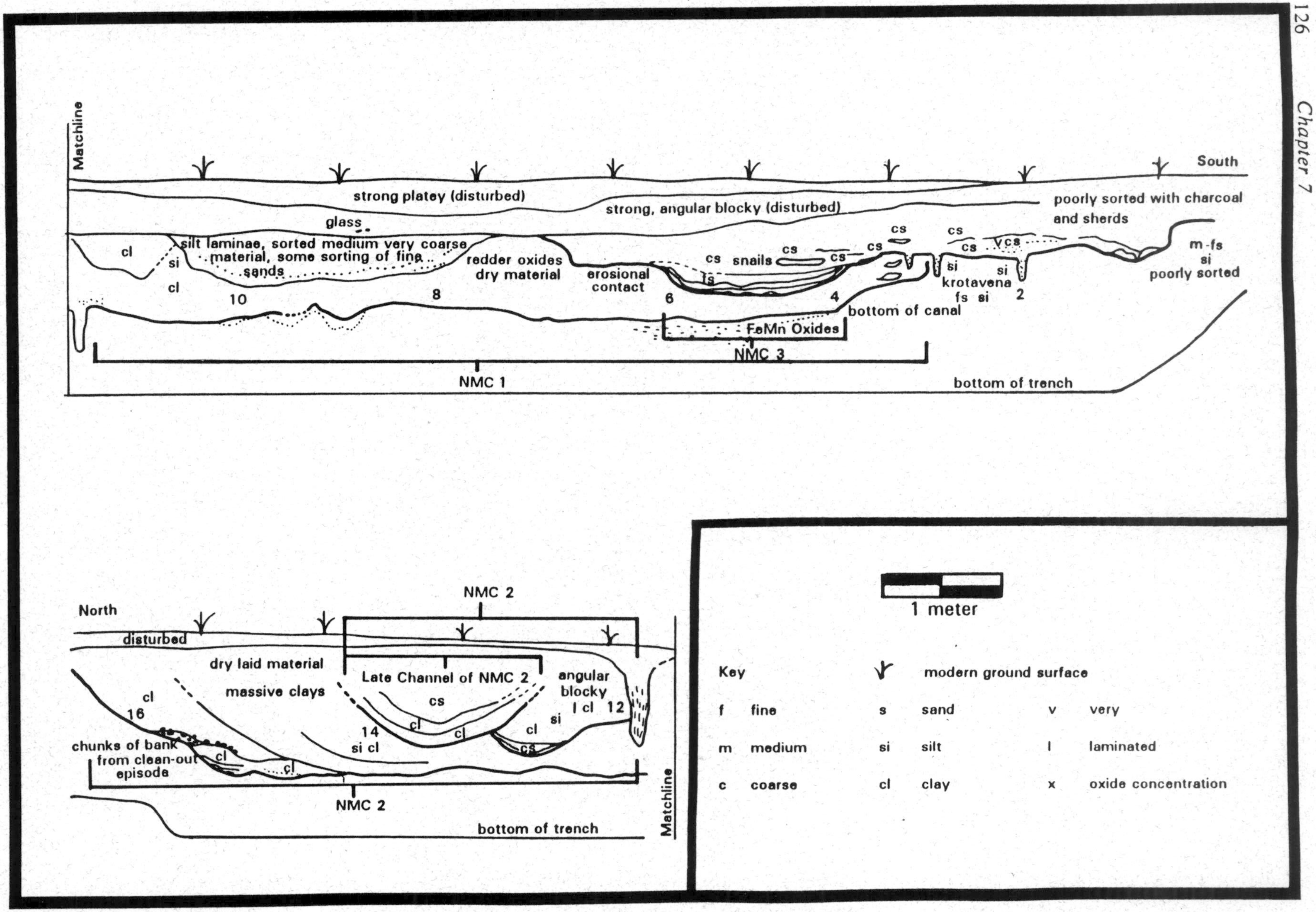

Figure 7.3. Area 6, east profile of Trench 1144 showing the widening of Alignment 8001.

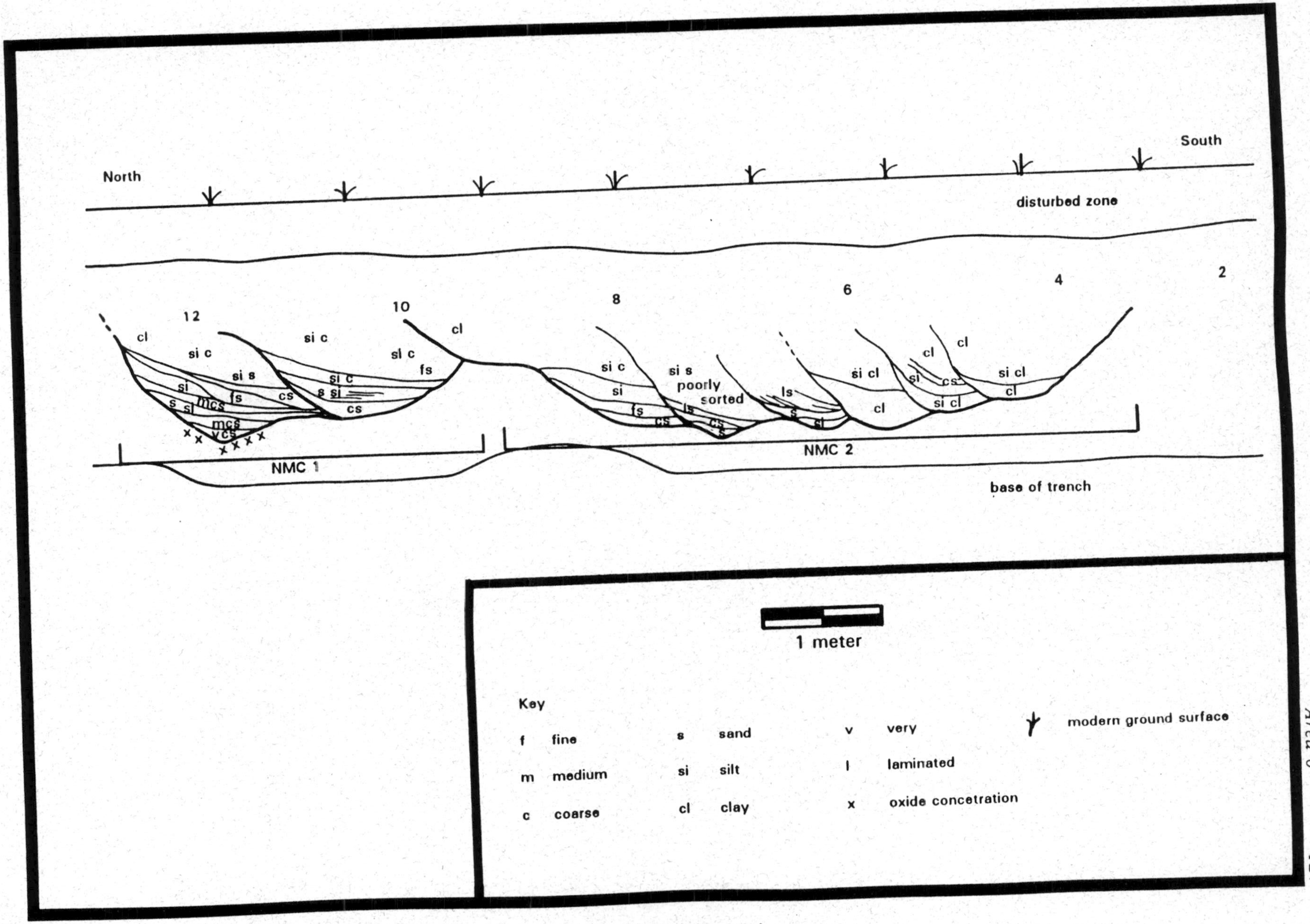

Figure 7.4. Area 6, east profile of Trench 1144b downstream of the bifurcation.

alignment by eroding the channel to the south, the downslope side. Although bank erosion was extreme, the base of the channel in this location was actually shallower than in adjacent profile exposures (compare Figure 7.3 with Figures 7.2 and 7.4). Hydraulically, the change in the force of the water after the decrease in canal gradient was too great to allow for velocity reduction and channel reorientation, which resulted in either bank erosion or channel migration, or both. Once past this curve, the force of the water achieved equilibrium, the channel returned to its typical profile, and the depth of the channel increased to that observed in Trench 1145.

The decrease in depth of the channel from Trench 1145 to Trench 1144 and the increase in depth at Trench 1144b were inconsistent with the normal hydraulic gradient. This may have resulted from intentional excavation of the channels to a shallower depth to slow and control the water, or, because of the gradient of the canals in the project area, the flow velocity may actually have cut other portions of the canals until velocity was regulated, reaching a natural equilibrium. Although the second explanation seemed unlikely, investigators commonly observed basal channel erosion in North Main Canals 1 and 2. In Trench 1144, lateral rather than basal erosion occurred because of the change in direction. Nevertheless, the first alternative seemed the most likely, given the consistent depth of North Main Canal 1 across the project area.

Alignment 8001b—North Main Canal 2

The early channel of North Main Canal 2 in Area 6 was very similar in morphology to that of North Main Canal 1, including similar deposits and the broad parabolic shape. Later inner channels of North Main Canal 2 exhibited steeper sides and finer sediments, attributes common to North Main Canal 3. Within and downstream of Area 6, North Main Canal 2 exhibited two main channels that were not present upstream (east) of Area 6. Extensive investigations of the later channel provided the only supportive evidence that Turney's bifurcation might have channeled water in two directions, but sequentially, not simultaneously. This later channel could have been used just prior to the construction of North Main Canal 3, or North Main Canal 3 could have been constructed after the flow to the northwest was discontinued. The similarities in sedimentation between the later channel of North Main Canal 2 and North Main Canal 3 also suggested that flow velocity and volume had been greatly reduced or, perhaps, more regulated, than in the earlier channels in Alignment 8001.

In the profiles of Trenches 1144 and 1145, North Main Canal 2 appeared as a bifurcated channel. The southern channel predated the northern channel in Trench 1144 but was later in Trench 1145. Sediments in the earlier channel consisted of fine materials, principally laminated silty clays and clays, in Trench 1144 but generally were coarser in Trench 1145. This indicated that velocity was reduced in the vicinity of Trench 1144 prior to the change in channel orientation. Because these sediments were well sorted, indicating water regulation and stability of flow, and since channel migration was not a problem at this time, investigators surmised that water control methods had improved by the time North Main Canal 2 was in use.

Alignment 8002—North Main Canals 2 and 3

The later channel of North Main Canal 2, which was probably a re-engineering of the earlier alignment, followed a new orientation to the southwest, eventually connecting with the previously abandoned South Main Canal alignment (8003). Sediments associated with this channel compared with those of the earlier channel, becoming finer near the perimeter of the canal. Given the orientation of the channel, investigators expected higher gradients along this route, and coarse materials therefore should have been present in the canal. Since this was not the case, either this channel had carried lower volumes of water or some control device(s) had reduced water velocity. Investigations did not reveal water control features such as dams in any of the trench exposures or excavation units.

North Main Canal 3 meandered along the general route of the previous canals, often exhibiting both northward and southward channel migration. In Trench 1145 (Figure 7.2), the channel appeared to be stable, with only a minor amount of erosion along its northern side. In the area of Trench 1145, investigators determined that North Main Canal 3 had migrated from north to south; evidence of bank erosion in this area reflected this change in orientation (Figure 7.1). In Trench 1144, however, erosion had occurred on the south side of the channel, probably as a result of overbank flow. Here, again, the channel was stable between erosional episodes (possibly deliberate seasonal flushing of the channels) and had a broad parabolic shape. From this point west, the later channels of North Main Canals 2 and 3 represented a realignment and were designated as Alignment 8002.

Originally described as a crosscut channel in the system (Volume 1:Chapter 8), Alignment 8002 was characterized as a re-engineered canal during data recovery. North Main Canal 3, dating to the latter part of the Gila Butte phase, followed earlier channels, either directly inside the earlier channels or adjacent to them. Such was the case in Area 5, where North Main Canal 3 followed Alignment 8501 into the project area, then changed direction to follow Alignment 8001 until reaching Area 6. Investigators originally thought that the change in orientation of this canal in both Areas 5 and 6 represented distinct crosscut canals. During data recovery, however, they discovered that this was not the case; in fact, North Main Canal 3 traversed the project area by migrating southward in a steplike pattern (Figure 1.2).

As illustrated in Figures 7.2, 7.3, and 7.5, North Main Canal 3 retained its parabolic shape. The profile illustrated in Figure 7.5 was distorted due to the oblique angle of Trench 1309 to the alignment of the canal. Even with the distortion, North Main Canal 3 was uniform in shape and was consistently shallower and contained finer sediments than the other main canals. North Main Canal 3 appeared from the fill sequence to have functioned efficiently, although sediment accumulation would have necessitated periodic cleaning.

Alignment 8005

This channel originated in the South Main Canal (Alignment 8003) as an upper channel and entered Area 6 from the southeast. Its general trajectory appeared to follow that of North Main Canal 1 as it changed direction and headed to the northwest. Alignment 8005 was easily traced from the South Main Canal alignment to the North Main Canal alignment, but field crews could not locate it once it entered Alignment 8001, either north of Alignment 8001 or in other channels. As is illustrated in Figure 7.1, Alignment 8005 changed course immediately prior to intersecting North Main Canal 3, causing the northern bank to erode into a flared channel. Alignment 8005 had clearly been cut by North Main Canal 3.

Alignment 8005 was a small channel, measuring approximately 1.0 m in width and 0.5 m in depth. In general, the fill sequence was composed of fine- or medium-fine materials, although sands were present in the inner beds. Although the canal followed a typical gradient for early pre-Classic period canals, the finer texture of the canal sediments was probably directly related to low volume and low velocity. At first, investigators thought Alignment 8005 was an early alignment of the North Main Canal (in particular, to have supplied North Main Canals 1 and 2 with water,

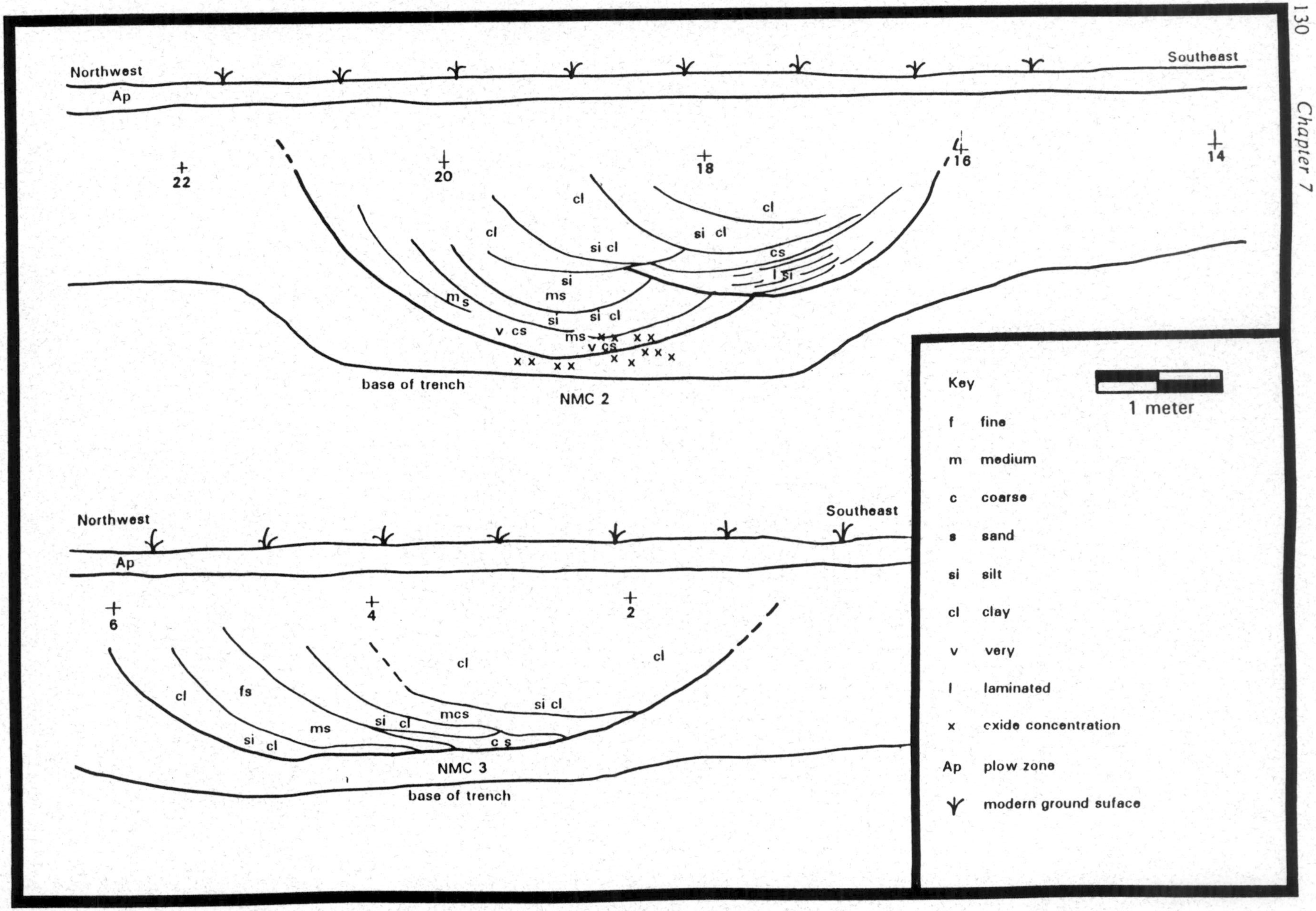

Figure 7.5. Area 6, northeast profile of Trench 1309 showing North Main Canals 2 and 3.

perhaps from the South Main Canal alignment). Investigations did not confirm this because as the channel entered Area 6, it was cut by North Main Canal 3 and could not be relocated in any of the excavation units. The base of Alignment 8005, because of its small size, was higher than that of any of the larger canals in Area 6. Therefore, preservation of its basal section would have been unlikely if any of the later three alignments had followed the same orientation. In a sense, its apparent lack of preservation supported the original notion that it was earlier than any of the canals in the North Main Canal alignment. This stratigraphic indication suggested that the South Main Canal was older than the North Main Canal. Alignment 8005 was a distributional canal (Nials and Gregory 1989:40), designed to deliver water from its associated main canal to areas where it would supply either field ditches (laterals) or the fields themselves.

MATERIAL CULTURE

The material culture assemblage from Area 6 was similar in size to samples from other areas at Dutch Canal Ruin and included ceramics, flaked stone, and one ground stone item. Canals commonly do not produce large frequencies of artifacts unless they are directly associated with larger habitation sites. No habitation loci appeared in direct association with Area 6.

Ceramics

The ceramic assemblage from Area 6 did not contain many temporally diagnostic types and was limited in diversity (Table 7.1). The presence of buffware types in the assemblage suggested a late pre-Classic period association. The absence of incised types, which are diagnostic of the Pioneer and early Colonial periods, further supported this notion. To add to the ceramic data base for the canals in Area 6, investigators examined the ceramic assemblage recovered from North Main Canals 2 and 3 during the investigations in the Squaw Peak Parkway corridor (D. H. Greenwald 1988c:129). These data are presented in Table 7.2.

Researchers found increased numbers of temporally diagnostic types in the canals within the Squaw Peak Parkway corridor, among them late Pioneer and early Colonial period types (Table 7.2). Investigators expected to locate these types in Area 6 but found none there, perhaps because no adjacent habitation area occurred along this portion of the canals. The ceramic assemblage recovered from Area 6 appeared to represent activities associated with the later use of the canals or with postabandonment activities. Analysts dated North Main Canal 3 from the late Gila Butte phase to the middle Santa Cruz phase, and the Area 6 ceramic assemblage provided some meager support for this association. Because North Main Canal 3 was the latest canal in Area 6, the ceramics recovered from the other canals could represent postabandonment activities or activities associated with North Main Canal 3.

Flaked Stone

The flaked stone assemblage consisted of 14 items, including 11 pieces of debitage (2 pieces of angular debris, 2 flake fragments, and 7 whole flakes), 2 cores, and a hammerstone. Basalt was the most common material, represented by 9 items, and the remainder were igneous other material (n=3), chert (n=1), and quartz (n=1). Use wear was present only on the hammerstone. In general, the small assemblage provided few insights into the types of activities associated with the canals. Although the inhabitants of the area may have produced flakes, both cores were of materials not represented in the debitage category. Core reduction had probably occurred, but because of the

Table 7.1. Area 6, Ceramic Frequencies

Ceramic Ware/Type	Absolute Frequency	Percent Ware	Percent Assemblage
Hohokam Plain Ware	**47**		**90.4**
Gila Plain, Salt Variety	32	68.1	61.5
Gila Plain, Gila Variety	6	12.8	11.5
Squaw Peak Plain	9	19.1	17.3
Hohokam Decorated Ware	**5**		**9.6**
Santa Cruz Red-on-buff	1	20.0	1.9
Salt Red-on-buff, indeterminate	1	20.0	1.9
Indeterminate Red-on-buff	2	40.0	3.9
Indeterminate Buff	1	20.0	1.9
Total	**52**		**100.0**

Table 7.2. Area 6 and the Squaw Peak Parkway Corridor, Ceramic Frequencies

Ceramic Ware/Type	Area 6	Canals within the Squaw Peak Parkway	
		NMC2	NMC3
Hohokam Plain Ware	**47**	**32**	**30**
Gila Plain, Salt Variety	32	30	25
Gila Plain, Gila Variety	6		
Squaw Peak Plain	9		
Hohokam Plainware, indeterminate		1	5
Salt Red		1	
Hohokam Decorated Ware	**5**	**8**	**16**
Santa Cruz Red-on-buff	1	2	
Salt Red-on-buff, indeterminate	1		
Snaketown Red-on-buff		1	
Gila Butte/Santa Cruz Red-on-buff		1	1
Indeterminate Red-on-buff	2	3	6
Indeterminate Red-on-buff with incising			1
Indeterminate Red-on-gray with incising		1	1
Hohokam Buff Undecorated with incising			2
Hohokam Buff	1		5
Total	**52**	**40**	**46**

small sample, the assemblage did not reflect these activities. Mechanical excavations also may have reduced the number of items recovered, as artifacts are not as readily visible with such techniques.

Ground Stone

Field crews recovered one fragment of ground stone, approximately one-half of an unfinished stone ring. It was made from vesicular basalt and was well formed, although not completely shaped. It appeared to have been broken in the process of producing the hole, a bidirectional perforation. Archaeologists do not know what function stone rings served; however, the Hohokam may have used them as digging-stick weights and may have been making this item for use during canal construction or cleaning.

CHRONOLOGY

Although investigators determined the chronological sequence of the canals because of frequent superpositioning, assigning date ranges to these canals and revising previous temporal assignments presented a greater challenge. For the current study, analysts used three approaches and re-evaluated previous data in an attempt to provide a more appropriate temporal reconstruction.

Researchers dated all three alignments of the North Main Canal included in the study focus of Area 6 to the pre-Classic periods. Two data sets (ceramic associations and archaeomagnetic dating) supported these temporal assignments. Investigators obtained much of their data during the investigations in the Squaw Peak Parkway corridor (Ciolek-Torrello and Greenwald 1988:158–164).

Ceramic associations came from both direct and indirect evidence. Direct evidence consisted of ceramics recovered from within deposits associated with the use of the canals. Indirect evidence was in the form of stratigraphic associations of other features dated by temporally diagnostic ceramics, archaeomagnetic dating, or radiocarbon assays. Date ranges for the canal features, based on either direct or indirect ceramic associations, indicated that the Hohokam had constructed and used Area 6 canals during the late Pioneer and Colonial periods. Table 7.2 presents the ceramic frequencies compiled from all areas investigated along Alignment 8001, including the Squaw Peak Parkway corridor. As indicated by these data, the ceramic assemblage for the North Main Canal alignment was dominated by Colonial period types. Alignment 8005 did not produce temporally diagnostic sherds, but investigators showed that it originated in the South Main Canal alignment, which was dated by ceramics and archaeomagnetic analysis to the late Pioneer and Colonial periods.

Stratigraphy

The stratigraphic relationships of the canals indicated that North Main Canal 1 and the earliest channel of North Main Canal 2 followed a trajectory that angled to the northwest. These were the earliest canals in Area 6, and North Main Canal 1 clearly predated North Main Canal 2. The latest channel of North Main Canal 2 and North Main Canal 3 followed a southwesterly trajectory from Area 6. Investigators had no question that North Main Canal 3 postdated North Main Canal 2. They did speculate, however, about the contemporaneity of the latest channel of North Main Canal 2 with the earliest channels of North Main Canal 2 and North Main Canal 3. North Main Canal 2 did not appear to have been intentionally bifurcated, and therefore the earliest and latest channels of North Main Canal 2 were not absolutely contemporaneous. Investigators were left with a twofold question: (1) what was the temporal association of the latest channel of North Main Canal 2 with the earliest channel of North Main Canal 2 and with North Main Canal 3; and (2) how much time elapsed from the initial channeling of water to the southwest in North Main Canal 2 to the construction of North Main Canal 3?

Alignment 8005 was cut by North Main Canal 3 and therefore predated it and may have predated North Main Canals 1 and 2 as well, since investigators could not locate it in any of the excavation units or trench profiles beyond where it was displaced by North Main Canal 3. As posited above, Alignment 8005 may have been the earliest alignment of Canal Patricio, originating from Alignment 8003, the South Main Canal.

Ceramics

The ceramic assemblage from Area 6 did not provide a solid foundation upon which to base any temporal conclusions. It did suggest affiliation with the pre-Classic Colonial period on the basis of a single Santa Cruz Red-on-buff sherd, a tenuous assignment at best. Other temporally diagnostic types were not represented (Table 7.2). For dating purposes, the Area 6 sample was compared to samples recovered from other locations of North Main Canals 2 and 3 (Table 7.2). These data supported late Pioneer and Colonial period use of the canals.

Absolute Dates

Field crews collected three archaeomagnetic samples from Area 6 during the current project, one from each of the three channels that constituted Alignment 8001 (North Main Canals 1, 2, and 3). Dr. Jeffrey Eighmy at Colorado State University analyzed the samples (Appendix A).

Sample CSUSED-77 from North Main Canal 1 produced two declination intervals, A.D. 625–1050 and A.D. 1175–1700, and three visual intervals, A.D. 900–1000, A.D. 1200–1475, and A.D. 1500–1700 (see Eighmy and Klein 1991 for the significance of the readings). Of the declination intervals, the second could be rejected because of its obviously late range. The first date interval fell within the expected range, especially the earlier part of the interval. All of the visual intervals appeared to be too late, perhaps by at least 200 years. Archaeomagnetic dating of North Main Canal 1 therefore did not provide much resolution to the current questions regarding the chronology of this canal. The quality of the sample was part of the problem; its alpha 95 reading of 5.07 suggested considerable variability in the sample, hence reduced reliability.

Sample CSUSED-78 from North Main Canal 2 produced one declination interval, A.D. 950–1200, and one visual interval, A.D. 1015–1300. Both date interpretations were too late when compared with previous results, stratigraphic position, and associated ceramics. Again, a long alpha 95 (4.86) reduced the reliability of this sample.

Sample CSUSED-79, collected from North Main Canal 3, produced two declination intervals, A.D. 600–1025 and A.D. 1450–1800, and three visual intervals, A.D. 600–700, A.D. 850–1015, and A.D. 1375–modern. The alpha 95 calculated for this sample was 7.87. The estimated age of North Main Canal 3 fell within the earliest declination date and the middle visual date. The other ranges were either too early or too late and could be rejected.

Each of these samples came from upper clay sediments, which in the past had produced better results. Later activities in this area perhaps had affected the declination of the samples, possibly by saturation of the sediments, enabling the metallic particles to become freely suspended and realign themselves with the current magnetic pole. The results of the archaeomagnetic sampling cannot simply be rejected, but as a larger data base is developed, refinement of these dates may be possible and the accuracy or inaccuracy of the results confirmed. At the present time, however, they do not offer much support to chronology building and the interpretation of Area 6.

SUMMARY

North Main Canals 1, 2, and 3 all functioned as main canals (Masse 1987:17, 1991:212), designed to distribute water from the intake or head (Nials and Gregory 1989:51) to various locations along the course of the canal. The general conception shared by many who have studied Hohokam canals (e.g., Herskovitz 1982; Masse 1981; Nials and Gregory 1989) is that distribution canals or branches carried water from the mains to areas where it would be distributed to villages and fields. Lateral canals or field laterals would have supplied water directly to the fields. Apparently, either the general system that extended through Dutch Canal Ruin (specifically, through Area 6) had no need for distribution canals or the canals identified as mains represented long segments of distribution canals that branched from the mains farther upcanal. Lateral canals were recorded during the data recovery phase of the Squaw Peak Parkway project, but they were not located in the Phoenix Sky Harbor Center, although field crews expended much effort in searching for them.

During the testing phase, investigators considered North Main Canal 3 to be a crosscut canal from the bifurcation to its intersection with the South Main Canal. In retrospect, no temporal association existed between North Main Canal 3 and the South Main Canal. The South Main Canal was abandoned for approximately 100 years, possibly longer, prior to the construction of North Main Canal 3. North Main Canal 3 simply followed a previous channel, at least from the point of intersection to the western project boundary, a distance of approximately 200 m. It functioned as a main canal, as defined here, although on a smaller scale than North Main Canals 1 and 2 and the South Main Canal.

Alignment 8005 also was designated a crosscut canal during the testing phase. A crosscut canal has been equated with the transfer of water from one canal to another; such a channel may have been constructed when the existing system changed. For example, if a segment of one canal needed extensive repairs, it would have been more efficient to route water around the damaged or nonfunctional section of the canal by supplying water from another source (canal). In addition, when a segment of a canal system was not being used, water could be delivered to other parcels from another canal via a crosscut.

When first found, Alignment 8005 was thought to be a lateral because of its small size. Upon tracing the channel, however, investigators determined that it was longer than most lateral canals. It also intersected the North Main Canal alignment, thereby connecting that main canal with the South Main Canal. By definition, Alignment 8005 should be categorized as a distributional canal (Nials and Gregory 1989:40).

Researchers have scrutinized Turney's bifurcation since the initial investigations in the Squaw Peak Parkway (Greenwald and Ciolek-Torrello 1988a). Discovery of a bifurcated alignment during the testing phase and its general spatial correlation to that mapped by Turney (1924) resulted in the selection of Area 6 for further investigation. Although investigators expected to see minimal historic disturbance around canal features in the project area due to their linear morphology (Greenwald and Chenault 1991), the effects of such disturbance on specific canal features can be as devastating as upon habitation features. Such was the case, at least in part, in a section of the North Main Canal 2 alignment. Historic disturbance in the form of a subfloor pit and septic system had removed a critical section of these channels. For this reason, investigators moved farther east, upcanal, in an attempt to define the relationship of the apparent bifurcated channel. Regarding Turney's bifurcation, they concluded that although North Main Canal 2 contained two separate channels, these channels were not related temporally and had not functioned as a bifurcated system. Instead, the sediment history of these two channels indicated that one channel had been in operation until it was no longer functional, and the other channel had then been constructed; therefore, the system had operated as a re-engineered canal.

The canals in Area 6 closely followed the natural grade in the project area. From 24th Street to 16th Street, the elevation changed as much as 3 m, resulting in steep gradients for the canals as long as they paralleled the natural slope. Altering the course of the canals to the northwest may have been prudent, given the erosive nature of the early alignments. A comparison of Figure 1.1 and Figure 1.2 shows that the first 500 m of this new canal trajectory would have positioned the canals at approximately a 45° angle to the natural grade. After that distance, the canals would have paralleled the natural contours, reducing erosion by reducing the gradient.

Diverting the canals to the southwest would have produced only slight variations in gradient. Water control devices or low volumes of water could have helped prevent channel erosion along Alignment 8002. North Main Canal 3 was consistently stable despite its gradient and might have been controlled simply through its volume and or its velocity.

Field crews uncovered few habitation features along the North Main Canal alignment. This suggested that the occupants of Area 6 preferred to locate their living areas not immediately adjacent to the canals but at some distance from them. The distribution of habitation loci during the late Pioneer and Colonial periods is illustrated in Chapter 18, Figures 18.1 and 18.2, which show that habitation areas were concentrated west of the bifurcation, along Alignment 8002.

Canal studies and their relevance for the entire project are discussed further in Volume 4. Analysts combined the information generated through investigation of Area 6 (as well as other research areas) with that from the testing phase to expand current understanding of Hohokam canal systems.

CHAPTER 8

DUTCH CANAL RUIN, AREA 7

M. Zyniecki
David H. Greenwald

Research Area 7 was one of two areas east of the Squaw Peak Parkway in the Phoenix Sky Harbor Center that archaeologists excavated during the 1990 data recovery phase (Figure 1.2). Field crews recorded the profiles of a burned pit structure and two canal channels during testing. Investigators decided to conduct data recovery excavations in Area 7 based primarily on the potential for information on canal morphology, canal feature function, and repair or maintenance activities (Volume 1:Chapter 10).

Area 7 was located at roughly N1825-1850, E1630-1690 within the project grid; its legal description was T1N, R3E, Sec. 10, NE¼NW¼SE¼ on the Phoenix, Arizona, 7.5 minute USGS topographic map (Figure 1.4). Area 7 was less than 100 m east of the freeway and less than 100 m south of the Southern Pacific Railroad tracks. The nearest research area, Area 5, was approximately 400 m to the east. An area investigated by the Museum of Northern Arizona as part of the Squaw Peak Parkway project (Greenwald and Ciolek-Torrello 1988a) was located approximately 300 m to the southwest.

Field crews opened about 67 m of backhoe trenches during the data recovery phase (Figure 8.1). Investigators excavated the burned pit structure in Area 7, based on the results of testing, because of its potential for providing information and the paucity of datable samples collected at other areas of Dutch Canal Ruin. Also, because of the way the structure had burned, the level of detection above the floor was greater than for any other pit structure excavated during the Phoenix Sky Harbor Center Project. Hand excavations in the canals revealed that a fire pit had been constructed after the pre-Classic period canal was abandoned. These excavations also uncovered an erosion control feature, a "baffle," along the downslope side of the pre-Classic period canal. This baffle was in a turning point of the canal, presumably to help redirect the flow of water and prevent erosion of the bank. Canal profiles of Alignments 8501 and 8537 (Canal Viejo and Canal Nuevo) drawn during testing exhibited stratigraphy associated with a breach in the canal, a headgate area, or stabilization or rebuilding of the canal bank through large-scale filling of the canal (Volume 1:Chapter 9). Excavation revealed that this complex stratigraphy was the result both of the natural meandering of the channels and of artificial modification of the canal's natural course.

RESULTS OF TESTING

The backhoe operator excavated seven backhoe trenches totaling 191 m in length in Area 7 during testing. Five of the trenches, a total of 114 m, were oriented north-south, and two more totaling 77 m were oriented east-west. Trench length ranged between 14 m and 50 m. Most of the trenches were in the western part of the area where investigators observed the profile of a pit structure. Field crews dug adjacent trenches in an attempt to determine whether more features existed in the vicinity but found no additional habitation features.

Investigators identified two parallel canal channels (Alignments 8501 and 8537) and a pit structure. They observed anomalies in the profiles of the canals that may have represented a headgate, a breach in the bank, or a massive filling of the canal by the Hohokam. During the

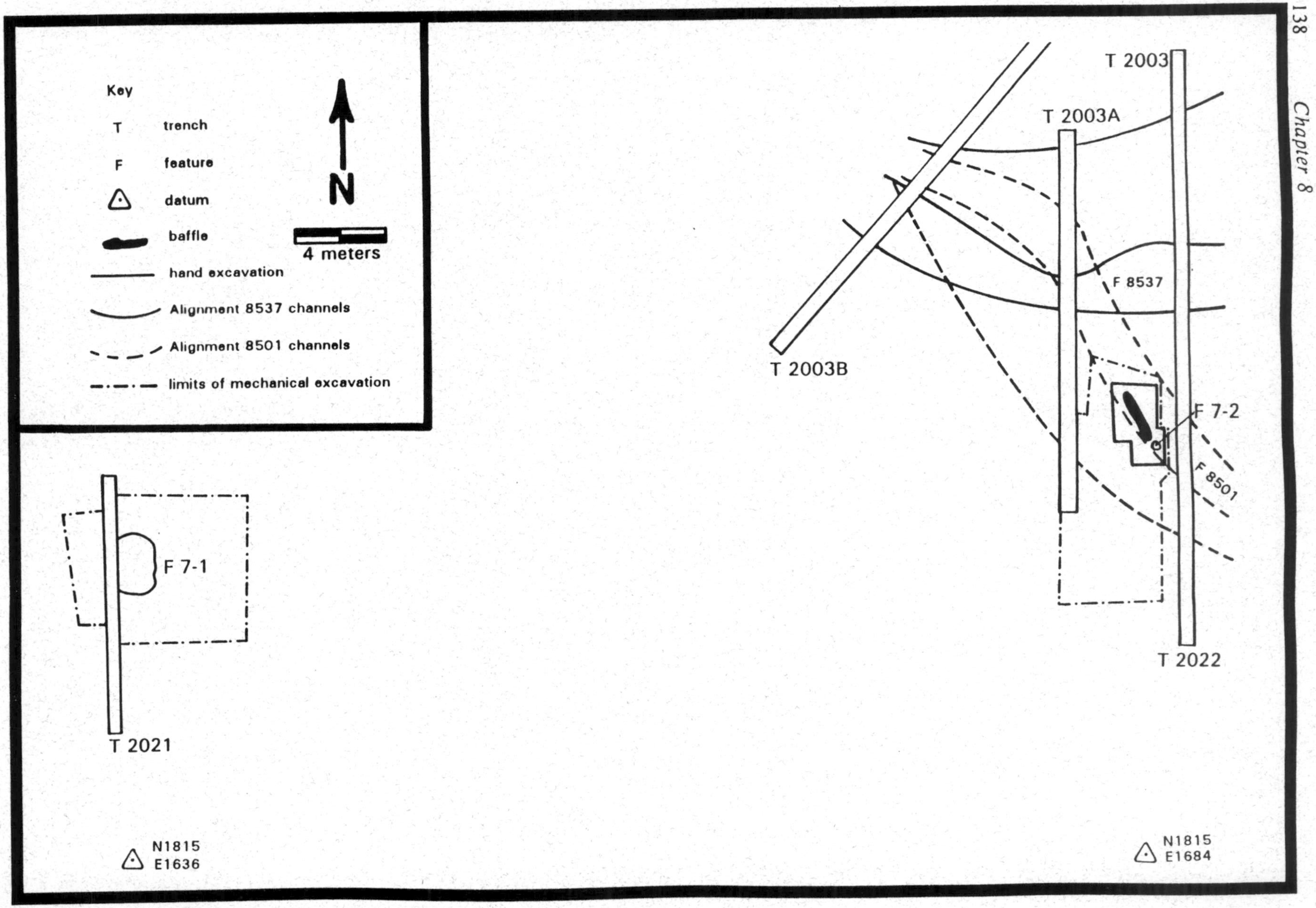

Figure 8.1. Area 7, Dutch Canal Ruin.

testing phase, they could not determine the ages of the two canals. Since previous researchers had dated neither during previous work at the site (Greenwald and Ciolek-Torrello 1988a), investigators considered the stratigraphic position of these two channels to be one means of defining their temporal association (Volume 1:Chapter 9). The pit structure was described as a well-defined pit containing a layer of ash and charcoal (Volume 1:Appendix A). The profile showed no evidence of floor or wall plastering or of floor features.

RESEARCH OBJECTIVES

The problem domain that bore most directly on the research in Area 7 concerned Hohokam canal irrigation technology (Volume 1:Chapter 10). Investigations revolved around questions related to canal morphology, comparison of Area 7 canals to canals of the same age, and analysis of changes over time and the effect of natural topography on canal construction and maintenance. Other areas of inquiry involved canal features and the chronological sequence. Because only the one known noncanal feature was excavated, some questions were limited in scope. Questions pertaining to the pit structure involved feature chronology and morphology, as well as those concerned with the broad patterns across Dutch Canal Ruin and among other sites.

FIELD METHODS AND STRATEGIES

Backhoe Trenching

Field crews partially reopened three of the original trenches in Area 7 trenches during data recovery (Figure 8.1). They re-excavated approximately 11.5 m of Trench 2021 to expose the profile of the pit structure, Feature 7-1, and about 1.0 m of Trench 2022 and the entire length of Trench 2003. They also excavated 54 m of new trench. This included the 17-m interval between Trenches 2003 and 2022, Trench 2003A (17.5 m), and Trench 2003B (19.5 m). Investigators expected the new trench excavations would shed some light on the anomalies observed in the canal profiles.

Mechanical Stripping

Data recovery began with mechanical stripping around the pit structure and over the canals between Trenches 2003 and 2003A (Figure 8.1). Field crews stripped roughly 46 m^2 near the pit structure to expose possible extramural features. They also stripped just over 45 m^2 in the vicinity of the canals to expose the limits of the canals. Within the mechanically stripped area, crews stripped an additional 5.75 m^2 and then excavated to identify the cause(s) of the anomalies observed in the canal profiles.

Hand Excavations

Most hand excavation techniques resembled those for habitation features in the project area. The fire pit at the top of the canal fill deposits in Alignment 8501 was partially destroyed during the excavation of the upper portion of that channel. Because of the small size of the fire pit, field crews excavated the intact portion as a whole and retained the entire fill for a macrobotanical sample. They also took a pollen sample from the base of the feature.

The purpose of the excavation of Alignment 8501, the pre-Classic period canal, was to uncover the various stratigraphic levels in order to discover the cause of the anomalies observed in the profiles. Excavations continued until the base of the canal was exposed. Field crews did not use screens during the excavation but instead collected artifacts as these were observed.

STRATIGRAPHY AND ASSOCIATION

Investigators recorded two natural strata in Area 7. The uppermost, Stratum 1, was a blocky clay loam that had been disturbed by historic and modern activities near the top and extended to a maximum depth of 35 cm below the present surface. This stratum had been developed by flooding of the Salt River and historic or modern irrigation farming practices. Stratum 2 extended below Stratum 1 to the bottom of the exposures in the trenches. This stratum was a silty loam that was deposited by Salt River flooding. The features in Area 7 all lay in Stratum 2.

CULTURAL FEATURES

Structures and Extramural Features

Feature 7-1

Testing Feature: 2037
Trench Number: 2021
Horizontal Provenience: N 1827.22–1829.67; E 1635.12–1637.51
Vertical Provenience: 2.77–3.13 mbd
Feature Type: Pit Structure

This subrectangular pit structure measured 2.44 × 2.38 m. The shorter axis was truncated by the backhoe trench, which removed the western edge of the structure. A possible entry (Subfeature 7-1-1) was excavated on the west side of the trench (Figure 8.2). The remaining portion of the possible entrance consisted of a step entry, subrectangular in outline; a comparison of similar structures in the project suggested that the entry probably had been appended to the structure. The remnant of this possible entrance measured 42 × 25 cm and was located slightly to the north of the east-west central line through the structure. Walls of the entrance were almost vertical, between 12 cm and 13 cm in height. The floor of the entrance was level, unprepared, compacted through use, and 3 cm higher than the floor of the structure. The walls of the structure also were not formally prepared and varied in height between 31 cm and 38 cm. The soil of the walls was oxidized around the entire structure, with the areas of heaviest oxidation along the north and south walls. A layer of charcoal-stained soil was next to the wall in many places, making the definition of the walls even easier. The floor of Feature 7-1 had been leveled and compacted through use, though not formally prepared. There was a slight basin shape to the floor, with the center 1 cm lower than the edges. The floor was well defined by a layer of charcoal-stained soil immediately above it.

Field crews observed no interior features. They did examine three areas of charcoal and charcoal-stained soil, between 15 cm and 40 cm in diameter, on the floor of the structure, as well as an oxidized area with a diameter of 10 cm. The charcoal-stained and oxidized areas were probably the remnants of informal fire areas. Burned daub and pieces of charcoal in the fill indicated that the structure had had a post and brush superstructure that had been covered with mud or adobe.

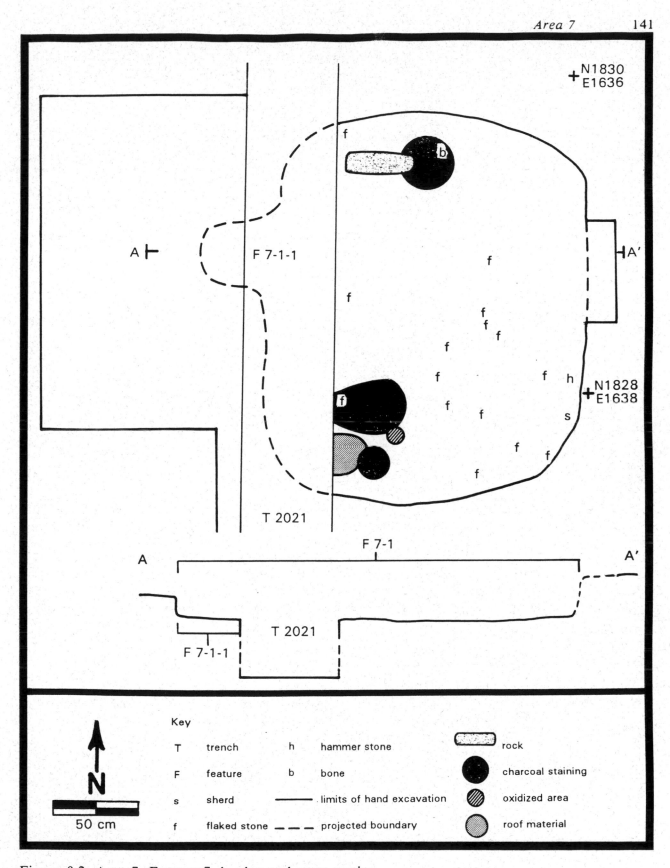

Figure 8.2. Area 7, Feature 7-1, plan and cross section.

Apparently, Feature 7-1 had been abandoned for some time before it burned. Field crews found a piece of unburned roof material, several areas of sterile sandy silts, and the layer of charcoal-stained soil in contact with the floor. The roof material and silts were deposited after abandonment of the structure. The charcoal-stained soil most likely had resulted from informal fire areas. In addition, approximately 12 cm of fill were between the floor and the burned daub and charcoal that constituted the remains of the burned superstructure. These factors supported the idea that Feature 7-1 had been abandoned for a period of time before burning. After burning, the remaining depression was filled with alluvium, which contained several fire-cracked rocks. The rocks might have been intentionally discarded in the depression left by the collapsed structure, or they might have been deposited through natural processes.

Feature 7-2

Horizontal Provenience: N 1833.05–1833.15; E 1684.47–1684.52
Vertical Provenience: 2.86–2.89 mbd
Feature Type: Fire Pit

Feature 7-2 was a circular, basin-shaped fire pit discovered during investigations of one of the canals, Alignment 8501. A portion of it was removed before it was recognized. The remaining portion of the pit was 10 × 5 cm in area with a depth of 3 cm. The east edge of the pit was oxidized, and the fill contained some ash, charcoal, and oxidized clays. The pit fill was composed of blocky clays similar to those in which the feature had been constructed. The fire pit was superposed on Alignment 8501 and therefore had been used after abandonment of the canal.

Canals and Related Features

Investigators examined two canal alignments in Area 7: Alignments 8501 and 8537. Within Alignment 8501 (Canal Viejo), they discovered water- and erosion-control devices while excavating horizontal exposures along the west side of Trench 2003 (Figure 8.1). They initiated excavations here because field crews had observed anomalous stratigraphic units during testing. To reiterate, analysts posited that these anomalous units were due to the construction and use of a headgate in this general location, a breach in the canal bank that was subsequently filled, or massive filling of the canal for unknown reasons. These anomalous fill units lay between the two channels illustrated in Figure 8.3, where they are identified as Strata 14, 15, and 16. In Figure 8.3, the distinct parabolic shape of Alignment 8501 and the less obvious southern channel wall of Alignment 8537 (Canal Nuevo) appeared to contain water-laid deposits between the two channels. Among these water-laid deposits were massive anomalous units. Limited by the two-dimensional view provided by the backhoe trench wall, investigations during the testing phase suggested two explanations: (1) Strata 14, 15, and 16 represented an earlier channel that had been all but destroyed by Alignment 8537; or (2) some type of breach or outlet in this location, associated with Alignment 8537, had filled rapidly with massive sediments.

The Hohokam could have constructed a channel earlier than Canal Viejo (which could have contained Strata 14, 15, and 16) in this location, but investigators found no evidence for such a channel in any of the other trenches or during subsequent data recovery efforts. The basic problem with interpreting Strata 14, 15, and 16 as representing some type of opening in the bank of Canal Nuevo is that these strata appeared to have been truncated or buried by the sediments accumulated in Canal Viejo. Data recovery revealed that the chronological ordering of the two channels precluded the second scenario. Canal Viejo was positively dated to the pre-Classic period

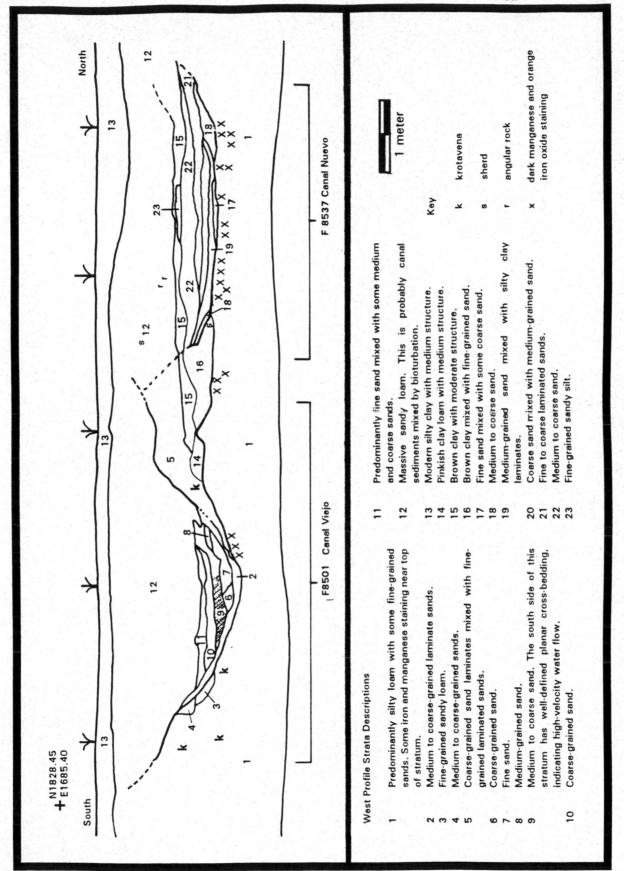

Figure 8.3. Area 7, west profile of Trench 2003 showing Alignment 8501 (Canal Viejo) and Alignment 8537 (Canal Nuevo).

occupation, and its stratigraphic association with North Main Canal 3 as observed in Area 5 placed it in the early Colonial period. Canal Nuevo was dated to the Classic period on the basis of Roosevelt Red Ware sherds in its fill. Trenches 2003A and 2003B revealed that Canal Nuevo was superimposed on Canal Viejo, conclusive evidence that Canal Viejo predated Canal Nuevo.

The confusion about the stratigraphy in Trench 2003 came about because a canal was trenched at a severely oblique angle. Investigators did not know at the time of testing that Canal Viejo turned sharply at this point in Trench 2003 and followed a north-northwest trajectory. Strata 14, 15, and 16 consisted of sediment units that had accumulated on the side of the canal bank as it began sloping upward. As Figure 8.3 shows, Canal Nuevo clearly cut Stratum 16, but a distinct channel boundary was not present above Stratum 16. Hence, the temporal association of these channels in terms of their stratigraphic relationship was unclear. Investigators only sorted out this problem when they opened Canal Viejo, exposing the strata horizontally. Instead of a gradually sloping canal bank as illustrated in Figure 8.3 and observed in the west profile of Trench 2003, Canal Viejo retained a uniform, parabolic profile even as it changed course. Canal Viejo could be traced only a short distance (about 2.5 m), however, because Canal Nuevo maintained a west-northwest trajectory, cutting through the earlier canal.

Defining Canal Nuevo was difficult because of the fine sediments accumulated in the channel. These sediments were weakly developed in areas adjacent to the canal banks and may have originated from erosion of the banks, which left the upper banks poorly defined. Although this initial explanation may be partly accurate, the sediments probably were also related to increased efficiency of canal technology during the Classic period, when flows were regulated more consistently and the variability in the size of particles being transported was more uniform. The lighter color of the sediments indicated that organic matter was greatly reduced compared to sediments in earlier canals—the result, possibly, of continuous or nearly continuous flows and the removal of vegetation along the canal's banks. The horizontal deposition of the sediments in Canal Nuevo (Figure 8.3), in contrast to Canal Viejo and other pre-Classic period canals, suggested that the flow and depositional history of this Classic period canal was more uniform and less erosive than that of the earlier channels. Field crews did not observe evidence of scouring events per se except in the extreme lower portion of the channel, as in Stratum 19. In comparison, Canal Viejo contained an entire series of "scour" deposits, including Strata 2, 7, 9, 10, and 11.

Baffle

Pre-Classic period canals at Dutch Canal Ruin frequently exhibited areas of erosion, usually in the basal areas and along the banks. Channel migration, usually downgrade (to the south), was evident in most canal exposures, and intrachannel scouring, whether intentional or a result of canal gradient, was present in nearly all canals in the project area.

The natural slope of the first terrace was to the west-southwest, and the canals followed a similar trajectory, occasionally turning northward, reducing the gradients. Bank erosion occasionally led to breaches of the canals (D. H. Greenwald 1988a:65-70) and eroded areas. In Canal Viejo, the Hohokam constructed a device to retard erosion along the southern canal bank where the canal turned sharply to the north-northwest. Referred to as a baffle, this feature was constructed inside the canal after the canal had been in operation for some time, as it was built on canal sediments. The baffle consisted of an embankment formed from canal clays and reinforced by rocks and artifacts placed along its top and the upper portion of the inside, similar to the way that engineers use rip-rap today to prevent water erosion in drainages and along roadcuts. The feature extended northward from the curved bank for 0.75 m, intruding into the canal channel, and had a width of

0.13–0.50 m (Figure 8.4). Although it may have extended farther along the canal bank, its excavated length might have been sufficient to reduce the threat of erosion at this location. By extending into the canal a short distance, the baffle moved water away from the canal bank and produced an area of slow-moving water, or an eddy.

Researchers have reported features with similar functions elsewhere in the Hohokam area. For example, cobble concentrations that included large fragmented artifacts and wooden structures placed in canals were interpreted as erosion and velocity control devices at La Cuenca del Sedimento, located on the south side of the Salt River in Tempe (Ackerly, Kisselburg, and Martynec 1989:155–163). Although of a different morphology and associated with larger canal channels, the La Cuenca intrachannel cobble concentrations probably reduced erosion through channel stabilization and reduction of water velocity before the stream reached a canal juncture (Ackerly, Kisselburg, and Martynec 1989:162). Intermixed with the cobble concentrations at La Cuenca were numerous postholes. Investigators likewise observed this association, although on a much smaller scale, in Area 7. At La Cuenca, lines of posts were set perpendicular to the canal. Their orientation suggested that these posts were probably not associated with dams or headgate construction. Rather, in combination with the cobble concentrations, they probably reduced water velocity in relation to the cobbles, forming a baffle system. The postholes in Alignment 8501 at Area 7 may have functioned similarly, while the "rip-rap" feature may have protected the canal bank at this radius in the canal alignment. Alternatively, the postholes in the bottom of the channel may have been associated with some type of brush dam. This inference was supported by the presence of two shallow depressions, one upcanal and the other downcanal from the postholes. Such depressions can be caused by a sudden increase in water velocity as it flows over or around an obstruction. Known as plunge pools, eroded features such as these are common in irrigation canals and natural streams. Rocks in the downcanal depression indicated that it was open when the baffle was built or in use. The other depression contained a single rock; if it was associated with a retention device, any evidence of the presence of postholes had probably been removed by Trench 2003.

At this point, where the course of the canal was directed to the northwest, the velocity of flow could have been slowed, but the change in direction would have resulted in downslope bank erosion unless preventive measures were taken. Within a distance of 10 m, Alignment 8501 returned to roughly its original trajectory, as can be seen by its location in Trenches 2003A and 2003B (Figure 8.1). The change in canal direction at this location may have been directly related to controlling water velocity, perhaps necessary if a turnout or an area susceptible to erosion existed beyond this point or if laterals were located downcanal.

Postholes

The two postholes in Canal Viejo, originating at the base of the channel and extending to a maximum depth of 20 cm, appeared to be related to some form of water control. Subfeature 7-3-2 was the larger of the two postholes, approximately 30 cm in diameter. Field crews recovered four river cobbles, each less than 10 cm in diameter, from the fill of this posthole: one from the base, the others from the middle levels. The cobbles appeared to have been used in some manner to stabilize the post, the way chinking is used in masonry. Subfeature 7-3-3 was adjacent to Subfeature 7-3-2 and measured about 18 cm in diameter and 15 cm in depth. Both postholes were filled with canal sediments, indicating that the posts had been removed or had decayed while the canal was in operation.

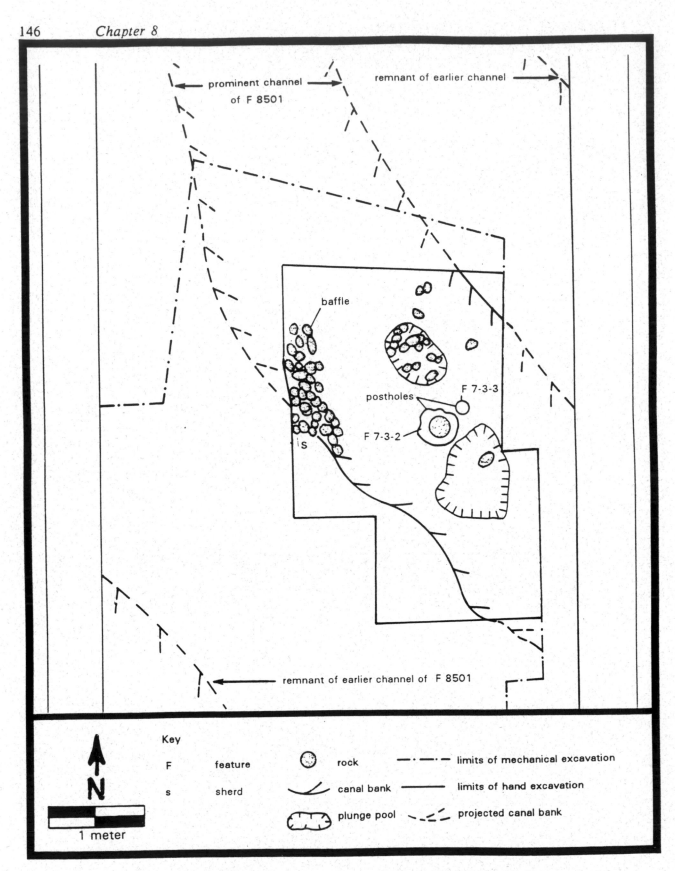

Figure 8.4. Area 7, canal baffle and postholes in Alignment 8501 (Canal Viejo).

MATERIAL CULTURE

The number of artifacts recovered from Area 7 was as high as or higher than in most pre-Classic period areas of Dutch Canal Ruin, although only three features were excavated. This detail was accounted for, in part, by excavation of the canal, which contained discarded artifacts. However, the quantity of material culture remains collected from Feature 7-1 was larger than for almost any other pre-Classic period structure. The large number of artifacts recorded for this feature could have been a result of investigators' ability to define the outline of the structure at a higher elevation above the floor due to its oxidized walls, because of the structure's relatively long occupation period, or because this structure was the locus of more activities than other pit structures at Dutch Canal Ruin.

Ceramics

Field crews recovered 78 sherds during data recovery in Area 7 (Table 8.1). All but one were Hohokam wares, the exception being a Gila Polychrome sherd. The majority of the sherds, 79.5%, came from Canal Viejo, Alignment 8501. Other than the Gila Polychrome sherd, only one other temporally diagnostic sherd, a Santa Cruz/Sacaton Red-on-buff, was recovered.

Table 8.1. Area 7, Ceramic Frequencies

Ceramic Ware/Type	Feature				Total
	7-1 Pit Structure	7-2 Fire Pit	A8501 Canal	A8537 Canal	
Hohokam Plain Ware					
Gila Plain, Salt Variety	6		37		**43**
Gila Plain, Gila Variety	7	2	9		**18**
Squaw Peak Plain			8		**8**
Hohokam Decorated Ware					
Santa Cruz/Sacaton Red-on-buff			1		**1**
Indeterminate Buff			7		**7**
Roosevelt Red Ware					
Gila Polychrome				1	**1**
Total	**13**	**2**	**62**	**1**	**78**

In addition to the sherds recovered from Canal Viejo, 16.7% were from the pit structure, 2.6% were from the fire pit, and 1.3%, a single Gila Polychrome sherd, was from the profile of Canal Nuevo, Alignment 8537. Little can be said about temporal affiliations from two temporally diagnostic sherds, especially when those sherds were collected from canals.

Flaked Stone

The flaked stone assemblage from Area 7 totaled 142 artifacts (Table 8.2). Of these, 85 (59.9%) were from Canal Viejo, 54 (38.0%) were from Feature 7-1, 2 (1.4%) were from Feature 7-2, and 1 (0.7%) was from the overburden above Feature 7-1. Debitage, which included angular debris, flake fragments, platform-remnant-bearing (PRB) flakes, and whole flakes, accounted for the majority (131, or 92.3%) of the assemblage. Of the remaining 11 artifacts, 8 (5.6%) were tools with use wear, and 3 (2.1%) were cores. The tools consisted of two hammerstones, four core/hammerstones, a used flake, and a flake biface.

Table 8.2. Area 7, Flaked Stone Morphological Type by Material Type

Morphological Type	Material Type					Total
	Quartz	Chert	Basalt	Igneous Other	Quartzite	
Angular Debris		3	1	5	1	10
Flake Fragment	2	15	29	5	6	57
PRB[1] Flake		4	17	9	3	33
Whole Flake	2	4	19	5	1	31
Used Flake		1				1
Core			1	1	1	3
Core/Hammerstone			2	2		4
Hammerstone			2			2
Flake Biface			1			1
Total	4	27	72	27	12	142

[1]Platform-remnant-bearing

Raw material types in Area 7 included basalt, chert, igneous other material, quartzite, and quartz, in descending order of frequency. This assemblage was slightly unusual in that it contained 27 (19.0%) chert artifacts. Although not large in absolute terms, this was the greatest number, as well as the highest percentage, of chert artifacts in the assemblages from all the pre-Classic period areas of Dutch Canal Ruin.

Feature 7-1 also was unusual in several ways when compared to the other pre-Classic period structures at Dutch Canal Ruin. First, it contained more artifacts generally (n=54). Second, it had more artifacts (n=39) on the floor than any other structure. Third, it contained 23 of the 27 chert artifacts from Area 7, not only more than any other pre-Classic period feature in both absolute and relative terms, but also more than in all other areas combined.

Of the 54 artifacts in Feature 7-1, 52 (96.3%) were debitage, and the remainder consisted of a used flake of chert and a quartzite core. Material types included 23 (42.6%) pieces of chert, 16 (29.6%) of basalt, and 8 (14.8%) of quartzite, with quartz and igneous other materials accounting for 4 (7.4%) and 3 (5.6%) items, respectively. The 39 artifacts collected from the floor constituted 72.2% of the total, a uniquely high quantity among the pre-Classic period structures at Dutch Canal Ruin. The floor contact artifacts all were debitage except for the quartzite core. Of the 23 chert artifacts from Feature 7-1, 15 (65.2%) lay on the floor; 4 (17.4%) were in the floor fill, including the used flake; and the feature fill contained another 4 (17.4%). The one artifact collected from the overburden above Feature 7-1 was a chert PRB flake.

The fire pit, Feature 7-2, contained one piece of chert debitage and another of basalt.

Field crews retrieved 85 flaked stone artifacts during the excavation of Canal Viejo, 76 (89.4%) of which were debitage. Material types for the debitage and other morphological categories were predominantly basalt (64.7%) and igneous other material (28.2%), with the remainder being quartzite and chert. One core, two core/hammerstones, two hammerstones, and the flake biface were basalt, and a core and two core/hammerstones were composed of igneous other material.

Six (75.0%) of the eight tools within Area 7 were hammerstones or core/hammerstones. The remaining two tools were a used flake with scraping use and a flake biface with scraping and battering. Apparently, the flake biface originally had been part of a core/hammerstone. All of the hammerstones and core/hammerstones from Area 7, as well as two of the three cores, were from Canal Viejo, suggesting the possibility that their presence in the canal was linked to the construction or maintenance of the canal or both. The presence of these artifacts, whether associated with canal activities or not, in conjunction with the debitage, indicated that reduction of flaked stone tools had taken place within Area 7. The used flake and the use wear on the flake biface suggested other activities as well.

Ground Stone

Investigators recovered three pieces of ground stone during the excavations in Area 7: two from Canal Viejo and one from Feature 7-1. A handstone made from a greenstone river cobble was collected from the floor of the structure. One of the ground stone artifacts in Canal Viejo was a fragment of a metate not further specified (NFS) made from vesicular basalt. The other item from the canal was a piece of unmodified, water-worn Squaw Peak schist.

The handstone, a river cobble, was minimally shaped, with a small amount of pecking; grinding use wear was present on both sides. In addition, both ends of the cobble showed evidence of battering, indicating that at some point this handstone had been used as a hammerstone or possibly a hand-held maul.

The presence of the handstone and the metate in Area 7 indicated that the inhabitants had processed botanical resources, either wild or domesticated. In addition, the results of the investigations suggested that the site inhabitants had expended minimal energy on tool production, although they may have curated tools for later reuse. The handstone appeared to be a river cobble of appropriate size and shape that the user had originally selected as a hammer; subsequently, the tool was used for grinding. This sequence suggested that artisans had made no effort to produce formal tools for specific functions.

CHRONOLOGY

The chronological evidence from Area 7 suggested that Canal Nuevo, Alignment 8537, was a Classic period canal and Canal Viejo, Alignment 8501, was a pre-Classic period canal. The fire pit, Feature 7-2, may have been a Classic period feature associated with Canal Nuevo. Convincing evidence for the date of the pit structure, Feature 7-1, was lacking, although investigators thought that this structure most likely dated to the pre-Classic periods.

Stratigraphy

The fire pit, Feature 7-2, lay in the uppermost layers of fill in Canal Viejo, the pre-Classic period canal, and therefore must have postdated that canal. This fact alone does not prove that it dated to the Classic period, and investigators found no further chronological evidence.

The linear nature of canals often permits reconstruction of temporal associations on the basis of their stratigraphic positioning in relation to each other. In Area 7, investigators traced Canal Viejo and Canal Nuevo through three trenches during the data recovery phase to address this question. In Trench 2003, the two channels were spatially separated; they merged in Trench 2003A, 3.0 m to the west. Six meters farther to the west in Trench 2003B, the channels were superposed, with Canal Nuevo positioned above Canal Viejo. This relationship clearly indicated that Canal Nuevo was the younger of the two alignments.

Field crews also observed Canal Viejo below North Main Canal 3 in Area 5 (Chapter 6). North Main Canal 3 was confidently dated to the late Colonial period elsewhere in the project area (Chapters 3 and 18). The relationship of the two canals in Area 5 indicated that Canal Viejo had been abandoned by the time North Main Canal 3 was built, possibly by the beginning of the Santa Cruz phase.

The stratigraphic position of Feature 7-1, the pit structure, compared closely with that of other pre-Classic period structures. The structure was dug through the clayey silts into the upper levels of the poorly sorted sands. Although buried by the dark clays associated with the lower levels of the Holocene soils unit, the feature was not filled with these soils. These stratigraphic indicators suggested that Feature 7-1 was more likely associated with the pre-Classic than the Classic period occupation.

Ceramics

Area 7 yielded two temporally diagnostic sherds. One was a Santa Cruz/Sacaton Red-on-buff sherd in the fill of Canal Viejo, which was consistent with, and supported, the dating of this canal to the pre-Classic period. The second sherd, a Gila Polychrome, appeared in the lower portion of one of the exposures for Canal Nuevo, supporting a Classic period date for this canal. Although two sherds did not provide conclusive evidence for the age of these canals, they did not contradict conclusions based on other data.

The ceramic assemblage from Feature 7-1, the pit structure, provided few insights concerning its age. Given the types present, the investigating team could have assigned Feature 7-1 to almost any period within the Hohokam chronology.

Absolute Dates

Field crews collected two chronometric samples from Area 7: a charcoal sample from Feature 7-1 and an archaeomagnetic sample from Canal Nuevo. The radiocarbon sample was from a charcoal concentration on the floor of the structure. The date for this sample (Beta-43287), provided by Beta Analytic, Inc., was computed using the AMS technique and adjusted by carbon 13 for total isotope effect. The adjusted date was A.D. 320–430, 1575±75 (Appendix B). The calibrated date (Stuiver and Becker 1986) for this sample was A.D. 413–542. Both dates were for a standard deviation of one sigma. Using the Hohokam chronology presented by Dean (1991), this date would place Feature 7-1 in the early Pioneer period. Although no other chronological data exist for this structure, a date in the Pioneer period seemed too early given the evidence from the other excavated pre-Classic period areas of Dutch Canal Ruin. Using the same arguments proposed in Chapter 4, investigators thought this early date resulted from the use of old wood.

The archaeomagnetic sample from Canal Nuevo (CSUSED-76) was not given visual date ranges (Appendix A; Eighmy and Klein 1991:Table 4), the preferred method for dating samples (Eighmy and Baker 1991:9). The declination date ranges for the canal sediments were A.D. 650–850, 925–1025, and 1275–1700. The presence of a Gila Polychrome sherd in the canal below the level at which the sample was taken made it highly unlikely that the two earlier ranges were correct. The later range indicated that the canal was in use after A.D. 1275, but it also seemed unlikely that the canal was in use as late as A.D. 1700. A more appropriate ending date would seem to be toward the end of the Hohokam sequence, ca. A.D. 1500. The more reasonable date for the canal would, therefore, be A.D. 1275–ca. 1500. This would place the earliest possible age of the canal in the late Soho phase, with the latest possible age in the Polvorón phase.

SUMMARY

The presence of a single structure would seem to indicate that the occupation of Area 7 had been brief, consisting of periodic use for perhaps only a few years. The presence of the pre-Classic and Classic or post-Classic period canals, however, indicated that the Hohokam had used Area 7 intermittently over a long period of time. The small number of features pointed to the low intensity of use of this area even while inhabitants were building and managing canals. Initial construction of the canals and subsequent maintenance would have required higher levels of social structure during those brief periods.

Activities in Area 7 included the reduction of lithic raw material, the grinding of foodstuffs, canal construction and maintenance, and possibly agricultural pursuits during the pre-Classic periods. Analysts found evidence for lithic technology and the processing of subsistence plants in the flaked and ground stone assemblages from Feature 7-1. They inferred canal maintenance from the presence of the baffle, Subfeature 7-3-1, in Canal Viejo. They based their inference of agricultural pursuits on the location of Feature 7-1 in the floodplain of the Salt River, the proximity of the pre-Classic period canal, and analogy to other pre-Classic period areas of Dutch Canal Ruin, which would indicate that Feature 7-1, the pre-Classic period occupation of Area 7, was a field house settlement.

As a field house locus, Area 7 was probably occupied by a subnuclear family during specific periods of the year. Because investigators did not conduct botanical analyses in Area 7, they did not address the issue of seasonality. However, the spatial association of the structure to the canals suggested that the occupants had practiced agriculture there, and investigators therefore expected that the occupation had occurred during summer.

Aside from the intrusive fire pit in Canal Viejo, the only evidence of activities associated with Classic period use in Area 7 was related directly to the construction and maintenance of the canal. The fire pit may have been associated with the Classic period; however, this inference could not be supported at this time. Assuming this association was correct, the fire pit could have been related to construction, maintenance, or management of the Classic period canal or of nearby field areas.

Analysts used archaeomagnetic dating to place the age of Canal Nuevo between the late Soho phase and the end of the Hohokam sequence. The presence of a Gila Polychrome sherd in the canal's fill supported the premise of a late Classic period affiliation. As with many of the areas investigated at Dutch Canal Ruin, the physical remains and material culture assemblages were extremely limited, precluding more definite conclusions about this location.

CHAPTER 9

DUTCH CANAL RUIN, AREA 8

Mark L. Chenault
Sarah L. Horton
David H. Greenwald

Area 8 within Dutch Canal Ruin consisted of a small hamlet. Architectural morphology and material culture remains indicated that Area 8 dated to the Polvorón phase (A.D. 1350 to 1450) (Sires 1984) of the post-Classic period. In this area, archaeologists found and excavated 24 prehistoric cultural features, consisting of 1 pit house, 1 pit structure, 2 cremations, 19 pit features, and what appeared to be an adobe wall segment.

Area 8 represented one of the southernmost loci at Dutch Canal Ruin. It was located within N1400–1450, E1070–1110 of the project grid and in T1N, R3E, Sec. 10, NW¼SE¼SW¼ on the Phoenix, Arizona, 7.5 minute USGS topographic map (Figure 1.4). The actual area investigated measured 22 × 34 m (Figure 9.1).

Excavation and analysis of the Area 8 remains showed that it was a small habitation site occupied year-round, probably by a nuclear family or small extended family. Pollen and macrobotanical remains suggested that farming activities performed by the inhabitants of the site included the cultivation of domesticated crops such as cotton and corn. Analysis also indicated that the inhabitants had exploited a variety of wild plant foods.

Located north of Hadley Street, south of Sherman Street and an alignment of the former Dutch Ditch, and east of 18th Street (Figures 1.1 and 1.2), Area 8 was situated in or near an area Midvale (1934) identified as a mound. However, modern disturbance of this area was extensive, and investigators found no evidence of a mound feature. The upper portions of most of the cultural features in Area 8 were truncated by the modern plow zone, and several recent pits occurred in and around the features.

RESULTS OF TESTING

Five backhoe trenches excavated in Area 8 during the testing phase of the Phoenix Sky Harbor Center project revealed six prehistoric features. All appeared in Trenches 1320 and 1321, including Feature 1067, recorded as a trash deposit; Feature 1068, a pit with evidence of burning; Features 1069 and 1070, pits; and Features 1072 and 1073, fire pits (Volume 1:Appendix A). The excavations revealed that Feature 1067 was actually a pit structure filled with trash and that Feature 1070 was a pit house and not merely a flat-bottomed pit. The results of the testing effort suggested the possibility of up to six structures and numerous associated features. Excavation showed that the occupation in Area 8 was smaller than anticipated. Nevertheless, it was the most substantial component of the portion of Dutch Canal Ruin within the project area.

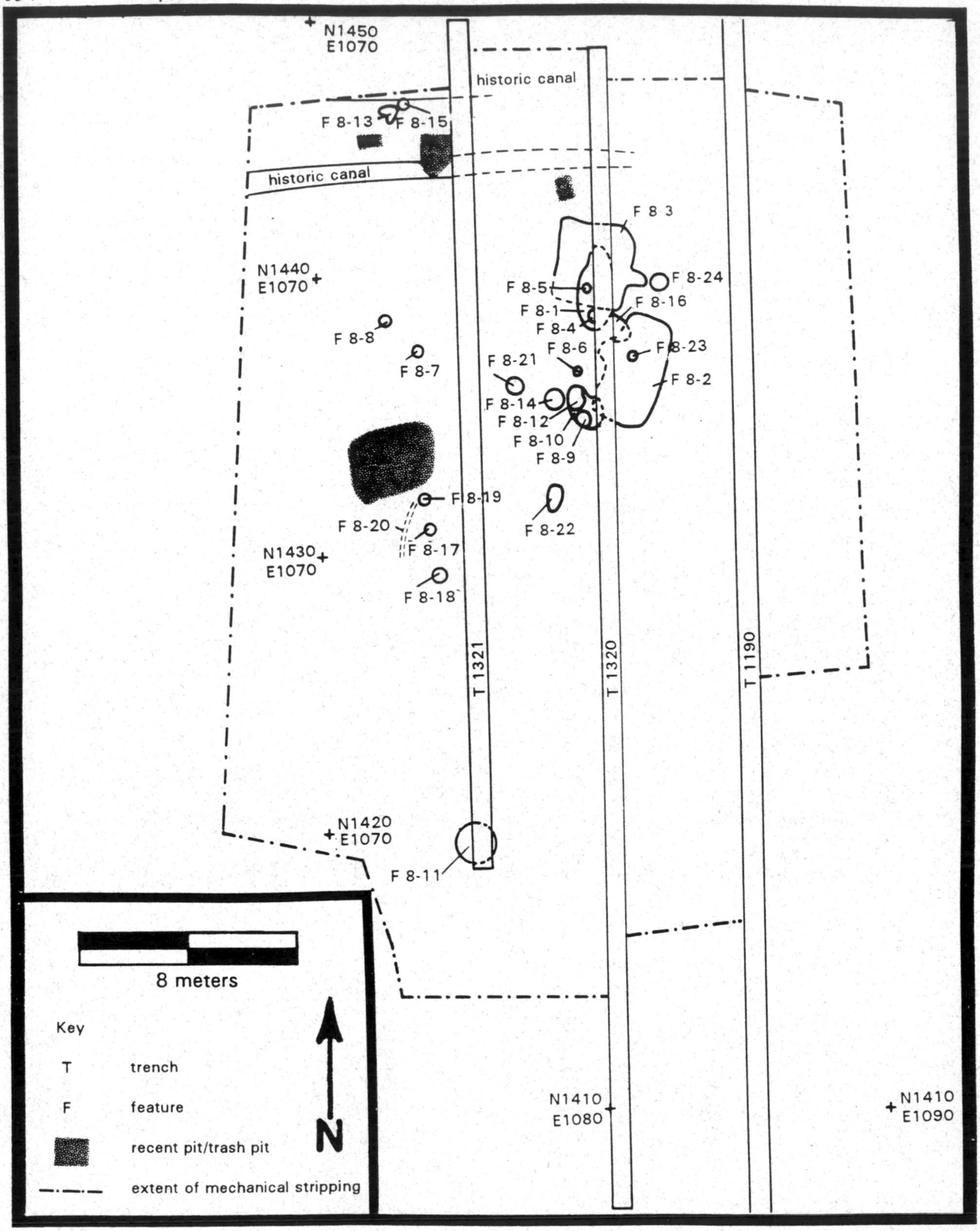

Figure 9.1. Area 8, Dutch Canal Ruin.

RESEARCH OBJECTIVES

Research in Area 8 was guided by the generalized problem domains and research questions outlined in the Phoenix Sky Harbor Center Project research design (Volume 1:Chapter 10): (1) site structure, settlement composition, and settlement typology; (2) spatial, functional, and temporal variation in settlement across Canal System 2; (3) factors influencing Hohokam settlement and land use in the geologic floodplain; (4) Hohokam canal irrigation technology; and (5) chronology building and refinement. Also of interest to the research program were subsistence, demographics, social structure, and intra- and interregional exchange and interaction, although these issues were not formalized as problem domains.

Initial data recovery efforts indicated the likelihood that Area 8 was associated with the post-Classic period. Investigators then developed additional research topics, one of them specific to Area 8: the hypothesis that the prehistoric remains discovered could be assigned to the proposed post-Classic period Polvorón phase (Sires 1984) (see discussion in Volume 4, Chapter 8).

Researchers have proposed the Polvorón phase (A.D. 1350–1450) to encompass the final recognizable episode of the Hohokam occupation of the Phoenix Basin; Sires (1984:301) considered it a post-Classic period manifestation. The characteristics of the Polvorón phase include a pattern of dispersed, small settlements, possibly located along small canal segments (Sires 1984:324). This pattern could take two different forms: first, the continued occupation of rooms within Civano phase compounds; and second, the construction of jacal pit structures forming house clusters or courtyard groups. Polvorón phase pit structures are generally rectangular with rounded corners. They are true pit houses, rather than houses in pits, in that the pit walls form the lower walls of the house.

Polvorón phase material culture consists of ceramics dominated by plainware and redware types, with decorated types consisting almost entirely of Roosevelt Red Ware. Intrusive ceramics in Polvorón phase assemblages include Hopi yellowware. In lithic assemblages, the relative frequency of obsidian is higher than in the Civano phase, but the number of shell artifacts is lower, especially those made from *Glycymeris*. Treatment of human remains appears to have included both inhumation and cremation. Subsistence strategies did not change drastically from earlier periods, with most of the traditional wild and domestic foods still used (Gasser and Kwiatkowski 1991; Sires 1984).

FIELD METHODS AND STRATEGIES

Mechanical Excavations

Field crews during data recovery investigations reopened Trenches 1190, 1320, and 1321 (excavated during the testing phase) to gain access to identified cultural features but excavated no additional backhoe trenches. Examination of profile drawings produced during testing aided in identifying and relocating these features. They then stripped off overburden using the backhoe to expedite feature excavation. The reopened trenches totaled 116 m in length, and the area stripped measured 620 m^2.

Hand Excavations

Investigators excavated the prehistoric cultural features described below by hand, using the controlled methods described in the work plan (Volume 1:Chapter 11). Because of the difficulty in defining features within this portion of the project area, field crews initiated excavation in 2 × 2-m grid squares above suspected features. Once they identified an occupation surface, they followed it to define the extent of the surface and any structure walls. They then excavated architectural features stratigraphically in quadrants after defining the features as structures. They excavated the 5 cm of fill directly above the floor of a structure down to the floor as a level, by quadrant.

STRATIGRAPHY AND ASSOCIATION

The prehistoric cultural features in Area 8 of Dutch Canal Ruin lay just below the plow zone and in most cases had been truncated by recent activity. Analysts classified soils in this part of the site as Gilman Loam, a fine sandy loam (Volume 1:Appendix B), which was underlain by a fine sand with a massive structure. All of the features originated in the upper portion of the undisturbed stratigraphy, although a few extended into the lower level of fine sand. This association was markedly different from those of surrounding areas at Dutch Canal Ruin, as cultural features in those areas originated at the base of the fine sand level.

CULTURAL FEATURES

The 24 features excavated and recorded in Area 8 comprised 1 pit house, 1 pit structure, 2 cremations, 2 trash pits, 6 fire pits, 1 posthole, 1 adobe wall segment, and 10 pit features of unknown function (Figure 9.1). Each feature is described below.

Structures

Feature 8-2

Testing Feature: 1070
Trench Number: 1320
Horizontal Provenience: N 1434.60–1438.60; E 1079.78–1082.81
Vertical Provenience: 3.67–3.81 meters below datum (mbd)
Feature Type: Pit House

Feature 8-2 was a roughly rectangular pit house, 4.00 × 3.03 m in plan and 14 cm in depth (Figure 9.2). The backhoe trench removed three-fourths of the west wall, including most of the entry. A substantial amount of historic and recent disturbance in the area had removed the upper portion of the structure as well (Volume 4:Chapter 2), leaving only its shallow base. This feature was described as a flat-bottomed pit during testing.

Feature 8-2 was a true pit house, in which the sides of the pit form the base of the structure walls. These lower walls were lined with adobe. At the time of excavation numerous gaps appeared in the wall lining, and what was present was very thin and not substantially constructed. The wall-floor juncture formed an obtuse angle, with the walls originating at the level of the floor. The floor

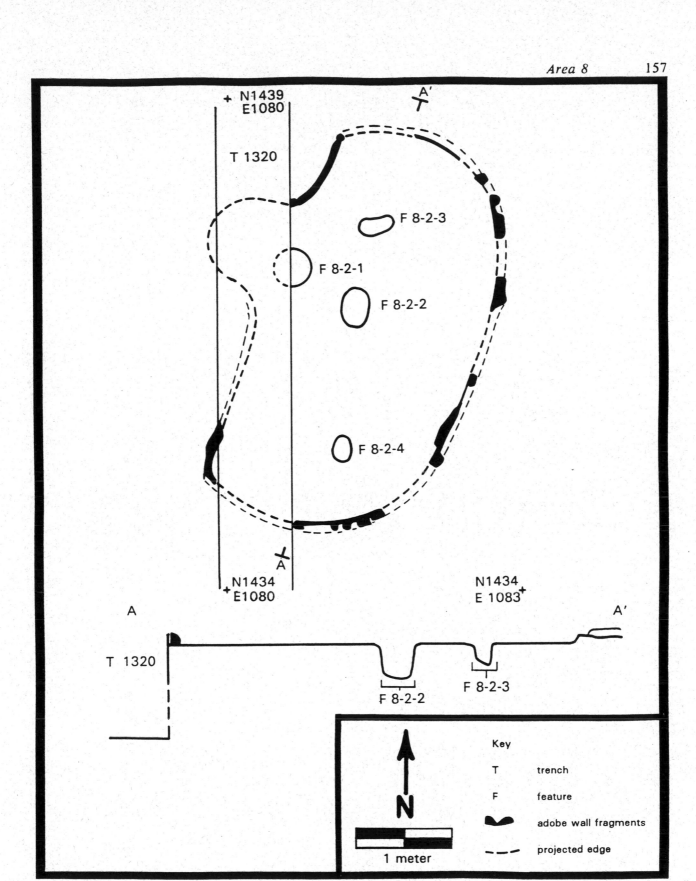

Figure 9.2. Area 8, Feature 8-2, plan and cross section.

was a level, unplastered, hard-packed surface. Originally the pit house probably had had a superstructure of jacal; however, field crews found no direct evidence for a superstructure.

Some conclusions concerning the roof may be derived from the patterning of floor features. One large, centrally located posthole (Subfeature 8-2-2) indicated that a primary roof support had been used. Two smaller pits (Subfeatures 8-2-3 and 8-2-4) were situated to either side of the center support and would have formed a fairly straight line of posts along the north-south axis of the structure. These latter floor features were probably the postholes for secondary roof supports.

Although most of the entry was missing, one small corner remained at the juncture with the sidewall of the structure. From this fragment, investigators surmised that the entry had opened to the west.

The fire pit (Subfeature 8-2-1) was located north of the center of the structure along the north-south axis and west of the center along the east-west axis, just inside the structure and in front of the entry. Trenching removed a large portion (approximately one-half) of the fire pit. Reconstructed dimensions indicated that the fire pit was 36 cm in diameter and 15 cm in depth. In profile, the feature had steep sloping sides and a slightly curved base, and both the sides and base were oxidized from exposure to heat. The fire pit had been dug below the floor surface into the silty clay substratum and was extensively disturbed by rodent activity.

Feature 8-2 was located stratigraphically at about the same level as the other pit structure in Area 8, Feature 8-3. Feature 8-2 appeared to have been abandoned intentionally, with artifacts removed by the inhabitants. The fill of the structure was a light grayish brown silty loam, moderately compacted, containing a high density of cultural materials and charcoal flecks, suggesting that it had been used for trash disposal.

Feature 8-3

Testing Feature: 1067
Trench Number: 1320
Horizontal Provenience: N 1438.70–1442.21; E 1078.35–1081.83
Vertical Provenience: 3.66–3.84 mbd
Feature Type: Pit Structure, Not Further Specified (NFS)

Feature 8-3 was an irregularly shaped pit structure, 3.51 × 3.48 m in plan (including the entry) and 18 cm in depth (Figure 9.3). The structure, bisected by Trench 1320, was identified as a trash deposit during testing (Volume 1:Appendix A). The fire pit within the structure was recorded during testing as Feature 1068 (Volume 1:Appendix A). The shape of the structure in plan view, east of the trench, was fairly regular (subrectangular), with an entry extending 78 cm from the east wall. The west side of the structure, however, was somewhat irregular. Historic or recent disturbance was a factor in the poor preservation of this feature; numerous historic pits and other forms of historic disturbance surrounded it. One of these pits, a posthole, was excavated into the wall and floor of the structure entry. The structure was situated 40 cm below the modern surface, which probably accounted for its ephemeral appearance and disturbed state.

The extant walls of Feature 8-3 were considerably less substantial than those of Feature 8-2; no intact adobe wall segments were visible. However, field crews found concentrations of what appeared to be melted adobe along the southern portion of the east wall and around the south and east portion of the entry. The edges of the feature were defined by the boundaries of the feature

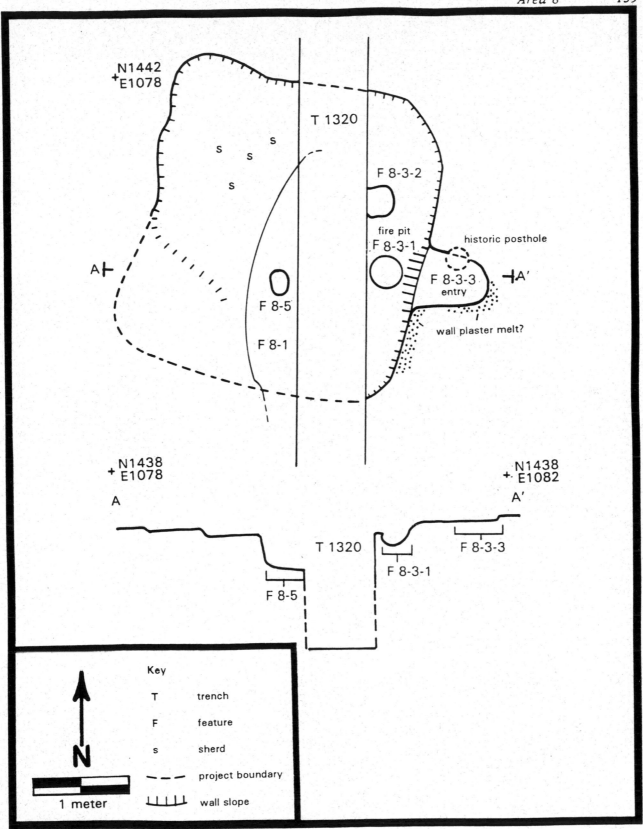

Figure 9.3. Area 8, Feature 8-3, plan and cross section.

fill, and flat-lying artifacts and a slight compaction difference identified the structure floor. The floor was unprepared and had originally been dug into the clayey substratum, providing a natural surface that would not necessarily have required preparation.

The entry (Subfeature 8-3-3) opened to the east. Beginning at a slight step by the fire pit (Subfeature 8-3-1), the entry floor gently sloped up to the outside. Field investigators observed no plaster on the walls or floor, although an accumulation of melted wall plaster was directly outside the entry.

The fill of Feature 8-3 was a medium grayish brown silty clay loam. Because of the difficulty in defining the floor, field crews did not separate the 10-cm floor fill level from the rest of the fill or excavate it as a separate level, as dictated by standard project procedure. Instead, they removed the entire fill of the structure as one level. The fill contained more than 2000 artifacts, mostly ceramics. In comparison, crews found fewer than 10 sherds on the floor.

Only two floor features, a fire pit (Subfeature 8-3-1) and a pit NFS (Subfeature 8-3-2), were present. The fire pit was located in the east half of the structure directly in front of the doorway. It was circular in plan and somewhat basin shaped in cross section. The fire pit was 37 cm in diameter and 23 cm in depth and contained grayish silt with inclusions of burned adobe, charcoal, ash, numerous ceramics, flaked stone, and shell. The sides and base of the fire pit were oxidized.

Subfeature 8-3-2 was a small, circular pit NFS, 30 cm in diameter and 9 cm in depth. In cross section, it was basin shaped. The pit fill was basically the same color as the surrounding fill, but the texture was clearly different, consisting of a well-sorted, well-consolidated loam with a lower percentage of clay. Whereas the surrounding soils appeared blocky, the pit fill was much finer. Both ceramics and flaked stone were collected from the pit. No function could be assigned to this subfeature.

The southeast corner of Feature 8-3 was 50 cm due north of the northwest corner of Feature 8-2 (Figure 9.1). The proximity and orientation of the two pit structures indicated that they probably had not been occupied simultaneously. Feature 8-3 had been casually abandoned and subsequently had been used for trash disposal.

Cremations

Feature 8-13

Horizontal Provenience: N 1446.1; E 1072.7
Vertical Provenience: 3.72–3.92 mbd
Feature Type: Secondary Cremation

This feature was a cremation placed within a Gila Plain, Salt Variety jar. Field crews could not define the pit in which the cremation was buried. The fill immediately around the jar was a light grayish brown silty loam, moderately compacted and sorted and disturbed by rodent activity and containing fragments of burned daub, charcoal, and a few artifact fragments. The cremation jar was approximately 30 cm in diameter and 20 cm in height. The top of the jar, however, had been truncated by historic disturbance processes, so investigators could not determine the exact height of the vessel. The fill inside the jar consisted of a light grayish brown ashy silt, loosely compacted, containing fragments of burned, calcified human bone. Kathryn L. Wullstein conducted a detailed analysis of the human remains contained in this cremation (see below, under "Human Remains"), and additional information on the cremation vessel is provided in Chapter 12.

Feature 8-15

Horizontal Provenience: N 1446.50; E 1073.15
Vertical Provenience: 3.72–3.89 mbd
Feature Type: Secondary Cremation

This cremation was placed in a jar and covered with a bowl; both vessels were Gila Plain, Salt Variety. The edges of the pit into which the cremation was placed could not be distinguished from the surrounding matrix. The fill immediately around the jar was a light grayish brown silty loam, moderately compacted and sorted, with small inclusions of burned daub and charcoal. The jar containing the cremated remains was approximately 27 cm in diameter and 20 cm in height. The fill within the jar was a loosely compacted, light grayish brown silty loam. It contained inclusions of charred and calcified human bone. Detailed analysis of the human remains contained in this cremation are presented below; details of the analysis of the cremation vessel are in Chapter 12.

Extramural Features

Feature 8-1

Trench Number: 1320
Horizontal Provenience: N 1438.00–1441.10; E 1079.35–1079.90
Vertical Provenience: 3.78–4.18 mbd
Feature Type: Pit NFS

Feature 8-1 was a large, somewhat oval-shaped pit, 3.10 × 0.55 m in plan and 40 cm in depth. The exact shape was impossible to determine because Trench 1320 had removed at least one-half of the feature. In cross section, the walls of the feature were vertical, and the base was fairly level. Feature 8-1 was exposed while field crews were excavating a 2 × 2-m control excavation unit. The fill, composed of loose silty soil with random charcoal flecks, was removed in one level. Extensive rodent disturbance occurred in the lower portion of the feature. This feature apparently intruded into the fill of Feature 8-3, extending through the south wall and floor. In addition, Features 8-4 and 8-5 had been dug into the fill of Feature 8-1.

Feature 8-4

Trench Number: 1320
Horizontal Provenience: N 1438.38–1438.68; E 1079.72–1079.86
Vertical Provenience: 3.85–3.98 mbd
Feature Type: Pit NFS

Feature 8-4 was a small pit, approximately 30 × 14 cm in plan (the backhoe removed an unknown portion) and 13 cm in depth. It was circular in plan and basin shaped in cross section. This feature was situated within the fill of Feature 8-1 and contained a grayish sandy-textured fill.

Feature 8-5

Trench Number: 1320
Horizontal Provenience: N 1439.40–1439.60; E 1079.48–1079.66
Vertical Provenience: 3.82–3.88 mbd
Feature Type: Pit NFS

Feature 8-5 was a small pit, 18 × 20 cm in diameter and 6 cm in depth. The pit was circular in plan and basin shaped in cross section. The feature fill was a dark silty loam. Feature 8-5 intruded into the fill of Feature 8-1.

Feature 8-6

Horizontal Provenience: N 1436.40–1436.66; E 1078.16–1078.44
Vertical Provenience: 3.77–3.88 mbd
Feature Type: Trash Pit

Feature 8-6 was a small trash pit, 26 × 28 cm in diameter and 11 cm in depth. The pit was circular in plan, with a basin-shaped cross section. The feature fill consisted of a very hard and compact soil with inclusions of caliche, ceramics, flaked stone, and oxidized daub. It was uncovered during the excavation of a 2 × 2-m control excavation unit. Although field crews defined no prehistoric ground surface in the area surrounding Feature 8-6, the presence of nearby pit features demonstrated that the trash pit was located in a prehistoric activity area. Investigators could not ascertain what types of activities other than trash deposition had taken place there.

Feature 8-7

Horizontal Provenience: N 1437.30–1437.80; E 1073.10–1073.52
Vertical Provenience: 3.76–3.86 mbd
Feature Type: Pit NFS

Feature 8-7, a small pit uncovered during backhoe stripping, was 42 × 50 cm in diameter and 10 cm in depth. It was circular to slightly oval in plan, and in cross section it was basin shaped. The fill was a light grayish brown silt, moderately compacted and evenly sorted with a few charcoal flecks, ceramics, and flaked stone. The upper portion of the feature appeared to have been truncated by historic disturbance. Feature 8-8, another pit feature, was 0.5 m to the northwest of this feature.

Feature 8-8

Horizontal Provenience: N 1438.15–1438.55; E 1072.21–1072.65
Vertical Provenience: 3.76–4.10 mbd
Feature Type: Pit NFS

Feature 8-8 was a small pit, 40 × 44 cm in diameter and 34 cm in depth. It was circular in plan; in cross section, the upper part of the feature was basin shaped and the lower portion was U-shaped. The fill of Feature 8-8 consisted of a light grayish brown silt, moderately compacted and evenly sorted, containing artifact and charcoal inclusions. The upper portion of the feature appeared to have been truncated by historic disturbance.

Feature 8-9

Horizontal Provenience: N 1433.44–1435.00; E 1079.12–1079.63
Vertical Provenience: 3.76–3.83 mbd
Feature Type: Pit NFS

Feature 8-9 was a small pit, 51 × 56 cm in diameter and 7 cm in depth. In plan it was circular, and in cross section it was a shallow basin. This feature was uncovered during the excavation of a 2 × 2-m control excavation unit. The fill consisted of a hard-packed, cement-like matrix with ceramics, flaked stone, and shell inclusions. Although the feature fill contained some caliche, the feature was not lined. The 2 × 2-m unit also contained three other pit features (Features 8-10, 8-12, and 8-14). Stratigraphically, Feature 8-9 was situated above the other three features, postdating them.

Feature 8-10

Trench Number: 1320
Horizontal Provenience: N 1434.64–1436.00; E 1078.80–1079.90
Vertical Provenience: 3.85–4.05 mbd
Feature Type: Fire Pit

Feature 8-10 was a large pit, 1.36 × 1.10 m in plan and 20 cm in depth. The plan of the feature was irregular, whereas the cross section of the pit was basin shaped. The fill was composed of a compact, silty soil and contained charcoal, burned adobe chunks, ceramics, flaked stone, bone, and shell. Although the base of this feature showed no sign of burning, the side walls were reddened, suggestive of oxidation. The fill above the pit contained a concentration of artifacts and burned materials. The area around the feature had been intensively used, and Feature 8-10 was intruded on by Features 8-9 and 8-12.

Feature 8-11

Trench Number: 1321
Horizontal Provenience: N 1418.80–1420.15; E 1074.35–1075.65
Vertical Provenience: 3.90–4.43 mbd
Feature Type: Trash Pit

Feature 8-11 was a large pit measuring 1.30 × 1.35 m in diameter and 53 cm in depth. The pit was circular in plan and basin shaped in cross section. The fill was a grayish brown silty clay containing charcoal, burned daub, fire-cracked rock, ceramics, and flaked stone. Although ash occurred across the base of the pit, the feature showed no indication of burning. Feature 8-11 was approximately 13 m south of the main feature cluster in Area 8 and was the southernmost feature in the area (Figure 9.1).

Feature 8-12

Horizontal Provenience: N 1433.15–1434.00; E 1078.80–1079.60
Vertical Provenience: 4.05–4.46 mbd
Feature Type: Fire Pit

Feature 8-12 was a medium-sized pit, 80 × 85 cm in diameter and 41 cm in depth. It was circular in plan and bowl shaped in cross section. The fill was a grayish soil with occasional pockets of sand and inclusions of charcoal flecks, burned adobe, fire-cracked rock, ceramics, bone, flaked stone, and shell. The walls were reddish, the result of exposure to heat. Feature 8-12 cut into a portion of Feature 8-10 and appeared to postdate it.

Feature 8-14

Horizontal Provenience: N 1433.15–1434.00; E 1078.00–1078.80
Vertical Provenience: 3.87–4.24 mbd
Feature Type: Fire Pit

Feature 8-14 was a medium-sized pit, 85 × 80 cm in plan and 37 cm in depth. It was irregular in plan and basin shaped in cross section. The fill was a grayish soil with inclusions of charcoal fragments, pockets of sand and ash, ceramics, flaked stone, and bone. The side walls and base of the pit were reddened, with indications of caliche content. The reddish color was probably a result of oxidation from some type of burning in the pit. Feature 8-14 was located within a 2 × 2-m area that contained four pits (Features 8-9, 8-10, 8-12, and 8-14), three of which appeared to have experienced some exposure to fire.

Feature 8-16

Testing Feature Number: 1069
Trench Number: 1320
Horizontal Provenience: N 1437.65–1438.60; E 1080.60–1081.15
Vertical Provenience: 3.77–4.44 mbd
Feature Type: Pit NFS

Feature 8-16 was the remains of a pit 95 × 55 cm in plan and 67 cm in depth. A portion of the feature was removed by Trench 1320. In plan it was irregular, and in cross section it was U-shaped. Field crews encountered considerable difficulty while trying to follow the outline of the sides and base of the pit. The pit fill consisted of a silty and sandy matrix containing ceramics, flaked stone, bone, shell, and charcoal flecks. Feature 8-16 did not seem to have been intentionally filled with refuse and exhibited no signs of burning. Investigators could not determine during excavation whether Feature 8-16 intruded into the pit house (Feature 8-2) or was truncated by it.

Feature 8-17

Horizontal Provenience: N 1430.80–1431.15; E 1073.45–1073.75
Vertical Provenience: 3.82–4.20 mbd
Feature Type: Posthole

Feature 8-17 was a posthole 30 × 35 cm in diameter and 38 cm in depth. The feature was circular in plan and had straight walls with a curved base in cross section. The fill was a grayish brown silty loam with a heavy ash and charcoal concentration and a small number of ceramics and flaked stone artifacts. Several patches of oxidized soil occurred around the perimeter of the posthole. This, in addition to ash and charcoal in the fill, indicated that a fire had burned within the feature. Investigators could not determine the original depth of the posthole because the upper portion was removed by historic disturbance.

The posthole appeared isolated because it was not associated with, or incorporated into, any observable structure. However, other feature types were located nearby. Feature 8-19, a fire pit, was within 1.0 m to the north, and Feature 8-18, a pit NFS, was 1.0 m south of Feature 8-17 (Figure 9.1). Also, a feature that appeared to be an adobe wall segment (Feature 8-20) occurred 1.0 m due west. This cluster of features may have been part of an extramural activity surface and use area, such as a ramada or courtyard.

Feature 8-18

Horizontal Provenience: N 1429.00–1429.50; E 1073.85–1074.35
Vertical Provenience: 3.83–3.91 mbd
Feature Type: Fire Pit

Feature 8-18 was a small, shallow pit, 50 cm in diameter and 8 cm in depth. In plan it appeared circular, and in cross section it had a shallow basin shape. The feature fill was a grayish brown silty loam, moderately compacted and sorted, containing ceramic and charcoal inclusions. The interior surface of the pit was uneven and moderately oxidized. This was the southernmost feature thought to be associated with the possible extramural activity surface that included Features 8-17 and 8-19.

Feature 8-19

Horizontal Provenience: N 1431.80–1432.10; E 1073.25–1073.55
Vertical Provenience: 3.74–3.93 mbd
Feature Type: Fire Pit

Feature 8-19 measured 30 cm in diameter and 14 cm in depth. In plan it was circular, and in cross section it was basin shaped. The feature contained gray and reddish brown fill with ash and charcoal inclusions. The top portion of the fire pit was removed by historic disturbance. This feature was considered to be part of the cluster of features (Features 8-17, 8-18, and 8-19) forming a possible extramural activity surface.

Feature 8-20

Horizontal Provenience: N 1430.00–1432.00; E 1072.55–1072.65
Vertical Provenience: 3.85 mbd
Feature Type: Wall Segment

Feature 8-20 was a north-south–oriented wall segment, 2.0 m in length, averaging about 10 cm in width. The wall was constructed of a grayish brown clay that was distinct from the surrounding browner and less-compacted clay substrate. The wall segment was not necessarily part of a structure but might have been a windbreak or courtyard wall. Because of its proximity to Features 8-17, 8-18, and 8-19, this wall segment may have been associated with those features.

Feature 8-21

Horizontal Provenience: N 1435.63–1436.22; E 1076.60–1077.20
Vertical Provenience: 3.96–4.26 mbd
Feature Type: Fire Pit

Feature 8-21 was a small pit, 59 × 60 cm in diameter and 30 cm in depth. It was an irregular circle in plan and had an irregular, slightly conical bowl shape in cross section. The feature fill was composed of a grayish silt containing charcoal inclusions, daub, fire-cracked rock, ceramics, shell, and bone. Evidence of heat exposure was present in the walls, which appeared to be oxidized. This pit was situated stratigraphically at about the same depth as three other fire pits (Features 8-10, 8-12, and 8-14), all located within a 2 × 2-m area to the west.

Feature 8-22

Horizontal Provenience: N 1431.60–1432.40; E 1078.05–1078.65
Vertical Provenience: 3.86–4.04 mbd
Feature Type: Pit NFS

Feature 8-22 was a shallow, medium-sized pit, 80 × 60 cm in plan and 18 cm in depth. It appeared in plan as oval to subrectangular and in cross section as a basin. The fill was a light grayish brown soil, moderately compacted, containing shell, ceramics, and flaked stone. An unknown amount of the upper portion of the feature was truncated by historic disturbance and possibly by backhoe stripping.

Feature 8-23

Horizontal Provenience: N 1437.25; E 1081.25
Vertical Provenience: 3.83–3.91 mbd
Feature Type: Pit NFS

Feature 8-23 was a small pit, 33 × 49 cm in plan and 8 cm in depth. The plan was subrectangular, with a small, outward protrusion (8 cm) on the east side of the pit. In cross section, it had steeply sloping sides with a flat base. The fill was a grayish brown silty loam, moderately compacted, sorted, and slightly disturbed by rodent activities. Additional inclusions of charcoal and caliche were present. Artifacts in the fill consisted of ceramics and flaked stone. This feature was very shallow and appeared to have been truncated by later activities. Feature 8-2, a pit house, was situated directly above the pit, indicating that Feature 8-23 predated the structure.

Feature 8-24

Horizontal Provenience: N 1439.37–1440.00; E 1081.87–1082.45
Vertical Provenience: 3.66–3.90 mbd
Feature Type: Pit NFS

Feature 8-24 was a small pit, 58 × 63 cm in diameter and 24 cm in depth. In plan it was circular, and in cross section it was basin shaped with a steep west wall and a sloping east wall. The fill was a grayish brown silty loam, moderately compacted, sorted, and slightly disturbed by rodent activity.

MATERIAL CULTURE

Ceramics

The excavations in Area 8 yielded 9843 sherds, 81.6% of the entire Dutch Canal Ruin ceramic assemblage. Analysts categorized the ceramics by ware: Hohokam undecorated (plainware and redware), decorated (buffware and Roosevelt Red Ware), and intrusive wares (Table 9.1).

Hohokam Undecorated Ceramics

Undecorated Hohokam ceramics constituted 95.0% (n=9352) of the ceramic assemblage for Area 8. Both plainware and redware ceramics were included in this ware category. The plainware ceramics dominated the assemblage, making up 81.2% of the Hohokam undecorated ceramics and 77.2% of the total ceramic assemblage for Area 8. The analysts divided the ceramics by temper type and then assigned them to traditional Hohokam ceramic types based on the temper. Hohokam Red Ware made up the second-largest ceramic category, constituting 18.8% of the Hohokam undecorated ceramics and 17.9% of the total ceramic collection for the area.

Decorated Ceramics

Analysts identified two types of decorated ceramics at Area 8: Hohokam Buff Ware and Roosevelt Red Ware. They analyzed 44 buffware sherds from Area 8, 14.2% of the decorated ceramics, and classified the buffware using traditional type categories. Salt Red-on-buff is a newly defined ceramic type with plainware paste but buff slips (Chapter 12). This type was not common in the Phoenix Sky Harbor Center assemblage, but researchers at other projects in the Phoenix Basin (e.g., Abbott and Gregory 1988:23-25) have recorded a similar ceramic type.

When buffware sherds exhibited no design elements, they were categorized as indeterminate buffware. Most of the buffware from Area 8 fit into this category. Seven sherds had small portions of design that were insufficient to allow type assignment; these sherds were categorized as indeterminate red-on-buff.

Roosevelt Red Ware, including the Salado Polychromes, constituted the majority (266, or 85.8%) of the decorated ceramics in Area 8 and 2.7% of the entire Area 8 ceramic assemblage. As with the buffware, analysts used traditional type definitions to classify these ceramics. Sherds with no distinguishing design elements were categorized as indeterminate polychrome.

Archaeologists currently do not know whether the Hohokam produced Gila and Tonto polychromes in the Phoenix Basin. Crown (1991:152) has stated that they did, and another study indicated numerous production loci (Crown and Bishop 1987). However, another study, based on temper analysis (Peterson and Abbott 1991), suggested that Roosevelt Red Ware was produced outside the Phoenix Basin. The latter theory would place the polychromes in the category of an intrusive ware.

Table 9.1. Area 8, Ceramic Frequencies

Ceramic Ware/Type	Absolute Frequency	Percent Ware	Percent Assemblage
Hohokam Plain Ware	**7593**		**77.2**
Gila Plain, Salt Variety	7337	96.6	74.6
Gila Plain, Gila Variety	129	1.7	1.3
Squaw Peak Plain	96	1.3	1.0
Wingfield Plain	30	0.4	0.3
Phoenix Plain	1	0.0	0.0
Hohokam Red Ware	**1759**		**17.9**
Salt Red	1716	97.6	17.4
Gila Red	19	1.1	0.2
Squaw Peak Red	16	0.9	0.2
Wingfield Red	6	0.3	0.1
Indeterminate Hohokam Red	2	0.1	0.0
Hohokam Buff Ware	**44**		**0.4**
Indeterminate Buffware	33	75.0	0.3
Indeterminate Red-on-buff	7	15.9	0.1
Casa Grande Red-on-buff	3	6.8	0.0
Salt Red-on-buff, Indeterminate Style	1	2.3	0.0
Roosevelt Red Ware	**266**		**2.7**
Gila Polychrome	165	62.0	1.7
Indeterminate Polychrome	61	22.9	0.6
Tonto Polychrome	27	10.2	0.3
Phoenix Red	8	3.0	0.1
Pinto Polychrome	3	1.1	0.0
Salado Red	2	0.8	0.0
Intrusive Wares	**142**		**1.4**
Indeterminate Tucson Plain	97	68.3	1.0
Lower Colorado Buff Ware	35	24.6	0.4
Jeddito Black-on-yellow	4	2.8	0.0
Indeterminate Prescott Gray	2	1.4	0.0
Tanque Verde Red-on-brown	1	0.7	0.0
Prescott Black-on-gray	1	0.7	0.0
Mimbres Corrugated	1	0.7	0.0
Linden Corrugated	1	0.7	0.0
Unidentifiable	**39**		**0.4**
Total	**9843**		**100.0**

Intrusive Wares/Types

Intrusive sherds in Area 8 included Tucson Plain, Tanque Verde Red-on-brown, Lower Colorado Buff Ware, Prescott Black-on-gray, Indeterminate Prescott Gray Ware, Mimbres Corrugated, Linden Corrugated, and Jeddito Black-on-yellow (Table 9.1). These ceramics indicated that the inhabitants of Area 8 had participated, to some degree, in an extensive regional exchange system. Types equated with the Sinagua and greater Anasazi culture traditions were not present in the assemblage, although a few examples from the Tucson Basin, the Hopi mesas, and the Mimbres, Prescott, Mogollon, and Patayan culture areas were present.

Distributions

Aside from the ceramics recovered from nonfeature contexts (55.4%), the two structures produced the largest number of sherds in Area 8 (2646 sherds, or 26.9% of the assemblage). Fire pits (831 sherds, or 8.4%) and pits NFS (730 sherds, or 7.4%) produced the next largest assemblages. The high frequencies of sherds from these proveniences suggested that these locations had been areas of relatively higher activity, possibly including daily subsistence, food processing, and the discarding of unusable artifacts. The frequencies of sherds from the pit structures may have represented trash deposition, which could have occurred after abandonment. Because Area 8 was truncated by historic activities, it is possible that other structures originally were present but had not been preserved. Surface adobe structures would have been susceptible to historic disturbance and could have been entirely removed by plowing.

Pits NFS could not be assigned specific functions based on their morphology, preservation, or contents. Ten pits were placed in this category, with a mean distribution of 73 sherds per feature. The six fire pits averaged 138 sherds each. The size ranges of pits NFS and fire pits were similar, although two overly large features contained high proportions of the assemblages: Feature 8-1, a pit, measured 3.10 × 0.55 m in plan and 40 cm in depth, and Feature 8-10, a fire pit, measured 1.36 × 1.10 m in plan and 20 cm in depth. These features contained higher frequencies of all material categories except ground stone. The pit structures also could have accumulated high frequencies of ceramics and other artifact categories due to their large size, whether deposition was intentional or natural.

The two structures (Feature 8-2 and Feature 8-3) contained 28.7% (n=76) of the Roosevelt Red Ware sherds from Area 8, which constituted 2.7% of the total ceramic assemblage from structures at Dutch Canal Ruin. Also from the structures, Hohokam Decorated Ware, primarily consisting of buffware or buff-slipped types, accounted for 18.2% (n=8) of these types from Area 8 and 0.3% of the total ceramic assemblage from structures. Of the Hohokam Red Ware, which was dominated by Salt Red (97.6% of the ware), 23.1% (n=406) came from the structures, accounting for 14.6% of all ceramics from structures. This distribution did not readily allow a temporal assignment for the structures other than to the Classic period. The structures could have been occupied during the Civano phase and could have filled after abandonment with trash representing a wide temporal range. The presence of the Roosevelt Red Ware sherds suggested a late affiliation for these structures, a premise supported by the occurrence of Jeddito Black-on-yellow, a Hopi ware that does not commonly appear in the archaeological record before A.D. 1300 (Breternitz 1966:78). Investigators had little doubt that Area 8 had been occupied during the Polvorón phase, but its origins may have been during the Civano phase or even the late Soho phase, given the presence of the Squaw Peak and Wingfield varieties (Brunson 1989).

Flaked Stone

The flaked stone assemblage consisted of 3183 items (Table 9.2). The majority of items consisted of debitage (2939, or 92.3%), with basalt (1728, or 54.3%) and igneous other materials (872, or 27.4%) as the predominant material types. Researchers have documented a trend toward the increased utilization of obsidian by the Hohokam during the Classic and post-Classic periods in the Salt-Gila Basin (Doyel 1974:159; Sires 1984). Area 8 fit this pattern (obsidian is 5.5% of the flaked stone assemblage), although the frequency of obsidian here was lower than at El Polvorón (approximately 30%) as reported by Sires (1984:277).

In the flaked stone assemblage from Area 8, used flakes accounted for 56% of the items showing use. Hammerstones and core/hammerstones accounted for the second-highest frequency, 32%, followed by flake unifaces that made up 13% of all items exhibiting use wear. Scraping was the most common form of use, accounting for 68 (38.8%) of the items, which included those items that exhibited more than one type of use wear. Battering was present on 37.0% of the items, due to the number (n=36) of hammerstones and core/hammerstones in the assemblage. Cut/saw use wear appeared on 37 items (21.2%). The Area 8 flaked stone assemblage exhibited a greater diversity of use wear than did the assemblages from other areas of Dutch Canal Ruin, and nearly all tool types at Area 8 had some use wear, indicating that the occupants of the area had conducted a variety of subsistence activities there.

The high frequencies of tertiary flakes and other forms exhibiting no cortex or cortex on the platform only (63.7% of all debitage) indicated that inhabitants had produced flakes and tools. Core reduction was not extensive, as indicated by few "exhausted" cores. The high frequencies of tertiary flakes and items with minimal cortex present, however, suggested that tool manufacture had been common, although the assemblage did not contain many tools (Table 9.2), perhaps because tool production had taken place at the site, with tools used and discarded off-site. The amount of flake production at Area 8 can be measured from the ratio of tools and whole flakes to cores (13.6:1), indicating extensive flake production. The low number of exhausted cores indicated that flakes either had been produced elsewhere and brought to this locus or, more likely, had been produced by the reduction of other flakes.

Few items exhibited evidence of extensive ventral facial thinning or marginal retouch. When present, facial reduction was minimal. Two items exhibited high levels of ventral face modification, and 16 items (0.5%) exhibited some modification.

Tool completeness indicated a strategy of expediency. Investigators determined that 63 items (74.1%) were complete, which suggested that flaked stone items had been produced, used, and discarded at the site. Curation of tools did not seem to have been a common practice. A high proportion of items could have been broken after their discard, a proposition that at present cannot be evaluated. Furthermore, evidence for a strategy of expediency was supported by the high frequency of used flakes versus retouched tools, as well as the low levels of facial reduction.

Of the flaked stone assemblage, 1742 items (54.7%) came from nonfeature contexts, while structures accounted for 767 (24.1%), fire pits accounted for 407 (12.8%), and pits NFS accounted for 191 (6.0%) of the flaked stone items. All other features accounted for 2.4% of the assemblage. These data may appear to suggest that fire pits were one of the areas of increased flaked stone use, but this is somewhat misleading. Feature 8-10 contained 309 (75.9%) of the 407 items from fire pits, and three other fire pits produced between 22 and 49 items each. Although the relative frequencies of whole flakes and used flakes from Feature 8-10 were similar to those from the structures, a smaller proportion of the cores (3.1%) came from Feature 8-10. The structures

Table 9.2. Area 8, Flaked Stone Morphological Type by Material Type

Morphological Type	Material Type									Total
	Quartz	Chert	Obsidian	Sedimentary	Basalt	Igneous Other	Quartzite	Metasediment	Metamorphic Other	
Angular Debris	14	4	3	4	45	28	8	1	3	110
	*12.7	3.6	2.7	3.6	40.9	25.5	7.3	0.9	2.7	**3.5
	**20.0	4.7	1.7	21.1	2.6	3.2	7.1	4.0	3.1	
Flake Fragment	21	29	38	6	640	264	35	10	31	1074
	2.0	2.7	3.5	0.6	59.6	24.6	3.3	0.9	2.9	33.7
	30.0	33.7	21.8	31.6	37.0	30.3	31.3	40.0	32.0	
PRB[1] Flake	9	15	23		369	183	20	3	24	646
	1.4	2.3	3.6		57.1	28.3	3.1	0.5	3.7	20.3
	12.9	17.4	13.2		21.4	21.0	17.9	12.0	24.7	
Whole Flake	22	23	65	8	578	333	37	7	36	1109
	2.0	2.1	5.9	0.7	52.1	30.0	3.3	0.6	3.2	34.8
	31.4	26.7	37.4	42.1	33.4	38.2	33.0	28.0	37.1	
Used Flake	1	8	37		22	26	4	2		100
	1.0	8.0	37.0		22.0	26.0	4.0	2.0		3.1
	1.4	9.3	21.3		1.3	3.0	3.6	8.0		
Core	2	5	3	1	30	23	1			65
	3.1	7.7	4.6	1.5	46.2	35.4	1.5			2.0
	2.9	5.8	1.7	5.3	1.7	2.6	0.9			
Core Tool			1		1	3				5
			20.0		20.0	60.0				0.2
			0.6		0.1	0.3				
Core/ Hammerstone					16	2	2		1	21
					76.2	9.5	9.5		4.8	0.7
					0.9	0.2	1.8		1.0	
Hammerstone					9	4			2	15
					60.0	26.7			13.3	0.5
					0.5	0.5			2.1	
Cobble Uniface					1		1			2
					50.0		50.0			0.1
					0.1		0.9			
Flake Uniface	1				8	2	2			13
	7.7				61.5	15.4	15.4			0.4
	1.4				0.4	0.2	1.8			
Other Uniface								1		1
								100.0		tr
								4.0		
Cobble Biface					1					1
					100.0					tr
					0.1					

Table 9.2. Area 8, Flaked Stone Morphological Type by Material Type, continued

Morphological Type	Material Type									Total
	Quartz	Chert	Obsidian	Sedimentary	Basalt	Igneous Other	Quartzite	Metasediment	Metamorphic Other	
Flake Biface		1	1		3	1				6
		16.7	16.7		50.0	16.7				0.2
		1.2	0.6		0.2	0.1				
Other Biface		1								1
		100.0								tr
		1.2								
Notch			1							1
			100.0							tr
			0.6							
Projectile Point			2							2
			100.0							0.1
			1.1							
Large Primary Flake					4	2	2			8
					50.0	25.0	25.0			0.3
					0.2	0.2	1.8			
Indeterminate					1	1		1		3
					33.3	33.3		33.3		0.1
					0.1	0.1		4.0		
Total	70	86	174	19	1728	872	112	25	97	3183
	*2.2	2.7	5.5	0.6	54.3	27.4	3.5	0.8	3.0	100.0

*Row %
**Column %
[1]Platform-remnant-bearing

produced 30.7% of the whole flakes, 21.0% of the used flakes, and 16.9% of all cores. The frequencies of whole flakes and used flakes in the structures appeared to be high, which may be partially explained by the apparent trash accumulation in that portion of Area 8. In the distribution of obsidian, 46.0% was present in proveniences other than features. Feature 8-12, the pit house, accounted for 16.7%, and Feature 8-10 accounted for 17.2%.

The flaked stone assemblage from Area 8 contrasted markedly with those from other areas investigated at Dutch Canal Ruin, primarily in the frequencies. This difference could be attributed directly to the increased use of Area 8 over the seasonally or temporarily occupied field house loci. Expediency in flaked stone production and use seemed to be the strategy employed, with the focus on flakes. Area 8 was located within 400 m of one of the Salt River channels, where most raw materials could have been procured easily. Obsidian, however, had probably been acquired through exchange, and its lower frequencies at Dutch Canal Ruin compared to other Polvorón phase sites could be accounted for by the smaller size of Area 8 and by less interaction with the exchange network.

Ground Stone

Field crews recovered 186 ground stone artifacts from Area 8. Most were indicators of food-processing activity, including active and passive grinding implements and tabular tools (Table 9.3). Archaeologists believe the latter were used in processing agave and other economic resources. The majority of the ground stone items were produced from igneous rock, although 4 items were sandstone and 1 item was made from a conglomerate material. Phyllite and schist accounted for 155 (83.3%) items of the ground stone assemblage. Among these materials were 2 tabular tools and 17 items classified as ground stone NFS.

The frequency of schist and phyllite raw materials (140 items, or 75.3%) in the ground stone assemblage was high. These tabular or planar materials can be used for a number of purposes, including tabular knives, hoes, digging tools, palettes, ornaments, and temper for ceramics. Analysts assigned two raw material types to sources: Squaw Peak schist and Squaw Peak (?) phyllite. Other sources may have been represented but could not be identified. The presence of the Squaw Peak materials indicated that residents had traveled as far as Squaw Peak (10 km) to acquire them or had received them through some form of exchange. Aside from the tabular tools and perhaps as temper in ceramics, investigators recognized no other uses for these raw materials in Area 8.

Active grinding implements consisted of 2 rectangular manos made of vesicular basalt (1.1% of the assemblage), 1 handstone made of granite (0.5%), 7 grinders NFS (of these, 3 were made of sandstone and 1 each of scoria, conglomerate, metamorphic, and metasediment, making up 3.8% of the assemblage), and 1 polishing stone made of metasediment (0.5%). Passive grinding implements included only 1 trough metate (0.5%) and 1 metate NFS (0.5%), both made of quartz-bearing basalt.

Of the 39 items evaluated for completeness, only 4 were complete, including 1 handstone and the polishing stone. Another 23 items were classified as fragmentary because not enough of the item was present to make a type determination. Of these, however, most were probably < 1/4 complete. The fragmentary condition of the ground stone assemblage suggested that items had been discarded only after they were broken or considered too small to be used. When fragmented to the extent that they could not be reworked or modified into another form, they were discarded. Another factor that may have affected the condition of the ground stone assemblage was the stratigraphic position of the resources. The cultural deposits were truncated by historic plowing, which may have contributed to the fragmentary condition of the assemblage.

Field crews recovered 18 items from features, the majority (n=12) being ground stone NFS. They also recovered 2 grinders NFS from Feature 8-2, the pit house, and found 1 of the tabular tools in Feature 8-3, the pit structure. The schist and phyllite raw materials came from nine of the features, with 65 pieces from nonfeature contexts. The paucity of ground stone items from features, especially structures, suggested that the inhabitants had used ground stone tools in areas removed from features or in areas unassociated with features. Fragmentary items from feature contexts were probably a result of discard or, in the case of fire pits, had been used in association with the function of these features.

The ground stone assemblage neither provided much insight into the use of ground stone at Area 8 nor gave many clues to the types of activities conducted there. The presence of rectangular manos and trough metates is often equated with the grinding of cultigens such as corn. Tabular tools can be used for both cutting and sawing and may have been used in conjunction with agave procurement and processing or the procurement of other resources such as cactus and wood. Tabular tools also can be used for digging or as hoes, although the two specimens in this collection did not exhibit use-wear patterns consistent with these activities.

Table 9.3. Area 8, Ground Stone Morphological Type by Feature Type

Morphological Type	Feature Type						Total
	Pit House	Pit Structure	Pit NFS	Trash Pit	Fire Pit	None/Not Applicable	
Polishing Stone						1 *100.0 **1.1	1 **0.5
Handstone			1 100.0 8.3				1 0.5
Grinder NFS	2 28.6 7.7					5 71.4 5.5	7 3.8
Rectangular/Loaf Mano						2 100.0 2.2	2 1.1
Trough Metate						1 100.0 1.1	1 0.5
Metate NFS					1 100.0 4.2		1 0.5
Tabular Tool		1 25.0 3.1				3 75.0 3.3	4 2.2
Raw Material	22 15.5 84.6	27 19.0 84.4	8 5.6 66.7	1 0.7 100.0	19 13.4 79.2	65 45.8 71.4	142 76.3
Ground Stone NFS	2 8.3 7.7	4 16.7 12.5	2 8.3 16.7		4 16.7 16.7	12 50.0 13.2	24 12.9
Manuport						2 100.0 2.2	2 1.1
Other Ground Stone			1 100.0 8.3				1 0.5
Total	26 *14.0	32 17.2	12 6.5	1 0.5	24 12.9	91 48.9	186 100.0

*Row %
**Column %

Shell

The Area 8 shell assemblage consisted of 80 items. Within the assemblage, the presence of manufacturing debris, unworked fragments, and worked *Anodonta* items indicated that shell artifacts had been made on location. *Laevicardium* was the most common genus in the assemblage, represented by 34 items (42.5%) (Table 9.4). Of these, 33 were either unworked fragments or manufacturing debris, which indicated that shell artifact production had occurred at Area 8 but the produced items had not been retained at the site. *Anodonta*, a freshwater taxon, was represented by 12 fragmentary specimens. Its occurrence at the site may have been related more to subsistence than to shell artifact production, based on the presence of only one item that showed evidence of modification. After *Anodonta* is removed from its environment, the shell becomes brittle and nearly impossible to work, so the presence of *Anodonta* at Area 8 in both raw and modified forms suggested local acquisition and shell artifact production. *Conus* was also relatively abundant, represented by 11 specimens, with the greatest variability in artifact form among the various types present. Its natural shape lent itself to the production of shell beads, rings, and tinklers.

Area 8 also contained a variety of other shell types, including *Olivella, Nassarius, Argopecten, Chione*, and *Sphaerium*. *Nassarius, Argopecten, Chione*, and *Sphaerium* were each represented by one item and are grouped under "Other" in Table 9.4. Each of these specimens was a whole shell or worked item. The *Argopecten* specimen was a whole shell pendant. Other forms in the assemblage consisted of disc beads made from unidentified marine shell, bracelets made from *Glycymeris*, shell beads made from *Olivella* and *Conus*, and rings and tinklers made from *Conus*.

As has been a common distribution pattern for artifacts from Area 8 features, the structures (Features 8-2 and 8-3) and the large fire pit (Feature 8-10) produced the highest frequencies of shell items after nonfeature contexts. Field crews found 20 (25.0%) of the shell items in the two structures and 9 (11.3%) in the fire pit. Of the remaining shell items, no more than 3 (3.8%) were present in any one feature. Two items (2.5%)—a worked disc bead and a worked fragment of unknown form—were from Feature 8-13, one of the cremations. Fire pits and pits NFS produced the remainder of the shell from features.

The shell assemblage from Area 8 (Table 9.4) compared with that at El Polvorón (Sires 1984) in exhibiting a low percentage of *Glycymeris* and in lacking the elaborateness seen in earlier Hohokam shell work (Volume 4:Chapter 13). Through time, the general trend in Hohokam shell work was away from life forms and toward geometric design. The Classic period shell artifacts from Las Colinas were more stylized than shell from earlier contexts (Vokes 1988), and this trend also was evident in materials from the Sacaton and Soho phases of the Salt-Gila Aqueduct Project (Vokes 1984).

Worked Bone

Of the 121 bone items from Area 8, analysts classified 5 as bone tools, and another 5 items exhibited cut marks and burning. The 5 tools, all classified as awls, were from Feature 8-15, a cremation. The awls from Feature 8-15 probably represented grave goods. All were fragmentary, probably the result of intensive burning. They could be identified only as having been produced from the bones of large mammals.

Table 9.4. Area 8, Provenience of Shell Artifact Type by Shell Type

Provenience/ Artifact Type	Shell Type					Indeterminate Marine Shell	Other	Total
	Laevicardium	*Conus*	*Olivella*	*Glycymeris*	*Anodonta*			
General Site								*35
Unworked Fragment	12				4	1		17
Artifact in Process		1						1
Whole/Reconstructible Valve			1				1	2
Disc Bead						3		3
Plain Bracelet				3				3
Whole Shell Bead			2					2
Manufacturing Debris	3	1						4
Cylindrical Bead		2						2
Decorated Ring		1						1
Feature 8-1								*2
Unworked Fragment	2							2
Feature 8-2								*8
Unworked Fragment	1							1
Manufacturing Debris						1		1
Plain Ring		2						2
Worked Fragment, Unknown Form					1			1
Whole Shell Bead			1					1
Tinkler		1						1
Plain Bracelet				1				1
Feature 8-3								*12
Cylindrical Bead		1						1
Plain Bracelet				3				3
Whole Shell Pendant							1	1
Unworked Fragment	7							7
Feature 8-9								*3
Unworked Fragment	2				1			3
Feature 8-10								*9
Unworked Fragment	4				3			7
Cylindrical Bead		2						2
Feature 8-12								*2
Unworked Fragment					1			1
Plain Bracelet				1				1

Table 9.4. Area 8, Provenience of Shell Artifact Type by Shell Type, continued

Provenience/ Artifact Type	Shell Type					Indeterminate Marine Shell	Other	Total
	Laevicardium	*Conus*	*Olivella*	*Glycymeris*	*Anodonta*			
Feature 8-13								*2
Disc Bead						1		1
Worked Fragment, Unknown Form							1	1
Feature 8-16								*3
Whole/Reconstruc- tible Valve							1	1
Unworked Fragment	1				1			2
Feature 8-21								*2
Unworked Fragment					1			1
Worked Fragment, Unknown Form	1							1
Feature 8-22								*2
Whole Shell Bead			1					1
Unworked Fragment	1							1
Total	34	11	5	8	12	6	4	80

*Provenience total

Subsistence Remains

Macrobotanical Remains

Field crews took macrobotanical samples from Features 8-2, 8-12, and 8-14 and Subfeatures 8-2-1 and 8-3-1. The analysis results for Feature 8-2 and Subfeature 8-2-1 indicated that plant processing had occurred within the structure, further supporting the interpretation of the pollen analysis results. Although the fire pit (Subfeature 8-2-1) contained very little pollen, macrobotanical remains, mostly charred mesquite wood and agave fibers, were numerous. The macrobotanical analysis also identified corn. Two extramural fire pits (Features 8-12 and 8-14) contained a large archaeobotanical assemblage. In addition to being filled with trash, these features apparently contained secondary food-processing refuse (Chapter 15). The results of the flotation and macrobotanical analysis for Area 8 are presented in Table 9.5, and a detailed discussion of the macrobotanical analyses for Dutch Canal Ruin is presented in Chapter 15.

Table 9.5. Area 8, Macrobotanical Results

Feature Number	Feature Description	Context	Taxa Identified (common names)
8-2	pit house	floor fill	paloverde, mesquite, agave, Cheno-am, sprangletop type
8-2-1	fire pit	feature fill	mesquite, agave, Cheno-am, spurge, grass family, little barley grass
8-3-1	fire pit	feature fill	mesquite, agave, Cheno-am, cotton, canary grass, seepweed, corn
8-12	fire pit	feature fill	mesquite, agave, grama grass type, brome grass/eel grass type, Cheno-am, tansy mustard, grass family, pea family, sprangletop type, wild tobacco, canary grass, dropseed type, seepweed, corn
8-14	fire pit	feature fill	agave, Cheno-am, tansy mustard, sprangletop type, canary grass, seepweed, corn

Several trends were evident from the archaeobotanical remains. Agave fibers were the predominant plant remains in Area 8. Agave cultivation has been documented in the lower Salt River Valley beginning by the mid Colonial period (Fish et al. 1985; Chapter 15). Both agave and canary grass may have been cultivated by the inhabitants of Area 8. Cheno-am seeds, mesquite, corn, and seepweed appeared to have been important subsistence items, given their frequency in the area. Overall, a more diverse range of floral resources, both agricultural and wild plants, had been exploited by the Classic period occupants of Area 8 than by the pre-Classic period occupants of other parts of Dutch Canal Ruin.

Pollen Evidence

Field crews collected pollen samples during excavation to obtain evidence of economic activity necessary to understanding prehistoric subsistence strategies and how they related to function and land use in Area 8. They selected pollen samples from the two pit structures (Features 8-2 and 8-3) and from two fire pits (Features 8-12 and 8-14); Cummings (Chapter 16) analyzed the samples. A large quantity of Cheno-am pollen was present in Feature 8-2, along with small amounts of saguaro/fishhook cactus (*Carnegiea/Mammillaria*) and cattail (*Typha*) pollen. A small amount of Cheno-am pollen came from the hearth in Feature 8-2. Saguaro was an important subsistence resource for the Pima and Papago (Castetter and Bell 1942:63) and also was used for shelter. The use of cattail as a food source also has been reported ethnographically (Chapter 16). Two fire pits (Features 8-12 and 8-14) yielded Cheno-am pollen and cholla (*Cylindropuntia*) pollen, suggesting that Cheno-ams and cholla buds had been processed there.

Faunal Remains

The 121 faunal remains from Area 8 were present in both screened and unscreened contexts (Chapter 17 and Appendix D). Field crews found 86 items in screened contexts, which consisted of the fill from within features. In comparing the taxa represented in the two contexts, bone recovered from both screened and unscreened contexts could almost always be associated with prehistoric activities at the site.

Stratton (Chapter 17) observed that the large mammal assemblage from Area 8 amounted to 28% of the total assemblage. This percentage was in marked contrast to the pre-Classic period components from Dutch Canal Ruin, in which no large mammal faunal remains were present, suggesting that large mammals had been acquired by the post-Classic period residents of Area 8 and used variously for food, tools, and possibly shelter. This observation was significant in demonstrating the presence of large mammals in the faunal assemblage at permanently occupied habitation loci as opposed to habitation loci that were seasonally or temporarily occupied. The results were similar for Area 10 (Appendix D), a Classic period locus. From these data, it appeared that residents of Area 8 had exploited a much larger faunal resource base than inhabitants of the adjacent pre-Classic period settlements, employing a subsistence base strategy apparently related to permanence of occupation.

Human Remains

Investigators recovered two cremations from Area 8 at Dutch Canal Ruin. Both were in the northern portion of the area, between two alignments of the historic Dutch Ditch (Figure 9.1). Kathryn Wullstein, who conducted the documentation of all human remains from the Phoenix Sky Harbor Center, provided the following descriptions. Analysts classified both burials as secondary cremations in which funerary vessels were used. The placement of these two burials and their proximity to each other suggested that occupants of the area had specifically used this location as a cemetery.

Feature 8-13

Feature 8-13, a secondary cremation, was in a Gila Plain, Salt Variety jar. The bone was highly burned and fragmentary, but some components were identifiable, including a right scapula, a right parietal, a left navicular, temporal bone fragments from both sides, a femur head, a left occipital, and the fragments from six vertebra bodies. In some instances analysts could not determine the side of the body from which the bone had come. No teeth were present. The fragmentation and limited amount of material also made it impossible to determine the sex of the individual. The remains indicated that the individual was an adult, but no further age determination could be made.

Feature 8-15

This secondary cremation, Feature 8-15, was also in a Gila Plain, Salt Variety jar. Portions of both femurs, the right scapula, four ribs, the ischium, the ilium, two thoracic vertebrae, a humeral distal end, the left scapula, and a manubrium were recovered. A portion of the right mandible was identifiable, and the second premolar was present. Alveolar resorption had occurred for the first and second molars and for the incisors. The third molar and first premolar were absent but, given the appearance of the teeth sockets, probably had been present at the time of death. The sex of the individual could not be determined. The remains indicated that the individual was an adult, but no further age determination could be made.

CHRONOLOGY

Architectural morphology and material culture attributes dating to the Polvorón phase (A.D. 1350–1450) (Sires 1984) indicated that Area 8 had been occupied late in the Hohokam sequence. Radiocarbon dates and the stratigraphic position of the Area 8 cultural deposits also supported a late date of occupation.

Stratigraphy

The cultural remains in Area 8 were situated high in the project area stratigraphic units, at a point above most other occupation areas of Dutch Canal Ruin. Cultural features in Area 8 were truncated by the plow zone, whereas plow zone activity in the other areas of the site usually did not extend deeply enough to adversely affect the prehistoric remains. The high stratigraphic position of the Area 8 remains argued for a more recent date of occupation than those assigned to components in other areas of the site.

Although the stratigraphic position of the remains might have been accounted for by an anomalous high topographic feature at this location, the natural stratigraphy in and around Area 8 indicated otherwise. The cultural remains that constituted Area 8 lay in the stratum defined as the Holocene soil zone (Volume 1:Chapter 3), and the stratigraphic position of Area 8 was consistently higher than that of the pre-Classic period components at Dutch Canal Ruin but similar to that of other Classic period loci. Both the cultural features and the upper portions of this stratum were truncated by historic use of the project area, principally in the form of plowing, but also through residential use.

Area 8 was one of only two research areas to contain multiple superimposed features and cultural deposits (Chapter 11). Internal stratification of deposits occurred when Feature 8-2 (a pit house) was constructed over Feature 8-23 (a pit NFS); when Feature 8-3 (a pit structure) was intruded on by Feature 8-1 (a pit NFS), which in turn was intruded on by Features 8-4 and 8-5 (pits NFS); and when Feature 8-10 (a large fire pit) was intruded on by Feature 8-9 (a pit NFS), by Feature 8-12 (a fire pit), and possibly by Feature 8-2 (the pit house). Given the stratigraphic relationships of these features, archaeologists can argue that the occupation at Area 8 was of longer duration than at most of the other investigated areas. Furthermore, the spatial arrangement of the two structures suggested that they were not absolutely contemporaneous. The longevity of occupation and the contemporaneity of the two structures is examined further in the following section.

Ceramics

Analysis of the ceramic assemblage from Area 8 indicated a strong association with the late Classic and post-Classic periods, especially supporting a Polvorón phase occupation. As mentioned above, Roosevelt Red Ware (Salado Polychromes) accounted for 86% of the decorated ceramics from the area. Salado Polychromes most frequently constitute 85–88% of the decorated ceramic assemblage in Polvorón phase components (Chapter 12; see also Peterson and Abbott 1991; Sires 1984) but have been recorded in frequencies as low as 66.9% (Abbott 1993). In four sherds of Jeddito Black-on-yellow, analysts found additional evidence that Area 8 dated to the Polvorón phase. Hopi yellowware ceramics often appear in association with Salado Polychrome types and occur in small numbers in Polvorón phase contexts (Sires 1984).

Hohokam Buff Ware types appeared in low frequencies in the assemblage (Table 9.1). The three buffware sherds that could be identified were Casa Grande Red-on-buff, a type generally equated with the Sedentary to Classic period transition, A.D. 1100–1150 (Dean 1991), which extended into the Soho phase. Earlier types could have been brought to the site well after they were no longer being produced, and curation of personal items could account for the presence of some Hohokam Buff Ware at a post-Classic period site. Nevertheless, the possibility of an earlier occupation cannot be ignored, although investigators found little evidence to support such a supposition.

Few distinctions were recognizable within the ceramic assemblages from individual features, including those that were superimposed. The frequencies of temporally sensitive types within stratified features were fairly evenly distributed. This pattern suggested that the features with temporally diagnostic types had been used within a narrow span of time, resulting in similar ceramic assemblages. Feature 8-5 contained no sherds, Feature 8-23 produced 1 plainware sherd, and Feature 8-4 contained 5 sherds. The other stratified features contained between 92 and 1626 sherds each. Although the types of ceramics present provided little help in identifying occupation events, the methods of disposal proved valuable. For example, Feature 8-3 (which was intruded on by Features 8-1, 8-4, and 8-5) had been abandoned long enough for a thin layer of clean sediments to cover the floor prior to an episode of trash deposition that included the accumulation of 1626 sherds. The other structure, Feature 8-2, was stratigraphically late and contained one of the four Jeddito Black-on-yellow sherds from Area 8. Feature 8-2 was filled, beginning at floor level, with cultural trash, suggesting that occupation had continued or a later occupation had immediately followed the abandonment of this structure. Although investigators inferred on the basis of fill sequence that a later occupation had occurred at Area 8 than was represented by excavated features, the presence of types such as late red-on-buff and the Wingfield varieties can be used to argue an earlier occupation. Investigators gained the general impression, however, that Area 8 had been occupied on a permanent basis for a short period of time. Enough time elapsed to allow at least two structures to be constructed, abandoned, and filled with trash. The investigating team contended that Area 8 had been occupied during the late Classic and post-Classic periods and that its features represented an unknown fraction of what had actually existed. Other Polvorón phase sites often contain surface adobe-walled structures without adobe footings (Greenwald and Ciolek-Torrello 1988b), and such features could have been destroyed by historic use of the area. The use of such adobe structures by residents of Area 8 could account for the postabandonment trash deposition within Features 8-2 and 8-3.

Absolute Dates

The preservation of the hearth in Feature 8-2 and the fire pit in Feature 8-3 was too poor to allow recovery of archaeomagnetic dating samples. However, analysts produced three sets of calibrated (Stuiver and Becker 1986) radiocarbon dates for Area 8. A sample from the fire pit in Feature 8-3 provided a calibrated age range of B.P. 670–545 (A.D. 1280–1405), one sigma; a calibrated range of B.P. 666–527 (A.D. 1284–1423), one sigma, was obtained from a sample taken from Feature 8-12, an extramural fire pit; and a calibrated range of B.P. 673–545 (A.D. 1277–1405), one sigma, was obtained from a sample from Feature 8-14, another fire pit (refer to Appendix B for the calibrated dates). The consistency of these date ranges suggested that the samples came from contemporaneous features.

The date ranges for the three samples fell into the Civano and Polvorón phases, and these samples could have originated during any interval associated with the late Classic period. Radiocarbon samples from Dutch Canal Ruin have consistently dated earlier than archaeomagnetic samples from the same loci and earlier than the ceramic assemblages suggest. One possible explanation for the disparity between radiocarbon dates and other dating methods from the geologic floodplain may be the introduction of organic matter through saturation, resulting from periods of flooding or historic and prehistoric irrigation. If carbon samples on the geologic floodplain were contaminated and the contamination was related to the duration that each sample was buried, then the date ranges for the samples from Area 8 should be earlier than their target date. Due to the disproportionate error, these late samples may need to be slightly adjusted to reflect a more accurate temporal association. Because the upper portions of these age ranges fell within the proposed temporal span for the Polvorón phase, they added some support to the placement of the Area 8 occupation in that time period.

Architectural Implications

One pit structure and one pit house, both true pit houses as defined by Howard (1988), constituted the architectural elements excavated in Area 8. The structures (Features 8-2 and 8-3) were very similar in morphology to those found at El Polvorón (Sires 1984:237–238) and at other sites with components assigned to the Polvorón phase (e.g., Greenwald and Ciolek-Torrello 1988b; Gregory 1988; Howard 1988). Sires described the pit houses at El Polvorón:

> All were rectangular with rounded corners. The average interior area was just over 15 square meters. Pit walls were generally 3 cm to 5 cm thick, constructed of a hard-packed material that is referred to here as adobe. However, even after prolonged exposure to the sun, it never achieved the hardness of true adobe, still crumbling easily. House pits were shallow, the average remaining height of walls being 29 cm. Pit walls were vertical or slanted outward slightly, curving at the base. There was no break between the pit walls and floors, which were constructed of the same adobe material. The floors were somewhat thinner than the walls, usually about 3 cm thick. The basic construction technique for house pits was to excavate a shallow depression and then coat the inside with adobe [Sires 1984:237].

Although poorly preserved, Feature 8-3 was subrectangular and had a ramped entry. Feature 8-2 was somewhat more D-shaped than rectangular but had thin adobe walls like the El Polvorón phase structures. In Features 8-2 and 8-3, the sides of the pit formed the lower walls. In summary, the structures in Area 8 compared with other Polvorón phase structures, and this observation was supported by their stratigraphic positions, ceramic associations, and absolute dates.

DISCUSSION

The two structures in Area 8 were situated close together, being separated by less than 1.0 m, with one structure facing east and the other facing west. In discussing possible Polvorón phase structures at Las Colinas, Gregory (1988:41) noted that the structures appeared to occur in pairs. Though this may have been the case with the Area 8 structures, their close proximity, the large amount of refuse in the fill of Feature 8-3, and the large pit intruding through Feature 8-3 suggested that the structures had not been occupied at the same time. Instead, the occupants of Area 8 had apparently occupied and abandoned Feature 8-3 first. Alternatively, since the structures were closer together than the usual Hohokam pattern, the inhabitants could have occupied the two structures contemporaneously but abandoned Feature 8-3 before Feature 8-2. The large number of artifacts in the fill of Feature 8-3 suggested that site inhabitants had used that structure as a trash receptacle after abandonment. Because Feature 8-2 was the only structure nearby and appeared to be the latest structure at Dutch Canal Ruin, the refuse and the intrusive pit (Feature 8-1) in Feature 8-3 had probably originated with the occupants of Feature 8-2. A third, speculative, alternative is the possibility that surface adobe structures existed in Area 8 but were completely removed by plowing.

Area 8 was the most complex of any of the loci investigated at Dutch Canal Ruin. At least one nuclear family and possibly an extended family occupied the area. Both of the structures were habitation use areas associated with domestic activities. Structures, although used variably throughout the year, served as focal areas for social activities at the family level. Use of structures, however, would have required that the occupants use an area larger than simply the interior of the structure. Many activities associated with structure use were probably initiated outside and transferred inside as necessary. Food processing and cooking were probably performed outdoors, based on the presence of extramural thermal features; food may have been consumed both indoors and outdoors. The structure also could have met sleeping and basic shelter requirements. Storage, especially of items of high value, probably occurred inside.

Investigators identified several additional use areas in association with the structures in Area 8. A cluster of fire pits (including Features 8-10, 8-12, and 8-14) near the southwest corner of Feature 8-2 suggested that food had been roasted in that area. Evidence from pollen analysis supported this interpretation. Analysts found pollen from Cheno-ams and *Cylindropuntia* (cholla) in samples from both Feature 8-12 and Feature 8-14, indicating that the occupants of the area had processed those plants in the pits. Both features also contained macrobotanical remains of numerous other plant types (Table 9.5 and Chapter 15), including agave and Cheno-ams. Trash from the occupation of Area 8 was discarded in a pit (Feature 8-11) approximately 15 m south of Features 8-12 and 8-14.

Another use area in the west-central portion of the excavated area contained Features 8-17, 8-18, 8-19, and 8-20. As described above, Feature 8-17 was a posthole, and Feature 8-20 was a thin adobe wall fragment. Those features appeared to have formed a windbreak or perhaps were part of a larger feature such as a ramada or courtyard, the evidence for which had been obscured or removed by modern disturbance processes and by Trench 1321. The presence of the fire pits suggested that this area may also have been used for food preparation.

A cremation area was located in the northwest portion of Area 8. Field crews found two cremations (Features 8-13 and 8-15) just south of a small ditch of historic age, which may have been associated with the historic Dutch Ditch. Investigators could not determine whether the cemetery extended north of those historic ditches. They found no evidence of primary cremation pits or crematoria in the area. Remains of a funeral pyre or primary cremation pit are generally distinctive features at sites. The apparent absence of such a feature at Area 8 may indicate that both individuals were cremated at another site and only portions of the remains were interred in Area 8.

Area 8 of Dutch Canal Ruin had elements of two of the settlement types described by Gregory (1991) for the Phoenix Basin Hohokam: farmsteads and hamlets. Farmsteads are defined as

> settlements established primarily for the purpose of agricultural and related subsistence pursuits and . . . most often occupied by a single social group. This settlement type is conceived of as having been occupied seasonally, perhaps for only a few seasons, or as the repeated seasonal locus of residence for small social groups with continuing ties to one or more larger settlements [Gregory 1991:163].

Archaeologists seem to have some confusion about the term, however, because Wilcox (1978:26) defined farmsteads as year-round habitations and hamlets as seasonally occupied sites. This study followed Gregory's definition. The inhabitants of Area 8 probably formed a single social unit, such as a nuclear or small extended family. Agriculture and related subsistence pursuits were almost certainly primary criteria for the selection of the area for habitation. Evidence in the form of large amounts of refuse, cremation burials, and macrobotanical and pollen remains indicative of seasonal variation suggested that the area had been occupied year-round rather than seasonally. In this respect, Area 8 also had characteristics of hamlets, defined as small population aggregates (fewer than 100 people) that were occupied throughout the year (Gregory 1991:162). Therefore, investigators identified Area 8 of Dutch Canal Ruin as a hamlet occupied by a small social group as a permanent habitation in which a broad range of subsistence and social activities took place.

Characterizing the occupation in Area 8 as a hamlet was consistent with the evidence. Part of the definition of a farmstead implies the existence of larger occupation aggregates for which farmsteads were "functional extensions" (Gregory 1991:163). During the post-Classic period Polvorón phase, large population aggregates apparently did not exist (Crown and Sires 1984; Doyel 1991; Sires 1984), as populations became dispersed across the landscape in a fashion similar to Pioneer period settlements. Even at sites with as many as six to eight Polvorón phase structures, such as El Polvorón (Sires 1984), Brady Wash (Greenwald and Ciolek-Torrello 1988b), and Las Colinas (Gregory 1988), all structures were probably not occupied simultaneously. Small, dispersed hamlets were the dominant settlement type during the Polvorón phase.

SUMMARY

Area 8 of Dutch Canal Ruin was a small, single-component hamlet assigned to the Polvorón phase. Architectural morphology and material culture attributes and frequencies, rather than absolute dates, were the primary criteria for this temporal assignment. Radiocarbon dates did, however, support the assignment of Area 8 to the time period A.D. 1350–1450, as defined for the Polvorón phase by Sires (1984).

The architectural remains supported the notion of an occupation by a nuclear or extended family and compared with late pit architecture from other sites (Greenwald and Ciolek-Torrello 1988b; Gregory 1988; Sires 1984). The use of the structures did not appear to be absolutely contemporaneous.

Artifacts from Area 8 also indicated a Polvorón phase occupation. The strongest evidence appeared in the ceramic assemblage. Roosevelt Red Ware (Salado Polychromes) accounted for 85.8%

of the decorated ceramics. Researchers have found similar percentages of polychromes at other Polvorón phase components, such as Pueblo Grande (66.9%) and El Polvorón (88.6%). Frequencies of Roosevelt Red Ware at several late Classic and post-Classic period sites are presented by Motsinger (Chapter 12) and by Peterson and Abbott (1991).

Preserved pollen and macrobotanical remains from Area 8 suggested that it had been occupied year-round as a habitation site. Analyses of the floral remains indicated that the inhabitants of Area 8 had exploited a wide variety of wild plant food sources in addition to cultivated crops such as corn, agave, and cotton. Additional evidence for permanent occupation at Area 8 appeared in the fairly dense trash deposits associated with the features in this locus. Field house loci usually do not contain large amounts of discarded items or mixed, diverse artifact assemblages, and this pattern was supported by the lack of extensive trash deposits in the pre-Classic period occupations of Dutch Canal Ruin. The occupation in Area 8 resulted in the deposition of numerous decorated sherds in addition to plainware sherds. This diversity of ceramics, along with objects of shell, flaked stone, and ground stone, indicated that the occupants had performed a wide range of activities in the area.

In spite of their limited numbers, the inhabitants of Area 8 were part of a network of interaction that reached outside of the immediate area and beyond the Phoenix Basin. Intrusive ceramics in Area 8 included sherds from the Tucson Basin and the Hopi, Prescott, Mimbres, and Lower Colorado areas. Although investigators have not determined the mode of exchange, the intrusive ceramics indicated participation in an exchange network linking a broad area of the Southwest. Exchange was, of course, not limited to ceramics; marine shell and obsidian were both obtained from distant sources.

In conclusion, Area 8 contained the remains of a small habitation site dating to the final stage of the Hohokam occupation of the Phoenix Basin, the Polvorón phase. The inhabitants conducted a wide range of domestic activities, including farming, collection and processing of plant foods, processing of faunal materials, and nonsubsistence activities such as shell artifact production, obsidian procurement, and the cremation and interment of the dead. The cultural remains in Area 8 indicated the continued pursuit of livelihood at this site following the general collapse of Classic period Hohokam society.

186

CHAPTER 10

DUTCH CANAL RUIN, AREA 9

David H. Greenwald

Investigators discovered Area 9 during the monitoring phase of construction of the remote parking facility that lies immediately west of the north runway of Phoenix Sky Harbor International Airport (Figure 1.2). It was outside of the Dutch Canal Ruin boundaries as defined by Midvale (1934) and in an area for which BRW (1986) recorded no surface remains. Area 9 was in T1N, R3E, Sec. 10, NE¼SE¼SE¼ on the Phoenix, Arizona, 7.5 minute USGS topographic map (Figure 1.4), in the extreme southeastern portion of the site, and within the area that extended from N1280 to N1330 and from E1940 to E2050 (Figure 10.1) within the project grid. On Phoenix street maps, Area 9 lies between 23rd and 24th streets and between Buckeye Road and Sherman Street. (Sherman and 23rd streets no longer exist in the project area but can be found on various extant maps of the Phoenix area.)

FIELD METHODS AND STRATEGIES

Archaeologists investigated this portion of the project area by monitoring only, because construction of the remote parking facility would not cause more than shallow disturbance. Although investigators had previously reported no cultural remains in this location, surface blading and leveling operations exposed two features. Their location was designated Area 9. Two other forms of subsurface disturbance occurred in the remote parking lot. First, construction crews installed a storm drain that traversed the southern portion of the lot in a north-south direction. At the extreme southern boundary, this drain then ran west until it intersected the Squaw Peak Parkway. Second, crews installed a series of light poles with underground wiring to provide lighting for the parking lot. The trenches dug for the electric lines and the holes dug for pole installation all were examined for archaeological remains. Field crews observed no subsurface remains in any of the trenches or light pole holes within or adjacent to Area 9, although they did find remains farther north in the portion of the site designated Area 10 (Chapter 11).

When cultural resources were encountered, archaeologists directed construction activities elsewhere while they recorded and mapped the remains, using hand trenches to further delineate artifact concentrations. They drew feature plan maps to scale and prepared detailed records. Investigators recovered a sample of artifacts from Features 9-1 and 9-2 through general feature collection and the collection of 5 × 5-m units and hand trenches. They located all artifacts, collection units, and shovel trenches according to the project grid system. Feature 9-1, an irregular, basin-shaped pit in the profile of the storm drain, was drawn in profile and described in detail as part of the recording process. After construction crews removed the asphalt surface associated with an earlier parking lot, investigators mapped the surface extent of Feature 9-2, an extensive scatter of prehistoric artifacts. A profile drawing of the west profile of the storm drain illustrated how the feature was preserved below the zone of historic and modern disturbance.

Once investigators completed hand trenching without finding additional features, they allowed grading operations to continue. They directed the heavy equipment operators to remove thin cuts with each pass in an attempt to locate and define additional features, particularly structures. Monitoring continued until culturally sterile soils were exposed and all remnants of the artifact scatter were removed.

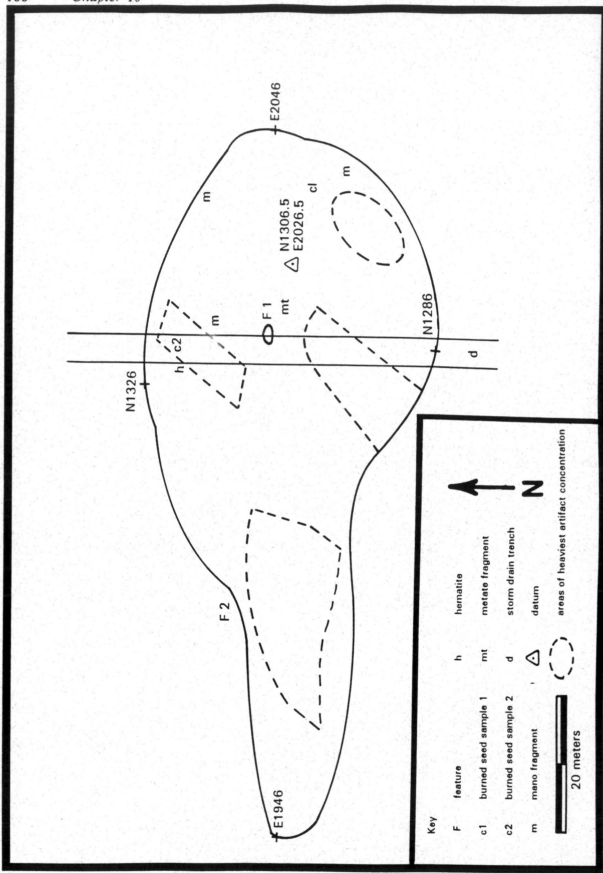

Figure 10.1. Area 9, Dutch Canal Ruin.

CULTURAL FEATURES

Feature: 9-1
Horizontal Provenience: N 1308.5–1309.8; E 2015.0–2016.5
Vertical Provenience: 2.35–2.97 meters below datum (mbd)
Feature Type: Pit, Not Further Specified (NFS)

This feature consisted of an irregular, basin-shaped pit, measuring 1.3 × 1.5 m, within a large construction trench. Its east half appeared in plan as a circular outline. The pit was filled with sandy silts with occasional inclusions of charcoal flecks, with a concentration of charcoal flecks in the basal portion of the pit. At this location, the pit extended approximately 32 cm below the base of Feature 9-2, a trash deposit. No artifacts were recovered from the pit, and other than the charcoal flecks, no evidence such as oxidation of pit walls was present to indicate that this pit had contained a fire.

Feature: 9-2
Horizontal Provenience: N 1286.0–1326.0; E 1946.0–2046.0
Vertical Provenience: 2.35–2.65 mbd
Feature Type: Trash Deposit

Feature 9-2 consisted of an irregularly shaped area containing varying densities of artifacts. The long axis of this trash feature ran east-west (100 m east-west by 40 m north-south), with the heaviest concentrations of materials contained within its southern half. The deposit was deepest in its northern portion, although this apparent increase in depth was gradual. This feature may have represented the remnants of a trash mound that originated in a slight depression. The trash consisted of a dense deposit of cultural materials, including ceramics, flaked and ground stone items, and macrobotanical remains. After the asphalt was removed from above the intact portion of the feature, investigators recognized that the trash deposit had been severely truncated by modern use of the area. Only about 10 cm of deposits remained in situ, although a horizontal variation of nearly 30 cm was recorded for the feature from north to south. The discovery of buried remnants of historic house foundations, utility lines, and septic features suggested that at least two house lots had covered Feature 9-2 prior to the leveling and paving of the area. Commercial buildings were present in association with the paved parking lot prior to development of the area as part of the Phoenix Sky Harbor Center. Field crews carefully monitored areas adjacent to the trash deposit for any evidence of prehistoric architecture but observed none. Earlier leveling of this area may have removed any surface structures or shallow pit structures.

MATERIAL CULTURE

The artifact collection from Area 9 amounted to a judgment sample. Nevertheless, investigators used systematic collection methods and attempted to augment the samples by collecting diagnostic ceramics and complete or nearly complete tools. Whenever botanical remains were observed, they were collected as well. Field crews often observed remains after the heavy equipment removed a level of cultural fill, exposing clean cuts to be examined for features and walls of structures.

Ceramics

The 589 sherds collected from Area 9 (Table 10.1) represented a wide variety of Hohokam types, with only two intrusive types present. Area 9 was the only area at Dutch Canal Ruin that contained sherds representative of all phases from the late Pioneer period (Snaketown phase) to the late Classic period (Civano phase, or perhaps as late as the Polvorón phase). It was the only area that supported an occupation during the Sedentary period, indicated by the presence of Sacaton Red-on-buff, Santa Cruz/Sacaton Red-on-buff, and Sacaton/Casa Grande Red-on-buff types, as well as an occupation during the early Classic period (Soho phase), indicated by the presence of Casa Grande Red-on-buff and Sacaton/Casa Grande Red-on-buff types.

Table 10.1. Area 9, Ceramic Frequencies

Ceramic Ware/Type	Count	Percent Ware	Percent Assemblage
Hohokam Plain Ware	**430**		**73.0**
Gila Plain, Salt Variety	171	39.8	29.0
Gila Plain, Gila Variety	41	9.5	7.0
Squaw Peak Plain	215	50.0	36.5
Wingfield Plain	3	0.7	0.5
Hohokam Red Ware	**70**		**11.9**
Salt Red	58	82.9	9.8
Gila Red	8	11.4	1.4
Squaw Peak Red	4	5.7	0.7
Hohokam Decorated Ware	**64**		**10.9**
Snaketown Red-on-buff	2	3.1	0.3
Gila Butte Red-on-buff	9	14.1	1.5
Santa Cruz Red-on-buff	9	14.1	1.5
Sacaton Red-on-buff	6	9.4	1.0
Casa Grande Red-on-buff	8	12.5	1.4
Santa Cruz/Sacaton Red-on-buff	4	6.3	0.7
Sacaton/Casa Grande Red-on-buff	2	3.1	0.3
Salt Red-on-buff, Indeterminate Style	1	1.6	0.2
Indeterminate Red-on-buff	5	7.8	0.9
Indeterminate Buff	18	28.1	3.1
Roosevelt Red Ware	**22**		**3.7**
Phoenix Red	17	77.3	2.9
Gila Polychrome	5	22.7	0.8
Intrusive Wares	**3**		**0.5**
Sells Plain	1	33.3	0.2
Lower Colorado Buff Ware	2	66.7	0.3
Total	**589**		**100.0**

In the decorated ceramic assemblage, 30 of the recovered sherds were associated with the pre-Classic period, and 13 were associated with the Classic period. Another 17 sherds, all Phoenix Red, have been included under the category of Roosevelt Red Ware because they appeared to be the undecorated portions of Salado Polychrome vessels. As a result, the assemblage was equally split between pre-Classic and Classic period types, suggesting that the Hohokam had continuously used Area 9 from the Snaketown phase to at least the Civano phase. Because of the lack of architectural features, investigators could not determine whether Area 9 had been occupied on a seasonal or a permanent basis. The numbers of pre-Classic period ceramic types suggested that the inhabitants may have used the area for longer periods than most other pre-Classic period loci at Dutch Canal Ruin, and it may have functioned as a farmstead.

Flaked Stone

Field crews recovered three flaked stone items from Area 9, two of them flake fragments and one a whole flake. These items were made from fine-grained basalt and igneous other material.

Ground Stone

The seven ground stone items from Area 9 included three loaf-shaped manos, one trough metate fragment, one complete basin metate, and one piece of hematite. Two of the three manos exhibited use wear on two sides, while the third mano had been used on only one side. Quartz-bearing basalt was the most common material represented, with only one item made from vesicular basalt. Analysts classified all ground stone items from Area 9 as coarse grained, except the hematite. Although small, the assemblage indicated that occupants of Area 9 had processed food, primarily using grinding tools, in this portion of the site. Hematite was often associated with pigment used for producing a red paint and could have been used in association with pottery decoration or body adornment. This particular specimen had 11 grinding facets and had been extensively reduced.

Subsistence Remains

The identified subsistence remains from Area 9 consisted solely of charred corn kernels and cob fragments. Field crews collected all that they recognized in the field. Investigators made no attempt to analyze any flotation or pollen samples from this area, since they found only one pit and a shallow trash feature. The presence of corn remains in this location supported the premise that agricultural activities were associated with the settlement.

CHRONOLOGY

The ceramic assemblage provided the only method with which to address the temporal associations of Area 9. The wide variety of Hohokam ceramic types suggested a long period of use, spanning the late Pioneer through Classic periods. This was the only locus at Dutch Canal Ruin to contain such an array of ceramic types and the only area to contain a Sedentary and early Classic period (Soho phase) component. Archaeologists have difficulty distinguishing Soho phase components in mixed contexts from other phases because of the generally lower frequencies of decorated types, which has been related to a decline in painted ceramics during that time. Therefore, other areas at Dutch Canal Ruin quite possibly could have contained Soho phase components. Possible use of the site during the Soho phase was demonstrated by the presence of burials in other areas of Dutch Canal Ruin that contained Hohokam Red Ware vessels. These burials, however, were generally not associated with areas of habitation but were intrusive into canals or abandoned structures.

The variety of diagnostic types represented at Area 9 was far greater than in any of the other areas at Dutch Canal Ruin (see Table 12.5). Investigators cannot state unequivocally that the small ceramic assemblage from Area 9 represented a Hohokam continuum from the late Pioneer period through the Classic period, but given the variability within the assemblage, this suggestion deserves consideration.

Analysts recognized no stratigraphic separation that signaled differential occupations. Normally, deposits associated with a span of 800 years are separated vertically, especially in an area subject to alluviation. An example of such vertical separation occurred in the elevations between Area 8, a late Classic period settlement, and nearby pre-Classic period settlements, where field crews recorded differences of as much as 1.3 m. Examination of the storm drain trench and systematic shovel trenches indicated that a maximum depth of 30 cm of cultural deposits was preserved at Area 9, although no vertical separation could be defined between the earliest and latest cultural materials.

SUMMARY

Because of the paucity of features, investigators could only speculate on the function of Area 9. The areal extent of the trash deposit and the amount of materials suggested long-term use of the area, yet the lack of structures was cause to question this interpretation. Historic modification of the landscape, in the form of excavation and leveling, may have removed many of the features. Alternatively, the absence of vertical separation between the earliest and latest ceramic types suggested that this was not an area of rapid alluvial accumulation and that use of the area during each period represented had been brief. The apparent mixture of early and late materials also may have resulted from periodic erosion of the area, as it was located proximate to Turney's Gully, one of the braided river channels of the Salt River (Turney 1929). The degree of slope of the general area gave some support to this interpretation.

Archaeologists more easily assigned the temporal use of Area 9 to a period that spanned the latter two-thirds of the Hohokam continuum, due to the recovery of ceramics from each recognized phase. After drawing an analogy with other areas at Dutch Canal Ruin, investigators determined that the pre-Classic period occupation was probably seasonal and related to agricultural pursuits. Analysts found it difficult to assess the Classic and post-Classic period occupations because of the limited information. The presence of corn remains, in the form of kernels and cob fragments, supported the notion of an agricultural association, at least during a portion of the area's occupation. Further, investigators speculated that Area 9 represented a farmstead locale and was a favored location that had been reused during each successive phase of the Hohokam sequence from the late Pioneer period through the Civano phase or, perhaps, the Polvorón phase.

CHAPTER 11

DUTCH CANAL RUIN, AREA 10

David H. Greenwald

Area 10 was a concentration of prehistoric features discovered during the monitoring of construction for the remote parking facility that was installed west of the north runway of Phoenix Sky Harbor International Airport (Figure 1.2). This area lay along the southeastern boundary of Dutch Canal Ruin as defined by Midvale (1934; n.d.; see Figure 1.5). The greatest concentration of features occurred north of Grant Street, with only two features to the south (Figure 11.1). Area 10 was located in T1N, R3E, Sec. 10, SE¼NE¼SE¼, on the Phoenix, Arizona, 7.5 minute USGS topographic map, along the former route of Grant Street in the extreme eastern portion of the project area (Figure 1.2) and within an area that extended from N1550 to N1650 and from E1915 to E2050 within the project grid.

In a survey of the Phoenix Sky Harbor Center, researchers recorded two loci of surface artifacts in the general location of Area 10: Locus I, which contained fewer than 10 surface artifacts and was located between Sherman and Grant streets, and Locus 17, with more than 10 artifacts, north of Grant Street (BRW 1986:Map 4). The Locus I artifact scatter bordered the southern edge of Area 10; the Locus 17 artifact scatter continued to the north of Area 10 and west of Area 5. Although archaeologists did not use systematic methods to quantify artifact density, surface materials in this area were visibly more concentrated than in most areas of Dutch Canal Ruin. The artifacts north of Area 10 may have been deposited either by prehistoric activities associated with a scatter of features in that location or by redeposition from historic leveling of Area 10. Given the direct spatial association of surface materials to Area 10, archaeologists surmised that these materials were associated with the features in Area 10.

FIELD METHODS AND STRATEGIES

Because researchers had not previously observed surface remains in this area and construction of the remote parking lot would result in only shallow disturbance, archaeologists monitored this area during construction rather than including it within the testing and data recovery phases for the greater part of Dutch Canal Ruin. Monitoring resulted in the definition of 11 features plus the identification of a canal alignment approximately 100 m south of the features. In addition to surface blading and leveling, the parking lot area had been disturbed in two other ways. First, construction crews built a storm drain oriented northwest-southeast through the extreme southern portion of Area 10; field crews found Feature 10-2, a pit structure, in this area and found a canal, Alignment 8555, in another section of the storm drain system. Second, construction crews installed a series of light poles with underground wiring to provide lighting for the parking lot. Investigators monitored each trench dug for the electric lines and each hole dug for the light poles as part of the monitoring process. Despite the intact cultural features in the extreme northern portion of the parking lot, they found no cultural features in any of the trench exposures or pole emplacements, probably because historic disturbance had left a modern surface that was below the prehistoric cultural deposits in this portion of the site.

Field crews made every effort to record all cultural features observed at Area 10 without delaying construction, which occasionally required that the heavy equipment be directed to another portion of the remote parking area to allow sufficient time for further investigation. Investigators

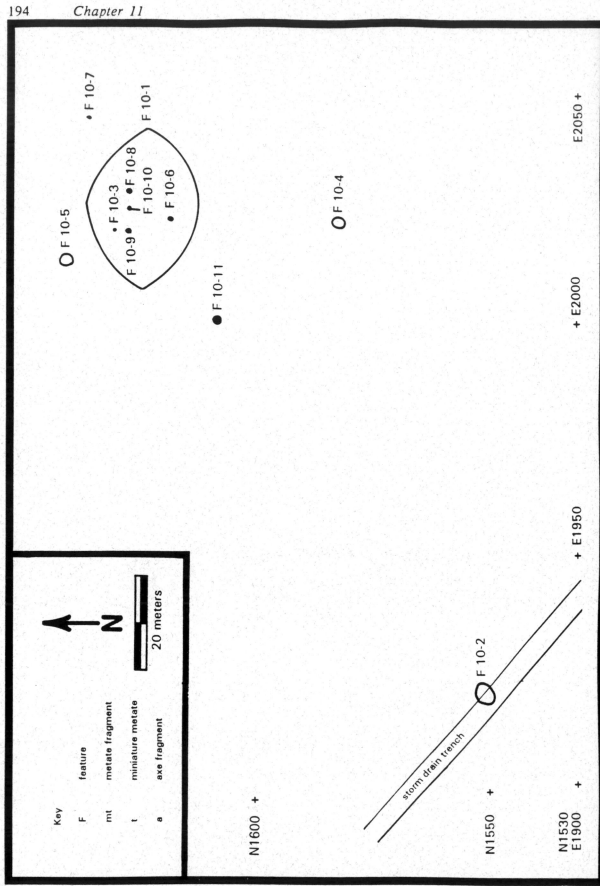

Figure 11.1. Area 10, Dutch Canal Ruin.

also attempted to delineate the preserved extent of the artifact scatter or trash deposit (Feature 10-1) associated with the features and defined two boundaries based on artifact density. The densest concentration measured 31 m east-west by 12 m north-south; the lighter concentration, which surrounded and encompassed it, measured 34 m east-west by 23 m north-south. Investigators defined the two boundaries subjectively, on the basis of visual observation, without the aid of any type of quantification. They plotted each feature according to its exact grid provenience.

Excavation and recording methods followed those outlined in the plan of work (Volume 1:Chapter 11), with some variations to expedite data recovery and prevent construction delays. Field crews recorded each feature as it was exposed. Surface stripping by heavy equipment achieved horizontal exposure, and vertical exposure occurred in a construction balk left in place with a grade (elevation) stake set in for monitoring the depth of the excavations. Investigators found three features in the construction balk profile. They closely examined storm drain trenches while construction crews were digging the trenches and again before crews installed the drain pipe. The trenches also produced vertical exposures and resulted in the identification of a pit structure (Feature 10-2) and Canal Barranca (Alignment 8555). Field crews drew profile maps of both of these features and recorded all information regarding their locations and morphologies. They found Feature 10-4, an horno, during surface stripping and mapped its plan. To expedite the excavation of this feature, the backhoe operator dug a north-south trench through its east half, leaving a profile exposure. Investigators then prepared a profile drawing and recovered botanical and radiocarbon samples from the exposure. Field crews cross-sectioned all other features (with the exception of Feature 10-5, a poorly preserved horno) and removed one-half of the fill. They collected all artifacts found in the excavated portions of the features and took botanical samples.

CULTURAL FEATURES

Structures and Extramural Features

Feature 10-1

Horizontal Provenience: N 1613.5–1636.5; E 2008.0–2042.0
Vertical Provenience: 1.65–1.95 meters below datum (mbd)
Feature Type: Trash Deposit

This feature consisted of a concentration of cultural materials truncated by modern and historic activities conducted in this portion of the site. In situ trash deposits had a maximum depth of 35 cm. Intrusive features extended to a maximum depth of 45 cm below the base of the trash deposit. Culturally sterile soils occurred beneath the trash deposit in the form of compacted silty clay. The plan of these deposits was oval, with the long axis oriented east-west. At its greatest extent the feature measured 34 m east-west by 23 m north-south, with its center at N1625, E2025. Profile exposure resulted from the creation of the grade balk used during construction. Feature 10-1 contained five pit features (10-3, 10-6, 10-8, 10-9, and 10-10) that either intruded into or were contemporaneous with the trash deposit.

Feature 10-2

Horizontal Provenience: N 1551.20–1552.30; E 1918.95–1920.50
Vertical Provenience: ca. 2.35–2.55 mbd
Feature Type: Pit Structure

Feature 10-2, in the north profile exposure of the storm drain trench, was a pit structure. A profile drawing was the only effort expended to investigate and record this feature (Figure 11.2). The floor surface was easy to identify from the feature's western extent to an area adjacent to a fire pit or hearth. Toward its western edge the floor appeared to slope upward, forming a gradually curving wall. The floor surface was somewhat irregular, undulating between the wall and the fire pit. East of the fire pit the floor was difficult to follow, and field crews identified no east wall. The structure was filled with dark sandy silts that contained charcoal inclusions and pieces of oxidized soil, which were probably remnants of the superstructure. The fire pit contained white ash, pieces of charcoal, and pieces of oxidized soil.

Feature 10-3

Horizontal Provenience: N 1630.0–1631.0; E 2020.2–2021.8
Vertical Provenience: 1.65–1.95 mbd
Feature Type: Pit, Not Further Specified (NFS)

Feature 10-3 was a shallow, basin-shaped pit that was exposed as oval in plan view by grading and leveling operations and measured 1.0 m north-south by 1.6 m east-west. This pit was filled throughout with dark, sandy silts containing sherds and charcoal flecks. It was located near the center of and intruded into Feature 10-1, a trash deposit. Ceramics, charcoal, and one piece of flaked stone were present in the pit. It was 30 cm in depth, extending to the base of Feature 10-1.

Feature 10-4

Horizontal Provenience: N 1581.30–1583.25; E 2021.00–2023.30
Vertical Provenience: Not recorded
Feature Type: Horno

When first exposed by blading operations, this feature was identified as an horno because of its dark brown and black outline. It measured 1.95 m north-south by 2.30 m east-west. After investigators drew the plan view, they had the backhoe operator bisect the feature to expedite recording. The north-south trench revealed a basin-shaped pit filled with large amounts of fire-cracked rock, burned adobe chunks, and river cobbles interspersed with ash and charcoal. The upper portion of the southern side of the pit was heavily oxidized, while the remainder of the pit edge was blackened, containing carbon deposits. Oxidized soils extended as much as 20 cm beyond the pit edge into the sterile substrate, indicating the intensity of the fires contained in the pit. A layer of water-laid silts covered the ash deposits resting in the very bottom of the pit, indicating that the pit had been left open for a period of time before being filled with the fire-cracked rock and charcoal-rich matrix that composed most of the feature fill. Field crews collected charcoal for dating purposes and flotation and pollen samples from the feature's fill.

Feature 10-5

Horizontal Provenience: N 1639.0–1641.0; E 2013.0–2015.3
Vertical Provenience: Not recorded
Feature Type: Horno

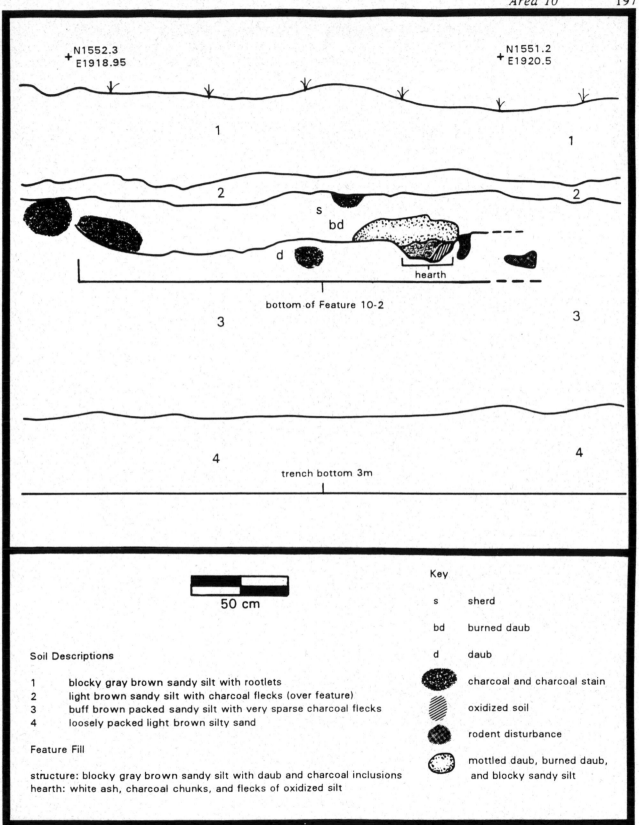

Figure 11.2. Area 10, Feature 10-2, profile.

Feature 10-5, the remnants of an horno, was exposed by grading activities and had been severely disturbed by historic residential activities at the site. Only small portions of the blackened rim were preserved. This feature was northwest of Feature 10-1, the trash deposit (Figure 11.1). Given the preserved segments exposed by blading, investigators concluded that the feature measured approximately 2.0 × 2.3 m. Due to its poor preservation, this feature was not investigated further.

Feature 10-6

Horizontal Provenience: N 1618.0; E 2023.5
Vertical Provenience: 1.65–ca. 1.95 mbd
Feature Type: Flaked Stone Concentration

This deep, basin-shaped pit was filled with various flaked stone items and tools, including 2 hammerstones, 1 core tool, 2 cores, 4 whole flakes, and 1 bifacially prepared flake tool. The pit was exposed by grading operations at 1.65 mbd and extended approximately 30 cm below its level of detection. It was located in the extreme southern portion of Feature 10-1 and intruded into the trash deposit. This feature appeared to be an artifact cache, because it contained only items related to core reduction that apparently did not represent refuse disposal.

Feature 10-7

Horizontal Provenience: N 1635.0; E 2046.5
Vertical Provenience: 1.65–1.95 mbd
Feature Type: Ceramic Concentration

Feature 10-7 was a partially complete, in situ plainware vessel with associated Roosevelt Red Ware sherds. The ceramic concentration was 30–50 cm below the modern ground surface. It was northeast of Feature 10-1, the trash deposit, and lay directly beneath the disturbed zone, as did many of the other features in Area 10.

Feature 10-8

Horizontal Provenience: N 1627.5; E 2029.0
Vertical Provenience: 1.75–2.45 mbd
Feature Type: Trash Pit

This pit measured 60 cm in diameter and 70 cm in depth and was filled with ash, charcoal, flaked and ground stone, sherds, and numerous fire-cracked rocks. It was in the profile exposure of the grade balk left during the construction of the parking lot. This feature lay near the center of Feature 10-1, the trash deposit, and appeared to be contemporaneous with Feature 10-1, as its upper portion began below the top of the trash deposit.

Feature 10-9

Horizontal Provenience: N 1627.5; E 2020.2
Vertical Provenience: 2.00–2.15 mbd
Feature Type: Mixed Artifact Concentration

Feature 10-9 was a basin-shaped pit in the north profile of the grade balk exposure left during construction of the parking lot. It appeared to originate at the base of or within the lower 5 cm of the trash deposit, Feature 10-1. It was 20 cm in diameter and 15 cm in depth. The concentration of sherds and flaked stone items in the pit was more dense than that observed in Feature 10-1.

Feature 10-10

Horizontal Provenience: N 1627.5; E 2025.6
Vertical Provenience: 1.65–1.92 mbd
Feature Type: Trash Pit

Feature 10-10 was a basin-shaped pit filled with sherds, flaked stone items, some faunal remains, and charcoal flecks. It lay in the profile exposure of the construction balk and was 70 cm in diameter and 30 cm in depth. It was near the center of Feature 10-1, the trash deposit, and appeared to intrude into the trash deposit.

Feature 10-11

Horizontal Provenience: N 1608.0; E 2001.5
Vertical Provenience: 1.92–2.25 mbd
Feature Type: Fire Pit

This fire pit contained fire-cracked and heat-altered rock and ground stone items. It was southwest of Feature 10-1, the trash deposit, and measured 36 × 45 cm and was 33 cm deep. The altered rock and ground stone appeared to have been used in conjunction with the fire in the pit.

Canal Barranca, Alignment 8555

Alignment 8555 was discovered as a result of excavation of the storm drain trench within the project grid at N1449.6–1471.0 and E1897.3–1906.8 (Figure 1.2). The storm drain trench cut the canal at a severe oblique angle, prohibiting accurate measurements of its width. Within this trench investigators delineated three channels: one as an isolated channel and the other two as slightly superimposed channels. Investigators profiled the canal despite the oblique angle, because it had not been located in any of the systematic or geomorphic trenches dug elsewhere in the project area. They calculated its trajectory, based on its exposure in the storm drain trench, and located it again west of 22nd Street after extending Trench 2081 to the south. This exposure was also at an oblique angle, so the backhoe trench operator dug Trench 2120 farther to the southwest and at an angle to the project grid to provide a perpendicular exposure (Volume 1:Figure 8.5). Investigators included Canal Barranca with the Area 10 features because they discovered it during monitoring and its projected course passed directly south of Area 10.

Canal Barranca is discussed in detail in Volume 1, Chapter 8, and in Volume 4, Chapter 6, and is described only briefly here. The multiple channels of Canal Barranca were parabolic in shape. The combined width of these channels was about 8 m. The best-preserved channel, however, was only 3.75 m in width, with a maximum preserved depth of 0.70 m. The section of Canal Barranca west of 22nd Street in Trench 2120 represented the terminal point of the canal alignment, for its southwesterly course had caused it to enter Turney's Gully, a small river channel that formed the southern boundary of Dutch Canal Ruin. Although the canal's age has not been determined through absolute dating techniques, it may date to the Classic period, given its shallow stratigraphic position and its spatial association with areas of Classic period occupation.

MATERIAL CULTURE

Ceramics

Investigators recovered 1066 sherds from Area 10 (Table 11.1). The variety of types indicated two periods of occupation: 5 sherds represented the Colonial period, and 259 sherds represented the late Classic period. The Classic period occupation was indicated by the dominance of Salt Red (235, or 99.6%) within the Hohokam Red Ware assemblage. Ceramic cross dates for Salt Red have been given at A.D. 1300–1450 (Wood 1987).

Table 11.1. Area 10, Ceramic Frequencies

Ceramic Ware/Type	Count	Percent Ware	Percent Assemblage
Hohokam Plain Ware	**702**		**65.9**
Gila Plain, Salt Variety	676	96.3	63.4
Gila Plain, Gila Variety	3	0.4	0.3
Squaw Peak Plain	22	3.1	2.1
Wingfield Plain	1	0.1	0.1
Hohokam Red Ware	**235**		**22.0**
Salt Red	235	100.0	22.0
Hohokam Decorated Ware	**9**		**0.8**
Snaketown Red-on-buff	1	11.1	0.1
Santa Cruz Red-on-buff	4	44.4	0.4
Santa Cruz/Sacaton Red-on-buff	1	11.1	0.1
Indeterminate Buff	3	33.3	0.3
Roosevelt Red Ware	**24**		**2.3**
Phoenix Red	6	25.0	0.6
Pinto Polychrome	1	4.2	0.1
Gila Polychrome	9	37.5	0.8
Tonto Polychrome	4	16.7	0.4
Salado Red	1	4.2	0.1
Indeterminate Polychrome	3	12.5	0.3
Intrusive Wares	**96**		**9.0**
Sells Plain	78	81.3	7.3
Lower Colorado Buff Ware	18	18.8	1.7
Total	**1066**		**100.0**

The percentage of intrusive types (9.0%) in Area 10 was considerably higher than the percentage for Dutch Canal Ruin as a whole and higher than at any other area at the site. One Sells Plain sherd collected from Area 9 was the only other example from Dutch Canal Ruin. However, the frequency of Lower Colorado Buff Ware was higher in Area 8 (35 sherds) than in Area 10 (18 sherds). Sells Plain spanned a period from A.D. 800 to post-1700 (Wood 1987), and Lower Colorado Buff Ware generally spanned a period from pre-A.D. 1000 to post-1850 (Waters 1982a:287–291). As a result of these wide temporal spans, the occurrence of either or both wares could be associated with either the Colonial period or the Classic period occupation. The only other area at Dutch Canal Ruin that had both Sells Plain and Lower Colorado Buff Ware was Area 9, which also contained (along with Area 10) one of the two Classic period components found at Dutch Canal Ruin.

Investigators examined pit features containing ceramics in terms of their stratigraphic positions. Feature 10-3 contained six Lower Colorado Buff Ware sherds and intruded into Feature 10-1, the trash deposit. Field crews recovered the remaining 12 Lower Colorado Buff Ware sherds from Feature 10-1. Feature 10-7, a ceramic concentration, contained a partially reconstructible Sells Plain vessel with three Tonto Polychrome sherds and was at the same elevation as Feature 10-3. The stratigraphic positions of these features and the direct association of Tonto Polychrome sherds with Feature 10-7 suggested that the Lower Colorado Buff Ware and Sells Plain sherds were associated with the Classic period occupation at Dutch Canal Ruin. Therefore, archaeologists may argue (with some circumspection) that the co-occurrence of Sells Plain and Lower Colorado Buff Ware at Dutch Canal Ruin represented a Classic period association. In addition, since Area 8 represented a post-Classic period occupation, these two wares may also have been associated with the Polvorón phase. Few data recovered from Dutch Canal Ruin supported this very tentative conclusion, but such an association may have been present at other late or post-Classic period sites in contexts of higher integrity than found here.

Flaked Stone

Field crews recovered 90 flaked stone items, representing 11 morphological types, from Area 10 (Table 11.2). They recovered 89 of these items from features but only 1 from Feature 10-1, the trash deposit. Most of the flaked stone items (64, or 71.1%) were debitage. Core tools and hammerstones (16, or 17.8%) made up the second-largest component of the assemblage.

Nineteen items exhibited identifiable use wear. Battering, the most frequently identified, was related to the high number of hammerstones in the tools. Battering can result from the pounding or pulverizing of foods such as legumes, greens, or meat. The remaining items showed evidence of scraping (n=2), scraping and battering (n=2), cutting/sawing (n=1), and cutting and scraping (n=1).

With the exception of the large primary flake, all morphological types were represented by fine-grained basalt (Table 11.2), indicating that the area's inhabitants had undertaken basalt core reduction and tool production. The accumulation of trash in Area 10, the number of features observed, and the amount of use wear present in the flaked stone assemblage suggested extensive use of Area 10. Variations within the range of feature types supported the inference that the flaked stone assemblage was related to subsistence activities, including both procurement and processing.

Table 11.2. Area 10, Flaked Stone Morphological Type by Material Type

Morphological Type	Material Type					Total
	Chert	Basalt	Igneous Other	Quartzite	Metamorphic Other	
Angular Debris	1	5			1	7
	*14.3	71.4			14.3	**7.8
	**33.3	7.5			100.0	
Flake Fragment	1	22	10			33
	3.0	66.7	30.3			36.7
	33.3	32.8	55.6			
PRB[1] Flake	1	3				4
	25.0	75.0				4.4
	33.3	4.5				
Whole Flake		19	1			20
		95.0	5.0			22.2
		28.4	5.6			
Used Flake		1				1
		100.0				1.1
		1.5				
Core		5				5
		100.0				5.6
		7.5				
Core Tool		3	1			4
		75.0	25.0			4.4
		4.5	5.6			
Core/Hammerstone		2				2
		100.0				2.2
		3.0				
Hammerstone		6	4			10
		60.0	40.0			11.1
		9.0	22.2			
Biface		1	1	1		3
		33.3	33.3	33.3		3.3
		1.5	5.6	100.0		
Large Primary Flake			1			1
			100.0			1.1
			5.6			
Total	3	67	18	1	1	90
	*3.3	74.4	20.0	1.1	1.1	100.0

*Row %
**Column %
[1]Platform-remnant-bearing

Ground Stone

Field crews recovered 31 ground stone items from Area 10 in a variety of forms and material types (Table 11.3). The assemblage was dominated by ground stone NFS (12, or 38.7%), followed by formal food-processing tools, including manos (10, or 32.3%) and trough metates (5, or 16.1%).

Feature 10-1, the trash deposit, contained 18 of the 31 items from Area 10, including 5 of the manos and all of the trough metate specimens. Feature 10-8, the trash pit, contained 6 of the items, including 4 of the manos. The remainder of the recovered items were from Features 10-4, 10-10, 10-11, and 10-12.

The ground stone assemblage from Area 10 was well represented by grinding implements. Food processing was probably a major activity, particularly the processing of cultigens such as corn. Field crews recovered charred and vitrified corn kernels and cobs from Feature 10-4, an horno, providing an association of rectangular manos and trough metates with this important food resource.

Shell

Marine and freshwater shell were represented by three items recovered from Feature 10-1. Two were unworked *Anodonta* sp., and the other was an unworked fragment of *Laevicardium*. *Anodonta* requires fresh, continuously moving water. Although *Anodonta* may have reached the project area by being carried down high-velocity canals, this bivalve had probably been gathered from the Salt River as a food source and transported to Area 10. Investigators observed no evidence of modification on either specimen. *Laevicardium*, a Pacific Ocean species, was imported to the Hohokam region for manufacture, primarily into jewelry. This fragmentary specimen, even though not worked, may have been debris from shell artifact production.

Subsistence Remains

Area 10 yielded little information on subsistence. Recovery of charred corn cobs with kernels from Feature 10-4 indicated that the occupants of the area may have built hornos for roasting corn at this location. In this particular situation, some of the corn was burned within the pit, indicating that an attempt at roasting corn was not completely successful. The ground stone assemblage suggested that the inhabitants may have processed foodstuffs by grinding. Rectangular manos and trough metates have often been equated with the grinding of corn, although the Hohokam could have processed a wide variety of other resources, including faunal resources, with these tools (Bostwick 1988:162–163). The presence of corn remains at Area 10 supported the notion that this settlement was associated with agricultural production. Area 10 was located near a Classic period canal, Canal Nuevo (Alignment 8537), 220 m to the north, and Canal Barranca (Alignment 8555), less than 125 m to the south. Residents of Area 10 probably made use of either or both of these canals.

Faunal resources recovered from Area 10 included 6 specimens recovered through screening and 115 specimens recovered from unscreened contexts (Appendix D). None of the 121 specimens exhibited evidence of having been worked or retained cut marks, although the 3 specimens from a mixed artifact concentration were burned. Field crews recovered 97 unidentifiable fragments from Feature 10-1, the trash deposit; the fragmentary nature of these items precluded identification as to taxa and element (bone type). Of these specimens, 7 represented medium-sized animals, while

Table 11.3. Area 10, Ground Stone Morphological Type by Material Type

Morphological Type	Vesicular Basalt	Quartz-Bearing Basalt	Basalt	Tuff/Diorite	Granite	Sandstone	Metasediment	Greenstone	Schist/Phyllite	Total
Round/Oval Mano						1 *100.0 **33.3				1 **3.2
Rectangular/Loaf Mano	3 33.3 60.0	3 33.3 50.0	1 11.1 50.0	1 11.1 50.0			1 11.1 25.0			9 29.0
Trough Metate	1 20.0 20.0	2 40.0 33.3	1 20.0 50.0				1 20.0 25.0			5 16.1
Maul								1 100.0 33.3		1 3.2
Haftable Item								1 100.0 33.3		1 3.2
Blank								1 100.0 33.3		1 3.2
Ground Stone NFS	1 8.3 20.0	1 8.3 16.7			2 16.7 100.0	2 16.7 66.7	2 16.7 50.0		4 33.3 100.0	12 38.7
Passive Ground Stone NFS				1 100.0 50.0						1 3.2
Total	5 *16.1	6 19.4	2 6.4	2 6.4	2 6.4	3 9.7	4 12.9	3 9.7	4 12.9	31 100.0

*Row %
**Column %

17 represented large mammals, principally deer or other ungulates. Contrary to their expectations, analysts identified only large mammal remains from screening. The relative frequency of large mammals in the assemblage, 14.05%, compared favorably with the results from Area 8, where large mammals accounted for 27.9% of the entire assemblage. Large mammals were consistently absent from the pre-Classic period field house and farmstead loci, suggesting that residents of these temporarily occupied sites had not exploited large mammals there. In contrast, Area 8 had been occupied year-round, allowing the capture and use of large mammals. The presence of large mammals in the Area 10 assemblage further supported the notion that this had been a locus of permanent habitation during the late Classic period.

The nearby wooded riverine zone along the Salt River would have provided a wide range of economically important plants for the occupants of Area 10 and would have been a refuge for a variety of faunal resources as well, including mammals, waterfowl, fish, amphibians, and reptiles. With numerous large mammals, especially ungulates, residents may have successfully hunted deer in the riverine zone, avoiding treks to the distant uplands. The limited representation of cacti in the botanical record from the Classic period loci suggested that residents of Dutch Canal Ruin had not extensively used the uplands but had instead acquired the bulk of their subsistence base from farming and exploiting resources local to the first terrace and riverine zone.

CHRONOLOGY

Reconstruction of Area 10 chronology was limited to ceramic data and stratigraphic associations. The ceramic assemblage provided information on periods of use, while stratigraphic relationships among different features permitted the relative ordering of cultural remains.

Investigators tentatively assigned Feature 10-2, a pit structure southwest of the main area, to the Colonial period. Because of a general paucity of sherds, they based its pre-Classic period assignment on its stratigraphic association. Its level of detection and associated prehistoric occupation surface was lower than that of any other feature in Area 10. The ceramic assemblage supported the contention that two periods of use had occurred at Area 10 (Table 11.1). Analysts identified a first occupation during the Santa Cruz phase of the Colonial period, A.D. 850–900 to 950–1000 (Dean 1991:90), based primarily on the presence of Santa Cruz Red-on-buff. The limited assemblage of pre-Classic period diagnostic types, which indicated a brief occupation of the area during this period, resembled that of other pre-Classic period settlements at Dutch Canal Ruin.

The second period of use occurred during the late Classic period, beginning in the Civano phase. During this time, trash began to accumulate in Feature 10-1, and Features 10-3, 10-6, and 10-10 later intruded into this refuse feature. Feature 10-9 appeared to have originated at the bottom of Feature 10-1 or within the lower 5 cm of the trash deposit, making it contemporaneous with Feature 10-1. Feature 10-8 was also contemporaneous with Feature 10-1, as its upper limits were within the central deposits of the trash. Features 10-3, 10-8, and 10-10 contained ceramic assemblages that supported Classic period use. Field crews recovered no ceramics from Feature 10-6, the artifact cache, but, because it was intrusive, its stratigraphic association indicated that it may have postdated Feature 10-1.

Feature 10-7, a ceramic concentration consisting of an in situ pot break with other miscellaneous sherds, lay outside the limits of Feature 10-1. Feature 10-7 contained three Tonto Polychrome sherds and the Sells Plain vessel, and its higher elevation supported the premise of a late association. Researchers have dated Tonto Polychrome to A.D. 1265–1385 (Breternitz 1966:97) and Sells Plain to between A.D. 800 and post-1700 (Wood 1987:169).

Similarly, although Feature 10-11 did not contain any ceramics, its stratigraphic placement suggested a late association. The levels of detection for the hornos, Features 10-4 and 10-5, were not available for comparison; however, field crews observed both during the early phases of construction blade work. Crews recovered 9 Salt Red sherds from Feature 10-4, implying a late association for this horno. Wood (1987:169) placed Salt Red at approximately A.D. 1300–1450, a late Classic period (Civano phase) or post-Classic period association.

The occurrence of other ceramic types in Area 10 further supported the premise of a late temporal association for Feature 10-1. Among the types recovered were Pinto Polychrome, Gila Polychrome, and Salado Red—all Classic period ceramic types. Phoenix Red, a type considered to represent the undecorated portion of Roosevelt Red Ware or Salado Polychrome vessels, was also present in Feature 10-1. Given these ceramic and stratigraphic data, investigators assigned the majority of cultural activities conducted at Area 10 to the late Classic period and the Civano or Polvorón phases.

SUMMARY

On the basis of ceramic evidence, investigators concluded that early use of Area 10 had occurred during the Colonial period. The limited number of pre-Classic period remains was, therefore, probably related to the activities conducted in this location rather than to preservation factors. The pre-Classic period component was probably associated with canal irrigation and agricultural production, being similar to other scattered field house loci during this time at Dutch Canal Ruin. Pre-Classic period use of the site was sporadic and ephemeral, resulting in the construction of crude structures that exhibited a lack of formality and low energy investment in their construction. Archaeologists have recognized few activities other than agriculture at other field house loci, a result of limited and poor preservation of ephemeral features. As a field house locus, Area 10 would be expected to contain few artifacts, poorly defined pit features, and small, crudely constructed structures, a general pattern throughout the site for pre-Classic period occupation.

The level of use of Area 10 during the Classic period greatly increased over that of the pre-Classic period. The increase in activity expressed by the trash accumulation and the number and variety of features may have been related either directly to habitation activities or to processing activities in connection with nearby fields. The amount of cultural debris and stratigraphic evidence suggested that the Hohokam had reused the area. Although investigators found no architectural features in this area, the level of activity represented by the features and artifactual materials suggested that structures had at one time existed but had apparently not been preserved. If the area's occupants had built surface structures, historic and modern activities in this location probably removed all evidence. Historic plowing, in particular, may have destroyed stratigraphically high features and surface architecture. Feature 10-1 was truncated by plowing, as were several of the other features, and historic residential use contributed further to the disturbance here. Area 10 appeared to resemble Area 8 in its level of use and variety of activities. Only the architectural features were absent. Although archaeologists often determine residential functions and longevity of use on the basis of the presence, size, and construction attributes of architectural features, they may also use other factors (such as numbers and types of extramural features and quantities of artifacts and refuse accumulation) to address the level and complexity of occupation at sites. The absence of architectural features in Area 10 during its Classic period use precluded interpretations concerning settlement type and function. However, the variety and types of features and the quantity of refuse supported the idea that its occupation was permanent, possibly as a farmstead locale.

CHAPTER 12

THE DUTCH CANAL RUIN CERAMIC ASSEMBLAGE

Thomas N. Motsinger

The ceramic assemblage recovered from Dutch Canal Ruin during the excavations of 1989 and 1990 included 12,059 sherds representing 12 different ware categories. Thomas N. Motsinger and Sarah L. Horton completed the analysis of the assemblage during the spring and summer of 1991, with a host of individuals and institutions volunteering assistance during several phases of the project. The research goals, methods, and ceramic type descriptions are reviewed below, followed by a discussion of the results and their implications.

Analysts used traditional type names to identify the categories of pottery. However, they devised a new name, "Salt Red-on-buff," to describe Gila Plain, Salt Variety ceramics that were covered with a buff slip and decorated with red paint. This shorthand method of describing a type previously described as a Hohokam Buff Ware with a brown paste and sand temper (Abbott and Gregory 1988) was intended to identify the type as a buff-slipped version of the local variant plainware.

The pattern for intrusive wares at Hohokam sites, suggested by Crown (1985b), seemed to hold for the Dutch Canal Ruin assemblage. Field crews recovered intrusive wares only from Classic and post-Classic period contexts, and types from the Tucson Basin were the single largest category. In addition, the presence of pottery scrapers provided evidence for local ceramic production.

RESEARCH OBJECTIVES

The main function of ceramic analysis has traditionally been to provide chronologic information concerning the occupation of a site or a portion of a site. Temporal information that may be provided by ceramic analysis is of particular importance in the Hohokam region, where researchers lack the luxury of tree species that lend themselves to dendrochronological dating. At Dutch Canal Ruin, the ceramic assemblages from structures, canals, trash deposits, cremations and inhumations, activity areas, and other features assisted in assigning cultural phases within the Hohokam chronology. Archaeomagnetic and radiocarbon dating techniques, when applicable, served to further bolster chronologic arguments, and, in turn, investigators used ceramic data to confirm or question conclusions drawn from the results of such techniques.

Another application of the ceramic analysis related to chronology building at Dutch Canal Ruin was the definition and temporal refining of the Polvorón phase (Sires 1984) component in Area 8. Researchers regard a high proportion of Roosevelt Red Ware polychrome sherds compared to other decorated types as one indicator of a Polvorón phase presence (Peterson and Abbott 1991). Investigators expected that the numbers and types of ceramics present in the Polvorón phase component at Dutch Canal Ruin would help to shed light on this recently defined post-Classic period Hohokam phase.

A secondary and somewhat traditional objective of the ceramic analysis was to provide information concerning cultural contacts with groups outside the Salt-Gila Basin Hohokam culture area. Achieving this objective, of course, involved the identification of intrusive ceramics within the local assemblage. The Dutch Canal Ruin assemblage, for example, included pottery produced by the Tucson Basin Hohokam, the Mimbres of southwestern New Mexico, the Prescott tradition, the White Mountain Mogollon, the Patayan of southwestern Arizona, and the early Hopi.

Another aim of the analysis of ceramic items from Dutch Canal Ruin was to provide information on some of the functions of the site and the differentiation of these functions across the site and through time (Heacock 1988:100). Relatively sparse ceramic and other material culture deposition in an area with numerous structures would suggest a temporary, perhaps seasonal, occupation of those structures. A higher density of ceramic items in an area with few structures may indicate that the structures were occupied more intensively, perhaps on a permanent basis for an extended period. Also, a high proportion of plainware ceramics suggests a primarily utilitarian function, whereas an unusually large number of decorated or intrusive ceramics, or exotic forms such as effigy vessels, could be associated with burial contexts or occupation of the site by individuals of high socioeconomic status (Wilcox 1991:268).

Recently, ceramic analysis in the Phoenix Basin has taken a new turn toward the study of intervillage exchange and resource procurement. As part of the analysis of ceramics from Pueblo Grande, Abbott (1991) identified temper sources within the Phoenix Basin and found that the temper from ceramics produced in the Phoenix Basin, when viewed microscopically by a properly trained analyst, can usually be assigned to one of these sources. A major focus of the second phase of the project analysis was to analyze a sample of the plainware and redware ceramics (using Abbott's data) to determine where the inhabitants of the site had been procuring their tempering material and whether the predominant sources of such procurement activities had changed through time. Such analysis could also determine whether the plainware and redware types had been tempered with material from the same source, or whether one ware or the other had been imported to the site from another village specializing in such production. Eventually, the results from Dutch Canal Ruin can be compared with those of nearby communities that were occupied contemporaneously, as well as with communities that shared Canal System 2.

This same technique, when applied to the Roosevelt Red Ware assemblage from the post-Classic period component of Dutch Canal Ruin, was expected to help determine if such types had been produced in the Phoenix Basin. Recent chemical analysis suggested such wares had been produced locally (Crown, Schwalbe, and London 1988). Analysis of the temper material promised to indicate whether the Roosevelt Red Ware types had been produced at the site or by specialists elsewhere in the region.

ANALYTICAL METHODS AND APPROACHES

The methods employed for the analysis of the Dutch Canal Ruin ceramic assemblage included a variety of approaches drawn from previous work (e.g., Abbott and Gregory 1988; Crown 1981; Gladwin et al. 1937), as well as some more novel techniques that analysts developed and tested for use in this and perhaps future undertakings. The analysis consisted of two basic parts: (1) a complete typological analysis that included the recording of several attributes of each sherd; and (2) a detailed temper source analysis to help study procurement strategies, intervillage exchange, and specialization of ceramic production. The details of the second phase of this ceramic analysis are presented in Volume 4, Chapter 11.

The use of a strictly typological approach to ceramic analysis has been appropriately criticized as inadequate by itself (e.g., Abbott and Gregory 1988:5), although a typological analysis is fundamental as a means of communicating and comparing certain aspects of ceramic variation, especially with respect to large bodies of ceramic data. For this reason, analysts classified each sherd in the Dutch Canal Ruin assemblage into ware and type categories. In the case of Hohokam plainware and redware types, however, they first recorded each sherd as belonging to a particular temper category, based on the types and relative proportions of temper present. Analysts conducted this recording as a challenge to the traditional type categories of Hohokam undecorated ware categories that are based largely on temper type. The procedure allowed the analysis to proceed objectively until analysts made a final determination regarding the assignment of type names to the temper categories. Although most of the traditional type names were eventually retained, the analysis yielded considerable insight into the actual proportions of various temper types necessary for classification into particular categories. This information is presented in greater detail later in the ware and type descriptions.

To better quantify the variation represented in the ceramic assemblage, analysts recorded several other variables for each sherd recovered. The recording of the primary and secondary surface treatments on both the interior and exterior of each sherd established the frequency of various treatments for particular ceramic types. To examine the range, frequency, and distribution of vessel form, analysts recorded the type and the portion of the vessel represented by each sherd when these data could be determined. Finally, archaeologists weighed all ceramic samples so that raw counts for any subset of the total assemblage could later be standardized.

After coding and recording all variables for each sherd, analysts entered the data into the dBASE III+ computerized data base system for storage and basic data manipulation. They performed all higher-level manipulation and analysis of the ceramic data using the SPSS/PC+ 4.0 statistical software package on an IBM-compatible personal computer equipped with an Intel 80386 processor.

CERAMIC TYPES AT DUTCH CANAL RUIN

The recent burgeoning of public land development in the Phoenix area has allowed for archaeological investigations of unprecedented size and frequency at such important Hohokam sites as Las Colinas (Abbott et al. 1988), Casa Buena (Cable and Gould 1988), the Grand Canal Ruins (Mitchell 1989a), Pueblo Grande (Soil Systems, Inc. 1993), and Dutch Canal Ruin (Greenwald and Ciolek-Torrello 1988a). The ceramic analyses from these investigations have led to important revisions of the ceramic typological system that archaeologists have long accepted for the Hohokam region, particularly within the Phoenix Basin. For the present study, analysts incorporated the recent changes that they thought would be most logical and most useful to future Hohokam research. In doing so, they decided not to incorporate some recently proposed typological changes (e.g., Abbott, personal communication 1991; Wood 1987) into the Dutch Canal Ruin analysis, although such changes were deemed useful in the context of the research to which they were applied and may prove to be useful to future investigations in the Phoenix area. In the following discussion, ceramics from Dutch Canal Ruin have been divided into four categories: Hohokam Plain Ware, Hohokam Red Ware, Hohokam Decorated Ware, and intrusive wares.

Recent work by Abbott (1991) called into question the validity of the typological scheme that has long served to categorize Hohokam undecorated wares, which include both plainware and redware. The distinction between Gila Red and Salt Red, for example, has traditionally been based on such attributes as temper type, slip color, presence or absence of polishing striations, and patterning and size of fire clouds (Crown 1981; Gladwin et al. 1937; Haury 1976). Abbott and Gregory (1988) argued that the attributes that archaeologists believed were particular to one type or the other actually were present in both types, at least in the Las Colinas assemblage, to such a degree that a single type name, "Gila-Salt Red," is more useful than the separation of the two. A similar question has arisen regarding the proper typological sorting of the plainware types, given the recent ability of researchers to more finely separate sherds of differing temper types (Abbott, Schaller, and Birnie 1991; Volume 4:Chapter 11). Analysts have at least three ways to use the newly available temper sourcing information in defining Hohokam undecorated ware types: (1) they can assign separate type names to ceramics from each narrowly defined source; (2) they can use these temper sources to define varieties within the traditional types; or (3) they can apply the new information to refine the definitions of the existing types, without altering the structure of the typological system.

Although each of these options offered a valid approach to typological sorting, analysts chose to apply the third, more conservative approach to the Dutch Canal Ruin analysis for several reasons. First, type categories are traditionally based on more than one ceramic attribute. Although temper may be the most important diagnostic attribute of a ceramic type, a temper difference must usually be coupled with a strong *tendency* toward a difference in paste, slip, surface treatment, vessel wall thickness, or other attributes to justify a separation of types. Although all of the types are defined narrowly by their gross temper attributes, the type separations are justified by a tendency for the temper types to be associated with other attributes, described in Tables 12.1 and 12.2. The Hohokam Plain Ware and Red Ware definitions, although similar to much previous work, were developed particularly for the Phoenix Sky Harbor Center project; they may be useful for future work but are specific to this ceramic analysis. Until archaeologists establish relationships between the finer temper categories and other ceramic attributes, the use of the important information on fine-grained temper sources should be restricted to a level below that of type or variety names.

Table 12.1. Hohokam Plain Ware Types from Dutch Canal Ruin

Plain Ware Type	Dominant Temper	Typical Surface Finish	Typical Wall Thickness
Gila Plain, Salt	arkosic sand	smoothed, untreated	thick (4–9 mm)
Gila Plain, Gila	micaceous sand	smoothed, untreated	thick (4–9 mm)
Squaw Peak Plain	schist	smoothed, sometimes floated	medium (3–5 mm)
Wingfield Plain	phyllite	floated, often polished	thin (2–4 mm)

Table 12.2. Hohokam Red Ware Types from Dutch Canal Ruin

Red Ware Type	Dominant Temper	Typical Slip Color	Interior Smudging	Polishing Striations
Salt Red	arkosic sand	orange-red	yes	no
Gila Red	micaceous sand	red	yes	yes
Squaw Peak Red	schist	orange-red	no	no
Wingfield Red	phyllite	red-brown	no	no

Hohokam Plain Ware

Plainware types that are considered indigenous to the Hohokam of the Salt and Gila river valleys accounted for just over three-fourths (76.4%) of the sherds and vessels recovered from Dutch Canal Ruin. Analysts separated all of these plainware types into the two varieties of Gila Plain, the recently named Squaw Peak Plain, and Wingfield Plain. The main differences among these types are their dominant temper type, surface finish, and wall thickness, summarized in Table 12.1. However, although temper type is the most discrete distinguishing attribute of these types, considerable overlap exists among the other attributes of the Hohokam Plain Ware types.

Gila Plain

Gladwin and his coresearchers (1937) first defined Gila Plain as the trademark plainware pottery of the Hohokam. Archaeologists now recognize two distinct varieties of Gila Plain (Salt Variety and Gila Variety) and distinguish them largely on the basis of temper composition. Though both varieties were present in the Dutch Canal Ruin ceramic assemblage, Salt Variety, presumed to have been made at the site, was by far the more abundant of the two.

Gila Plain, Salt Variety

This type constituted the vast majority of the sherds recovered from Dutch Canal Ruin, just over 70% of the total assemblage. This variety is tempered with sand that includes an abundance of quartz and some feldspar. A few fine muscovite particles are sometimes visible as well, but these appear to be a natural constituent of the sand that was selected as temper and not a purposeful addition. By the definition developed for the Dutch Canal Ruin analysis, these free mica particles account for less than 5% of the total temper material and are typically present infrequently on the surface of vessels, often only microscopically. Sometimes other metamorphic rock such as schist or phyllite is included in the temper, but this never amounts to more than 50% of the total temper. Rarely, a small amount of crushed sherd, often redware, is a visible part of the temper of Gila Plain, Salt Variety, but the arkosic sands always remain the predominant temper material. Surfaces of vessels are typically smoothed but otherwise left untreated. Bowl interiors are smudged on about 5% of the samples analyzed, and exteriors of bowls and jars are infrequently well finished with a floated surface.

Gila Plain, Gila Variety

This second variety of Gila Plain accounted for only 2.5% of the plainware assemblage recovered from Dutch Canal Ruin. Gila Plain, Gila Variety is distinguished from its sibling Salt Variety by an abundance of mica and micaceous schist intentionally added to the sand temper. By the analytical definitions developed for the Dutch Canal Ruin assemblage, Gila Variety contains over 5% micaceous schist and free mica particles, with the remainder of the temper consisting of the same sands found in Gila Plain, Salt Variety. Although some researchers (e.g., Haury 1976; Howard et al. 1990) have described this variety as containing more mica than sand, the Dutch Canal Ruin analysis revealed that, while more mica than sand is visible on the surfaces of vessels of this variety, sand usually makes up more than half of the temper. During manufacture the free mica particles tend to "float" to the surface of the vessel, resulting in the distinctive glittery appearance of Gila Plain, Gila Variety sherds. As with Gila Plain, Salt Variety, the surfaces of Gila Plain, Gila Variety are normally left smoothed but untreated, with smudging occurring only infrequently on the examples analyzed from Dutch Canal Ruin. Polishing striations are sometimes visible as patterns on the interiors or exteriors of bowls or both and on the exteriors of jars.

Squaw Peak Plain

Lane (1989) first named this Hohokam Plain Ware type to distinguish the schist-tempered plain ceramics common in the Phoenix Basin from both the phyllite-tempered Wingfield Plain and the sand-tempered Gila Plain. Crown (1981), Wood (1987), and Brunson (1989) identified Gila Plain, Wingfield Variety as a schist-tempered Hohokam Plain Ware variety, but this designation is commonly confused with the distinctively thin-walled, well-finished Wingfield Plain, whose phyllite temper is also often mistaken for schist. Although sand is still often present in the temper of Squaw Peak Plain, schist platelets make up over half of the total temper and often account for nearly all of the temper in individual examples. Under a stereoscope, the schist particles appear as heterogeneous plates of quartz separated by dark bands. Sherds from the interiors of jars often have been left raw or are eroded, making the plates of crushed schist readily visible. Vessel walls are usually thicker than those of Wingfield Plain but thinner than Gila Plain walls. Surfaces usually exhibit a poorer finish than those of Wingfield Plain, but well-smoothed or floated surfaces are more common than with either of the Gila Plain varieties. One or both surfaces of Squaw Peak Plain vessels often oxidize to a dull to bright orange color. Though such vessels earned the separate type designation "Squaw Peak Orange" in Lane's (1989) report, the lack of a formal slip or even a finely floated surface on these examples indicates that the orange color was probably due more to differing firing conditions or different clay sources than to a purposeful surface color change.

Wingfield Plain

Phyllite is the dominant temper in this Hohokam Plain Ware type, normally constituting well over three-fourths of the total temper in any one example. Macroscopically, the phyllite of Wingfield Plain is a distinctive matte gray color, in contrast to the silvery sheen of the quartz-laden schist found in Squaw Peak Plain. Under a stereoscope, the phyllite appears to be the same color as pencil lead but has slightly less sheen. Like Squaw Peak Plain, the temper plates are readily visible on untreated surfaces and may be up to 5 mm in length. Wingfield Plain vessel walls are typically the thinnest of the Hohokam Plain Ware types, and surfaces usually have a more homogeneous, floated finish than those of either Gila Plain or Squaw Peak Plain. The floated surfaces are always the same color as the paste just below, distinguishing the floating on Wingfield Plain from the application of a red or red-firing slip on Wingfield Red.

Hohokam Red Ware

The redware ceramics recovered at Dutch Canal Ruin and at other Hohokam sites roughly mirrored their plainware counterparts in terms of temper, paste, and method of construction. Only the additional surface treatments added to these vessels (particularly the addition of a red slip and sometimes smudging and polishing) set the Hohokam Red Ware types apart from the Hohokam Plain Ware types. Table 12.2 summarizes typical attributes of the four Hohokam Red Ware types identified at Dutch Canal Ruin. Temper is the only one of these attributes that remains constant within each type; however, exceptions to the three surface treatment attributes occur within each type.

Salt Red

This was the dominant redware type recovered from Dutch Canal Ruin, accounting for over 97% of all Hohokam Red Ware sherds analyzed. This type is characterized by the same sand temper found in Gila Plain, Salt Variety. The exterior surfaces of Salt Red bowls and jars contain a thick orange to orange-red lustrous slip. The interiors of bowls are usually smudged and polished to create a smooth, often highly lustrous surface. Occasionally, the interiors of jars are also smudged but are usually not polished to the same degree as bowl interiors. Polishing striations are conspicuously absent from both surfaces of Salt Red bowls and jars. On whole vessels, exterior fire clouds are typically small and sometimes arranged in symmetrical patterns around the lower portions of vessels.

Gila Red

Like Gila Plain, Gila Variety, Gila Red is characterized by micaceous, glittering surfaces with between 5% and 50% mica in the core temper. The slip on Gila Red is more of a true red than the orange-red slip found on Salt Red, and smudging is nearly ubiquitous on bowl interiors and common on jar interiors. Typically, both the surfaces of bowls and the exteriors of jars exhibit readily apparent, patterned polishing striations. Fire clouds on the exteriors of vessels are larger than those on Salt Red and are usually unpatterned.

Squaw Peak Red

Aside from an obvious difference in temper, this type appears to possess many of the characteristics of Salt Red but exhibits less care given to its constructed form. Like Squaw Peak Plain, Squaw Peak Red is predominantly schist tempered, though the paste of Squaw Peak Red is often orange rather than gray (Lane 1989). Smudging of interior surfaces is common, though less so than with Salt Red, and both the slipped and the smudged surfaces usually lack the smooth, lustrous sheen that is the result of careful polishing on Salt Red vessels. Although the Squaw Peak Red slip can be quite thick, more often than not it is much thinner and more easily eroded than Salt Red slips. Slipped surfaces sometimes exhibit extensive, unpatterned fire clouding, the result of less controlled firing methods than were employed for Salt Red.

Wingfield Red

Wingfield Red has been confused with well-finished examples of Wingfield Plain to the extent that at least one researcher (Bruder 1983) has even challenged the type's existence. At Las Colinas, however, Abbott and Gregory (1988) identified a distinct red-slipped type—otherwise exhibiting the same characteristics as Wingfield Plain—that they are confident represents a true Wingfield Red. While heeding the warnings of Bruder (1983), the Dutch Canal Ruin analysts identified a small but distinctive assemblage of phyllite-tempered ceramics that reflected the description of Wingfield Red developed by Abbott and Gregory (1988). The red slips on Wingfield Red vessels are thinner and more fugitive than those of other Hohokam Red Ware types, and evidence of wiping and polishing is rare. Bowls, jars, and scoops may be slipped on one or both surfaces, with unslipped surfaces apparently receiving no smudging or other special treatment. Aside from the red slip, Wingfield Red is identical to Wingfield Plain in all respects.

Hohokam Decorated Ware

Hohokam Buff Ware

Hohokam Red-on-buff is the hallmark of the Hohokam ceramic tradition. Changes in the design styles of the buffware types have long served as the temporal markers for the beginnings of chronological phases in the widely recognized Hohokam cultural sequence. The buffware recovered from Dutch Canal Ruin covered the range in stylistic types from Gila Butte through Casa Grande, and the typological characteristics of the assemblage did not differ from those that are well documented in previous publications (e.g., Gladwin et al. 1937; Haury 1945, 1976). Hohokam Buff Ware is characterized by a light tan- to buff-colored paste and predominantly micaceous schist and free mica temper. The potters often applied a buff to white slip to vessel surfaces, over which they painted a deep maroon, purple, red, or pink design. Researchers have identified and described a brown-paste, sand-tempered variant of this ware at some sites in and north of the Salt River Valley (Abbott and Gregory 1988; Antieau 1981; Crown 1981). Analysts of the Dutch Canal Ruin assemblage named this alternate Hohokam Buff Ware type "Salt Red-on-buff."

Salt Red-on-Buff

The formal naming of Salt Red-on-buff serves two purposes. First, the name associates the type with the Salt River, the region where it has been most commonly identified. Second, because the type was previously given such cumbersome designations as "Hohokam buff, brown-paste, sand-tempered variant" (Abbott and Gregory 1988), the name properly affiliates the type with Gila Plain, Salt Variety, which shares paste and temper characteristics with Hohokam Buff Ware. The brown paste and sand temper of Salt Red-on-buff warrants the separate ware category that analysts created for it. Also, the formal naming and description of the type may help to avoid the misidentification of Salt Red-on-buff as one of the Tucson Basin Red-on-brown types, which has occurred in the past. Although the design styles of Salt Red-on-buff changed through time, the division of Salt Red-on-buff into separate types would be counterproductive, given the wide recognition of the type names associated with the traditional Hohokam Buff Ware design styles. Because the changes in design style mirror those of traditional Hohokam Buff Ware, analysts recommend that the Hohokam design style in which individual examples of Salt Red-on-buff were decorated be expressed after the type name (e.g., Salt Red-on-buff, Santa Cruz style or Salt Red-on-buff, Casa Grande style).

As noted above, Salt Red-on-buff is essentially locally produced Gila Plain, Salt Variety to which the potter added a buff-colored slip and red decoration. The paste color (brown) and temper (composed of angular to subrounded, predominantly quartz and feldspar sand) is indistinguishable from that of its sibling plainware. The slip is slightly darker than slips found on traditional Hohokam Buff Ware, though the design style and execution of the Dutch Canal Ruin examples appeared to be nearly identical. Abbott and Gregory (1988) reported that the paste and temper of Salt Red-on-buff remains remarkably constant from the Gila Butte through the Casa Grande styles. Salt Red-on-buff may represent attempts by Salt River Hohokam communities to supplement the supply of buff-paste, micaceous buffware imported from the Gila River Valley with similar, locally produced vessels.

Roosevelt Red Ware

Although generally not considered true Hohokam ceramics because of their apparent origin with the Salado in the Tonto Basin, Roosevelt Red Ware types are common in Civano and Polvorón phase contexts in Hohokam sites. The ware traditionally has been called Salado Red Ware (Haury 1945) or Salado Polychrome (Wood 1987) by researchers who have assumed that it was produced only by the Salado and exchanged widely. However, evidence indicates that potters produced at least some examples of these types within the Salt River Valley (Crown 1981, 1991; Crown, Schwalbe, and London 1988). Alternatively, consistency within the paste and temper materials supports the premise that production occurred within a common location. Archaeologists have not recognized the paste, in particular, as having been locally available, suggesting that potters had produced Roosevelt Red Ware vessels outside of the Salt-Gila Basin (Volume 4:Chapter 11). In any case, analysts for this assemblage chose the category "Roosevelt Red Ware" to allow for the possibility of manufacture by groups other than the Salado.

Roosevelt Polychromes

Roosevelt polychrome types, which include the earlier Pinto Polychrome and the later Gila and Tonto polychromes, were common within Classic and post-Classic period contexts at Dutch Canal Ruin, though investigators noted only four examples of Pinto Polychrome. These polychrome types generally consist of red-slipped, sand- and hornblende-tempered bowls and jars that are overlain with a white slip and decorated with black carbon or iron-carbon paint. In form and design characteristics, the examples of Roosevelt polychromes recovered from Dutch Canal Ruin conformed to previous descriptions of the three types (Crown 1981; Haury 1945; Wood 1987).

Pinto Polychrome

Pinto Polychrome, the earliest of the three polychrome types, is characterized by very fine black lines on a white background and an abundance of bold scrolls filled with fine, well-executed hachure, similar in style to the White Mountain Red Ware types. The design is confined to the interiors of hemispherical bowls, though archaeologists have also encountered rare examples of jar forms.

Gila Polychrome

Much broader brushwork, sloppier execution, and a wider variety of motifs typify the black-on-white designs found on Gila Polychrome bowls and, more rarely, jars. Bowl forms are commonly hemispherical with slightly everted rims. A band of decoration sometimes appears just below the rim on the exteriors of bowls, in addition to the interior design.

Tonto Polychrome

Tonto Polychrome incorporates the red-slipped background as an integral part of the design pattern, usually resulting in a more expansive layout. This design style is usually confined to the exteriors of bowls and jars. Sometimes, a Tonto-style design on the exterior of a bowl is coupled with an interior decorated in the Gila Polychrome style.

Phoenix Red

Chemical analyses of Roosevelt polychromes and Phoenix Red from Las Colinas supported the notion that the latter is simply an unpainted version of the polychrome types. Also, petrographic analysis of Phoenix Red has shown it to be identical to other Roosevelt Red Ware types in temper composition (Volume 4:Chapter 11). For these reasons, analysts included this type in the Roosevelt Red Ware category.

Phoenix Red represents a type that had been previously identified and described (Crown 1981) but was not named until the analysis of the 1982–1984 Las Colinas ceramic assemblage (Abbott and Gregory 1988). Investigators identified a few examples of the type in the Dutch Canal Ruin assemblage.

Phoenix Red is very similar to, and possibly indistinguishable from, unpainted sherds of Roosevelt polychromes. X-ray fluorescence analysis of the Las Colinas assemblage strongly suggested that the clay source for the polychrome types and the Phoenix Red recovered from the site was the same, and probably local (Crown, Schwalbe, and London 1988). All the Phoenix Red sherds recovered from Dutch Canal Ruin had pronounced dark carbon streaks at the center of the core, with tan or brown paste at each margin. The temper is a well-sorted arkosic sand, typically much smaller and less abundant than in any of the Hohokam Red Ware types. The slip color is a distinctive bright or raspberry red, though poor firing conditions result in some muddy brown examples. Both surfaces of bowls and the exteriors of jars are slipped, and unslipped surfaces are usually a light tan color. Smudging is never present on either bowls or jars.

Salado Red

This type was represented by only three sherds at Dutch Canal Ruin, all of which were essentially obliterated, indented corrugated versions of Phoenix Red jar sherds (Wood 1987). Temper, paste, and slip color were otherwise identical to such attributes of Phoenix Red. Wood (1987) described this type as having originated outside the Tonto Basin, in the Sierra Ancha range to the east, though the type is very common at core-area Salado sites.

Intrusive Wares

Although analysts identified nine ceramic types from outside the Phoenix or Tonto basins in the Dutch Canal Ruin assemblage, only two intrusive wares (Tucson Plain and Lower Colorado Buff) occurred in sufficient quantity to warrant formal descriptions. These two wares accounted for about 96% of the intrusive ceramics recovered from Dutch Canal Ruin.

Tucson Plain Ware

Just over half of the sherds classified as Tucson Plain Ware were unidentifiable as to type. Most of these sherds closely fit Wood's (1987) definition of Gila Plain, Tucson Variety, though this type varies across its production area and has thus been given many names, for example, "Palo Parado Plain" and "Ramanote Plain" (Wood 1987). The variety most commonly found at Dutch Canal Ruin is characterized by finely sorted quartz sand temper, a dense brown paste with a crisp break, and floated and often polished surfaces. Many of the better-finished examples may simply be sherds from the undecorated portions of Tucson Basin Red-on-brown vessels. Sells Plain, which made up about one-third of the intrusive ceramics analyzed, was the only Tucson Plain type that was identifiable in the Dutch Canal Ruin assemblage.

Sells Plain was produced over a broad area in southern Arizona, and many local variations have been noted (Wood 1987), but the examples recorded for the present analysis shared several attributes. Vessel walls are thick (6-10 mm), and the temper is coarse, with quartz sand dominating, though analysts have noted some biotite, schist, and phyllite. The paste is most often red-orange to gray, though surfaces are always gray, untreated, and usually quite rough. Often, the red-orange core of the sherds stands in marked contrast to the dark gray surfaces at either margin.

Lower Colorado Buff Ware

This southwestern Arizona Patayan ware, which accounted for about 23% of the intrusive wares from Dutch Canal Ruin, is very poorly understood and has been separated inconsistently into a variety of types (Beckwith 1988; Colton 1939; Rogers 1945; Waters 1982b). Because of these inconsistencies, and because the variation within the ware was not characterized by apparent breaks along type lines, analysts made no attempt to separate types within this ware classification. The surfaces of most examples of this ware from Dutch Canal Ruin ranged in color from orange to buff to salmon, often with violet and pink bands visible in the core cross section. The paste is usually gritty, and finely crushed sherd, grog, and sometimes calcium carbonate constitute the temper. Both the surfaces of bowls and the exteriors of jars sometimes exhibit a wiped appearance, with bold, dark striations suggesting the use of a corn husk or similar soft material as a smoothing tool. A fugitive, translucent red slip is sometimes applied to the visible surfaces of vessels.

THE DUTCH CANAL RUIN ASSEMBLAGE

The assemblage of sherds recovered from Dutch Canal Ruin was not particularly large or diverse, but it did provide a good view of what was left at this riverine settlement, which had probably been occupied both seasonally and permanently at times from the late Pioneer period through the post-Classic period. Table 12.3 presents the total type frequencies of all ceramics recovered from the site.

Ceramic Frequencies by Weight

Analysts recorded the weights of all ceramic items in an attempt to standardize the count totals for each ceramic type and to provide a comparative base for future research that may investigate the use of weight totals in ceramic analysis. The usefulness of the weight data presented in Table 12.4 is limited, however, by certain methodological considerations. Although weight totals for ceramic categories initially may seem to reflect the relative abundance and importance of the types more accurately than individual sherd counts, a closer look at the analytical significance of each type of ceramic total revealed that comparison by weight was confounded by certain factors that caused plainware types to be overrepresented in the assemblage. Plainware vessels were generally larger than their decorated or redware counterparts, and thus the sherds that resulted from the breakage of these vessels were larger in both surface area and thickness. Comparison by weight, then, was really a comparison of the total mass of ceramic material that was in use at the site, while comparison by sherd count was a better indicator of the frequencies of whole vessels. If the goal of ceramic type comparison is to determine the relative frequencies of different types of ceramic vessels that were in use prehistorically, sherd count is the better of the two comparisons.

Table 12.3. Ceramic Frequencies from Dutch Canal Ruin

Ceramic Ware/Type	Absolute Frequency	Percent Ware	Percent Assemblage
Hohokam Plain Ware	**9,214**		76.4
Gila Plain, Salt Variety	8,539	92.7	70.8
Gila Plain, Gila Variety	228	2.5	1.9
Squaw Peak Plain	409	4.4	3.4
Wingfield Plain	37	0.4	0.3
Phoenix Plain	1	0.0	0.0
Hohokam Red Ware	**2,068**		17.1
Salt Red	2,012	97.3	16.7
Gila Red	27	1.3	0.2
Squaw Peak Red	21	1.0	0.2
Wingfield Red	6	0.3	0.0
Indeterminate Hohokam Red	2	0.1	0.0
Hohokam Decorated Ware	**184**		1.5
Snaketown Red-on-buff	3	1.6	0.0
Gila Butte Red-on-buff	10	5.4	0.1
Santa Cruz Red-on-buff	35	19.0	0.3
Santa Cruz/Sacaton Red-on-buff	6	3.3	0.0
Sacaton Red-on-buff	6	3.3	0.0
Sacaton/Casa Grande Red-on-buff	2	1.1	0.0
Casa Grande Red-on-buff	11	6.0	0.1
Indeterminate Red-on-buff	27	14.7	0.2
Salt Red-on-buff, Indeterminate	3	1.6	0.0
Indeterminate Buff Ware	81	44.0	0.7
Roosevelt Red Ware	**313**		2.6
Pinto Polychrome	4	1.3	0.0
Gila Polychrome	180	57.5	1.5
Tonto Polychrome	31	9.9	0.3
Phoenix Red	31	9.9	0.3
Salado Red	3	1.0	0.0
Indeterminate Roosevelt Polychrome	64	20.4	0.5
Intrusive Ware/Type	**241**		2.0
Sells Plain	79	32.8	0.7
Tanque Verde Red-on-brown	1	0.4	0.0
Lower Colorado Buff Ware	55	22.8	0.5
Prescott Black-on-gray	1	0.4	0.0
Mimbres Corrugated	1	0.4	0.0
Linden Corrugated	1	0.4	0.0
Jeddito Black-on-yellow	4	1.7	0.0
Indeterminate Tucson Plain Ware	97	40.2	0.8
Indeterminate Prescott Gray Ware	2	0.8	0.0
Unidentifiable Ware/Type	**39**		0.3
Total	**12,059**		100.0

Table 12.4. Ceramic Frequencies from Dutch Canal Ruin, by Weight

Ceramic Ware/Type	Absolute Weight (grams)	Percent Ware	Percent Assemblage
Hohokam Plain Ware	93,024		78.8
Gila Plain, Salt Variety	85,778	92.2	72.7
Gila Plain, Gila Variety	1,991	2.1	1.7
Squaw Peak Plain	4,849	5.2	4.1
Wingfield Plain	391	0.4	0.3
Phoenix Plain	15	0.0	0.0
Hohokam Red Ware	19,058		16.2
Salt Red	18,037	94.6	15.3
Gila Red	295	1.5	0.3
Squaw Peak Red	336	1.8	0.3
Wingfield Red	67	0.4	0.1
Indeterminate Hohokam Red	323	1.7	0.3
Hohokam Decorated Ware	1,132		1.0
Snaketown Red-on-buff	26	2.3	0.0
Gila Butte Red-on-buff	58	5.1	0.0
Santa Cruz Red-on-buff	234	20.7	0.2
Santa Cruz/Sacaton Red-on-buff	48	4.2	0.0
Sacaton Red-on-buff	62	5.5	0.1
Sacaton/Casa Grande Red-on-buff	7	0.6	0.0
Casa Grande Red-on-buff	89	7.9	0.1
Indeterminate Red-on-buff	158	14.0	0.1
Salt Red-on-buff, Indeterminate	23	2.0	0.0
Indeterminate Buff Ware	427	37.7	0.4
Roosevelt Red Ware	1,751		1.5
Pinto Polychrome	20	1.1	0.0
Gila Polychrome	902	51.5	0.8
Tonto Polychrome	231	13.2	0.2
Phoenix Red	245	14.0	0.2
Salado Red	69	3.9	0.1
Indeterminate Roosevelt Polychrome	284	16.2	0.2
Intrusive Ware/Type	2,682		2.3
Sells Plain	1,211	45.2	1.0
Tanque Verde Red-on-brown	2	0.1	0.0
Lower Colorado Buff Ware	696	26.0	0.6
Prescott Black-on-gray	8	0.3	0.0
Mimbres Corrugated	11	0.4	0.0
Linden Corrugated	7	0.3	0.0
Jeddito Black-on-yellow	14	0.5	0.0
Indeterminate Tucson Plain Ware	723	27.0	0.6
Indeterminate Prescott Gray Ware	10	0.4	0.0
Unidentifiable Ware/Type	329		0.3
Total	117,976		100.0

Spatial, Temporal, and Functional Implications

Spatial Implications

The small number of ceramic items associated with the structures and other features in Areas 1-7 supported the contention that Dutch Canal Ruin was a farmstead site that had been seasonally occupied through the Santa Cruz phase of the Colonial period. After an apparent lack of activity, or at least a drastic decline in activity, within the project area during the Sedentary period, permanent occupation of the site during the Classic period was indicated at Area 8 and possibly at Areas 9 and 10. However, investigators uncovered no structures in the latter two loci to support this notion. Area 8 yielded a particularly large amount of ceramic material and two pit houses, indicating a permanent Polvorón phase settlement during that time. Table 12.5 presents ceramic frequencies for all of the research areas.

Temporal Implications

The appearance of Roosevelt Red Ware types, particularly polychromes, in sites within the Hohokam core area has traditionally been the indicator of the onset of the Civano phase of the Classic period. Recently, the use of Roosevelt Red Ware in making temporal inferences has advanced beyond a simple presence/absence approach, as relative percentages of Roosevelt Red Ware ceramics in Hohokam contexts have proved useful in the identification of Polvorón phase sites, areas, and features (Peterson and Abbott 1991). At Dutch Canal Ruin, percentages of Roosevelt Red Ware ceramics helped confirm the Polvorón phase designation given to Area 8 and strongly suggested a Polvorón phase component at Area 10, though investigators identified no diagnostic Polvorón phase features there. The percentage of this ware at Area 9, the only other locus of Dutch Canal Ruin to yield more than a trace of Roosevelt Red Ware, appeared to be more consistent with a Civano phase occupation, although a ceramic mixture from several phases could mask a Polvorón phase association.

One important methodological consideration enters into the usefulness of percentages of Roosevelt Red Ware. Calculating Roosevelt Red Ware as a percentage of the total ceramic assemblage from a given context mixes functional types, thus confounding the results. In this analysis, archaeologists treated decorated ceramics, including both Hohokam Buff Ware and Roosevelt Red Ware, as a separate functional class from undecorated ceramics, including redware and plainware types; however, an overabundance of plainware ceramics resulting from a largely utilitarian function of a context could artificially deflate the apparent importance of Roosevelt Red Ware. The percentages from the three Dutch Canal Ruin areas and other known Polvorón and Civano phase contexts presented in Table 12.6 indicate the relative amount of Roosevelt Red Ware pottery present among decorated ceramics only.

Another trend that became apparent during the analysis of the ceramics from Dutch Canal Ruin and Pueblo Salado was an inverse relationship between the frequencies of schist-tempered Squaw Peak Plain and Roosevelt Red Ware. As Table 12.7 shows, the frequency of Squaw Peak Plain as a percentage of all plainware types decreased as the frequency of Roosevelt Red Ware as a percentage of all decorated wares increased. Results of the Pueblo Salado ceramic analysis supported this trend from the Soho phase through the Civano and Polvorón phases (Volume 3:Chapter 5). Presumably, the manufacture of schist-tempered plainware ceramics decreased from the Soho through the Polvorón phases, due perhaps to reduced access to schist sources near Squaw Peak. This decrease in Squaw Peak Plain also reflects the general Classic period trend toward the use of sand

Table 12.5. Ceramic Frequencies from Dutch Canal Ruin, by Area

Ceramic Ware/Type	1	2	3	4	5	6	7	8	9	10
Hohokam Plain Ware										
Gila Plain, Salt Variety	13/72.2*	100/64.9	62/43.4	85/89.5	20/95.2	32/61.5	43/55.1	7337/74.7	171/29.0	676/63.4
Gila Plain, Gila Variety		12/7.8	11/7.7	8/8.4		6/11.5	18/23.1	129/1.3	41/7.0	3/0.3
Squaw Peak Plain		24/15.6	33/23.1	2/2.1		9/17.3	8/10.3	96/1.0	215/36.5	22/2.1
Wingfield Plain			3/2.1					30/0.3	3/0.5	1/0.1
Phoenix Plain								1/0.0		
Hohokam Red Ware										
Salt Red		2/1.3			1/4.8			1716/17.4	58/9.8	235/22.0
Gila Red								19/0.2	8/1.4	
Squaw Peak Red			1/0.7					16/0.2	4/0.7	
Wingfield Red								6/0.1		
Indeterminate Hohokam Red								2/0.0		
Hohokam Decorated Ware										
Snaketown Red-on-buff									2/0.3	1/0.1
Gila Butte Red-on-buff			1/0.7						9/1.5	
Santa Cruz Red-on-buff	5/27.8		16/11.2						9/1.5	4/0.4
Santa Cruz/Sacaton Red-on-buff						1/1.9	1/1.3		4/0.7	1/0.1
Sacaton Red-on-buff									6/1.0	
Sacaton/Casa Grande Red-on-buff									2/0.3	
Casa Grande Red-on-buff								3/0.0	8/1.4	
Salt Red-on-Buff, Indeterminate						1/1.9		1/0.0	1/0.2	
Indeterminate Red-on-buff		7/4.5				2/3.9		7/0.1	5/0.8	
Indeterminate Buff Ware		9/5.8	10/7.0			1/1.9	7/9.0	33/0.3	18/3.1	3/0.3
Roosevelt Red Ware										
Pinto Polychrome								3/0.0		1/0.1
Gila Polychrome							1/1.3	165/1.7	5/0.8	9/0.8
Tonto Polychrome								27/0.3		4/0.4
Phoenix Red								8/0.1	17/2.9	6/0.6
Salado Red								2/0.0		1/0.1
Indeterminate Roosevelt Polychrome								61/0.6		3/0.3

Area

Table 12.5. Ceramic Frequencies from Dutch Canal Ruin, by Area, continued

Ceramic Ware/Type					Area					
	1	2	3	4	5	6	7	8	9	10
Intrusive Wares										
Sells Plain									1/ 0.2	78/ 7.3
Tanque Verde Red-on-brown								1/ 0.0		
Lower Colorado Buff Ware								35/ 0.4	2/ 0.3	18/ 1.7
Prescott Black-on-gray								1/ 0.0		
Mimbres Corrugated								1/ 0.0		
Linden Corrugated								1/ 0.0		
Jeddito Black-on-yellow								4/ 0.0		
Indeterminate Tucson Plain Ware								97/ 1.0		
Indeterminate Prescott Gray Ware								2/ 0.0		
Unidentified Ware/Type								39/ 0.4		
Total	18	154	143	95	21	52	78	9843	589	1066

*Absolute frequency/percentage

Table 12.6. Roosevelt Red Ware Frequencies from Selected Hohokam Sites

Site Locus	Roosevelt % of Decorated	Roosevelt % of Total	Phases Represented	Total Assemblage
Dutch Canal Ruin/Area 10	75.0	2.3	Santa Cruz–Polvorón (?)	1,066
Dutch Canal Ruin/Area 8	85.8	2.7	Polvorón	9,843
Pueblo Grande (Abbott 1993)	66.9	2.8	Polvorón	17,372
El Polvorón (Sires 1984)	88.6	5.2	Polvorón	82,889
Dutch Canal Ruin/Area 9	25.6	3.7	Snaketown–Civano	589
Pueblo Grande (Abbott 1993)	29.1	0.8	Civano	16,873
Pueblo del Monte (Weaver 1977)	31.7	0.4	Civano	9,374
Casa Buena (Cable and Gould 1988)	24.2	0.3	Soho/Civano	14,967

After Peterson and Abbott 1991
Note: Data from Dutch Canal Ruin is from eastern (predominantly Classic period) portion of site

Table 12.7. Roosevelt Red Ware and Squaw Peak Plain Ware as Percentage of Wares, by Area

Area	Roosevelt Red % of Decorated Wares	Total Decorated Wares	Squaw Peak Plain % of Plain Wares	Total Plain Wares
8	85.8	310	1.3	7593
9	25.6	86	50.0	430
10	75.0	33	3.1	702

temper in locally produced undecorated wares. In any case, the percentage of Squaw Peak Plain may also serve as a temporal indicator through the post-Classic period, at least at these two sites.

Functional Implications

One goal of the Dutch Canal Ruin ceramic analysis was to provide information regarding the function of the site and the features found within the site. Investigators defined three approaches to accomplishing this goal: (1) comparison of ceramic types deposited within various feature types at the site; (2) comparison of vessel forms deposited at these feature types; and (3) comparison of vessel forms deposited at Dutch Canal Ruin with vessel forms recovered from similar sites.

The total bowl/jar ratio from Dutch Canal Ruin, shown in Table 12.8, was very similar to those for other known Hohokam farmstead sites, which have been defined as sites that may have been seasonally or permanently occupied and consisted of one or two contemporaneous structures and associated extramural features and activity areas (Heacock 1988). The Baccharis Site, for instance, located in the northern Salt River Valley, yielded a bowl/jar ratio of 0.7:1 and had other characteristics of a farmstead (Heacock 1988). When analysts separated this ratio by feature type, as shown in Table 12.8, they perceived that areas within structures and in extramural features shared a similar bowl/jar ratio.

Table 12.8. Vessel Form Frequencies for Dutch Canal Ruin by Feature Type

Feature Type	Bowl	Jar	Bowl/Jar Ratio
Structure	184	221	0.8:1
Extramural Pit	73	108	0.7:1
Canal	10	7	1.4:1
Cremation	·8	2	4.0:1
Other/No Feature	102	188	0.5:1
Total	**377**	**526**	**0.7:1**

Analysts could define no appreciable difference among the types of ceramics deposited within various contexts at Dutch Canal Ruin, as shown in Table 12.9. Decorated wares, plainwares, redwares, and intrusive wares seemed not to have been used differentially for specific functions associated with the various feature types. Also, the types of ceramics found at Dutch Canal Ruin were similar to the types found at Hohokam sites of all types within the Salt River Valley.

Intrusive Ceramics

Given the ceramic assemblage from Dutch Canal Ruin, analysts assigned 241 sherds to one of seven intrusive ware categories. An additional 39 sherds, unidentifiable to either ware or type category, were probably intrusive as well. Two geographic areas, the Tucson Basin and western Arizona, accounted for nearly 96% of all intrusive ceramics identified. This stands in marked contrast to the percentages of intrusive ceramics from other Hohokam sites (Table 12.10), where a large proportion were ceramics from northern Arizona. The counts presented in Table 12.10 may not be entirely comparable, particularly in the category of "western Arizona," in which Lower Colorado Buff Ware has been quantified only at Las Colinas (Beckwith 1988) and at Dutch Canal Ruin. Four sherds, all Jeddito Black-on-yellow, represented Anasazi influence at Dutch Canal Ruin, whereas other Hohokam intrusive assemblages from the Salt and Gila basins yielded between 17% and 80% Anasazi whiteware and yellowware pottery.

Crown (1985b) proposed that ceramic interaction between the Hohokam and other cultures of the Southwest was strongly influenced by the Gila and Salt rivers, which served as barriers to ceramic exchange during portions of the Hohokam sequence. Although Doyel's (1991) compilation of more recent data indicated that these rivers had not served as boundaries to the degree that Crown's data suggested, the basic hypothesis persists. Colonial and Sedentary period sites north of the Salt River contain the largest percentage of northern Arizona pottery, sites south of the Gila

Table 12.9. Ceramic Type Frequencies for Dutch Canal Ruin by Feature Type

Ceramic Type	Feature Type							
	Structure		Extramural Pit		Canal		Cremation	
	N	%	N	%	N	%	N	%
Hohokam Plain Ware	2221		1334		162		44	
Gila Plain, Salt Variety	2106	76.4	1288	73.6	117	65.4	43	91.5
Gila Plain, Gila Variety	47	1.7	20	1.1	14	7.8	1	2.1
Squaw Peak Plain	59	2.1	17	1.0	31	17.3		
Wingfield Plain	9	0.3	9	0.5				
Hohokam Red Ware	431		316		2		2	
Salt Red	406	14.7	297	17.0	2	1.1	2	4.3
Gila Red	1	0.0	3	0.2				
Squaw Peak Red	22	0.1	10	0.6				
Wingfield Red			6	0.3				
Indeterminate Red	2	0.1						
Hohokam Decorated Ware	21		19		15			
Gila Butte Red-on-buff	1	0.0						
Santa Cruz Red-on-buff			11	0.6				
Casa Grande Red-on-buff			1	0.1				
Indeterminate Red-on-buff	9	0.4	1	0.1	2	1.1		
Indeterminate Buff	11	0.4	6	0.3	13	7.3		
Roosevelt Red Ware	74		39					
Pinto Polychrome	1	0.0	1	0.1				
Gila Polychrome	52	1.9	25	1.4				
Phoenix Red	6	0.2	4	0.2				
Indeterminate Roosevelt Polychrome	15	0.5	9	0.5				
Intrusive Ware/Type	31		41				1	
Indeterminate Tucson Plain	23	0.8	23	1.3				
Jeddito Black-on-yellow	1	0.0	1	0.1				
Lower Colorado Buff Ware	7	0.3	16	0.9			1	2.1
Tanque Verde Red-on-brown			1	0.1				
Total	2778		1749		179		47	

River contain the most southern Arizona pottery, and sites between the rivers exhibit intrusive ceramics from both northern and southern Arizona (Crown 1985b:443). From the beginning of the Civano phase of the Classic period and later, the pattern changes. Hopi yellowware appears as an intrusive ware for the first time, and types from the Tucson Basin predominate over other southern Arizona types, irrespective of a site's relationship to drainages in the basin (Crown 1985b:443). Investigators recovered no intrusive wares from pre-Classic period contexts at Dutch Canal Ruin during the data recovery phase, but the pattern of these wares for the Classic period (and later) supported the distribution Crown identified. The pattern of intrusive wares within Dutch Canal Ruin was supported by the earlier work in the Squaw Peak Parkway corridor (Greenwald and Ciolek-Torrello 1988a:Table 10), in which 1.7% of the ceramic assemblage was represented by southern Arizona types. These data compared favorably with the current work, in which 1.5% of the assemblage was composed of types from southern Arizona (Table 12.3).

Table 12.10. Intrusive Ceramics from Selected Hohokam Sites

Ware			Site			
	Dutch Canal Ruin (N=241)	Dutch Canal Ruin[1] (N=12)	Snaketown[2] (N=824)	La Ciudad[3] (N=171)	Pueblo Grande[4] (N=368)	Las Colinas[5] (N=5622)
Kayenta/Tusayan			26	47	30	17
Hopi	2			10	9	tr
Little Colorado				12	4	1
Cibola			1	11	7	tr
White Mountain Red Ware				1	5	tr
Mogollon/Mimbres	tr		58			1
Corrugated Brown Ware	tr		1	4	18	1
Southern Arizona	73	100	2	9	17	4
Western Arizona	23				*	74
Prescott	1				2	1
Other			13	6	7	1

After Doyel 1991:235
[1]Greenwald and Ciolek-Torrello 1988a
[2]Haury 1976:328; includes 1935 and 1964–1965 excavations
[3]Wilcox 1987
[4]Doyel 1991
[5]Beckwith 1988; 1982–1984 excavations
tr = trace
*Ceramic type later identified in assemblage but not quantified

Miscellaneous Ceramic Objects

Spindle Whorls

Field crews at Dutch Canal Ruin recovered one object similar to an ellipsoidal hand-molded spindle whorl (Haury 1976:348) but with the perforation extending only partway through the piece. The object was 3.6 cm at its maximum diameter and 3.4 cm in height. The function of this object was unknown.

Field crews also recovered two fragments of perforated sherd spindle whorls. Both were from Salt Red vessels and were biconically drilled. One measured 3.9 cm in diameter and the other, though less than half complete, was projected to have been 4.6 cm in diameter.

Worked Sherds

The Dutch Canal Ruin assemblage also included fragments of three sherds purposely worked into discs. The three would have ranged in diameter from 4.2 cm to 6.8 cm. The types represented by the discs were Gila Plain, Salt Variety; Salt Red; and Gila Polychrome. Five other worked sherds appeared to have functioned as pottery scrapers to smooth vessel walls (Haury 1945:121). All of these sherds were Gila Plain, Salt Variety.

Figurine

A tiny fragment of a hand-molded ceramic figurine was the only example of this artifact type recovered from Dutch Canal Ruin. The piece appeared to represent the neck area of a human form, with a series of small holes, possibly a representation of a necklace, pressed into the front.

SUMMARY

Analysts made two general observations regarding the Dutch Canal Ruin ceramic assemblage. First, a rather distinctive dichotomy existed between the pre-Classic periods and the later period assemblages; second, the assemblage was heavily weighted by Classic and post-Classic period types collected principally from Areas 8 and 10. Although the intrusive assemblage was dominated by wares from western and southern Arizona, these intrusive ceramics appeared almost exclusively in Classic and post-Classic period contexts. This pattern suggested that the inhabitants of Dutch Canal Ruin had had more contact with other areas (whether through trade or other spheres of interaction) in the later periods than in the pre-Classic periods. Moreover, the early and late occupation of Dutch Canal Ruin represented entirely different settlement strategies. The general tendency has been for field house and farmstead sites such as those of the pre-Classic periods at Dutch Canal Ruin to produce small artifact assemblages; thus, the intrusive ceramic assemblage from these types of sites has always been limited. This pattern was present at Dutch Canal Ruin. The ceramic assemblage from the permanent settlements during the Classic and post-Classic periods exhibited higher frequencies and greater variability, including increased frequencies of intrusive wares.

The 10 areas investigated during the data recovery phase at Dutch Canal Ruin produced ceramic assemblages that were consistent with the results of the testing phase. The ceramic assemblage indicated that the site had been occupied by the latter part of the Pioneer period, as represented by Snaketown Red-on-buff ceramics (Greenwald and Ciolek-Torrello 1988a). The main period of occupation prior to the Classic period was during the Colonial period. The Sedentary period witnessed a marked decline in site use, with only Area 9 exhibiting ceramic evidence that suggested occupation during this time. Investigators could not determine exactly when Dutch Canal Ruin had first been occupied during the Classic period. The lack of diagnostic types during the Soho phase certainly did not help clarify this issue; however, Areas 8 and 10 both appeared to represent single occupation episodes during the Civano or Polvorón phases. Although a complete hiatus in occupation of Dutch Canal Ruin during the Sacaton and Soho phases was not immediately evident from the ceramics recovered, the results of the ceramic analysis indicated that the use of the project area had markedly declined during that time.

On the whole, the Dutch Canal Ruin ceramic assemblage contained limited quantities of sherds and ceramic artifacts. As a site that apparently was dominated by seasonal field house and farmstead loci, the size and variability of the assemblage was as expected. Activities at the various habitation loci would have focused on routine, daily subsistence practices centered on agriculture.

The ceramic assemblage did not indicate craft production and specialization at the site, although the inhabitants seemed to have practiced ceramic manufacture, as indicated by the presence of pottery scrapers, polishing stones, and pigment on grinding tools (Chapter 14). In summary, the ceramic assemblage indicated that residents, whether occupying the project area on a seasonal or a permanent basis, had focused on basic activities associated with subsistence and day-to-day economics.

CHAPTER 13

THE DUTCH CANAL RUIN FLAKED STONE ASSEMBLAGE

Kimberly Spurr
Dawn M. Greenwald

Flaked stone analysis can provide much information concerning past human activities such as reduction technology, tool use and function, and resource procurement and utilization strategies. Earlier analyses concentrated on formal tools and unique or spectacular items. More recently, archaeologists have noted (Collins 1975; Phagan 1976; Proper 1990; Sullivan and Rozen 1985) that the debitage produced by lithic reduction at a site can provide as much information, if not more, about prehistoric flaked stone technology and function. Through the use of statistics, analysts can reconstruct stages of manufacture and address reduction strategies used by different groups and for different purposes. Because debitage is the by-product of lithic reduction, it is not likely to have been removed from the site of tool manufacture and can provide information on production locations and use areas. An understanding of the reduction technology and the overall lithic assemblage thus can help reconstruct prehistoric behaviors. A study of the distributional patterns of debitage within a site may also clarify functional interpretations of site features.

Although the flaked stone assemblage from Dutch Canal Ruin was not extensive, it was large enough to provide information on general technological and functional trends. Additionally, because both pre-Classic and Classic period occupations were represented at the site, analysts were able to compare the flaked stone from these two distinct time periods. They analyzed all of the flaked stone assemblage to locate patterns pertinent to the research questions posed at the beginning of the investigation.

RESEARCH OBJECTIVES

The analysis of the Dutch Canal Ruin flaked stone assemblage addressed several research questions. One of the objectives was to describe functional attributes of the site and its components (Volume 1:Chapter 11), specifically, to determine if patterned social or functional differences existed in the use of space and if these differences were related to occupational changes through time. Spatial distribution patterns of debitage and tool types, raw material, and kinds of use wear contributed to this determination. Once analysts identified these patterns for Dutch Canal Ruin, they could compare the flaked stone assemblage to assemblages from similar sites and other sites within Canal System 2. They could then address broader questions about the temporal and spatial range of settlement types in the Hohokam culture (Volume 4:Chapter 12).

Another research goal was to use diagnostic lithic artifacts to aid in making temporal designations of features and site areas and help determine settlement and abandonment episodes. Although efforts to link Hohokam projectile point styles with particular periods or phases have met with little success, researchers have noted general trends (Bernard-Shaw 1988; Haury 1976; Huckell 1981). Alternatively, lithic materials may be recovered from contexts that can be dated by ceramics, and these associations can contribute to refinement of Hohokam lithic assemblage chronologies.

Investigators expected to distinguish general trends over time in material types, morphological types, and use-wear patterns in the different areas of Dutch Canal Ruin. Specifically, if the assemblages from Area 8 (Polvorón phase), from Area 10 (primarily Civano phase), and from the other areas (late Pioneer and Colonial periods) differed, analysts hoped to determine whether changes in lithic technology had occurred through time.

ANALYTICAL METHODS

Scott Kuhr and Kimberly Spurr conducted the analysis of the 3801 artifacts in the flaked stone assemblage between May 2 and July 22, 1991. Dawn M. Greenwald, who acted as a consultant, developed the analysis format and variables list.

The analysts used type collections for materials and grain size to ensure consistency. Using 10X hand lenses and low-power binocular microscopes (10–30X), they examined any artifact that macroscopically exhibited possible edge damage. They also inspected any questionable use-wear patterns or artifact classifications.

Although researchers have used a number of criteria in various Hohokam lithic analyses (Elstein 1989; Huckell 1981; Mitchell 1988; Proper 1990), the set of variables designed for this analysis specifically addressed the research questions listed above. Morphological type, tool completeness, and use-wear patterns contributed to goals pertaining to function. Material type, grain size, artifact weight and size, and debitage reduction stages contributed to describing the technological system of the site inhabitants. Ordinal-level evaluations of facial thinning and marginal retouch measured the general level of effort that had been expended on tool manufacture.

Morphological Type

Each artifact, whether a flake or a tool, was classified by morphological type, which is a combination of item form, function, and technology. Whenever possible, morphological types emphasized technological attributes rather than functional types. For example, a flake that had been retouched or purposefully modified on only one face was classified as a uniface rather than as a scraper. The exceptions to this pattern were the categories of projectile point and hammerstone, both functional designations. The presence of facial or marginal retouch, as well as the original "blank" form, were indicators of tool morphological types.

Debitage

Debitage is the debris produced during the manufacture and maintenance of tools and the reduction of cores. The analysis defined several subcategories of debitage, similar to those used by Sullivan and Rozen (1985), to produce data comparable with Sullivan and Rozen's core reduction and tool manufacture groupings (components of their model for interpreting prehistoric technological variability). For example, the category of angular debris refers to pieces of stone that have no complete flake morphology although they may show evidence of partial flake scars. This category also includes "shatter" and chunky residue from reduction and is comparable to Sullivan and Rozen's "debris" category. Flake fragments are flakes that have definable ventral face flake morphology but have incomplete length or width dimensions, as well as incomplete platforms. This category includes sheared flakes, those that break longitudinally along the axis of percussion during reduction. Platform-remnant-bearing (PRB) flakes have a complete platform as well as clear flake morphology but do not have a distal termination. Flakes that have step terminations due to inadequate striking force were included in this type because they could not be distinguished from flakes that had broken distal terminations. PRB flakes are similar to Sullivan and Rozen's "broken flakes" category. Whole flakes are complete, or nearly complete, flakes that have retained their entire platform and enough of their ventral surface that the maximum length and width dimensions are discernible.

Used Flakes

This category includes any type of debitage that exhibits use wear without evidence of production modification. Like retouched tools, used flakes were evaluated for tool completeness and use wear. Analysts considered them the products, rather than the by-products, of the reduction process. Hohokam flaked stone tool assemblages tend to be dominated by used flakes rather than formal tool types (Huckell 1981; Masse 1980), and this was certainly the pattern for the Dutch Canal Ruin assemblage.

Cores

This category includes any artifact with two or more flake scars that was used as a source of flakes. Two general groups of cores were evident in the assemblage. The first group, which included most of the cores, had been used to produce only a few flakes. The second group consisted of small, exhausted cores, with flake scars overlapping on every face. This latter group was less common but indicated some intensive use of resource material.

Small remnants of bipolar cores, usually of obsidian, were evidence of bipolar technology. The obsidian was evidently procured through trade, as no known sources exist within the immediate area.

Core Tools

This morphological type refers to cores that have some kind of secondary modification other than battering, including cores that exhibit evidence of use wear on one or more edges. These tools differed from cobble unifaces and bifaces in that they had functioned originally as cores, with flake scars usually in locations other than on the tool edge. Use wear was evident on both modified and unmodified edges. Most of the core tools were not considered exhausted cores.

Core/Hammerstones

The most common type of use wear exhibited on cores was battering, and those cores that subsequently had been used to batter other materials were classified as core/hammerstones. Often these tools exhibited both negative flake scars and areas with heavy crushing and step fracturing. Cores probably were used as hammerstones only when they were no longer required as cores, because battering use would have produced cones of force and stress points within the material that would have made it unsuitable for flake production.

Hammerstones

A hammerstone is a cobble used to batter some other material and is distinguished from core/hammerstones by the absence of intentional flake removal. The used edges and faces of hammerstones exhibited step fracturing, crushing, and surface pitting. A few showed extensive use wear, but the majority had evidence of only light use. Most of the artifacts in this category were alluvial cobbles of basalt or other dense material.

Hammerstones probably were used for different tasks than were core/hammerstones. Hammerstones have been associated with food processing and have been identified as percussors in flaked stone reduction, while core/hammerstones have been associated with ground stone maintenance and manufacture (Dodd 1979:238-239).

Unifaces

A uniface is a tool with flakes removed from one face. Analysts for the Dutch Canal Ruin assemblage divided unifaces into three categories, depending on the original morphology of the artifact. Cobble unifaces were modified cobbles that retained cortex on both sides and had a general cobble shape. Flake unifaces were modified flakes with some flake morphology on one or both faces and were generally thinner in profile than cobble unifaces. A third group (other unifaces) included items that did not fit into either of the first two categories, such as those without flake morphology or cortex. Analysts created this third category mainly to preserve the homogeneity of the cobble and flake uniface categories.

Bifaces

A biface is a tool with flakes removed from both faces. Analysts divided biface subcategories in the same manner as unifaces (using cobble, flake, and other categories), with cortex and flake morphology criteria also the same. In the Dutch Canal Ruin assemblage, most of the bifaces were flake bifaces and were more intensively shaped than unifacial tools.

Notched Tools

Notched tools are flakes that were intentionally notched by the removal of one or more flakes but that have no other modification. In this analysis the notch referred to a work facet on the tool rather than modification to expedite hafting. The one notched tool recovered from the site had evidence of scraping wear and may have functioned as a spokeshave, for the purpose of smoothing or shaping wood (Figure 13.1a).

Projectile Points

Field crews recovered three projectile points from Dutch Canal Ruin. One was a small, serrated basalt point, and the other two were small, obsidian side-notched points. None of the highly stylized Hohokam projectile points that have been associated with pre-Classic period habitation sites (e.g., Haury 1976) was recovered.

Large Primary Flakes

Large primary flake tools are classified separately in analyses because of the distinctive reduction technique required to produce them (Huckell 1981:178) and because of their assignment

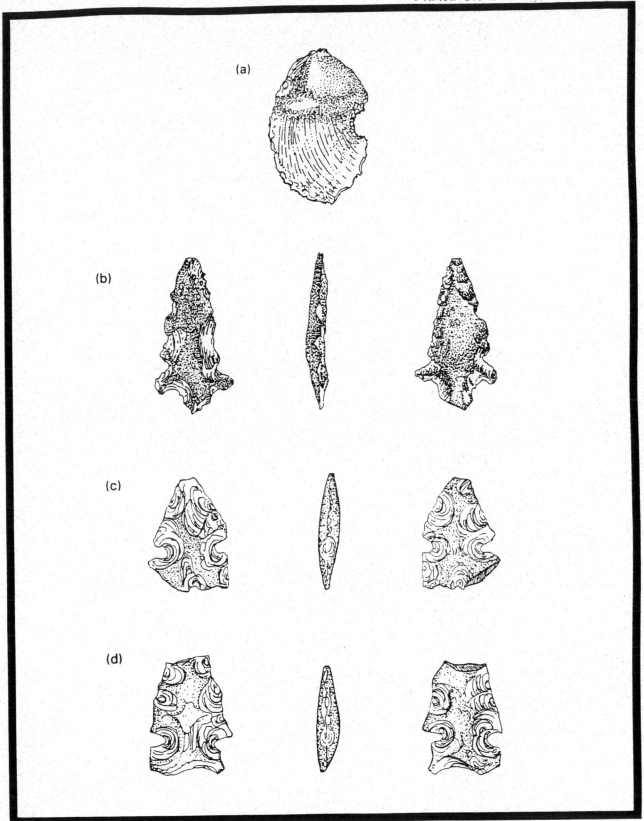

Figure 13.1. Flaked stone, Dutch Canal Ruin: (a) notched tool (length, 1.9 cm); (b–d) projectile points.

as the functional equivalent of tabular tools (Huckell 1981:180; Mitchell 1988:199; Osborne 1965). Analysts for the Dutch Canal Ruin assemblage classified all primary flakes larger than 8 cm² as large primary flakes regardless of use wear. These artifacts were composed of quartzite or igneous material; most were oval and had a plano-convex profile.

Indeterminate Tools

This miscellaneous category was for tools that did not fit into any of the above categories. Many of these artifacts were fragmentary or exhibited morphological characteristics of more than one tool type.

Material Type

The goal of the material type analysis was to provide an accurate and consistent identification of materials at a macroscopic level. Specific mineralogy, therefore, was not included. The analysts did not expect that a detailed identification of the composition of raw materials would contribute to information on raw material selection and use. Other archaeological studies have documented that the Hohokam derived most of their flaked stone from local sources and collected raw material as alluvial cobbles from the river systems within the Phoenix Basin (Elstein 1989; Hoffman and Doyel 1985; Huckell 1981). The Dutch Canal Ruin assemblage followed this pattern.

Information regarding the relative and absolute quantity of raw material types can be useful in determining lithic procurement and trade systems. In addition, correlation between morphological type and material is useful for investigating lithic technological strategies. The Dutch Canal Ruin flaked stone assemblage revealed that the inhabitants of the site had utilized a variety of raw materials, both local and nonlocal.

Chert was the most common sedimentary rock type in the collection. Sandstone occurred occasionally, often as fragments of ground stone that had been reworked by flaking. Only one piece of limestone, classified as debitage, was recorded.

Basalt was the most common igneous rock type in the assemblage, followed by andesitic basalt, rhyolite, and diorite. Obsidian constituted 4.6% of the collection. A very small quantity (< 0.1%) of another igneous material, vesicular basalt, was present, and this only as debitage.

Metamorphic rock types in the assemblage included quartzite, schist and phyllite, metasediment, and several igneous materials that had been altered to varying degrees. Quartzite was the most common type, constituting 3.8% of the assemblage.

Grain Size

This variable represented a relative measure of the grain size, texture, and flakability of each item. By correlating grain size, material, and morphological type, analysts could infer patterns of lithic preference and production. For example, certain classes of tools may have been produced from certain classes of material. Similarly, aspects such as use-wear patterns and tool breakage and discard may have been related to the texture of the material used.

Analysts evaluated grain size using four size classes. Size 1, micro- to cryptocrystalline, included materials such as obsidian and most cherts. Materials in this category had no macroscopically visible grains and had very good flaking characteristics, often showing conchoidal fracture. Size 2, fine grained, was represented by materials such as coarse cherts, fine quartzites, basalts, and other igneous materials in which the grains are visible but are not large enough to cause poor flaking characteristics. The fine granular composition of these materials makes flake morphology easy to recognize. Fine edge retouch is another characteristic of items included in this class. Size 3, medium grained, included basalts, quartzites, sandstones, and crystalline quartz, materials that are composed of grains visible to the naked eye and that create a rough surface. Flake morphology on these materials is less obvious, and use wear is often difficult to see. Natural fracture planes in the rock often cause abnormal flake terminations. Size 4 included coarse materials that contain large crystals or vesicles and show very poor flaking patterns. This size class was represented by porphyritic or vesicular igneous rocks or amorphous quartz. Both flake morphology and use wear are difficult to identify on these materials.

Analysts did not necessarily assign items of a specific material type to the same grain size category but instead evaluated grain size for each artifact, reflecting the range of variation within a single material type. For example, basalt appeared in nearly equal proportions in Sizes 2 and 3. The evaluation depended on the size of the inclusions and crystals in each piece of rock, as well as the matrix. At the beginning of the study, analysts chose type specimens for grain size and used them as references throughout the analysis.

Dorsal Face Evaluation

Dorsal face evaluation is a reduction stage variable, adopted from Phagan's analysis (1976), that allows a consistent description of artifact form through "a ranked evaluation of energy investment into facial regularization or flattening" (Phagan 1976). Facial classification provides a tool evaluation without applying typologies that imply function, such as "blank" or "preform" (Muto 1971), and can be used, along with margin evaluation, to assess the energy expended in the production of each tool.

Face evaluation for the Dutch Canal Ruin flaked stone assemblage included the amount, size, and pattern of retouch; the amount of cortex that remained; the regularization of form (plan outline); and the extent of thinning or flattening (section symmetry). Analysts noted eight levels of facial regularization, ranging from unworked with cortex to highly stylized. Explicit criteria determined each level. The dorsal face was identified by flake morphology (e.g., the presence of previous flake scars), the presence of cortex, or the convexity of the face; for artifacts that were completely reworked, the analyst arbitrarily defined the dorsal face.

Ventral Face Evaluation

Ventral face evaluation, the counterpart to dorsal face evaluation, used the same eight levels of regularization. A ventral face was determined by the presence of morphological attributes such as a bulb of percussion and a concave to straight longitudinal section (curvature). For artifacts that were completely retouched, the analyst arbitrarily defined the dorsal and ventral faces.

Margin Evaluation

Margin evaluation is a relative measure of the degree to which a tool's margin has been retouched and applies only to those tools with manufactured edges. Analysts of the Dutch Canal Ruin assemblage divided margin evaluation into four levels: (1) < 1/4 of the margin retouched; (2) 1/4–1/2 of the margin retouched; (3) 1/2–3/4 of the margin retouched; and (4) all or nearly all of the margin retouched. Analysts used the category of indeterminate tool if they could not determine the original extent of the retouch because of the fragmentary condition of the tool.

Weight

When compared within tool classes, the weight variable can add information on tool breakage and discard patterns. For example, within a given tool type, complete tools generally weigh more than incomplete tools. Evaluation of the weight of complete tools may indicate a preference for material weight or size characteristics. Analysts weighed all recovered tools and cores to the nearest tenth of a gram.

Size

Using a series of squares at intervals of 1 cm, analysts measured all debitage and used flakes to provide an ordinal measure of size. For example, if a flake fit into the third size interval, its maximum dimension lay between 2 cm and 3 cm. Items with a maximum dimension greater than 8 cm had a value of 9. Analysts based this technique on one proposed by Patterson (1982), which used square size templates. Flake size was useful in investigating patterns of tool preference, for example, whether the Hohokam had selected flakes of a certain size for use as tools.

Tool Completeness

This attribute measured the relative amount of a tool that was present. Analysts used four levels in this determination: (1) < 1/4 of the tool present; (2) 1/4–1/2 present; (3) 1/2–3/4 present; and (4) a complete or nearly complete tool. Tools that were too fragmentary for analysts to assess their form completeness were assigned to the category of indeterminate tool.

Use Wear

Analysts inspected any artifact with possible edge damage for evidence of use wear that could provide information on tool function. They made the determination of use-wear type on the basis of edge damage, such as striations, polish, flake scars, attrition, surface rounding, and the continuous pattern of these attributes. They then examined artifacts under 10–30X magnification to evaluate wear patterns. When they observed more than one type of use wear on a tool, they recorded both patterns and noted whether one pattern apparently preceded another. Analysts adopted Neusius's (1988) descriptions of use-wear patterning on tools during the analysis because they are clear, concise, and accurate and can be easily and consistently applied to determine wear type.

Cut/Saw

Cutting/sawing use wear indicates a linear motion of an edge against material to slice the material. Evidence of this type of use may include scarring on one or both faces and striations, near and parallel to an edge:

> The utilized edge will often exhibit a denticulated appearance due to alternate scarring. The majority of scars will be at angles other than perpendicular to the edge, but some perpendicular scars may occur, depending upon the tool morphology and angle of use [Neusius 1988:215].

Scrape

Scraping use refers to a pulling motion transverse to the working edge (Odell and Odell-Vereecken 1980:99), often to remove a layer or portion of material without cutting it. According to Neusius (1988), this action on a tool will produce generally unifacial scarring on much of the tool margin. Any striations or polish are perpendicular to the edge of the tool "and occur primarily on the surface opposite the scars" (Neusius 1988:215).

Shave/Plane

Shaving/planing use is similar to scraping but involves a low-angle pushing motion with the intention of removing small amounts of material without cutting it. Neusius (1988) describes this action as intermediate to cutting and scraping "due to the low angle of the tool during use." Less frequent bifacial scars "tend to be rolled over, almost snapped" (Neusius 1988:215). Polish from abrasion is more likely on the surface in contact with the material, and wear tends to be from attrition rather than scarring (Neusius 1988:215).

Hammer/Batter

This action has the potential to produce a variety of use-wear patterns. Materials may be crushed and broken or may simply be reworked through battering. In any type of hammering, pounding, or battering, the resulting use wear tends to produce pitted and cracked use surfaces and edges and "a series of overlapping, large step fractures although the actual nature of the removals may not be observable due to the extreme crushing which often results" (Neusius 1988:216). Polishing is generally absent.

Chop

Chopping is an activity intermediate between cutting and battering. This type of wear likely represents processing activity. It tends to produce polish, striations, and bifacial scarring "with both hinge and step terminations" (Neusius 1988:216). The striations on chopping tools may be diagonal to the edge.

Pierce/Drill/Grave

All of these types of use wear occur on tool projections. The projections could be produced deliberately, as on drills and burins, or could simply be naturally protruding sections of a flake or tool. Piercing wear is attritional, appearing on "the tip and adjoining margins of the item with decreasing intensity from the tip down" (Neusius 1988:215), with edge scars similar to those produced by cutting. "Both point contact and bending initiations can occur" (Neusius 1988:215).

Drilling use will produce a roughened tip with crushing and either bifacial or unifacial scars. "Usually, these scars are small and have a bending rather than point contact initiation" (Neusius 1988:215).

Neusius also noted,

[Graving] microwear traces are dependent upon the direction of motion. If the tool is used longitudinally, the result will be the type of wear discussed for cutting. If it is used transversely, wear will be similar to that discussed for scraping [Neusius 1988:215].

None

This value was assigned to formal tools that had retouched edges but exhibited no evidence of edge damage due to use. This is a separate category from indeterminate use wear (see below).

Indeterminate

This category applied to artifacts that had edge damage but for which analysts could not accurately determine the type of wear. Often these items had characteristics of several different types of use, or they may have had patterns atypical of the types described above.

Coded Comments

Coded comments provided additional information about the artifacts and included debitage reduction stage evaluations of primary, secondary, or tertiary. Analysts recorded the amount of cortex as 100% (primary flake), as less than 100% (secondary flake), or as none present (tertiary flake). In addition, they also recorded the presence of cortex on the platform of tertiary flakes and noted whether a flake or core had been produced by bipolar reduction.

TECHNOLOGY

Procurement of Raw Materials

In any flaked stone technology, two of the controlling variables are the availability and the quality of raw material. In the Hohokam region, high-quality material was often rare and had to be obtained through some system of exchange. The Hohokam needed to balance the cost of obtaining nonlocal material against the cost of developing a technology utilizing local resources of lower quality. For the more common, expedient flake tools, lower-quality material generally sufficed, but a few specialized tool forms (e.g., projectile points) required high-quality material.

In their discussion of expedient flake tool technology, Parry and Kelly (1987) proposed that in mobile groups, flaked bifaces and other formal tools provide a complete tool assemblage during travel. The multifunctional bifaces decrease the number of tools that must be carried and can serve as cores, allowing the production of flake tools even in areas where suitable raw material is not available. Sedentism, however, decreases the need for such tools, since tools for most specific tasks can be produced expediently from materials of lesser quality that may be found locally. This proposed pattern fits the Hohokam, who used local materials for the majority of tool types.

The following section describes materials that were available to the inhabitants of Dutch Canal Ruin and the patterns that were evident in their use (Table 13.1). The information on material types was compiled with the help of Kirk C. Anderson, Consulting Geomorphologist.

Sedimentary

The most common type of sedimentary rock in the Dutch Canal Ruin assemblage was chert (3.3%). Most of the chert was probably alluvial in origin, as was indicated by the polished cortex on many of the debitage items, and locally available. Chert reported from other Hohokam lithic assemblages (Elstein 1989; Hoffman and Doyel 1985) also commonly appeared as alluvial cobbles. Chert in the Dutch Canal Ruin collection was generally very fine grained, with 83.9% of the material classified as micro- to cryptocrystalline. More granular varieties were present but in smaller quantities. Chert colors varied from white, the most common color, to yellow and reddish brown.

Most of the small amount of sandstone in the collection consisted of fragments from ground stone production and maintenance. One core and 10 pieces of debitage, 0.3% of the assemblage, were of sandstone. Much of this material was heavily burned and may have been used to line hearths. One piece of limestone debitage was recorded, as were a few pieces (0.2%) of undifferentiated sedimentary material, also debitage.

Igneous

The majority (82.4%) of the flaked stone from Dutch Canal Ruin consisted of alluvial cobbles that were igneous in origin. Basalt was the single most common material, constituting 54.9% of the collection. The colors ranged from black to dark green. Texture also varied from very coarse, with obvious crystals of olivine and pyroxene, to a fine, almost aphanitic ground mass. Even coarse basalts were utilized for all types of tools except projectile points. Analysts thought it likely that the basalts were of local origin, probably from the mountain rages surrounding the Phoenix Basin.

Related to the basalts is andesitic basalt that grades into andesite. Although macroscopically andesitic basalt often resembles basalt, it is slightly less mafic than basalt. The andesitic basalt in the project assemblage was probably pre-Cambrian and had been metamorphosed mainly by pressure, evident in the presence of greenish crystals (plagioclase minerals altering to chlorite and epidote) in the darker matrix. These rocks range from very fine to medium grained and often contain inclusions. This material probably corresponded to the "metabasalt" described by Elstien (1989:656).

Obsidian made up 4.6% of the assemblage and included two of the three projectile points and 35.9% of the used flakes. Almost one-third (30.1%) of the obsidian showed evidence of bipolar

Table 13.1. Flaked Stone Morphological Type by Material Type

Morphological Type	Material Type									Total
	Quartz	Chert	Obsidian	Sedimentary	Basalt	Igneous Other	Quartzite	Meta-sediment	Metamorphic Other	
Angular Debris	15	8	3	4	65	46	10	1	6	158
	*9.5	5.1	1.9	2.5	41.1	29.1	6.3	0.6	3.8	**4.2
	**18.8	6.5	1.7	21.1	3.1	4.4	6.8	3.9	5.9	
Flake Fragment	25	47	38	6	777	328	49	11	32	1313
	1.9	3.6	2.9	0.5	59.2	25.0	3.7	0.8	2.4	34.5
	31.3	37.9	21.7	31.6	37.3	31.4	33.6	42.3	31.7	
PRB[1] Flake	10	20	23		409	204	26	3	24	719
	1.4	2.8	3.2		56.9	28.4	3.6	0.4	3.3	18.9
	12.5	16.1	13.1		19.6	19.5	17.8	11.5	23.8	
Whole Flake	25	33	66	8	697	385	47	7	36	1304
	1.9	2.5	5.1	0.6	53.5	29.5	3.6	0.5	2.8	34.3
	31.3	26.6	37.7	42.1	33.4	36.8	32.2	26.9	35.6	
Used Flake	1	9	37		24	26	4	2		103
	1.0	8.7	35.9		23.3	25.2	3.9	1.9		2.7
	1.3	7.3	21.1		1.2	2.5	2.7	7.7		
Core	2	5	3	1	40	28	2			81
	2.5	6.2	3.7	1.2	49.4	34.6	2.5			2.1
	2.5	4.0	1.7	5.3	1.9	2.7	1.4			
Core Tool			1		5	5				11
			9.1		45.5	45.5				0.3
			0.6		0.2	0.5				
Core/ Hammer-stone					24	3	2		1	30
					80.0	10.0	6.7		3.3	0.8
					1.2	0.3	1.4		1.0	
Hammer-stone					19	10			2	31
					61.3	32.3			6.5	0.8
					0.9	1.0			2.0	
Cobble Uniface					1	1	1			3
					33.3	33.3	33.3			0.1
					tr	0.1	0.7			
Flake Uniface	1				9	2	2			14
	7.1				64.3	14.3	14.3			0.4
	1.3				0.4	0.2	1.4			
Other Uniface								1		1
								100.0		tr
								3.9		
Cobble Biface					2					2
					100.0					0.1
					0.1					

Table 13.1. Flaked Stone Morphological Type by Material Type, continued

Morphological Type	Material Type									Total
	Quartz	Chert	Obsidian	Sedimentary	Basalt	Igneous Other	Quartzite	Metasediment	Metamorphic Other	
Flake Biface		1	1		4	2	1			9
		11.1	11.1		44.4	22.2	11.1			0.2
		0.8	0.6		0.2	0.2	0.7			
Other Biface		1			1					2
		50.0			50.0					0.1
		0.8			tr					
Notched Tool			1							1
			100.0							tr
			0.6							
Projectile Point			2		1					3
			66.7		33.3					0.1
			1.1		tr					
Large Primary Flake					5	4	2			11
					45.5	36.4	18.2			0.3
					0.2	0.4	1.4			
Indeterminate	1				2	1		1		5
	20.0				40.0	20.0		20.0		0.1
	1.3				0.1	0.1		3.9		
Total	80	124	175	19	2085	1045	146	26	101	3801
	*2.1	3.3	4.6	0.5	54.9	27.5	3.8	0.7	2.7	100.0

*Row %
**Column %
¹Platform-remnant-bearing

reduction. The source of this material is small nodules, known as Apache tears, which are available from a number of volcanic sources in the Southwest and northern Mexico (Shackley 1988:757–760). Most of the obsidian in the collection was transparent, with some pieces showing opaque black bands within the transparent black matrix.

Igneous other materials constituted 27.5% of the Dutch Canal Ruin assemblage and included several rock types common in the mountains surrounding the Phoenix Basin and as alluvial cobbles along the local major watercourses. Two types of local rhyolite were present in the collection. One was composed mainly of alkaline feldspar (e.g., sanidine) and quartz, with small biotite and amphibole crystals. The material was fine grained and appeared to have been slightly metamorphosed. The second rhyolite was actually a latitic rhyolite. Latite differs from rhyolite in the composition of the feldspars it contains; latite has sodic and calcic plagioclase, while rhyolite contains potassium feldspar and sodic plagioclase. The material in the collection appeared to contain all of these feldspars, although the potassium and sodium varieties were most prevalent. The matrix was fine grained, grading to aphanitic, with small crystals of quartz, feldspar, and amphibole present throughout.

Also included in the igneous other category was a very coarse material that was probably diorite, the intrusive equivalent of andesite. The major minerals in this rock are plagioclase, biotite, quartz, and hornblende. The andesitic basalt and diorite in the collection could have come from the same source. However, the diorite did not appear to have been altered, and it may have originated from Tertiary deposits.

Another type of igneous rock in the assemblage was vesicular basalt in the form of flakes and flake fragments that had probably derived from ground stone manufacture and maintenance. This material type constituted only 0.1% of the flaked stone collection.

Metamorphic

Quartzite made up 3.8% of the assemblage and was the most common metamorphic rock. Most of the quartzite came from alluvial cobbles, and 65.8% of the flakes still retained dorsal cortex. Grain size ranged from very fine to very coarse. Most of the quartzites were rather dirty, although pure quartz sand quartzite and an arkosic variety also were present.

Another type of metamorphic rock, metasediment, is a metamorphosed sedimentary rock of unspecified origin. This material (0.7% of the assemblage) tends to be very fine grained and is usually yellowish brown in color with small iron concretions. Cortex on dorsal surfaces indicated that these items were derived from alluvial cobbles.

The miscellaneous category of metamorphic other contained a small amount (2.7%) of material. The majority of metamorphic rock in this category was igneous material that had been altered to various degrees, producing a coarse rock.

Two flaked stone artifacts were made of schist. Some smaller pieces collected in the field may have been flaking debris, but due to the difficulty of identifying flake morphology on platy material such as schist, analysts did not include these pieces in the flaked stone analysis.

Other

Quartz, which appeared mainly as angular debris, constituted 2.1% of the assemblage. Both well-developed crystals and amorphous quartz were present, neither of which exhibit good flaking characteristics. The noncrystalline forms were probably procured as small nodules that had weathered out of veins in local metamorphic formations.

Reduction Technology

Hohokam lithic reduction technologies have been summarized by Huckell (1981:197–199). After his investigation of Las Colinas lithic materials, Huckell proposed that direct percussion had been the most common reduction technique and had probably been used both in the reduction of cores and in the modification of cobbles to produce tools. Bipolar reduction of small obsidian and chert nodules is a specialized percussion technique that uses an anvil as support for the cores (Crabtree 1972:42; Hayden 1980:4; Spears 1975). The production of large primary flake tools also is a special technique, involving reduction of large riverine cobbles using "massive percussion" (Huckell 1981:198) or cone splitting (Haury 1976:293). The Hohokam employed pressure flaking on very few tools, mostly projectile points, but it constituted another major technique of tool manufacture.

The lithic assemblage at Dutch Canal Ruin produced evidence for all reduction techniques outlined by Huckell that archaeologists have generally accepted for the Hohokam. Tool makers procured the vast majority of their lithic raw material as alluvial cobbles from local water courses. They split some cobbles to produce large primary flakes (0.3% of the Dutch Canal Ruin lithic assemblage) and reduced others through direct or bipolar percussion to produce flakes of various sizes and shapes. Some of these flakes were, in turn, used as tools. Of the retouched tools in the assemblage, 11% were either cobble unifaces or bifaces (alluvial cobbles flaked on one or both sides to produce a working edge). Although little evidence existed in the assemblage for formal biface and uniface production, analysts noted indications of pressure flaking on projectile points.

The site assemblage contained a few bipolar flakes (with evidence of simultaneous percussion and some crushing at both ends) and cores, representing 1.3% of the collection. Over three-fourths (84.8%) of these were obsidian. Because of the small size of these flakes and cores and the type of reduction required to produce them, Apache tear nodules were probably the primary type of obsidian used in the Hohokam core area. Quartz and basalt (two and three items, respectively) were the only other materials that showed evidence of bipolar reduction.

Debitage attributes also were informative for interpreting the reduction technology at Dutch Canal Ruin. Within the debitage, analysts could decipher some patterns of morphological type using Sullivan and Rozen's (1985) model of debitage variability. The assemblage produced nearly identical frequencies of whole flakes and flake fragments (34.3% and 34.5% respectively), approximately half that amount of PRB flakes (18.9%), and a very low frequency of angular debris (4.2%) (Table 13.1). The core frequency was 3.2%. This pattern of debitage distribution matched Sullivan and Rozen's Group IB1, a core reduction pattern with additional evidence of tool manufacture. The Dutch Canal Ruin assemblage varied only in the low percentage of angular debris recovered. Because angular debris is assumed to increase with the amount of core reduction, this particular activity probably was not intensive at the site. This finding correlated with an analytical evaluation that the majority of cores were not intensively reduced.

Table 13.2 presents data on the distribution of cortex. The frequency of tertiary reduction was the highest for all flake types, supporting indications of tool manufacture in Sullivan and Rozen's Group IB1. These data were puzzling, however, because few retouched tools occurred in the Dutch Canal Ruin assemblage. PRB flakes had the highest percentage of tertiary stage (66.5%) reduction and the smallest percentage of primary stage (6.8%) reduction. Percentages of secondary flakes were nearly the same for flake fragments, PRB flakes, and whole flakes (26.6–29.3%), and somewhat higher for used flakes (39.0%).

Debitage was distributed over a wide range of sizes (Table 13.3). Nearly two-thirds of the debitage fell into the 2 or 3 size range (30.4% and 29.2%, respectively), and the number of artifacts steadily decreased with increasing size. PRB flakes were, in general, larger than flake fragments and angular debris. Whole flakes were almost equally distributed among Sizes 2, 3, and 4, demonstrating a wide variation in the sizes of the items undergoing reduction and evidence for tool manufacture as well as core reduction.

Table 13.2. Debitage and Used Flake Reduction Stages

Debitage Class	Primary	Secondary	Tertiary	Total
Flake Fragment	135	385	792	1312
	*10.3	29.3	60.4	**38.5
	**40.4	39.6	37.6	
PRB[1] Flake	49	191	477	717
	6.8	26.6	66.5	21.0
	14.7	19.7	22.7	
Whole Flake	140	357	785	1282
	10.9	27.8	61.2	37.6
	41.9	36.7	37.3	
Used Flake	10	39	51	100
	10.0	39.0	51.0	2.9
	3.0	4.0	2.4	
Total	334	972	2105	3411
	*9.8	28.5	61.7	100.0

Note: Does not include products of bipolar reduction
*Row %
**Column %
[1]Platform-remnant-bearing

SITE ASSEMBLAGE

Analysts described and evaluated the flaked stone artifacts from Dutch Canal Ruin to note variability in form that would be meaningful technologically and functionally. They determined that attribute descriptions of morphological types were valuable information that should be documented and added to the small data base that presently exists in the Hohokam region. Intra- and intersite organizational studies would benefit from an adequate data base providing the foundation for further comparative research.

Debitage

Debitage included all material types except schist (Table 13.1). As previously discussed, field crews may have recovered schist debitage from the site, but because this material did not exhibit discernible flake morphology, analysts could not classify it as such. Limestone, sedimentary other materials, and vesicular basalt were present only as debitage. Of the other material categories, obsidian constituted the lowest percentage of debitage (74.3%), followed by chert (87.1%). All other materials had similar proportions of debitage that ranged from 90.4% to 93.8%. Two factors may have combined to account for the lower obsidian and chert frequencies. First, as obsidian was the

Table 13.3. Debitage and Used Flakes by Size Classification

Size	Angular Debris	Flake Fragment	PRB[1] Flake	Whole Flake	Used Flake	Total
1	5	23	6	13		47
	*10.6	48.9	12.8	27.7		**1.3
	**3.2	1.8	0.8	1.0		
2	63	443	206	350	30	1092
	5.8	40.6	18.9	32.0	2.7	30.4
	39.9	33.8	28.7	26.9	29.1	
3	42	402	237	348	22	1051
	4.0	38.2	22.5	33.1	2.1	29.2
	26.6	30.6	33.0	26.7	21.4	
4	19	247	127	299	17	709
	2.7	34.8	17.9	42.2	2.4	19.7
	12.0	18.8	17.7	22.9	16.5	
5	12	123	78	157	12	382
	3.1	32.2	20.4	41.1	3.1	10.6
	7.6	9.4	10.8	12.0	11.7	
6	7	45	45	69	12	178
	3.9	25.3	25.3	38.8	6.7	5.0
	4.4	3.4	6.3	5.3	11.7	
7	5	23	14	43	5	90
	5.6	25.6	15.6	47.8	5.6	2.5
	3.2	1.8	1.9	3.3	4.9	
8	4	6	4	16	4	34
	11.8	17.6	11.8	47.1	11.8	0.9
	2.5	0.5	0.6	1.2	3.9	
9	1		2	8	1	12
	8.3		16.7	66.7	8.3	0.3
	0.6		0.3	0.6	1.0	
Total	158	1312	719	1303	103	3595
	*4.4	36.5	20.0	36.2	2.9	100.0

*Row %
**Column %
[1]Platform-remnant-bearing

most expensive material to procure, tool makers had probably taken more care not to waste it. The handling of cherts would have been similar because of their micro- to cryptocrystalline grain and good knapping quality. Second, due to core reduction and field sampling techniques, field crews would normally recover less debitage from these materials. Obsidian and chert were derived from small nodules, appropriate for bipolar core reduction, and a fair amount of the small debitage pieces (less than 1/4-inch screen mesh size) produced by this technique will escape analysis. Flenniken's (1981) investigation of bipolar technology at the Hoko River site on the Northwest Coast documented that the number of "microliths" produced by bipolar reduction was low. Flenniken recorded 332 microliths (nonfunctional and potentially functional) and pieces of debitage and 270 cores, producing a ratio of 1.2:1 microliths to cores, not including material from bulk samples. However, almost 60.0% of the debitage recovered from bulk soil samples was composed of items 5 mm less or in length; these bulk sample statistics exhibited a ratio of 4.3:1 microliths and debitage to cores. Evidence of bipolar products in the Hohokam region, therefore, should be interpreted with these cautions in mind. Obviously, a large proportion of bipolar products and by-products were not recovered by the usual field recovery techniques, and very few had actually been produced during bipolar reduction.

Dutch Canal Ruin debitage, in general, was composed of somewhat coarser material than tools, a trend also noted by Proper (1990:306) for the Pine Creek assemblage. Fine, medium, and coarse grain sizes had similar debitage frequencies: 92.1%, 94.0%, and 90.5%, respectively. Grain Size 1, micro- to cryptocrystalline, had a lower percentage of debitage (78.4%) due to the low frequencies of obsidian and chert.

Used Flakes

Used flakes also were present in a wide range of materials (Table 13.1). Over one third (35.9%) were obsidian, followed by igneous other materials and basalt (25.2% and 23.3%, respectively). Very few used flakes were made of the coarser materials (e.g., metamorphic other). The majority were micro- to cryptocrystalline (43.7%) or fine grained (36.9%), corresponding somewhat to the elevated use of obsidian in this artifact class. Analysts expected that good quality material such as obsidian would have been used to produce technologically expensive tools, but they did not expect that obsidian would account for such a large percentage of used flakes.

Used flakes were distributed over the entire range of sizes (Table 13.3), with the exception of Size 1. The number of used flakes gradually decreased with increasing size, a trend similar to that noted for debitage. However, the used flake category showed a higher frequency of large flakes than the debitage category (Table 13.3). For Sizes 2 through 5, the frequencies were comparable between used flakes and all types of debitage. For Sizes 6 through 9, used flakes occurred in a much higher frequency than debitage. The Hohokam probably selected larger flakes as tools to achieve greater ease of manipulation.

Approximately half (49.0%) of the used flakes had cortex on the dorsal face (Table 13.2). Of the tertiary flakes, 37.0% had cortex on the platform, and 57.4% had no cortex at all. The remaining 3 flakes were bipolar, and reduction stage was not recorded. According to Table 13.2 frequencies, used flakes had more cortex retention than debitage, as seen in a larger proportion in the secondary reduction stage. These results agreed with those of other researchers (Elstein 1989:687; Hoffman and Doyel 1985:637; Proper 1990:306), who found that cortex-bearing flakes were more commonly used as tools than those without cortex.

Just over half (51.5%) of the used flakes showed evidence of scraping use wear (Table 13.4), and another third (29.7%) exhibited cutting/sawing use wear. The other 19.0% of used flakes were fairly evenly divided among the other types of use wear (Table 13.4), with no more than 5.0% exhibiting any one type of use wear.

Cores and Hammerstones

Modified and unmodified cores were made of seven different material categories (Table 13.1), including quartz, chert, obsidian, sandstone, basalt, igneous other, metamorphic other, and quartzite. All grain sizes were represented, although core/hammerstones were generally larger grained and included the only coarse-grained materials. Hammerstones also consisted of larger-grained materials, with the medium-grained category significantly higher (71.1%) than for either modified or unmodified cores (43.9% and 38.3%, respectively). Because tool makers did not select hammerstones for the quality of their internal material, texture/flakability would not be as important a factor in material procurement as the size, weight, and shape of the cobble. Hammerstones from Dutch Canal Ruin had a mean weight of 298.6 g, with a standard deviation of 234.3 g. The large standard deviation was due to three outliers: one large (1036.4 g) hammerstone and two small (less than 50 g) items. Core/hammerstones, also used for battering, had the highest mean weight of all cores (174.1 g), with a standard deviation of 165.8 g. The cores and core tools exhibited a wide range of weights, 1.0 g to 510.0 g for core tools and 0.8 g to 774.6 g for unmodified cores. This variability was due to differences in the nodule size of particular material types. Obsidian cores, derived from small Apache tear nodules, had a mean weight of 0.7 g. Chert cores, also found in small nodules, had a mean weight of 40.2 g. Larger, heavier cores were usually from dense igneous and metamorphic alluvial cobbles, appropriate for direct percussion reduction. Basalt cores had a mean weight of 164.5 g, indicative of this larger cobble size. Five cores had been reduced by bipolar techniques, and one of these subsequently had been used as a scraping tool. In the Hohokam area, bipolar cores do not usually exhibit use wear, but use wear on bipolar cores has been documented elsewhere (Hayden 1980:3).

The most common use wear on core tools was scraping (66.7%), followed by chopping and cutting/sawing (16.7% each) (Table 13.4). Of all battering wear exhibited on tools in the assemblage, most (48.4%) was on core/hammerstones rather than hammerstones (40.6%). All of the core/hammerstones exhibited battering wear, and two (6.1%) had chopping wear as well. Core/hammerstones most often have been associated with the maintenance and manufacture of ground stone tools (Dodd 1979:239; Hoffman 1988:363; Mitchell 1988:192–193), based on studies of use wear and characteristics of material and form. Researchers have determined that other possible functions, such as plant processing and use as a percussor, are more appropriate for the nonflaked hammerstones. Archaeologists need controlled and detailed analyses of hammerstone types in the Hohokam region to interpret variability adequately between these two common but important tool classes.

Unifaces

The majority of unifaces (14, or 77.8%) were produced from flakes (Table 13.1). Various materials were represented, including quartz, basalt, igneous other, and quartzite. Most (78.6%) were fine grained. Of the others, 7.1% were micro- to cryptocrystalline and 14.3% were medium grained. Most of the flake unifaces had had only a minimal amount of effort invested in their production. Analysts noted that 75% had retouch on less than one-half of their margins. Although one uniface was retouched on 1/2 to 3/4 of its margin, and one was retouched around its entire margin, none

Table 13.4. Flaked Stone Morphological Type by Use-Wear Type

Morphological Type	Use Wear									Total
	None	Cut/Saw	Scrape	Shave/ Plane	Batter	Chop	Grave	Indeter- minate	Multi- use	
Used Flake	1	30	52	2	2	3	1	4	6	101
	*1.0	29.7	51.5	2.0	2.0	3.0	1.0	4.0	6.0	*45.5
	**3.4	88.2	81.3	50.0	3.1	60.0	100.0	57.1	42.9	
Core	2				1					3
	66.7				33.3					1.4
	6.9				1.6					
Core Tool		1	4				1			6
		16.7	66.7				16.7			2.7
		2.9	6.3				20.0			
Core/Hammer- stone					31				2	33
					93.9				6.1	14.9
					48.4				14.3	
Hammer- stone					26				2	28
					92.9				7.1	12.6
					40.6				14.3	
Cobble Uniface			1		1				1	3
			33.3		33.3				33.3	1.4
			1.6		1.6				7.1	
Flake Uniface	5	1	5	2		1				14
	35.7	7.1	35.7	14.3		7.1				6.3
	17.2	2.9	7.8	50.0		20.0				
Other Uniface					1					1
					100.0					0.5
					1.6					
Cobble Biface					1				1	2
					50.0				50.0	0.9
					1.6				7.1	
Flake Biface	5	1						2	1	9
	55.6	11.1						22.2	11.1	4.1
	17.2	2.9						28.6	7.1	
Other Biface	1		1							2
	50.0		50.0							0.9
	3.4		1.6							
Notch			1							1
			100.0							0.5
			1.6							
Projectile Point	3									3
	100.0									1.4
	10.3									

Table 13.4. Flaked Stone Morphological Type by Use-Wear Type, continued

Morphological Type	Use Wear									Total
	None	Cut/ Saw	Scrape	Shave/ Plane	Batter	Chop	Grave	Indeter- minate	Multi- use	
Large Primary Flake	10 90.9 34.5	1 9.1 2.9								11 5.0
Indeter- minate	2 40.0 6.9				1 20.0 1.6			1 20.0 14.3	1 20.0 7.1	5 2.3
Total	29 **13.1	34 15.3	64 28.8	4 1.8	64 28.8	5 2.3	1 0.5	7 3.2	14 6.3	222 100.0

*Row %
**Column %

of the flake unifaces had facial thinning beyond the first stage of facial regularization, and 71.4% retained some dorsal cortex. Flake unifaces were used for a variety of tasks: 5 exhibited scraping wear, 2 had shaving/planing wear, 1 had chopping wear, and 1 had cutting/sawing wear.

Field crews recovered only three cobble unifaces from Dutch Canal Ruin, each made from a different material (Table 13.1). Two were fine grained, and the other was medium grained. Two had margin evaluations of < 1/4 retouch and retained most of their original cobble form, but the other was completely retouched around its margin and had undergone the first stage of facial regularization on its flattened surfaces. One cobble uniface showed evidence of scraping use, one had battering use, and the third exhibited both scraping and battering use.

One other uniface was made of schist and was the only coarse-grained tool in this category. The margin could not be evaluated, due to its indeterminate form, but battering wear was recorded.

Bifaces

Flake bifaces, like the unifaces, were made of a variety of materials, including chert, obsidian, basalt, igneous other, and quartzite (Table 13.1). All grain sizes except coarse were represented, with medium grain constituting the majority (55.6%). Unlike unifaces, however, most flake bifaces had had their margins completely or nearly completely retouched (60.0%), while 20% were retouched on less than half of their margins. The amount of effort invested in tool edge shaping, however, did not correspond to the level of effort put into item thinning. Since analysts found only the first stage of item regularization present, they assumed that flake bifaces had been, for the most part, thin and regular enough in section symmetry to adequately perform their function. Five of the nine flake bifaces had no observable use wear, and two had indeterminate use wear. Identifiable wear patterns occurred on one item with cutting/sawing wear and one that had multiple patterns of battering and scraping.

Field crews recovered only two cobble bifaces. Both were made of basalt, one fine grained and one medium grained. These tools exhibited use patterns similar to those on the cobble unifaces, with one showing battering and the other having multiple battering and scraping wear. When compared to unifaces, all bifaces had less scraping use wear but similar amounts of battering (Table 13.4).

Two bifaces, one made of chert and the other of basalt, were classified as other bifaces because analysts could not determine if they were produced from flakes or cobbles. One had scraping wear, and the other had no evidence of edge damage.

Notched Tool

Analysts recorded one notched tool, made of obsidian, during their investigation (Figure 13.1a). This small tool (size category 2) was unmodified except for the margin, < 1/4 of the total, which had been retouched to form a small notch. Both tool faces remained unworked, and the notched area showed evidence of scraping use wear. Huckell (1981:185) noted similar tools at Las Colinas.

Projectile Points

Field crews recovered three projectile points from excavations at Dutch Canal Ruin. The small sample and lack of comparable specimens limited the information gained from these artifacts.

In the Phoenix Sky Harbor Center flaked stone assemblage, analysts grouped projectile points into types based on notching and base shape characteristics (Volume 3:Chapter 6) and then compared these types with the literature to integrate the present collection with the general Hohokam tradition. None of the points from Dutch Canal Ruin fit into the types established for the project. Two of the points represented probable Polvorón phase contexts, and the third was associated with the pre-Classic period occupation. Each point from Dutch Canal Ruin is described below and is compared to similar items from previous excavations in the region.

When analysts encountered projectile points during the initial flaked stone analysis, they kept these artifacts separate and reanalyzed the points later using more detailed criteria. For each point they recorded maximum length, maximum width, maximum thickness, weight, blade length, stem length, length-to-width ratio, base shape, edge shape, transverse profile shape, location of edge retouch, completeness, presence and location of notches, presence of serrations, and whether the artifact had been heat treated or reworked. Analysts also considered attributes from the initial analysis, such as use wear and material type.

One of the points from Dutch Canal Ruin, made of fine-grained basalt, was stemmed and deeply serrated (Figure 13.1b). This point was small (1.9 cm in length by 1.0 cm in width by 0.3 cm in thickness) and rather rough in manufacture. The blade was 1.6 cm in length and irregular in outline. Stem length was 0.3 cm, and its shape was also irregular. The transverse profile was curved, and the flake scars on both faces were minimally invasive or patterned. No use wear was evident. This point was present in the fill of a pit not further specified (NFS) in Area 4, a limited activity locus from the pre-Classic period. It was very irregular in form, and researchers located no other points of this type in the literature. Haury (1976:297, Figure 14.39j) illustrated one vaguely similar point but did not assign it to a time period.

The second projectile point was small and tri-notched, although the basal notch was very small (Figure 13.1c). The tip and one side of the base were missing, so analysts recorded only a thickness measurement of 0.3 cm. The blade was approximately twice the length of the base and had straight edges. The base was 0.5 cm in length and had an irregular shape. The point, made of translucent obsidian, was well flaked and retouched on both faces. Analysts noted no evidence of use wear on the point, which field crews recovered from Feature 3 in Area 8, a pit structure NFS that dated to the Polvorón phase. A survey of the literature showed that tri-notched points were relatively rare in the Hohokam area. Huckell (1981:174, Figure 111e) described one similar point, although the current specimen had smaller notches and a somewhat wider blade. Huckell attributed this point type to the Classic period.

The final projectile point in the Dutch Canal Ruin collection was from Area 8. It also was made of translucent obsidian and was present in the fill of a pit with evidence of burning. This artifact was a small, side-notched point with a moderately concave base 0.5 cm in length and a blade with straight edges that was approximately twice as long as the base (Figure 13.1d). The tip and one side of the base were missing, so analysts measured only the thickness (0.4 cm). They noted no evidence of use wear on this point. Investigators did not date the pit that produced the point, so analysts could not define its temporal context. Because most of Area 8 was dated to the Polvorón phase, this feature probably corresponded to the Polvorón phase occupation of the site. Similar points have been reported by Landis (1989:429, Figure 9.6f) and Bernard-Shaw (1988:288, Figure 7.2p). Landis (1989:428) stated that this type was "common late in the Classic period, through the protohistoric, and into the historic Pima and Papago assemblages," while Bernard-Shaw (1988:289) placed the type in the Sacaton through Civano time periods.

Large Primary Flakes

Large primary flakes were made of basalt, igneous other material, and quartzite (Table 13.1). The vast majority of these artifacts were medium grained (90.9%), and the remainder (9.1%) were fine grained.

Ten of the 11 (90.9%) large primary flakes showed no evidence of use wear. The one used item showed cutting/sawing use wear. This pattern contrasted with those reported by previous researchers, although past studies generally have not classified both unused and used large primary flakes together. Usually, the classification has been "primary flake tools" (Hoffman 1988; Huckell 1981; Mitchell 1988). Most researchers have recorded similar use wear on large primary flake tools: cutting, scraping, shredding, and chopping. Following results documented by Lewenstein (1986), researchers have used these patterns, as well as other evidence, to associate primary flake tools with agave processing. Although field crews recovered relatively few of these artifacts from Dutch Canal Ruin, the presence of primary flake tools, as well as tabular tools (Chapter 14) and agave fibers (Chapter 15), suggested that the inhabitants of the site had processed agave there.

Indeterminate Tools

Indeterminate tools accounted for 0.1% of the assemblage. Most of these tools were so classified because they were fragmentary, and, as such, they had little in common. They were made of quartz, basalt, igneous other material, and schist (Table 13.1) and had grain sizes ranging from fine to coarse. Two of the indeterminate tools had no use wear, one showed evidence of battering wear, one scraping and graving, and one indeterminate use. Of the three artifacts in this group that were retouched, two had < 1/4 of the margin retouched, and one had an indeterminate portion of the margin retouched. Dorsal and ventral face evaluations ranged from facially unworked to the first stage of facial regularization.

DISTRIBUTION

Analysts further investigated variability in the flaked stone assemblage at Dutch Canal Ruin by plotting distribution synchronically and diachronically. They examined spatial distribution at the area and feature levels but limited temporal distribution to a definition of pre-Classic versus post-Classic periods.

Spatial Distribution

Feature Types

Analysts examined flaked stone distribution within feature types to gain information on functional patterns within Dutch Canal Ruin. They compiled data from different feature types, as well as from those that were similar in form, to discern discrete activity loci. However, because the structures in Area 8 were filled with trash, an adequate sample was not available from floor or associated contexts to interpret feature activities.

The lithic profile for the canals, however, showed a striking variation compared to other feature types. The heaviest types of stone artifacts, such as cores, core tools, core/hammerstones, and hammerstones, were present in these features in large quantities (approximately 14.0% of main canal lithics). Other researchers have suggested (Swidler 1989:169–170) that large stone items were used prehistorically to stabilize the canal beds against water damage at junctions. This was probably the case at Dutch Canal Ruin, where inhabitants had constructed a baffle in the canal at Area 7 to slow erosional damage (Chapter 8). Eight of the heavy canal artifacts, in fact, were recovered from Canal Viejo (Alignment 8501), the pre-Classic period canal alignment in Area 7 that contained the baffle feature.

Areas

Analysts examined lithic profiles to determine if functional assignments would be supported by assemblage characteristics and if the flaked stone would be able to provide additional information on area activities. Data presented in Table 13.5 support functional evaluations of project areas and furnish a comparative overview of the site's spatial distribution.

As expected, areas with habitations had the most diverse assemblages, accommodating seasonal or year-round occupation that incorporated multiple activities. Areas 3, 7, 8, and 10, the habitation locales, had the largest assemblages as well. Area 1 was the exception, with a total of 19 lithics and very few items other than debitage. Area 8, designated as a small hamlet, had more than twice the relative frequency of used flakes of any other area, and the used flakes from features occurred most often in pits with evidence of burning and pit structures NFS.

Area 10, a multicomponent locale, consisted of a pre-Classic period field house and a Classic period occupation, possibly a hamlet. The lithic profile for this area was distinctive, with a relatively low frequency of debitage and a high percentage of cores and heavy tools, including core tools, core/hammerstones, and hammerstones. This profile was deceptive, however, since data recovery in this area was augmented by mechanical stripping; although archaeologists monitored the work, artifacts recovered in this manner tend to be those that are large or otherwise distinctive. In fact, the only feature context represented in the flaked stone profile of Area 10 was a lithic cache, consisting of four whole flakes, two cores, one core tool, two hammerstones, and one flake biface.

Table 13.5. Flaked Stone Areal Distribution

Morphological Type	Area									
	1	2	3	4	5	6	7	8	9	10
Debitage	17 *89.5	80 93.0	138 93.2	66 93.0	45 100.0	11 78.6	131 92.3	2939 92.3	3 100.0	64 71.1
Used Flake		1 1.2					1 0.7	100 3.1		1 1.1
Core	1 5.3		3 2.0	2 2.8		2 14.3	3 2.1	65 2.0		5 5.6
Core Tool			1 0.7				1 0.7	5 0.2		4 4.4
Core/ Hammerstone		3 3.5	1 0.7				3 2.1	21 0.7		2 2.2
Hammerstone		2 2.3	1 0.7			1 7.1	2 1.4	15 0.5		10 11.1
Retouched Tool			3 2.1	3 4.2			1 0.7	30 0.9		3 3.3
Large Primary Flake	1 5.3		1 0.7					8 0.3		1 1.1
Total	19	86	148	71	45	14	142	3183	3	90

*Column %

Area 4, a limited activity locale, contained the highest relative percentage of retouched tools compared to the other areas and exhibited little diversity in lithic forms. Both of these factors supported its designation as a limited activity area and provided additional evidence for its function as a resource procurement and processing locale (D. M. Greenwald 1988c:268-270).

Temporal Distribution

Dutch Canal Ruin produced material from the pre-Classic, Classic, and post-Classic periods. Areas 1 through 7 dated to the pre-Classic periods, Area 8 to the post-Classic period, and Area 10 to the Classic period; Area 9 was a multicomponent pre-Classic and Classic period locus. Analysts compared the data from these broad but distinctive time periods to determine whether the flaked stone assemblage had undergone functional or technological changes at the site through time. Not all the lithic information from the site could be included, however, due to varying strengths of temporal association and assignment. The distribution shown in Table 13.6 consists of data recovered from single-component areas or from lithics associated with dated features from multicomponent areas. The pre-Classic period category consists of data from Areas 1 through 7,

Table 13.6. Flaked Stone Temporal Distribution

Variable	Period	
	Pre-Classic	Post-Classic
Material Type		
Quartz	10	70
	*2.3	2.2
Chert	31	86
	7.1	2.7
Obsidian	1	174
	0.2	5.5
Sedimentary		19
		0.6
Basalt	233	1728
	53.3	54.3
Igneous Other	129	867
	29.5	27.2
Quartzite	29	112
	6.6	3.5
Schist/Phyllite		2
		0.1
Metasediment	1	23
	0.2	0.7
Metamorphic Other	3	97
	0.7	3.0
Vesicular Basalt		5
		0.2
Total	437	3183
Grain Size		
Micro-Crypto	24	262
	5.5	8.2
Fine	161	1387
	36.8	43.6
Medium	249	1496
	57.0	47.0
Coarse	3	38
	0.7	1.2
Total	437	3183
Reduction Stage[1]		
Primary	51	277
	11.7	8.8
Secondary	77	848
	17.7	26.9
Tertiary	308	2027
	70.6	64.3
Total	436	3152

*Column %
[1]Reduction stages do not include items produced from bipolar reduction.

and the post-Classic period Polvorón phase category contains information from Area 8. Comparability, then, was limited by functional variation between the two categories. The pre-Classic periods exhibited more diverse site types (including field house, farmstead, limited activity, and canal locales) than the post-Classic period category, which comprised only a small hamlet.

Debitage types and most tool types were similar in the pre-Classic and post-Classic periods, except for the high frequency of used flakes in the post-Classic period (3.1%) compared to the count in the pre-Classic periods (1.9%). Without further instances of variability in assemblage profile, this particular difference may have been due to the functional variation between categories. Perhaps the relative abundance of used flakes at Area 8 illustrated the general-purpose nature of a permanent habitation as opposed to the more intermittent, specialized nature of activities represented at Areas 1 through 7.

The distribution of material types changed somewhat through time. Table 13.6 indicates that the inhabitants of Dutch Canal Ruin utilized more obsidian and metamorphic other material in the post-Classic period. The shift was countered by less use of chert, igneous other material, and quartzite. Material grain size was similarly affected, with a corresponding increase in micro- to cryptocrystalline and fine-grained materials. Additionally, evidence of bipolar reduction, the typical technique used with obsidian nodules, intensified in the Polvorón phase. Analysts expected to find an increase in obsidian due to the development through time of established trade networks (Findlow and Bolognese 1982; D. M. Greenwald 1988c:276; Upham and Rice 1980:84–86).

Another change in the flaked stone assemblage through time was cortical retention on debitage. The presence of cortex generally increased in the assemblage from the pre-Classic periods to the post-Classic period, although the incidence of primary flakes decreased slightly. The percentage of secondary flakes increased from 17.7% to 26.9%, and that of tertiary flakes decreased from 70.6% to 64.3%. Analysts could not determine the significance of this trend, since reduction stage values used by other flaked stone analysts are not used consistently (Sullivan and Rozen 1985:756) and other technological factors do not appear to have varied.

CONCLUSIONS

Analysts investigated the flaked stone assemblage at Dutch Canal Ruin for technological and functional information about the site. They retrieved technological data by examining debitage and tool morphological types, material source and attribute characteristics, and patterns of reduction. When they compared debitage types to Sullivan and Rozen's (1985) debitage classification scheme, they discerned patterns indicating a technology incorporating both nonintensive core reduction and tool manufacture. The lack of intensity of core reduction also became evident during analysis of assemblage cores, the majority of which had had very few flakes removed from them.

Although few retouched tools were recovered, tool manufacture had apparently been a significant reduction activity. Debitage consisted mostly of tertiary flakes (61.7%), and flake size indicated tool production as well as core reduction.

Reduction techniques at Dutch Canal Ruin reflected all of the varieties outlined by Huckell (1981:197–199) for the Hohokam region. These include direct percussion, indicated by the majority of cobble cores and debitage; bipolar reduction, evidenced by bipolar cores and their products and by-products; "massive percussion," distinguished by the form and ventral attributes of large primary flakes; and pressure flaking, noted on projectile points.

The site assemblage reflected a reliance on an expedient technology based on tools that were produced and used for tasks at hand. Tools such as unretouched and marginally retouched flakes required minimal effort to manufacture. Maintenance and recycling were not evident on assemblage tools, and suitable material had been readily available. These factors would have further reduced the need for a curation strategy. Nonlocal obsidian was even used to produce used flakes, a general-purpose, expedient type of tool. The production of obsidian used flakes occurred during the post-Classic period, at a time when trade networks in the Southwest had become well established.

Analysts examined intrasite distribution by feature types and areas. The canal assemblages were weighted by larger artifacts, such as cores, core tools, core/hammerstones, and hammerstones, a situation similar to that found at La Cuenca del Sedimento (Swidler 1989:169–170), which was interpreted as indicating an erosional deterrent. With the additional evidence of a baffle in the canal at Area 7 of Dutch Canal Ruin, investigators concluded that the large artifacts had probably served a similar purpose at this locale.

The investigation of flaked stone temporal distribution was limited to a comparison between the pre-Classic periods and the post-Classic period Polvorón phase. Analysts noted three general trends: (1) the Polvorón phase contained a larger frequency of obsidian and metamorphic other material, accompanied by smaller grain sizes; (2) cortical retention on debitage increased through time; and (3) the frequency of used flakes increased through time. The former trends have been documented during other archaeological projects in the Hohokam area that include both early and late components, but the latter trend has not been generally noted. Further lithic analysis will be needed to document and adequately interpret reduction stage variability through time.

CHAPTER 14

THE DUTCH CANAL RUIN GROUND STONE ASSEMBLAGE

Kathleen S. McQuestion
Dawn M. Greenwald

With the recent recognition of ground stone manufacturing quarries, particularly in the Hohokam culture area (Bostwick and Burton 1989; Bruder 1983; Burton 1976; Green 1989; Hoffman and Doyel 1985; McQuestion and Gibson 1987), interest in this major class of artifacts has increased dramatically. Now that archaeologists realize the potential for tracing the sources of raw materials and interpreting the behavior of selecting raw materials (Fratt and Biancaniello 1991; Schaller 1985, 1987, 1988), they are directing increased attention toward ground stone tools, especially those used for food processing. In addition, investigators have recently begun to research shaping techniques (Benitez 1991), motor habits (Adams 1991), use-wear patterns (Walker 1991), and residues (Greenwald 1990, 1991; Logan 1991) in Arizona, with intriguing results. Finally, after a long period of almost totally descriptive reporting on ground stone collections, data on this artifact class are beginning to add to our understanding of prehistoric social interaction as well as economic and subsistence strategies.

The ground stone assemblage at Dutch Canal Ruin exhibited patterns indicative of a settlement dominated by subsistence-related activities. Nonutilitarian items, except manuports, were nonexistent, and the most common specialized tools in the assemblage had been used for food processing. The focus of subsistence strategy changed significantly through time, from an emphasis in the pre-Classic periods on corn production and processing to an emphasis in the post-Classic period on agave farming and processing. This change was reflected in the ground stone assemblage by a decrease in the frequency of manos and metates and a concurrent increase in tabular tool frequency. Except in Areas 8 and 10, analysts noted little diversity in the assemblage, which indicated that most areas had been only seasonally occupied or had been inhabited for only a short period of time.

RESEARCH OBJECTIVES

Analysis of ground stone artifacts can significantly contribute to an understanding of functional attributes within sites and their spatial and temporal components. The analysis of the ground stone assemblage from the project area addressed several research questions pertaining to three problem domains outlined for the Phoenix Sky Harbor Center Project: site structure, settlement composition, and settlement typology; spatial, functional, and temporal variation in settlement within Canal System 2; and chronology building and refinement.

The relative frequency of ground stone within site loci and features can provide information about settlement size and composition. Determining the extent of utilitarian versus nonutilitarian items, trade versus local items, and general-purpose versus special-function tools can provide a basis for evaluating the site and site components. Morphological types and the variability apparent within each type, when combined with pollen and flotation information, can contribute evidence for specialization and possibly for seasonally related agricultural activities. Assemblage diversity also contributes to the definition of site typology. Seasonal occupation should produce less diversity in ground stone artifacts than year-round occupation. In addition, analysts expected that locales that represented limited activities, such as resource procurement and processing sites, would have less assemblage diversity than habitations or multiple activity sites (D. M. Greenwald 1988a:200–201). Such generalizations, however, should be used cautiously. Crown (1985a) reported, after examining ethnographic examples, that field house sites may produce a full complement of artifacts. Although curation of ground stone tools may be more common when major village sites are nearby, she found all types of household goods in ethnographically documented field house structures (Crown 1984:7). Household goods in association with field houses usually consisted of items of local origin; trade items occurred infrequently in these contexts.

Concerning spatial, functional, and temporal variation in settlement, ground stone items at Dutch Canal Ruin proved useful in detecting technological and functional changes within the site through time. Analysts compared the frequency and diversity of ground stone morphological types and attributes for the site's pre-Classic and post-Classic period components. Comparisons between the Dutch Canal Ruin and Pueblo Salado assemblages and with other sites along Canal System 2 appear in Volume 4, Chapter 12. The analysis also included examination of the distribution of artifacts and materials across feature types and areas for variation in function or resource access. Particular tools, such as manos, metates, pestles, and mortars, are directly related to food-processing activities. Although the size of the Dutch Canal Ruin ground stone sample was inadequate for archaeologists to interpret area and feature variability as planned, the distribution of the artifacts within the site as a whole provided information about general site function and resource access.

As of yet, few ground stone artifacts have been useful in defining temporal associations and in chronology building, although Hohokam ridged axe forms and stone palettes have proved to be somewhat temporally sensitive (Haury 1976:286, 291). Stylistic attributes that change through time are often good temporal indicators, and items used in ritualistic behavior or as ornamentation tend to exhibit greater variability in such attributes. Utilitarian ground stone artifacts, however, typically do not display great stylistic variability. The Dutch Canal Ruin assemblage consisted almost exclusively of utilitarian objects, preventing the determination and application of ground stone temporal style markers.

ANALYTICAL METHODS

Analysts designed this investigation to measure seven variables: morphological type; material type; material source; surface texture; condition; production investment; and comments. They did not assess all items analyzed in the ground stone assemblage for each variable. For instance, they assessed surface texture only for the implements used for grinding foodstuffs. The coded comments included as the seventh variable noted such observations as multiple use, pigment or other adherents, evidence of burning, and reshaping or reuse.

The morphological type assigned each item was in accordance with traditional Southwest terminology (Gladwin et al. 1937; Haury 1976), with a more general classification for those items that could not be identified to a particular tool category. For instance, if a metate fragment was too small for analysts to determine whether it was part of a slab or a trough metate, they classified it as a metate not further specified (NFS).

In most instances, analysts made material type designations macroscopically. If necessary, they used a 10X hand lens and a 10–30X microscope to identify mineral crystalline shape and to evaluate cementation in granular rocks. They specifically identified materials in the three major rock classes whenever possible. They grouped textures within material types more simply, for example, classifying porphyritic andesite as andesite. However, materials that could not be identified beyond their major rock class were categorized as materials not further specified (e.g., sedimentary NFS).

Because of the fine-grained mineralogy of igneous rocks, volcanic rocks are difficult to identify macroscopically with a high degree of accuracy. Examination with a microscope helps determine the presence or absence of minerals, but it is the proportions of those minerals that determine igneous rock classification. Geologists advise the use of petrographic analyses when precise identifications are necessary. Specific raw material identifications of quartz-bearing basalt made during this analysis should be considered preliminary. Quartz-bearing basalt, also called quartz basalt (Schaller 1987), is usually vesicular and may have been categorized as a vesicular basalt, the usual practice in past analyses. In making the distinction between these material types analysts hoped to provide a potential sample for future petrographic analysis of tools made from quartz basalt and to aid in estimating trends in the frequency of this material through time.

As a result of recent petrographic analyses, sourcing of specific types of schist and phyllite, used as ceramic tempering agents, is now possible (Abbott and Schaller 1990). Because schist does not occur naturally in the project area, its presence implied intentional transport to the locality. The raw material could have been procured from stream or drainage gravel deposits nearby or quarried from primary rock outcrop sources, such as Squaw Peak (Abbott 1991; Abbott and Schaller 1990). Mary-Ellen Walsh-Anduze of SWCA examined and identified a sample representing the variety of schist and phyllite materials collected during the Phoenix Sky Harbor Center excavations. She assigned preliminary identifications to 8 items collected from two areas of Dutch Canal Ruin (Areas 8 and 10), as well as 18 items from Pueblo Salado. Project analysts then identified the remainder of the project assemblage based on macroscopic comparisons to Walsh-Anduze's classified collection.

Evaluating cultural modification on schist and phyllite was more difficult than for other materials. Because of their layered character, these materials do not exhibit conchoidal fracture or flake morphology. In addition, platy cleavage can create the appearance of a smoothly ground surface, and fractured edges can appear almost serrated. Identification of cultural modification on objects in the collection was also limited by the size of the item present. Although investigators analyzed flaked tools of schist or phyllite as part of the flaked stone assemblage, they classified the remainder of these materials, possibly debitage, as raw material in the ground stone analysis. These materials were used prehistorically as ornamentation (pendants), as tabular tools (rasps, knives), as palettes, and as a sherd-tempering agent.

Analysts identified minerals specifically when possible and used a general mineral category for less common mineral types. None of the minerals in the assemblage occurred naturally in the project area, and all had probably been brought to the area from another location.

The raw material source or origin variable indicated whether materials had been procured from nearby streambed deposits or from nonlocal resources. Waterworn cortical surfaces indicated a riverine cobble source. Often the source could not be identified because the artifact had been well shaped or completely shaped and completely lacked its original cortical surface.

The lithics analysis assessed surface texture only for grinding implements that had been used to process foodstuffs and included only those surfaces that had been used for grinding. Active grinding implements comprised handstones, round/oval manos, rectangular/loaf manos, manos NFS, and grinders NFS. Passive forms of ground stone included basin, slab, and trough metates; metates NFS; grinding slabs; and passive ground stone NFS. In addition, analysts examined ground stone NFS and mano and metate blanks.

Although analysts subjectively judged the surface texture variable based on its granularity and frictional qualities, they also used an objective means of defining the texture. Artifacts were coarse if, as in the case of most igneous rocks, the vesicles or pecking scars were greater than 2 mm in diameter. Rocks with very large crystal inclusions, such as granite, could also be categorized as coarse textured. Implements with medium surface textures had vesicles or pecking scars that measured 2 mm or less in diameter. Smooth, low-friction surfaces with virtual absence of pecking were fine textured. Certain material types had surface textures that varied from item to item depending on their specific granularity. Quartzite tools, for example, could be either coarse or medium textured, and sandstone tools varied from medium to fine textured. If an implement was too fragmentary to be accurately characterized as to grinding surface texture, it was coded as not applicable.

Analysts based their evaluation of artifact condition on the amount of an item that was present. If an item was too fragmentary for them to estimate the percentage of the whole artifact it represented, it was coded as fragmentary. If enough of an incomplete item was present to make a judgment, analysts described it as $\leq$ 1/4, $\leq$ 1/2, or $\leq$ 3/4 complete. They measured items that were complete or nearly complete. Analysts considered fragments identified as reused as complete, because they had functioned as tools in their fragmentary state. Items such as minerals, raw material, and pigment were not evaluated for condition, since the attribute of completeness was neither appropriate nor informative.

The analysis also included determination of the amount of effort invested in the manufacture of ground stone tools and other artifacts based on evaluation of item form (as well as the roughening or refurbishing of grinding use surfaces) for evidence of production modification, such as flaking, pecking, grinding, drilling, or incising. Five stages of production investment were possible: (1) none, indicating that the item had no modification due to production; (2) minimal, noting the presence of modification that was insufficient to alter item shape; (3) shaped, denoting shaping on < 1/2 of an artifact, such as pecked, flaked, or ground edges; (4) well shaped, indicating that $\geq$ 1/2 of an item's form had been changed; and (5) completely shaped, identifying ground stone that had been totally transformed. However, analysts considered a trough metate completely shaped even if its underside had not been worked. An example of the minimal stage is a boulder slab metate with the use surface roughened by pecking but the natural form not changed in any way. Analysts determined the level of production investment only if an implement was $\geq$ 1/2 complete, so that they could reasonably estimate how much shaping the complete tool had undergone. If a tool had been reused or reshaped, the labor that went into its original production was included.

The seventh variable included in this analysis was coded comments, miscellaneous multiple comments that were coded when they were deemed analytically valuable. Any item that had pigment adhering to it or had been burned was noted. A tool was coded as recycled if it had originally had one function and was later used for another, such as an axe that had been reworked and later used as a maul. Analysts coded a tool as having had multiple uses, such as an axe that was also used as a hammerstone, when they could not determine whether the different uses were contemporaneous or which function was most recent. Another coded comment included observations regarding reshaping or reuse of broken tools that showed evidence of having been reworked into a tool of similar function. Examples are two-hand mano fragments that were shaped for use as one-hand manos, or a broken trough metate that was used later as a slab metate. Reshaping characteristically appeared as flaking or pecking on broken edges. These items may or may not have been reused after reshaping. Tool reuse was indicated by additional wear following tool reshaping, as described above. Reuse was apparent when analysts noted grinding use over freshly broken edges.

Further coded comments indicated whether grinding use occurred on one or two sides of certain grinding equipment, such as manos, grinders, metates, and grinding slabs. This did not include wear created indirectly through use on the undersides of metates. Incomplete use surfaces or surfaces that were too fragmentary or eroded were not assessed.

Analysts included additional comments if they made written descriptions or drawings of items or if they measured the artifacts. They measured for maximum dimensions using a metric board and used a hand-held tape to measure dimensions of grinding surface areas. They also used a vernier caliper to measure the diameter of very small objects, such as beads.

VARIABLE DESCRIPTIONS

Morphological Type

Analysts identified 21 morphological types in the ground stone assemblage from Dutch Canal Ruin and analyzed them as part of the data recovery phase in the Phoenix Sky Harbor Center. The assemblage included food-processing equipment; other tools, such as mauls and polishing stones; and raw material and ground stone that could not be identified as to type. Although many types of activities were represented, none of the items was nonutilitarian. Table 14.1 lists the frequencies for all morphological types recovered from the site.

Active Grinding Tools

Tools that are the hand-held member during the grinding process have been variously labeled as manos, handstones, grinders, rubbing stones, and polishing stones. They represent the active component of a system in which an active element is manipulated against another, passive component. In the case of the following tools, manipulation was in the form of grinding/smoothing use.

Polishing stones are small, finger-held, waterworn cobbles used in ceramic manufacture. The two recovered at Dutch Canal Ruin had some evidence of use wear, such as striations or facets, and exhibited a fine-grained surface. Field crews found one rhyolite polishing stone (6.3 × 3.0 × 2.1 cm) at floor contact in a pit structure (Feature 1-1). The stone had two abraded facets and was located near a grinding slab spotted with pigment. From overburden deposits in Area 8, crews recovered another polishing stone (5.9 × 4.2 × 2.7 cm) of metasediment. It had unidirectional striations on one flat surface and around the midsection and battering wear on one end.

Handstones are small, often irregularly shaped, active grinders that can be held within the palm of the hand. Field crews collected three handstones from three different areas at Dutch Canal Ruin. A quartzite handstone, the smallest of the three (Table 14.2), lay in overburden deposits in Area 3 and had striations on its single, fine-textured use surface. Another handstone with a fine surface texture, made of greenstone, came from floor contact in a pit structure (Feature 7-1). It was the largest of the three and had polish on both surfaces. This handstone had striations on only one surface and was battered on both ends from use as a hammerstone. The third handstone, of unshaped granite, was from a pit NFS (Feature 8-1) in Area 8. This was the only handstone with a coarse surface texture.

Round/oval manos are grinding stones larger than handstones that can be compared to one-hand manos (Table 14.2). They are usually ovoid or spherical riverine cobbles with minimal to no shaping

Table 14.1. Ground Stone Morphological Type by Material Type

Morphological Type	Vesicular Basalt	Quartz-Bearing Basalt	Granite	Other Igneous*	Conglomerate	Sandstone	Metasediment	Quartzite	Greenstone	Phyllite	Squaw Peak Phyllite	Squaw Peak (?) Phyllite	Squaw Peak Schist	Other Metamorphic*	Minerals	Unknown	Total
Polishing Stone				1			1										2
Handstone			1	1					1								3
Round/Oval				1				1	1								3
Mano	7	6	1	1		1											16
Rectangular/Loaf Mano				1													1
Mano NFS		1	1		3		1						1				7
Grinder NFS														1			1
Basin Metate		5	1		1		1		1								9
Trough Metate	2		2				1										5
Metate NFS	3		2														2
Grinding Slab									1								1
Passive Ground Stone NFS	4			4	13		4	2	2	10	5		2	2			51
Ground Stone NFS								2									2
Maul								1									1
Grooved Haftable Item								1									1
Blank				2					1			1					4
Tabular Tool											1						1
Ring	1																1
Raw Material									1		82	42	15	3	1		144
Manuport							1								2		3
Other							1									1	1
Total	17	13	2	8	12	1	17	9	4	7	93	47	18	6	3	1	258

*Includes andesite, diorite, rhyolite, scoria, tuff, igneous NFS

**Includes other schist, slate, metamorphic NFS

Table 14.2. Dimensional Data for Complete Active Grinding Tools

Active Grinding Tools	Field Specimen No.	Length	Width	Thickness
Round/Oval Mano	2-20	14.5	11.5	15.5
	3-163	13.7	10.3	5.5
	10-32	14.2	9.2	4.1
Handstone	3-162	6.9	4.6	3.8
	8-327	7.2	6.4	4.7
	7-31	10.5	7.4	3.6
Rectangular/Loaf Mano	10-1	19.0	9.8	5.9
	10-9	16.6	8.7	3.7
	10-27	15.4	9.1	5.2
	10-45	16.9	8.6	3.8
Grinder NFS	8-354	5.9	5.9	3.7

Note: all measurements in centimeters

required. Field crews recovered three of these tools from three different areas at Dutch Canal Ruin. A coarse-textured diorite round/oval mano that exhibited striations and use on two surfaces came from floor contact in a pit structure (Feature 2-2). Another round/oval mano, from Area 3, was lying on the prehistoric surface adjacent to a pit structure (Feature 3-16). This quartzite mano had two medium-textured use surfaces, and the ends were battered from hammerstone use. One complete sandstone specimen with two surfaces was from a trash pit in Area 10.

A grinder NFS is a tool that can be identified as the active partner in the grinding process but cannot be further classified due to its small size and, often, fragmented state. Field crews collected seven of these items, all in Area 8. Five of the seven lay in overburden deposits, and the other two came from the floor fill of Feature 8-2, a pit house. The latter two were reused pieces of fragmented sandstone grinding tools. Each displayed grinding over the fractured edges, indicating active use. Five materials were represented by the seven tools, including one each of scoria, conglomerate, metasediment, and metamorphic NFS, and three of sandstone material. Analysts could not determine the degree of production investment for four of the tools because they were too fragmentary. Three, however, had no evidence of production modification. All surface textures were represented: one coarse-textured material (scoria); one medium-textured material (sandstone); and five fine-textured materials. Only one of these items was complete enough to be measured (Table 14.2).

Rectangular/loaf manos are formal manos, usually of a size requiring two-handed use, that are shaped into distinctive rectangular or loaf shapes. They were the most numerous type of active grinding tools recovered and had the greatest amount of production investment. Most of these manos (62.5%) were completely shaped; the others ranged from minimally to well shaped. All but 2 of the rectangular/loaf manos had coarse textures, and 3 were complete (Table 14.2). The makers had selected almost exclusively igneous material for these manos (Table 14.1). Of 16 rectangular/loaf manos, 14 (87.5%) were basalt, including 7 vesicular and 6 quartz-bearing varieties. Four of these tools lay in overburden deposits, 2 each from Areas 2 and 8; 3 came from the fill of a trash deposit (Feature 9-2); and 9 were recovered during the monitoring of Area 10, with 5 having come from a trash deposit (Feature 10-1), 3 from a trash pit (Feature 10-8), and 1 from the main canal (Feature 10-12). Only 1 of the manos could be identified as having been derived from a river source.

One granite mano was too fragmentary for analysts to determine whether it was a rectangular/loaf or round/oval mano. This mano NFS came from floor contact in a pit structure (Feature 3-9). It had a riverine cobble cortex and possibly had been altered by pecking.

Passive Grinding Tools

The tools representing the passive element in an active-passive grinding system are the stationary member in the grinding process. They are used both as receptacles and as part of the grinding mechanism.

Basin metates are large, passive grinding tools that display multidirectional grinding wear on their generalized, circular or oval use surface. Field crews recovered one basin metate from a nonfeature context in Area 9. The complete specimen (29.0 × 23.1 × 6.7 cm) was of coarse-textured quartz-bearing basalt. The basin was 2.6 cm in depth and exhibited multidirectional striations. The tool was made from a naturally smooth boulder that may have had a natural concave surface, because the metate displayed no evidence of having been shaped. Grinding wear over the lip and upper edge of the depression indicated use with two different manos.

Trough metates are passive grinding tools that display unidirectional use patterns and have rectangular to subrectangular use surfaces, reflecting the grinding pattern of the long, rectangular mano used with this metate type. Nine trough metates were represented in the assemblage, including one miniature specimen. Single examples came from Areas 2, 5, 8, and 9, and five trough metates were present in Area 10. All were coarse textured and from an unknown source, except for one medium-textured example of metasediment that had been made from a waterworn boulder. Two were completely shaped, and seven were too fragmentary for analysts to determine the degree of shaping represented. The only trough metate that was almost complete was the miniature specimen. It measured 8.5 × 5.3 × 2.8 cm, with a use surface 1.2 cm in depth. Seven of the nine metates were located in features. One fragment was at floor contact in a pit structure NFS (Feature 5-1), and another came from a trash deposit (Feature 9-2). All metates from Area 10 were in a trash deposit (Feature 10-1). Most of the fragments were portions of trough walls with use surface depths ranging from 5.3 cm to 9.0 cm. Five exhibited arm wear (Adams 1991), and two were complete enough to be described as open-ended varieties. All of the trough metates had only one use surface.

If metates were too fragmentary for analysts to determine whether they were slab, basin, or trough types, they were classified as metates NFS. Field crews recovered five of these (three of vesicular basalt and two of granite, representing all texture types), all from features. One came from the South Main Canal (Feature 2-1), one from the canal in Area 7, one from a pit with evidence of burning (Feature 8-10), and two from Feature 3-9, a pit structure (one in the fill and the other on the floor). Two had riverine cortex, but analysts could not determine the source of the other three.

Grinding slabs are relatively flat passive grinding tools that do not have typical metate forms. They are larger than palettes but smaller than typical slab metates. Field crews recovered two grinding slabs from Area 1 at Dutch Canal Ruin. Both were complete, and neither was shaped. One grinding slab of igneous NFS material (25.8 × 21.8 × 9.3 cm) came from floor contact in a pit structure (Feature 1-1) and had hematite and possible black pigment adhering to its use surface. The other grinding slab, from another pit structure (Feature 1-2), was a small, flat, greenstone cobble (13.1 × 11.5 × 2.6 cm) exhibiting possible red pigment and polishing on one surface. Use was evident in a 10.5 × 7.5–cm area with a depth of 0.05 cm. The smaller surfaces and adhering pigments on these artifacts indicated that they had probably been used to grind substances other than, or in addition to, plants.

Passive ground stone NFS was a category for items that had evidence of passive use but for which further classification into a grinding slab or metate category was not possible. One such item, a burned fragment made of diorite, was located in a trash pit (Feature 10-8).

Other Ground Stone

Investigators at Dutch Canal Ruin recovered two complete mauls with rounded, battered working ends. Birnie and Walsh-Anduze (1991) have proposed that mauls are simply reworked axes. Although one of the two mauls in the Dutch Canal Ruin assemblage supported that theory, the other did not. Both mauls were made from greenstone river cobbles and had been completely shaped. The maul from the fill of Feature 10-1, a trash deposit, was made of an elongated, cylindrical cobble and measured 19.4 × 6.3 × 5.4 cm. The tool's working end may once have had an axe blade but had been used last as a maul. Its length would have allowed for the remaining portion of the tool to be resharpened into an axe blade. In fact, two flakes had been removed from the working edge, as if the intent was to sharpen the bit. The other maul came from the floor of Feature 3-1, a pit structure. Shaped from a flat, round, greenstone cobble, the tool was 11.6 × 8.7 × 4.7 cm and had a rounded working edge that measured 7.0 × 3.5 cm. It had a ridge below its groove, but cortex on the head and ridge had not been removed by pecking. The ridge appeared to have functioned in the hafting of the tool rather than as an attribute of style.

One grooved, haftable artifact of greenstone material that could not be identified as either an axe or a maul was recovered from a trash deposit (Feature 10-1). The head had fractured, and the remainder of the tool may then have been recycled for use as a hammerstone.

Blanks are items that do not have definable shapes but are in the process of manufacture and have no evidence of use wear. Area 10 at Dutch Canal Ruin yielded one small, waterworn, greenstone cobble (5.9 × 5.7 × 4.0 cm) that may have been an axe blank. The cobble, found in a trash deposit (Feature 10-1), had been altered by pecking, with cortex having been completely removed from one surface. Less than one-fourth of the blank was present.

A ring, or doughnut-shaped stone (Haury 1976:290), was the only ground stone item recovered from Area 6. The maker was bidirectionally perforating it by pecking when it broke. With one-half of the item present, analysts evaluated it as well shaped, although it had not yet been completely rounded to a doughnut form.

Tabular tools are thin, flat items that may exhibit ground, flaked, or backed edges and possibly serration. They have been referred to as tabular knives, agave knives, mescal knives, and hoes, and they are most often associated with agave processing (Bernard-Shaw 1983, 1984; Castetter, Bell, and Grove 1938; Russell 1975:70). Field crews recovered four from Area 8 at Dutch Canal Ruin, made from rhyolite, igneous NFS, phyllite, and Squaw Peak schist. All were fragmentary, and only two were shaped. Three came from nonfeature contexts; the fourth lay in the fill above the floor in a pit structure NFS (Feature 8-3).

The raw material category included both pigments and fragmentary items of schist and phyllite. Of the items in this category, none was local to the project area, so they must have been brought in prehistorically for some purpose. The schist and phyllite items, the most numerous (n=141), were not modified but were often naturally angular fragments. Some of the smallest pieces may have been by-products of production by flaking, but because of their platy structure analysts could not distinguish flake morphology. One piece of hematite recovered from Area 9 had 11 ground facets from use wear.

Ground stone NFS is a miscellaneous category for items that were modified but could not be classified into any of the above categories. For example, it included ground stone that was too fragmentary for analysts to determine whether it was passive or active, as well as modified schist and phyllite fragments that were too small to be identified as tabular tools, worked discs, or other forms. The majority (84.3%) of items in this category were < 1/4 complete, too fragmentary to determine item form.

Field crews recovered three manuports (unmodified items that did not occur naturally in their excavated contexts) from the site: two gypsum crystals and a small greenstone cobble. The cobble had a flat, rounded shape (13.7 × 12.9 × 4.5 cm) and came from the floor of a pit structure (Feature 1-2).

One artifact from a pit NFS in Area 8 was classified as other ground stone because it was unique. It was a complete, oblong cobble of metasediment that had four flattened faces. Although the object was of mano length, measuring 11.3 × 2.8 × 2.8 cm, grinding use on one of the thin faces followed the long axis of the cobble. If the artifact was an active grinder, it had been used differently than other handstones or manos.

Possible ground stone included items on which cultural modification could not be conclusively determined. The majority (83.3%) of the 24 items in this category were fragmentary. Three-quarters of the materials displayed riverine cortices on which ephemeral use might not have been evident. Distinguishing use wear on the materials in the assemblage, which included sandstone (29.2%), quartzite (29.2%), granite (20.8%), metasediment (8.3%), and greenstone, gneiss, and igneous NFS (4.2% each), was difficult. A large proportion (66.7%) of the possible ground stone items had been burned. Although 29.2% of these items were in overburden deposits and 25.0% were in feature fill proveniences, 45.8% came from the floor fill and floors of six structures. Because this category may not represent cultural modification, these items are not included in the site discussions that follow.

Material Type

The Salt River Valley has abundant material resources that have weathered out of predominantly volcanic and metamorphic rock outcrops. The stream bed and primary drainages feeding into the Salt River are filled with waterworn cobbles that have been transported many miles from these original outcrops. Igneous, metamorphic, and sedimentary rocks all were available prehistorically as alluvial cobbles and were exploited for use as ground stone tools in the project area.

The Hohokam also quarried raw material for ground stone tool production from various igneous outcrops within their culture area (Bernard-Shaw 1984; Bruder 1983; Burton 1976; Green 1989;

Hoffman 1988; Hoffman and Doyel 1985; McQuestion and Gibson 1987). Artifacts manufactured from some quarried materials have been documented at major village sites, such as Las Colinas and Snaketown (Schaller 1987).

Analysis of the Dutch Canal Ruin ground stone assemblage included both material type and source, as described under "Methods," above. Unless otherwise specified, raw material definitions used for this analysis were taken from Hamblin 1985.

Igneous

Igneous rocks, formed by the cooling and solidification of molten silicate minerals (magma), are classified on the basis of their mineralogical differences and textures. Textural descriptions of volcanic rocks include descriptions of the sizes and shapes of crystal inclusions in a fine-grained groundmass. Igneous rocks can also be distinguished as having vesicular textures, in which small holes, or vesicles, have formed as a result of gas bubbles being trapped in solidifying lava. Vesicular basalt, vesicular andesite, and scoria, a rock containing abundant vesicles, were present in the ground stone assemblage. Vesicle sizes ranged from less than 2 mm, as was common in the scoria, up to 25 mm in vesicular basalt with vesicles.

Analysts identified nine specific igneous materials in the Dutch Canal Ruin assemblage. They used igneous NFS (3, or 1.2%) as a general category for materials that could not be identified beyond their igneous origin. Of the ground stone artifacts collected from Dutch Canal Ruin, 52 (20.2%) were of volcanic material (Table 14.1).

Basalt is a dark-colored, fine-grained igneous rock composed of over 50% plagioclase and pyroxene. Basalt may or may not contain olivine as one of its mineral constituents. Olivine is one of the minerals present in the vesicular basalt outcropping in the Hohokam northern periphery (Bostwick and Burton 1989; Green 1989; Schaller 1985, 1987). Basalt material with quartz crystals in the groundmass has been identified as quartz basalt (Schaller 1987) or quartz-bearing basalt (Bostwick and Burton 1989). A total of 32 artifacts (12.4%) consisted of some type of basalt material (Table 14.1): 2 items (0.8%) were identified as basalt, 17 artifacts (6.6%) as vesicular basalt, and 13 items (5.0%) as quartz-bearing basalt.

Granite, a coarse-grained igneous rock composed of potassium feldspar, plagioclase, and quartz, has small amounts of ferromagnesian minerals, such as olivine, pyroxene, and amphibole. Analysts identified eight granite items (3.1%) in the ground stone recovered from Dutch Canal Ruin.

Andesite is a fine-grained igneous rock composed of plagioclase feldspar, amphibole (25–40%), and biotite. This rock type does not contain any quartz or potassium feldspar. Two ground stone artifacts (0.8%) were of andesite.

Rhyolite, another fine-grained volcanic rock, is composed of quartz, potassium feldspar, and plagioclase and is the extrusive equivalent of granite. Two artifacts (0.8%) in the site assemblage were of rhyolite material.

When ash flows solidify, glass, ash, and other hot inclusions in the flow fuse together and form welded tuff, a fine-grained rock. Tuff material also made up 0.8% (2 artifacts) of the assemblage.

Diorite is another coarse-grained igneous rock that consists mostly of intermediate plagioclase feldspar and pyroxene, with some amphibole and biotite. Analysts categorized two ground stone items (0.8%) as diorite.

An igneous rock containing abundant vesicles is scoria. The one scoria ground stone item (0.4%) collected from Dutch Canal Ruin was red.

Sedimentary

Sedimentary rocks, formed by the accumulation and consolidation of sediment, represented the smallest portion of the Dutch Canal Ruin assemblage (Table 14.1). This probably reflects the general sparsity of sedimentary deposits in the Salt River Basin.

A coarse-grained sedimentary rock composed of rounded fragments of pebbles, cobbles, or boulders is a conglomerate. Field crews recovered one artifact of conglomerate material (0.4%).

Sandstone is composed predominantly of sand-size particles, usually cemented by calcite, silica, or iron oxide. Sandstone artifacts totaled 17 items, or 6.6% of the total assemblage.

A sediment or sedimentary rock that has been subjected to a low level of metamorphism but is still identifiable as a sedimentary rock is a metasediment. The metasedimentary rocks in the Dutch Canal Ruin collection may represent sediments composed of very fine to very coarse grained clastic deposits that have undergone varying degrees of diagenesis. Although amounts of silica cementation differed between rocks assigned to this category, in-hand or microscopic examination revealed that individual, variable-sized grains were still observable within the matrix. These items showed evidence of fracturing between grains rather than along crystal faces. Most of the metasediment artifacts were rather dense and heavy, physical characteristics for which they probably had been selected. Analysts identified nine ground stone artifacts (3.5% of the total) as metasediment.

Metamorphic

Metamorphic material is rock formed from pre-existing rock that has been altered by pressure and temperature and by the chemical actions of fluids. Five identifiable metamorphic types were present in the assemblage and constituted the majority (65.5%) of ground stone materials from Dutch Canal Ruin (Table 14.1). Some metamorphic rocks have been so drastically altered that they can be distinguished only by the fact that they have undergone intense metamorphism. Such items that could not be identified to a specific type (6, or 2.3%) were placed in a general category of metamorphic NFS.

Metamorphic rocks are classified into two groups, foliated or nonfoliated, based on their structure and composition. Foliation is the planar feature of these rocks, produced by the secondary growth of minerals, and has either a layered or a nonlayered texture. Slate, phyllite, and schist are nonlayered types of foliated rocks. Nonfoliated rocks used as ground stone material included quartzite and greenstone. Quartzite is sandstone that has been recrystallized by metamorphism. Field crews recovered four quartzite artifacts (1.6% of the assemblage) from Dutch Canal Ruin.

The Hohokam commonly used a fine-grained, green metamorphic material called greenstone for hammerstones and for axe production. Its name derives from the fact that most of the minerals (such as epidote, chlorite, and actinolite) formed during metamorphism are green (Schaller 1987). Greenstone represented 2.7% of the assemblage (n=7).

Schist is a medium-grained metamorphic rock with strong foliation or schistosity resulting from parallel orientation of platy minerals such as mica, chlorite, and talc. When possible, analysts classified schist artifacts more precisely as Squaw Peak schist. Squaw Peak schist has been described by Abbott (1991) as having an appearance that grades from smooth and platy to more granular, depending on the amount of foliation. It has a waxy luster, due to microcrystalline grains of mica, and is typically silvery white but can have a grayish and, less frequently, a greenish tint. It may also have large crystals of quartz and magnetite inclusions that commonly appear as tiny black dots in the rock (Abbott 1991). It rarely contains biotite and hornblende, but it may have magnetite or large quartz crystal inclusions. Of the 21 items identified as schist, 18 (7.0%) were further classified as Squaw Peak schist.

Phyllites are similar to slates but are distinguished by a satinlike luster or sheen developed on the planes of foliation. They usually are the color of pencil lead, but other colors do occur (Abbott 1991). Phyllite constituted 140 items in the ground stone assemblage, 54.3% of the entire collection from Dutch Canal Ruin. Most (124) were raw material, that is, small pieces that did not exhibit evidence of cultural modification. Analysts tentatively identified some phyllite material as Squaw Peak (?) phyllite.

Slate is a fine-grained metamorphic rock with characteristic slaty-type foliation (slaty cleavage), resulting from the parallel arrangement of microscopic platy minerals such as mica and chlorite. Slate was commonly used for pendants, spindle whorls, and palettes (Haury 1976). Investigators recovered only one piece of worked slate (0.4%) from Dutch Canal Ruin.

Other Material

Analysts also investigated minerals as part of the ground stone assemblage from Dutch Canal Ruin. One specimen (0.4%) was hematite, an iron mineral of varied appearance identified in three especially common varieties: specular hematite, which has a metallic luster; earthy hematite; and oolitic hematite. All have a characteristic deep red streak. Two other items (0.8%) were small, fibrous gypsum crystals. Gypsum is a common evaporite mineral composed of calcium sulfate with water formed by the evaporation of sea water or saline lakes. Possibly the crystals were disaggregated from some rock that originated in this type of depositional environment. Gypsum does not occur naturally in the project area, and the two crystals found at Dutch Canal Ruin may have originated somewhere around the Verde Valley. It is not uncommon to find crystals on Hohokam sites that the prehistoric inhabitants curated; some, such as quartz crystals, were sometimes worked. Because of their fibrous quality, however, gypsum crystals did not lend themselves well to such modification.

Analysts classified as unknown those materials that could not be identified as either igneous, sedimentary, or metamorphic. They placed one item (0.4%) into this category.

Technology

Resource procurement patterns at Dutch Canal Ruin exhibited a combination of local resource exploitation and nonlocal acquisition. Nearby streambed deposits provided a significant portion of alluvial materials for ground stone tools. Items that still retained riverine cortex accounted for 16.3%

of the assemblage. Material procured from other sources could only be determined by specific material identification. Squaw Peak schist and Squaw Peak (?) phyllite represented 25.2% of the assemblage, and quartz-bearing basalt accounted for at least 5.0%. Although Squaw Peak schist, and possibly phyllite, outcrop in the area of Canal System 2 (Abbott and Schaller 1990), procurement of these material sources would still require special effort and planning, as they occur in discrete zones over 10 km north of Dutch Canal Ruin. Quartz-bearing basalt is located even farther away, in the New River area of the Hohokam northern periphery. Implications of a long-distance procurement strategy for ground stone include specialization of tool manufacture in the New River area (Hoffman and Doyel 1985) and a network for trade between the Hohokam northern periphery and the core area (Doyel 1985).

Evidence of tool production at Dutch Canal Ruin was limited. An axe blank and a nearly completed ring, or doughnut stone, indicated that these particular tool types had been locally produced. Other evidence was furnished by three tools that had been reshaped and recycled after breakage, as well as a small amount of obvious ground stone debitage analyzed from the flaked stone assemblage (Chapter 13). Maintenance of equipment, however, was evidenced by a large amount of use-surface rejuvenation.

The amount of production invested in a tool varied within the site assemblage. Without including items for which production attributes did not apply, such as raw material and manuports, the frequency of ground stone that had been completely transformed by production modification (33.3%) was similar to that of items that had been left in their natural form (38.0%). Minimal production (19.0%) had the next highest frequency, suggesting that makers had taken less effort to shape tools than to create an entire form. Tools that exhibited the greatest production investment were rectangular/loaf manos, trough metates, mauls, and the stone ring. Morphological types with the least amount of effort invested in manufacture were polishing stones, the basin metate, and grinding slabs. Rectangular/loaf manos and trough metates, high-investment tools, accounted for 84.6% of the quartz-bearing basalt. The combination of nonlocal material and a high level of energy input suggested that these tools had been valuable equipment and must have played an important role in household economics and in subsistence-related activities.

High values of production investment in the ground stone assemblage occurred most often on items that were less than one-half complete. The majority (66.7%) of well-shaped items and 50.0% of completely shaped items were < 1/2 complete. The reverse situation also occurred; tools with no or minimal production investment were usually unbroken. The majority (62.5%) of items with no modification and 50.0% of minimally shaped items were complete. Although some of this variability was reflected in the material types, with the majority of vesicular and quartz-bearing basalts < 1/2 complete, analysts did not find a significant correlation. Other material types, such as sandstone, quartzite, and granite, were frequently present as small fragments. Because tools made of alluvial cobbles generally retained the shape of the selected nodule, they rarely needed further modification. The majority of riverine-rounded cobbles occurred as a dense material protected by a resilient cortex covering and were not easily broken, as indicated by the high frequency of complete items from an alluvial cobble source (78.3%). Table 14.3 shows a high correlation between material derived from a riverine cobble source and levels of production investment. The eta statistic, 0.831, means that if one knows the level of production investment on a ground stone tool, one can predict whether or not the material source was a riverine cobble, with an 83% proportional reduction in error. Axes are an obvious exception to this pattern because, although they were usually produced from river cobbles, they required considerable production investment in their final form.

Table 14.3. Ground Stone Production Investment Levels by Material Source

Production Investment	Source	
	River Cobble	Unknown
None	14	2
	*87.5	12.5
	**58.3	11.1
Minimal	8	
	100.0	
	33.3	
Shaped		1
		100.0
		5.6
Well Shaped		3
		100.0
		16.7
Completely Shaped	2	12
	14.3	85.7
	8.3	66.7
Total	**24**	**18**

Note: Based on eta statistic (used to determine material sources at the 0.831 level of confidence)
*Row %
**Column %

DISTRIBUTION

The Dutch Canal Ruin ground stone assemblage was neither large nor diverse (Table 14.1). All of the tools and other stone from the site, with the exception of the manuports, were utilitarian items. Activities implied by morphological types included ceramic production, food processing, farming, pigment processing, and tool manufacture. The lack of ornamentation and ritual paraphernalia was apparent. Although Dutch Canal Ruin was located within a major canal system, the material culture did not argue for high social or economic status. Instead, the assemblage indicated that the site occupants gave primary attention to subsistence-related activities, with no evidence of differential status.

Special-function tools, those that are reserved for a particular task or set of tasks, were present in the Dutch Canal Ruin assemblage. They included polishing stones, tabular tools, the stone ring, and hafted tools and made up 3.9% of all ground stone. Food processing tools, such as the manos and metates, accounted for the majority (67.7%) of tool types. Analysts gained little information from the spatial distribution of these artifacts due to the small sample recovered. Two areas, 8 and 10, contained samples with more than 20 artifacts (186 and 31, respectively). The rest of the areas contained fewer than 20 artifacts each.

The samples from feature types also were small, so the evaluation of ground stone distribution among feature types was unproductive as well. The only interesting association was that rectangular/loaf manos and trough metates were most often recovered from contexts of primary refuse; 68.8% of rectangular/loaf manos and 66.7% of trough metates came from trash features at Dutch Canal Ruin.

Investigation of the temporal distribution of ground stone was more productive. When analysts combined data from Areas 1 through 6 and the pit structure in Area 7, they created an adequate sample of the pre-Classic period component. Another component with an adequate sample was the Polvorón phase of the post-Classic period, represented by Area 8. Table 14.4 compares data from these two groups. The large discrepancy in sample sizes between the pre-Classic and post-Classic period assemblages was caused by the large amount of raw material (schist and phyllite) in Area 8; most of this occurred in very small pieces, possibly temper material or fragments of tabular tools. If the raw material is not included in the Area 8 profile, the Polvorón phase is represented by 44 items, comparable to the sample size of the pre-Classic period profile (n=34). The third column in Table 14.4 presents the data for the post-Classic period without the raw material. It shows a large percentage of unidentifiable active tools, but the metate frequencies still are much lower than in the pre-Classic period assemblage. In addition, the schist and phyllite raw material that predominate the Polvorón phase profile may represent fragments of tabular tools, especially since schist-tempered ceramics decreased in frequency during this time (Chapter 12).

The presence of schist and phyllite in the post-Classic period assemblage was important, however, because it reflected a change through time in resource procurement strategies at Dutch Canal Ruin. The Polvorón phase, represented by Area 8, accounted for 100.0% of the Squaw Peak schist and 100.0% of the Squaw Peak (?) phyllite. Residents of Dutch Canal Ruin apparently did not exploit the Squaw Peak resources during the pre-Classic period. Quartz-bearing basalt, however, occurred throughout the occupation of Dutch Canal Ruin, implying that exchange networks for this material were in place early in the Hohokam sequence.

Another change through time expressed by the ground stone assemblage was an increase in the importance of agave farming. The pre-Classic period profile is dominated by manos and metates (Table 14.4). The Polvorón phase profile, instead, has relatively low frequencies of manos and metates and a larger percentage of tabular tools. This difference in food processing equipment reflected a change in subsistence strategies, also documented by the flotation and pollen analyses. The flotation and pollen records both indicated that the pre-Classic periods at Dutch Canal Ruin had been dominated by activities related to corn agriculture, and the flotation data for the post-Classic period features was dominated by charred agave remains (Chapter 15). These combined data formed a clear pattern of a dramatic shift in the subsistence base at Dutch Canal Ruin. An alternative explanation would be that residents of the site had conducted different subsistence activities at different loci. Lacking contemporaneous loci exhibiting evidence for different strategies, however, archaeologists had no evidence to support such an interpretation.

The shift in subsistence base must not have occurred until the Polvorón phase, since Area 10, occupied during the Civano phase, was dominated by corn-processing equipment and also yielded macrobotanical remains (Chapter 11). Trough metates and rectangular/loaf manos accounted for 45.2% of the entire ground stone assemblage at Area 10. Although investigators monitored this area during stripping operations, which creates a bias for the recovery of large items, more than half of the corn-processing equipment collected was $\leq$ 1/2 complete, similar to breakage patterns documented for the site as a whole.

Table 14.4. Ground Stone Temporal Distribution

Morphological Type	Period		
	Pre-Classic	Post-Classic	
		Entire Assemblage	*Excluding Raw Material*
Polishing Stone	1	1	1
	*2.9	0.5	2.3
Handstone	2	1	1
	5.9	0.5	2.3
Round/Oval Mano	2		
	5.9		
Rectangular/Loaf Mano	2	2	2
	5.9	1.1	4.5
Mano NFS	1		
	2.9		
Grinder NFS		7	7
		3.8	15.9
Trough Metate	2	1	1
	5.9	0.5	2.3
Metate NFS	4	1	1
	11.8	0.5	2.3
Grinding Slab	2		
	5.9		
Maul	1		
	2.9		
Tabular Tool		4	4
		2.2	9.1
Ring	1		
	2.9		
Raw Material		142	
		76.3	
Ground Stone NFS	15	24	24
	44.1	12.9	54.5
Manuport	1	2	2
	2.9	1.1	4.5
Other		1	1
		0.5	2.3
Total	**34**	**186**	**44**

*Column %

CONCLUSIONS

Investigation of ground stone provided information on site technology, function, and economic strategies. Although evidence for tool production was not overwhelming, the inhabitants of Dutch Canal Ruin had apparently followed a reliable pattern for the exploitation of material resources, the appropriateness of materials for particular tools, and the amount of production effort that was necessary to produce an adequate tool. Riverine cobbles, found in local streambed deposits, were used for many tools and would have often been applied to a task without any form of alteration. They required low investment and were less expensive to procure and produce than trough metates and rectangular/loaf manos. The latter tools, which were often made from quartz-bearing basalt, a nonlocal material, had high values of production investment. The frequent use of tools produced from riverine cobbles suggested that they had been selected for these characteristics, as well as for the attributes of the material itself. Analysis demonstrated that tools made from alluvial cobbles were more likely to remain complete than other tools. Trough metates and rectangular/loaf manos, by contrast, were usually recovered as small fragments.

Trough metates and rectangular/loaf manos must have been worth the expense of procuring the appropriate material and producing their well-shaped forms. Vesicular material, which was commonly used for these instruments, has a self-sharpening quality in its surface, requiring less resurfacing of grinding areas (Haury 1976:280; Teague 1981:202). These types of tools are also associated with corn processing (Goodyear 1975:88; Haury 1950:319) and are found more frequently at sites whose inhabitants engaged in corn cultivation (D. M. Greenwald 1988a:191-197). Apparently, then, a prehistoric cultural bias in the Hohokam region supported the use of trough metates and rectangular/loaf manos for corn processing even though the tool expense was high. This bias probably developed in the Pioneer period from which quartzite trough metates have occasionally been noted (Haury 1976:280), only to disappear almost entirely in favor of the vesicular variety by the late Pioneer period (D. M. Greenwald 1990:109).

Except for the manuports, all of the ground stone at Dutch Canal Ruin was utilitarian. Ornaments and ritual paraphernalia were virtually nonexistent. Most tools were associated with subsistence-related activities. Handstones, manos, metates, and tabular tools all were used in the processing of food. Ceramic production, pigment processing, and tool production activities were also represented in the assemblage.

A change in subsistence focus occurred during the post-Classic period. The ground stone profile for the pre-Classic periods was dominated by tools for processing corn and smaller seed crops, the Classic period assemblage exhibited strong association with corn processing, and the post-Classic period assemblage exhibited a mixed strategy of corn and agave processing. This pattern was also evident in the flotation and to some extent in the pollen analyses. However, the change may have been more dramatic than is apparent if the large frequency of schist and phyllite raw material in the post-Classic period profile actually represented tabular tool fragments.

A large proportion of the raw material that analysts assigned to the post-Classic period they identified as Squaw Peak schist and Squaw Peak (?) phyllite, sources that the Hohokam apparently had not exploited during earlier periods at Dutch Canal Ruin. The temporal variability of these materials represented a change in patterns of resource exploitation that probably had developed in response to the change in the subsistence base. Analysts noted no other temporal variability in the ground stone distribution but did document additional changes in material distribution through time (Chapter 13) that reflected changes in exchange networks within the Southwest. These patterns not only apply to Dutch Canal Ruin but also represent general trends throughout the Hohokam region related to changes in political, social, and economic complexity and interaction.

CHAPTER 15

FLOTATION AND MACROBOTANICAL RESULTS FOR DUTCH CANAL RUIN

Scott Kwiatkowski

This chapter presents the results of the analysis by Soil Systems, Inc. (SSI), of 19 flotation samples and one macrobotanical sample collected from six areas investigated at Dutch Canal Ruin during the data recovery phase of the Phoenix Sky Harbor Center Project. The samples from the pre-Classic periods, although consistently yielding charred corn, produced a less diverse and less dense charred plant assemblage than did the samples from the post-Classic period.

RESEARCH OBJECTIVES

The research design for this project specified that the archaeobotanical studies should address several topics (Volume 1:Chapter 10). Archaeologists were concerned with defining activities conducted within structures, house clusters, and other site components that were related to plant use; determining to what extent botanical resources had been a factor in site settlement and land use; inferring the importance and use of wild plant resources as they related to cultivated species and if such economically important species had been encouraged to grow in conjunction with crops; and comparing field house and farmstead components recorded earlier at Dutch Canal Ruin and at other sites with results from this project.

ANALYTICAL METHODS

Sample Processing

The field laboratory for SWCA in Phoenix processed the flotation samples prior to submitting them for analysis to SSI. Laboratory staff members poured sediment samples, generally of 4 liters, into a 5-gallon wash basin filled with water. A hose ran water into the bottom of the basin, and water running off the top of the basin was caught in fabric. This process lasted approximately five minutes, or until charcoal was no longer visibly suspended in the water of the basin. The heavy fractions, retained at the bottom, were screened through window screen. After both fractions were dried, laboratory staffers bagged them separately in preparation for analysis.

Analysis

The SSI analysts passed the light fractions from flotation through geological screens with 2.00-mm, 1.00-mm, 0.50-mm, and 0.25-mm mesh to facilitate sorting into similar size classes. They examined only light fractions because previous research at several sites in the Phoenix area (e.g., Kwiatkowski 1988a, 1988b, 1988c, 1989a, 1989b, 1990) demonstrated that heavy fractions rarely yield identifiable charred plant remains, and when they do, the taxa are usually the same as those in the light fractions. The analysts also omitted one or more of the larger screens if a light fraction was small and contained little large material

None of the samples contained more than seven pieces of wood charcoal large enough for identification. Analysts broke these fragments to expose a fresh cross section, examined them at

40X under a binocular microscope, and classified them by comparison to charred modern reference material. The SSI investigators then examined the remainder of the sample at 10–11X for the presence of nonwoody plant remains but occasionally used magnification up to 140X on problematic specimens. Analysts identified the remains using manuals (Knight 1978; Martin and Barkley 1961) and by comparing sample remains with modern and archaeological reference specimens. They quantified carbonized and uncarbonized materials separately and categorized plant parts as either whole or fragmentary (Appendix C).

SSI analysts examined the residue retained in the 0.25-mm mesh in the first eight samples (Appendix C). When they concluded that this fine material was basically unproductive, they did not further analyze it, and none of the material smaller than 0.25 mm is reported. The single productive sample from the small screen was 9 charred agave fibers from Feature 8-2 (Sample 8-110). However, analysts recovered 99 charred agave fibers in the larger screens of this sample and thus obtained no new information from the 0.25-mm screening.

Identification Issues

SSI analysts identified plant remains to the greatest level of specification possible. Table 15.1 is a list of taxa recovered, including common and scientific name equivalents. Fragmentary material exhibiting some key attributes of a particular taxon, yet lacking others normally present, received a tentative identification of "compares favorably" (cf.). Because cf. identifications could be in error, little significance should be attached to them. Plant parts classified as "indeterminate" are even less identifiable than cf. identifications, and "miscellaneous" refers to plant remains with little diagnostic value.

Researchers consider uncarbonized plant remains from open-air archaeological sites to be recent contaminants, for reasons discussed by Keepax (1977) and Minnis (1981). At Dutch Canal Ruin the uncarbonized plant remains recovered consisted entirely of weedy taxa, including one Old World native (Mediterranean grass), which could have been growing near the site recently (Table 15.1), whereas all of the domesticated plant remains were carbonized.

The identification of four nonwoody taxa (dropseed-type and sprangletop-type grains, agave, and little barley grass grains) merits further discussion. The small grass grains categorized here as dropseed type and sprangletop type are synonymous with the *Eragrostis-Sporobolus*-type grains and the *Leptochloa-Muhlenbergia*-type grains, respectively, that were described and illustrated by Kwiatkowski (1989b:503–504). Analysts based their identification of agave remains on the presence of fibers with trough-shaped cross sections. Archaeologists do not regard fibers of other shapes as identifiable beyond the class level (Bohrer 1987:72). Each little barley grass grain fragment lacked adherent paleas or lemmas; the current view is that such grains represent an incipient domesticate (Adams 1986, 1987; Bohrer 1984:252, 1987:86–88).

The wood charcoal identifications should be viewed as tentative for reasons discussed by Bohrer (1986:34). In addition to the usual problems (which include variability in wood structure due to the specimen's age at burning, its original position within the tree or shrub, and differing edaphic regimes), the charcoal from Dutch Canal Ruin tended to be small, making identifications even more difficult and less certain. Analysts identified two types of arboreal legume charcoal and separated them based on density; the cf. paloverde charcoal was less dense than the cf. mesquite charcoal.

Table 15.1. Plant Taxa in Flotation or Macrobotanical Samples

Common Name	Scientific Name	Condition and Parts Present*
Agave	*Agave* sp.	Fibers (C), fragments of round and trough-shaped fibers with attached parenchyma (C)
Arboreal legume cf. mesquite	cf. *Prosopis* sp.	Wood (C)
Arboreal legume cf. paloverde	cf. *Cercidium* sp.	Wood (C)
Brome grass–eel grass type	*Bromus-Elymus* type	Grain (C)
Canary grass	*Phalaris* sp.	Grain (C)
Cotton	*Gossypium hirsutum* var. *punctatum*	Seed (C)
cf. Creosotebush	cf. *Larrea tridentata*	Wood (C)
Desert broom	*Baccharis* sp.	Achene (U)
Dicot	Magnoliopsida	Leaf, stem fragment (U)
Dropseed type	*Sporobolus* type	Grains (C)
Goosefoot family or pigweed, Cheno-ams	Chenopodiaceae or *Amaranthus* sp.	Seeds (C,U)
Grama grass type	*Bouteloua* type	Grains (C)
Grass family	Gramineae	Culm fragment (C), grain (C)
Ground cherry	*Physalis* sp.	Seed (U)
Horse purslane	*Trianthema portulacastrum* L.	Seeds (C,U)
Little barley grass	*Hordeum* sp.	Grains (C)
Maize, corn	*Zea mays* L.	Cupules (C), cf. embryo fragments (C), glume fragments (C), cf. kernel fragments (C)
Mediterranean grass**	*Schismus arabicus* Nees., *Schismus* sp.	Floret (U), grain (U)
Mesquite	*Prosopis* sp.	Seeds (C), cf. pod fragments (C)
Pea family	Leguminosae	Seed (C)
Purslane	*Portulaca* sp.	Seed (U)
Seepweed	*Suaeda* sp.	Seeds (C)
Sprangletop-type grass	*Leptochloa* type	Grains (C)
Spurge	*Euphorbia* sp.	Seeds (C,U)
Sunflower family	Compositae	Achene (U)
Tansy mustard	*Descurainia* sp.	Seeds (C)
Wild tobacco	*Nicotiana* cf. *trigonophylla*	Seed (C)

*Condition codes: C = carbonized; U = uncarbonized
**Introduced taxon

Quantification

Two primary quantitative measures are used in this discussion to interpret the flotation data. The first, presence value, is the percent of occurrence in a suite of samples (Hubbard 1980). For example, if charred corn remains were found in 10 of 20 analyzed samples (including unproductive ones), their presence value would be 50%. The second measure, relative density, is the number of charred plant parts per liter in each sample, feature type, or time period expressed as a percent of the total. For example, 10 charred agave fibers in a 4-liter flotation sample equals 2.5 relative parts $(10 \div 4)$. This would constitute 1.25% of the total if 200 charred plant relative parts were recovered from the site.

RESULTS

The 20 analyzed samples came from the fill of extramural pits and fire pits, pit structures, a pit house, and a primary inhumation (Table 15.2). Thirteen of the features were pre-Classic (i.e., late Pioneer period to early Colonial period), whereas the others dated to the post-Classic period. Eight samples contained identifiable charcoal (Table 15.3), although only three wood types were present. Samples were considered unproductive if they did not contain nonwoody charred plant material identifiable at least to the family level. Using this criterion, analysts defined 12 samples as productive (Tables 15.4 and 15.5). Pre-Classic period pit structure floors tended to be the least productive. Over half (5 of 9, or 55.6%) of the samples analyzed from this context were unproductive (Table 15.2). Also, no identifiable charred plant remains were present in the samples for Areas 1 and 2 (Table 15.2).

Analysts identified 23 nonwoody plant taxa (Table 15.1). Of these, 14 yielded only charred specimens, 6 had only uncarbonized representatives, and 3 had both charred and uncharred examples (Table 15.1). Table 15.6 lists the actual number of charred plant parts recovered, the presence values, and the relative part percentages for each charred plant taxon, excluding cf. level identifications. Investigators for a previous study (Ruppé 1988) also analyzed 19 flotation samples from this site. This report is referred to below.

DISCUSSION

Evidence for Plant-Related Activities

This reconstruction attempts to identify the activities most likely to have produced the charred plant assemblage. Archaeologists consider a plant economically important when (1) its context of recovery fits known ethnographic uses (e.g., Castetter and Bell 1942; Castetter and Underhill 1935; Curtin 1984; Russell 1908); (2) they believe the plant was economically important to the Hohokam (e.g., Gasser 1982b; Gasser and Kwiatkowski 1991); and (3) they have no reason to believe that it might be an accidentally burned component of the prehistoric seed rain (Minnis 1978, 1981). Table 15.7 presents a summary of the economic activities inferred to have occurred in the analyzed features. These assessments should be viewed as speculative.

Pre-Classic Period Features

The pre-Classic period samples characteristically yielded few or no charred plant remains. Corn was by far the most common charred plant taxon recovered. In fact, with the exception of agave,

Table 15.2A. Productivity of Samples by Feature Type: Pre-Classic Period Features

Feature Type	Productive Samples		Unproductive Samples	
	No. Samples	Feature Nos.	No. Samples	Feature Nos.
Extramural Fire Pit, Burned	0	—	1	3-18
Pits, Burned	2[a]	3-3	0	—
Pits, Unburned	1	4-1	0	—
Pit Structures:				
Fire Pit, Burned	0	—	1	3-2-1
Floor Contact, Unburned	1	5-1	0	—
Floor Fill, Burned	1	3-1	4	1-1, 1-2, 2-2, 3-19
Floor Fill, Unburned	2	3-2, 3-9[b]	1	3-16[b]
Total	7		7	

Table 15.2B. Productivity of Samples by Feature Type: Post-Classic Period Features

Feature Type	Productive Samples		Unproductive Samples	
	No. Samples	Feature Nos.	No. Samples	Feature Nos.
Pits, Burned	2	8-12[b], 8-14	0	—
Primary Inhumation[c]	0	—	1	1-3
Structures:				
Pit House Floor Fill, Unburned	1	8-2[b]	0	—
Pit House Fire Pit, Burned	1	8-2-1	0	—
Pit Structure Fire Pit, Burned	1	8-3-1	0	—
Total	5		1	

Note: All samples are flotation unless otherwise indicated.
[a]One flotation and one macrobotanical sample.
[b]Refuse-filled feature.
[c]Although Feature 1-3 is undated, a "best guess" age estimate attributes it to the Soho phase.

Table 15.3A. Wood Charcoal Identified in Flotation Samples: Pre-Classic Period Features

| Context and Provenience | Arboreal Legume | | Perennial Shrub |
	cf. *Cercidium*	cf. *Prosopis*	cf. *Larrea tridentata*
Pit Structure F3-2 Floor Fill	—	1	—
Pit Structure F3-19 Floor Fill	—	—	1
Pit Structure F5-1 Floor Contact	—	6	—

Table 15.3B. Wood Charcoal Identified in Flotation Samples: Post-Classic Period Features

| Context and Provenience | Arboreal Legume | | Perennial Shrub |
	cf. *Cercidium*	cf. *Prosopis*	cf. *Larrea tridentata*
Pit House F8-2 Floor Fill	2	1	—
Pit House Fire Pit F8-2-1	—	6	—
Pit Structure Fire Pit F8-3-1	—	3	—
Pit F8-12	—	7	—
Pit F8-14	—	6	—

Table 15.4A. Distribution of Charred Nonwoody Plant Remains in Productive Flotation or Macrobotanical Samples: Pre-Classic Period Features

Context and Provenience	Sample Vol. (L)	Agave	Bouteloua Type	Bromus-Elymus Type	Cheno-am	Descurainia	Euphorbia	Gossypium hirsutum	Gramineae	Hordeum	Leguminosae	Leptochloa Type	Nicotiana	Phalaris	Prosopis	Sporobolus Type	Suaeda	Trianthema	Zea mays
PS F3-1 FF	2.0	—	—	—	1s	—	—	—	—	—	—	3g	—	—	—	—	—	—	—
PS F3-2 FF	4.0	—	—	—	—	—	—	—	—	—	—	—	—	—	—	—	—	—	2cf
PT F3-3	2.0	—	—	—	—	—	—	—	—	—	—	—	—	—	—	—	—	—	2c 60cf 5glf
PT F3-3	—*	—	—	—	—	—	—	—	—	—	—	—	—	—	—	—	—	—	5c 8cf 1glf
PS F3-9 FF	4.0	—	—	—	—	—	—	—	—	—	—	—	—	—	—	—	—	—	1cf
PT F4-1	4.0	—	—	—	—	—	—	—	—	—	—	—	—	—	—	—	—	1sf	—
PS F5-1 FC	4.0	4fi	—	—	2sf	—	—	—	—	—	—	—	—	—	—	—	—	—	1c
Total	20.0	4fi	—	—	1s 2sf	—	—	—	—	—	—	3g	—	—	—	—	—	1sf	8c 71cf 6glf

*Macrobotanical sample

FC = floor contact; FF = floor fill; FP = fire pit; PH = pit house; PS = pit structure; PT = pit

c = cupule; cf = cupule fragment; cmf = culm fragment; fi = fiber; frg = fragment; g = grain; gf = grain fragment; glf = glume fragment; s = seed; sf = seed fragment

Table 15.4B. Distribution of Charred Nonwoody Plant Remains in Productive Flotation or Macrobotanical Samples: Post-Classic Period Features

Context and Provenience	Sample Vol. (L)	Agave	Bouteloua Type	Bromus-Elymus Type	Cheno-am	Descurainia	Euphorbia	Gossypium hirsutum	Gramineae	Hordeum	Leptochloa Type	Leguminosae	Nicotiana	Phalaris	Prosopis	Sporobolus Type	Suaeda	Trianthema	Zea mays
PH F8-2 FF	2.5	99fi	—	—	2s 5sf	—	—	—	—	—	1g	—	—	—	—	—	—	—	—
PH FP F8-2-1	2.0	141fi	—	—	4s 8sf	—	1s 1sf	—	1cmf 1g	3gf	—	—	—	2g	2sf	—	1s	—	1cf
PS FP F8-3-1	2.0	463fi	—	—	1sf	—	—	1sf	—	—	—	—	—	1g 2gf	1sf	—	3s	—	3cf 1glf
PT F8-12	4.0	322fi 2frg	6g	1gf	8s 16sf	3s	—	—	2cmf	—	9g	1s	1s	21g 12gf	1s 7sf	21g 3gf	1s	—	20cf 2glf
PT F8-14	4.0	150fi	—	—	7s 11sf	1s	—	—	—	—	6g	—	—	2g 1gf	3sf	—	1s	—	2cf
Total	14.5	1,175fi 2frg	6g	1gf	21s 41sf	4s	1s 1sf	1sf	3cmf 1g	3gf	16g	1s	1s	26g 15gf	1s 13sf	21g 3gf	6s	—	26cf 3glf
Project Totals	34.5	1,179fi 2frg	6g	1gf	22s 43sf	4s	1s 1sf	1sf	3cmf 1g	3gf	19g	1s	1s	26g 15gf	1s 13sf	21g 3gf	6s	1sf	8c 97cf 9glf

FC = floor contact; FF = floor fill; FP = fire pit; PH = pit house; PS = pit structure; PT = pit
c = cupule; cf = cupule fragment; cmf = culm fragment; fi = fiber; frg = fragment; g = grain; gf = grain fragment; glf = glume fragment; s = seed; sf = seed fragment

Table 15.5A. Distribution of Charred Plant Remains in Productive Flotation Samples: Pre-Classic Period Features

Sample Locus	Actual No. Parts*	No. Relative Parts	Relative Parts (%)
Pit F3-3	67	33.5	87.58
Pit Structure F3-1 Floor Fill	4	2.0	5.23
Pit Structure F5-1 Floor Contact	7	1.75	4.58
Pit Structure F3-2 Floor Fill	2	0.5	1.31
Pit Structure F3-9 Floor Fill	1	0.25	0.65
Pit F4-1	1	0.25	0.65
Total	**82**	**38.25**	**100.00**

*Includes whole and fragmentary plant parts but not cf. identifications.

Table 15.5B. Distribution of Charred Plant Remains in Productive Flotation Samples: Post-Classic Period Features

Sample Locus	Actual No. Parts*	No. Relative Parts	Relative Parts (%)
Pit Structure Fire Pit F8-3-1	476	238.0	45.37
Pit F8-12	459	114.75	21.88
Pit House Fire Pit F8-2-1	166	83.0	15.82
Pit F8-14	184	46.0	8.77
Pit House Floor Fill F8-2	107	42.8	8.16
Total	**1392**	**524.55**	**100.00**

*Includes whole and fragmentary plant parts but not cf. identifications.

Table 15.6. Taxonomic Distribution of Charred Plant Remains in Flotation Samples

Taxon	Actual No. Parts			Relative Parts (%)[1]		Presence Value (%)[2]	
	Whole	Fragments	cf.	Pre-Classic	Post-Classic	Pre-Classic	Post-Classic
Agave	—	1181	—	2.61	87.72	7.69	83.33
Fibers	—	1179	—	2.61	87.62	7.69	83.33
Other Fragments	—	2	—	—	0.10	—	16.67
Cheno-am Seeds	22	43	26	2.61	3.77	15.38	83.33
Descurainia Seeds	4	—	8	—	0.19	—	33.33
Euphorbia Seeds	1	1	—	—	0.19	—	16.67
Gossypium hirsutum Seeds	—	1	3	—	0.10	—	16.67
Gramineae Culms	—	3	7	—	0.19	—	33.33
Gramineae Grains (except *Hordeum* and *Zea mays*)	73	19	42	3.92	4.56	7.69	83.33
Bouteloua Type	6	—	—	—	0.29	—	16.67
Bromus-Elymus Type	—	1	—	—	0.05	—	16.67
Leptochloa Type	19	—	8	3.92	0.79	7.69	50.00
Phalaris	26	15	14	—	2.19	—	66.67
Sporobolus Type	21	3	14	—	1.14	—	16.67
Indeterminate Type	1	—	6	—	0.10	—	16.67
Hordeum Grains	—	3	—	—	0.29	—	16.67
Leguminosae Seed	1	—	—	—	0.05	—	16.67
Nicotiana Seed	1	—	—	—	0.05	—	16.67
Prosopis	1	13	24	—	0.81	—	66.7
Pod Fragments	—	—	9	—	—	—	—
Seeds	1	13	15	—	0.81	—	66.67
Suaeda Seeds	6	—	—	—	0.48	—	66.67
Trianthema portula-castrum Seeds	—	1	—	0.65	—	7.69	—
Zea mays	3	97	158	90.20	1.62	30.77	66.67
Cupules	3	89	143	83.66	1.43	30.77	66.67
Embryos	—	—	1	—	—	—	—
Glumes	—	8	4	6.54	0.19	7.69	33.33
Kernels	—	—	10	—	—	—	—
Total	112	1362	268	99.99	100.02	—	—

Note: Relative parts and presence value percentages include whole and fragmentary plant remains, but not cf. identifications. Feature 1-3 is grouped with the post-Classic period features.

[1]Percentages are based on 38.25 pre-Classic period and 524.55 post-Classic period total relative parts.

[2]Percentages are based on presence in 13 pre-Classic period and 6 post-Classic period sample loci.

Table 15.7A. Summary of Inferred Plant-Related Activities: Pre-Classic Period Features

Feature	Remains						
	Agave (used for food, cordage, or fuel)	Corn (cobs used as fuel)	Secondary Refuse	Seeds (prepared for consumption)	Grass Culms (used as fuel)	Seepweed (used as flavoring agent)	Wood (used as fuel)
Pit Structure F3-2 Floor Fill	—	X	—	—	—	—	—
Pit F3-3	—	(X)	X	—	—	—	—
Pit Structure F3-9 Floor Fill	—	(X)	X	—	—	—	—
Pit Structure F5-1 Floor Contact	X	X	—	—	—	—	X

Note: Parentheses indicate activities that probably had occurred on-site but perhaps not within the features indicated, as these features may have contained secondary refuse.

Table 15.7B. Summary of Inferred Plant-Related Activities: Post-Classic Period Features

Feature	Remains						
	Agave (used for food, cordage, or fuel)	Corn (cobs used as fuel)	Secondary Refuse	Seeds (prepared for consumption)	Grass Culms (used as fuel)	Seepweed (used as flavoring agent)	Wood (used as fuel)
Pit House F8-2-1 Fire Pit	X	X	—	X	X	X	X
Pit Structure F8-3-1 Fire Pit	X	X	—	X	—	X	X
Pit F8-12	(X)	(X)	X	(X)	(X)	(X)	(X)
Pit F8-14	(X)	(X)	X	(X)	—	(X)	(X)

Note: Parentheses indicate activities that probably had occurred on-site but perhaps not within the features indicated, as these features may have contained secondary refuse.

the noncorn charred plant remains occurred so infrequently and in such low densities (Table 15.4) that they could have been inadvertently carbonized components of the prehistoric seed rain.

Pits

The single charred horse purslane seed fragment in Feature 4-1 could have been an accidentally burned component of the prehistoric seed rain, and therefore investigators could deduce nothing about this pit's function. The second pit, Feature 3-3, contained the highest density of pre-Classic period plant remains recovered during the current study: more than 60 charred corn cob fragments and additional macrobotanical corn remains (Table 15.4A). Although the sample from Feature 3-3 was dense compared to those of the other pre-Classic period features, each of the productive post-Classic period samples had even higher charred plant densities (Table 15.5). The material in Feature 3-3 may have been secondary refuse unrelated to the pit's original function, because the base and side walls of the pit were not oxidized, even though its fill was burned.

Pit Structures

The maximum density of charred plant remains in pre-Classic pit structures was 2.0 parts per liter (Table 15.5A). No sample yielded more than three charred plant taxa (Table 15.4A). Corn cupules occurred in three of the four productive samples (Table 15.4A). Although the charred sprangletop-type grains could have been remnants of a grass-covered superstructure in Feature 3-1, a burned pit structure, investigators thought this notion unlikely, because no burned grass culms were associated with the grains. The agave fibers, recovered from an unburned structure, may have represented discarded matting, cordage, or waste from the consumption of agave hearts or leaves. Because the wood charcoal in the burned structures could have been the residue either of hearth fuel or of burned superstructures, only the cf. mesquite charcoal from Feature 5-1, an unburned pit structure, was likely to be evidence of fuel (Table 15.7).

Seasonality

The macrobotanical data indicated that the pre-Classic period site components had been used during the harvest of corn. Noncorn remains in the pits and structures could have been stored products (e.g., agave), or they could have been inadvertently carbonized during any time of the year. The site could have been occupied during other times of the year, particularly during field preparation and crop growing seasons, but investigators currently have no direct macrobotanical evidence for such uses. Given historical analogues, analysts assumed that the inhabitants of Dutch Canal Ruin could have harvested two corn crops: one in late June and the second during October (Castetter and Bell 1942:179).

Processing, Consumption, and Storage

Although the charred corn cob fragments in pit structures had probably been burned either as fuel or to dispose of a by-product of corn harvesting, investigators could not determine whether they related primarily to corn processing prior to storage, discard after on-site corn consumption, or a combination of these two activities. If freshly harvested corn had been eaten near or within the structures, the cobs could have been disposed of in pit structure fire pits, which ultimately would have led to their carbonization. However, none of the corn recovered was from fire pits, and the single fire pit sampled was unproductive. Similarly, the only context within pit structures that produced charred corn remains during the previous site flotation study was floor fill (Ruppé 1988).

If, by contrast, the inhabitants of Dutch Canal Ruin had processed corn in bulk prior to storage by roasting ears in the husks and then shelling them near the fields, which is the practice of the historic Pima (Castetter and Bell 1942:180:183), investigators would expect to find some charred corn kernel fragments as the result of roasting accidents; the prehistoric agriculturists could have disposed of some of the leftover corn cobs by burning them as a source of fuel. A problem with this hypothesis of bulk processing is that field crews for the present study failed to recover burned corn kernel fragments, although kernel fragments were collected as macrobotanical samples from Areas 9 and 10. Ruppé (1988) did identify six charred corn kernel fragments in pit structure fill, although they were outnumbered by cupules and cupule fragments (n=13).

The charred corn remains in pit structures did not seem to be directly related to storage in the structures, because burned cob or kernel fragments would in all likelihood have been separated from the rest of the corn prior to storage. Corn could have been routinely stored in field houses, but the flotation and macrobotanical results did not support the notion of storage. Pollen analysis is better suited for detecting evidence of corn storage.

Post-Classic Period Features

Agave fibers dominated the charred plant remains in post-Classic period features, all located in Area 8 (Table 15.4B). Researchers have observed high values indicating the presence of agave at a number of other habitation sites located along Turney's Canal System 2, including the pre-Classic period sites of La Lomita Pequeña, El Caserío, and La Lomita (Kwiatkowski 1990) and the Classic period sites of Casa Buena (Gasser 1988) and the Grand Canal Ruins (Kwiatkowski 1989b). The Hohokam apparently cultivated agave in the lower Salt River Valley (Bohrer 1987:79; Fish et al. 1985; Gasser and Kwiatkowski 1991:425–426) by the mid-Colonial period (Gasser and Kwiatkowski 1991:441). Three other cultivated plants were represented at Area 8: corn, cotton, and little barley grass. Given the evidence from archaeological sites east of New Mexico, archaeologists believe that canary grass also could have been encouraged or grown (Cowan 1978, 1985:212–214). Grass grains in general were well represented in the post-Classic period features (Tables 15.4B and 15.6).

Six charred plant taxa were present in the majority of samples from the post-Classic period (Table 15.4B). Five of these—agave, Cheno-am seeds, canary grass grains, mesquite, and corn—had probably been important subsistence items to the residents of Area 8; the sixth, seepweed, had probably been a flavoring agent (Curtin 1984:71; Russell 1908:78). Their availability, however, would have varied seasonally. Canary grass ripens in early summer (Bohrer 1975:199), mesquite pods are available in midsummer, around August (Castetter and Underhill 1935:24), and corn was historically harvested in late June and October by the Pima (Castetter and Bell 1942:179). The various species of Cheno-ams ripen at different times of the year, although pigweed was harvested in September by the Tohono O'odham (Castetter and Underhill 1935:24). Even though agave leaves were available throughout the year, the more palatable and nutritious caudexes would have been available from fall through spring.

Given the consistent co-occurrence of plant remains available at different times of the year (and the higher density and greater diversity of material compared with that of the pre-Classic period features), investigators concluded that the Hohokam had used stored plant products at Area 8, had inhabited the area for a large portion of the year, had occupied the area more intensively or longer than the inhabitants of the pre-Classic period site components, or had combined these alternatives. In any case, the occupants of Area 8 had apparently conducted a greater range of plant-related activities than the occupants of any of the pre-Classic period loci (Table 15.7).

The Area 8 plant remains dated to the post-Classic period or the Polvorón phase (Chapters 9 and 18). The few (n=6) roughly contemporary flotation samples (attributed to the post-Classic period) analyzed from Las Colinas differed considerably from the Area 8 samples (Miksicek and Gasser 1989:111). Investigators recovered no agave remains or cotton seeds from these Las Colinas samples, but corn, mesquite, saguaro, and a number of weedy taxa were well represented. No grass family members other than corn appeared in these late Las Colinas samples (Miksicek and Gasser 1989:Table 7.9).

Pit Structure Fire Pit

Feature 8-3-1 yielded the highest density of charred plant remains recovered during the current project (Table 15.5). The wood charcoal probably represented fuel remnants, and the carbonized nonwoody plant remains were compatible with food preparation refuse deposited as de facto or primary refuse (Table 15.7B). Most of the identifiable carbonized material in this fire pit was charred agave fibers that could have been deposited as food refuse, as refuse from cordage making, or as fuel (Sheehy 1990). The presence of charred seepweed seeds in this and other features may indicate that the Hohokam favored it and used it as the Pima had, who referred to it as "black salt" and used it as a flavoring agent in cooking (Curtin 1984:71; Russell 1908:78).

Pit House

Feature 8-2 was an unburned pit house that had been filled with trash after abandonment. A flotation sample from its fire pit contained a more diverse and higher density of charred plant remains than a sample from its floor fill. Material in the fire pit may have been spread out into the floor fill by postdepositional disturbance processes (Table 15.4B). Wood charcoal, corn cob fragments, and a grass culm fragment recovered from the fire pit could have been fuel remnants, and most of the remaining charred nonwoody remains could have been food processing de facto or primary refuse. The charred spurge seed and seed fragment probably were not intentionally collected, because they are toxic (Curtin 1984:99–100; Felger and Moser 1985:295). The three charred "naked" little barley grain fragments were noteworthy because of a recent assertion that these grains are rare at Classic period sites in the Salt River Valley (Gasser and Kwiatkowski 1991:442–443).

Pits

Adjacent pits, Features 8-12 and 8-14, exhibited evidence of in situ burning. Feature 8-12 appeared to have been filled with trash (Chapter 9), and both features appeared to have been filled with secondary refuse from food processing activities (Table 15.7B). The two features had a number of archaeobotanical similarities and therefore might have been related (Table 15.4B). The presence of a charred wild tobacco seed in Feature 8-12 was noteworthy, because this taxon is rarely recovered from Hohokam sites. This seed could have been related to smoking, but because of its isolated occurrence, the possibility that it was an accidentally burned component of the prehistoric seed rain cannot be rejected.

Plant Use, Prehistoric Vegetation, and Site Settlement

Flotation and macrobotanical studies are not as well suited as pollen analysis for addressing paleoenvironmental research objectives. However, analysts made several observations by combining the archaeobotanical data with ecological information.

Because of the abundance of charred corn within the earlier features and current archaeological understanding of Hohokam field houses, Dutch Canal Ruin was probably near or surrounded by fields of corn during its pre-Classic period occupation. The presence of charred weeds was, however, minimal, perhaps indicating minor growth of disturbed-earth plants around the site due to agricultural and other activities.

Ecological studies provided relevant information about the motives for original site location. A reconstruction of the native vegetation of the Phoenix area (Turner 1974) indicated that the site had probably been located prehistorically in a desert saltbush series (*Atriplex polycarpa*). Land occupied by this vegetative association is well suited to agriculture (Masse 1991:196; Turner 1974; Turner and Brown 1982:194). In one sense, then, the pre-Classic period component of Dutch Canal Ruin may have been settled in a desert saltbush community because the vegetative association signaled the presence of arable land. However, this notion is not helpful for understanding the specifics of site location, because this association formerly constituted one of the most widespread plant communities in the Phoenix area (Shantz and Piemeisel 1924:788; Turner 1974).

An additional reason for site location related to vegetation may have been the propinquity of two other vegetative communities. The transition to a creosotebush-bursage (*Larrea-Ambrosia*) community occurred just north of the project area (Turner 1974), and Dutch Canal Ruin was also located approximately 2.2 km north of a deciduous riparian forest (Turner 1974), an important wildlife habitat. This is well within the 5-km catchment radius traditionally inferred for agricultural groups (e.g., Barker 1975; Brumfiel 1976; Jarman and Webley 1975; Rossman 1976; Vita-Finzi and Higgs 1970; West 1980). The inhabitants of the site would, however, have had to travel farther, about 6 km, to exploit the relatively dense cactus stands near present-day Papago Park (Turner 1974). Therefore, the site did not appear to have been settled primarily to exploit cactus products, a conclusion corroborated in the flotation and macrobotanical data; investigators recovered no charred cactus products from the site during either the current or the previous (Ruppé 1988) archaeobotanical study. This absence is somewhat unusual, because archaeologists commonly recover charred cactus products from Hohokam sites (e.g., Gasser 1982b; Gasser and Kwiatkowski 1991; Miksicek 1988).

The prehistoric irrigation of the area, combined with the presence of nearby braided stream channels of the Salt River, may have increased the density of mesquite near the site, especially in the later phases of occupation. Turner's (1974) map notes that mesquite bosques tend to grow in artificially created habitats "such as irrigation overflow areas or areas behind runoff retention dams." This idea was perhaps supported by the consistent occurrence of charred mesquite seeds in the post-Classic period features.

Lower Salt River Valley Field Houses

Miksicek (1987:229-230) was the first to point out that flotation samples from Hohokam field houses contain relatively few charred plant taxa per liter. Since then archaeologists have analyzed botanical samples from a number of lower Salt River Valley field houses and have developed ideas about the prehistoric use of these features. This section briefly reiterates the current conception of

Hohokam field house function, summarizes previous flotation studies of nearby field houses, and determines how closely the pre-Classic period pit structures analyzed during the current study compared to these data and ideas.

The Current Conception of Hohokam Field House Utilization

Implicit in the term *field house* is an agricultural function, that is, archaeologists believe these structures to have been associated with farming (Cable 1983b:159; Greenwald and Ciolek-Torrello 1987:5, 1988a:208; Wilcox 1978:26). Field houses were situated within or near agricultural fields (Wilcox 1978:26) and were intermittently occupied throughout the year, probably most intensively during crop harvests (Cable 1983a:202; Cable and Doyel 1983, 1985:284). A field house was probably used by a nuclear family, although not all family members would have necessarily occupied a structure at a given time (Cable 1983a:202). An auxiliary function for at least some of these features may have been as a base for the collection of wild plant resources (Cable and Doyel 1985:275). This model is consistent with ethnographic analogues (e.g., Castetter and Bell 1942:126; Ezell 1961:38) and with the archaeological characteristics of field houses, including small size and simple construction, uncomplicated site structure, a limited range and general dearth of artifacts and artifact types, and spatial associations with soils of high agricultural potential (Cable and Doyel 1985; Crown 1985a). The Hohokam used field houses in the lower Salt River Valley from at least the early Pioneer period through the Sedentary period (Henderson 1989) but may have replaced them with ramadas during the Classic period (Cable and Doyel 1985:281).

Comparative Field House Flotation Data

Flotation samples from at least five lower Salt River Valley field house sites have been analyzed: Casa Buena (Gasser 1988); AZ T:12:51(ASM) (Kwiatkowski 1988c); Pueblo Patricio (Gasser 1982a, 1983, 1984, 1985); La Cuenca del Sedimento (Miksicek 1989); and Dutch Canal Ruin (Ruppé 1988). The results of these analyses are compared below.

Casa Buena, Pueblo Patricio, and AZ T:12:51(ASM)

The flotation data from field houses at these three sites were summarized by Kwiatkowski (1988c). Investigators recovered charred corn remains from 12 (63.2%) of the 19 field houses at Pueblo Patricio, in 3 out of 4 field houses at AZ T:12:51(ASM), and within both of the field houses at Casa Buena. The Pueblo Patricio results included four structures without identifiable charred plant material. Corn was recovered more often than any other charred plant taxon, although mesquite pod fragments were relatively common at Pueblo Patricio.

The ubiquity of charred mesquite pod remains at Pueblo Patricio may have reflected an increased use of this resource due to its local availability; a large mesquite bosque nearby, illustrated on the earliest survey plats of Phoenix (U.S. Land Office 1870a, 1870b), presumably existed prehistorically as well. Mitchell (1989b:135–137), moreover, found that the frequency of occurrence of mesquite remains at several lower Salt River Valley sites, including Pueblo Patricio, increased near what would have been large stands of this resource.

The data from these three sites plus that from 8 field houses at three more distant sites (Casas Pequeñas, Ellsworth, and Chiadag) were compared to a sample of flotation data from 55 other structure types at six lower Salt River Valley sites using T tests (Kwiatkowski 1988c). The entire field house sample had an average of 1.29 charred plant parts per liter (s.d.=0.91), whereas the mean of the other structure types was 15.67 parts per liter (s.d.=27.61). Similarly, the field house sample averaged 0.64 taxa per liter (s.d.=0.39), whereas the mean for the other structures was 1.05 taxa per liter (s.d.=0.76). The field house sample contained significantly fewer charred plant parts per liter (p=.05), whether unburned structures were considered separately or were grouped with burned houses. Significant differences (p=.05) also occurred in the number of taxa recovered per liter. More charred plant taxa occurred in other structure types compared to the field houses, both when the data from unburned structures were considered alone and when they were combined with the data from burned structures.

This study (Kwiatkowski 1988c) found further that charred corn presence values within the field house sample did not statistically differ (p=.05) from those of the other structure types, either when only burned structures were considered or when both unburned and burned structures were considered together. However, the unburned field houses contained corn more often (p=.05) than did unburned structures of other types. These analyses should, however, be viewed with caution, because larger sample sizes could reverse some of the presence value results (Kwiatkowski 1988c:229).

The quantitative results supported the idea that field houses may have been used less intensively and for a more restricted range of activities than other Hohokam structures. Further, the primary subsistence-related activity occurring around these field houses seemed to have involved corn.

La Cuenca del Sedimento

Miksicek (1989) analyzed 28 flotation (76 total liters) and four macrobotanical samples from nine field houses at La Cuenca del Sedimento. Six of the structures had burned. These figures are based on information in the site feature descriptions (Henderson and Morgan 1989:Table 3.1), because Miksicek's chapter differs from other portions of the report as to which features were actually field houses.

Corn was the most common charred plant taxon, occurring in six (66.7%) of the structures. The next most common taxon was Cheno-ams, the seeds of which occurred in three (33.3%) field houses. Charred plant densities were low, varying from 0.00 to 3.00 parts per liter, except within a burned Sedentary–Early Classic period structure that contained 5.40 charred plant parts per liter. Four structures exhibited relatively high taxonomic diversities (≥ 1.00 taxa per liter). These features are problematical, however, because each was represented by a single 1-liter flotation sample. The other five more intensively sampled field houses contained fewer (0.09–0.69) taxa per liter.

Previous Work at Dutch Canal Ruin

Ruppé (1988) analyzed six flotation samples (22 total liters) from the floor fill of three field houses. Each sample contained charred corn, Cheno-am seeds, and other plant remains. Charred plant densities were low (1.00–3.00 parts per liter), as were the number of charred plant taxa per liter (0.33–0.75).

Comparisons with Other Studies

The results of the current study were similar to data from other Hohokam field houses of the lower Salt River Valley in their low density of charred plant remains but consistent occurrence of corn. The few differences from the results of prior studies seemed less important than the similarities. For example, although charred Cheno-am seeds were better represented in the pit structures previously analyzed from Dutch Canal Ruin (Ruppé 1988), the seed densities in both studies were so modest that they may not have represented intentionally gathered resources (Minnis 1981).

Further, charred corn kernel fragments were not identified in the present study, although they have been consistently recovered during other field house studies (Gasser 1983, 1984, 1988; Kwiatkowski 1988c; Miksicek 1989; Ruppé 1988). However, this inconsistency seemed unimportant, because charred corn kernel fragments may in fact have been present. Poorly preserved material resembling charred corn kernel fragments in Features 3-2-1 and 5-1 received cf. identifications (Appendix C). Area 9, located in the southeastern portion of the site, did contain charred corn kernels (Chapter 10). However, the Hohokam used Area 9 from the late Pioneer period through the Classic period, and due to general mixing of the deposits, analysts could not determine with which period or phase(s) the charred corn was associated.

SUMMARY AND CONCLUSIONS

The archaeobotanical data, combined with archaeological information, appeared to indicate that the pre-Classic period occupation at Dutch Canal Ruin was that of a field house locality oriented toward activities related to corn agriculture. The site had probably been occupied seasonally or intermittently, especially during the corn harvest, and only a restricted range of plant-related economic activities seemed to have been conducted there.

The character of occupation changed by the post-Classic period, at least in Area 8. The residents of this locus seemed to have engaged in a wider range of subsistence-related activities and may have lived at the site throughout the year.

One feature in favor of site location may have been the presence of arable land easily exploited for corn agriculture. Early irrigation practices at Dutch Canal Ruin may have fostered the local growth of mesquite trees and also may have resulted in other economically important resources (Volume 4:Chapter 7), while easy access to cactus products had probably not been an important determinant in site location. The archaeobotanical data appeared to support these speculations.

CHAPTER 16

POLLEN ANALYSIS FOR DUTCH CANAL RUIN

Linda Scott Cummings

Analysts at PaleoResearch Laboratories examined pollen samples recovered during the 1990 excavations at Dutch Canal Ruin from floor contacts and features in Areas 1, 2, 3, 5, and 8. The purpose of the pollen analysis was to obtain evidence for economic activities involving plants and to determine if and how botanical factors had played a role in site settlement and land use. The pollen record was a basis for defining the nature of the activities conducted within structures, house clusters, and other site components at Dutch Canal Ruin. The analysts then compared data recovered from these investigations with data documented in earlier studies of field house components at the same site and other sites in the region. In addition, this analysis presents a comparison between the pre-Classic and post-Classic period samples and addresses the issue of seasonal versus permanent occupation at Dutch Canal Ruin.

ANALYTICAL METHODS

A chemical extraction technique based on flotation is the standard preparation technique used by PaleoResearch Laboratories for the removal of the pollens from the large volume of sand, silt, and clay with which they are mixed. Analysts developed this process for extracting pollen from soils where preservation has been less than ideal and pollen density is low.

Laboratory analysts used hydrochloric acid (10%) to remove calcium carbonates present in the soil, after which they screened the samples through 150-micron mesh. Immediately after centrifugation to remove the dilute hydrochloric acid and water, the samples were mixed with zinc bromide (density 2.0) while still moist. All samples received a short (10-minute) treatment in hot hydrofluoric acid to remove any remaining inorganic particles. The samples were then acetolated for 3 minutes to remove any extraneous organic matter.

Analysts counted the pollen to a total of 100 to 200 pollen grains at a magnification of 500X using a light microscope. Pollen preservation in these samples varied from good to poor. Researchers then used comparative reference materials from the Intermountain Herbarium at Utah State University and the University of Colorado Herbarium to identify the pollen to the family, genus, and species level, where possible.

Aggregates (clumps of a single pollen type) were recorded during identification of the pollen. Aggregates may represent either pollen dispersal over short distances or the actual introduction of portions of the plant represented into an archaeological setting. In the pollen counts, aggregates were included as single grains, as is customary; their presence is noted by an "A" next to the pollen frequency on the pollen diagram (Figure 16.1).

The category "indeterminate pollen" includes pollen grains that were folded, mutilated, and otherwise distorted beyond recognition. Because these grains were part of the pollen record, they were included in the total count.

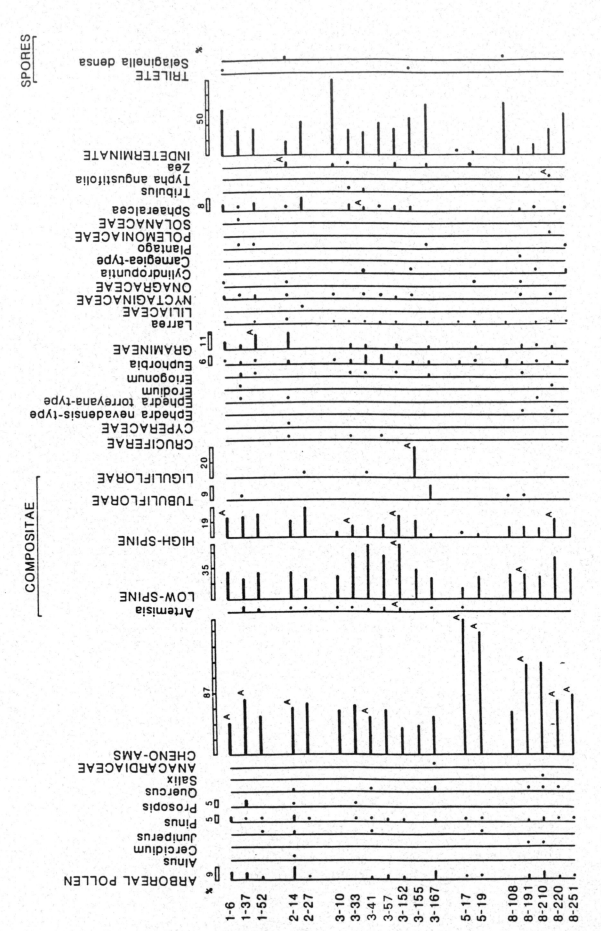

Figure 16.1. Pollen diagram, Dutch Canal Ruin.

DISCUSSION

Analysts examined samples from the floors of pit structures, a burial, fire pits, and pit fill (Table 16.1) to address subsistence activities at Dutch Canal Ruin. These samples represented five of the eight excavation areas defined during the 1990 excavations. Thirteen of these samples represented pre-Classic period contexts, while the remaining six were from the post-Classic period. Samples are discussed by area later in the chapter.

Preservation of the pollen recovered from this study was relatively poor. Indeterminate pollen regularly accounted for approximately 15% of the total pollen recovered, whereas pollen from a previous study at Dutch Canal Ruin exhibited less than 4% indeterminate pollen per sample (Cummings 1988:180–194). The two major pollen types recovered in this study were Cheno-ams and Low-spine Compositae, which probably represented bursage (Figure 16.1; Table 16.2). Levels of High-spine Compositae pollen also were frequently elevated. Previous analysis exhibited far more *Larrea* (creosotebush) pollen than was recorded in these samples, probably the result of the poor condition of the pollen recovered in this study, as poorly preserved *Larrea* pollen is relatively indistinct and would be included in the indeterminate pollen. The pollen signature recovered from these samples was consistent with that of a lowland desert creosotebush-bursage community, which analysts interpreted from the previous pollen study (Cummings 1988:180–194).

Ethnobotanical Review

Pollen types that probably represented economic activity at this site were varied and included *Prosopis* (mesquite), Cheno-ams, Compositae (sunflower family), Cruciferae (mustard family), *Euphorbia* (spurge), Gramineae (grasses), Liliaceae (lily family), *Cylindropuntia* (cholla cactus), *Carnegiea/Mammillaria*-type (saguaro/fishhook cactus), Solanaceae (potato/tomato family), *Typha* (cattail), and *Zea mays* (corn). The following brief review of ethnobotanic literature concerning use of these plants by historic Native Americans provides evidence of the exploitation of numerous plants, both by broad categories (such as greens, seeds, roots, and tubers) and by specific example (e.g., seeds parched and ground into meal that was formed into cakes and fried in grease). Repetitive evidence of the exploitation of resources indicates widespread utilization and strengthens the possibility that the same or similar resources were used in prehistoric times. Nevertheless, the ethnobotanic literature serves only as a guide to indicate the potential for prehistoric utilization, not as conclusive evidence that the Hohokam used the same resources. Pollen and macrofloral remains, when compared with the material culture (artifacts and features) recovered by archaeologists, become indicators of use. The pollen types observed in samples from Dutch Canal Ruin were exploited in historic times and may be indicators of prehistoric use.

Prosopis is a well-adapted xerophytic shrub. Mesquite wood is second in quality only to that of *Olneya* (ironwood) as a firewood in semidesert regions. Mesquite pods have long been an important food source among Native Americans. Kearney and Peebles (1960:402) observed that pinole flour made from the long, sweet mesquite pods was a staple food for the Pima. Mesquite pods have been described as an important food source among the ancient Pima, Papago (now Tohono O'odham), and other Native American groups from central Arizona (Castetter and Bell 1942:63; Curtin 1984).

The term *Cheno-ams* refers to a morphologically similar group of pollen representing the Chenopodiaceae family and the genus *Amaranthus* (pigweed). Cheno-ams, including a variety of plants such as *Chenopodium* (goosefoot), *Amaranthus*, *Suaeda* (seepweed), and *Atriplex* (saltbush), have been used as food and for processing other foods. The greens may be harvested and cooked

Table 16.1. Provenience Data for Pollen Samples

Period	Sample No.	Feature No.	Feature Description	Context	Pollen Counted
PC	1-6	1-1	Pit structure	Floor contact	100
PC	1-37	1-2	Pit structure	Floor contact	200
C	1-52	1-3	Primary inhumation (Soho phase)	Feature fill	200
PC	2-14	2-2	Pit structure	Floor contact	200
PC	2-27	2-2-2	Fire pit	Feature fill	100
PC	3-10	3-1-1	Fire pit	Feature fill	100
PC	3-33	3-2	Pit structure	Floor contact	200
PC	3-57	3-2-1	Fire pit	Feature fill	100
PC	3-41	3-3	Pit with burning (secondary refuse)	Feature fill	200
PC	3-152	3-18	Fire pit	Feature fill	100
PC	3-155	3-16	Pit structure	Floor contact	100
PC	3-167	3-19	Pit structure	Floor contact	100
PC	5-17	5-1	Pit structure	Floor contact, beneath metate	200
PC	5-19	5-1	Pit structure	Floor contact	200
C/pC	8-108	8-2-1	Fire pit	Feature fill	100
C/pC	8-191	8-2	Pit house	Floor fill/ refuse fill	200
C/pC	8-220	8-3	Pit structure	Floor fill	200
C/pC	8-210	8-12	Pit with burning (secondary refuse)	Feature fill	200
C/pC	8-251	8-14	Pit with burning (secondary refuse)	Feature fill	200

C = Classic
PC = Pre-Classic
pC = Post-Classic

Table 16.2. Observed Pollen Types

Scientific Name	Common Name
Arboreal Pollen	
Alnus	Alder
Cercidium	Paloverde
Juniperus	Juniper
Pinus	Pine
Prosopis	Mesquite
Quercus	Oak
Salix	Willow
Nonarboreal Pollen	
Anacardiaceae	Sumac family
Cheno-ams	Includes amaranth and pigweed family
Compositae:	Sunflower family
Artemisia	Sagebrush
Low-spine	Includes ragweed, cocklebur, etc.
High-spine	Includes aster, rabbitbrush, snakeweed, sunflower, etc.
Tubuliflorae	Includes eroded low- and high-spine
Liguliflorae	Includes dandelion and chicory
Cruciferae	Mustard family
Cyperaceae	Sedge family
Ephedra nevadensis-type	Mormon tea
Ephedra torreyana-type	Mormon tea
Eriogonum	Wild buckwheat
Erodium	Heron-bill
Euphorbia	Spurge
Gramineae	Grass family
Larrea	Creosotebush
Liliaceae	Lily family
Nyctaginaceae	Four o'clock family
Onagraceae	Evening primrose family
Cylindropuntia	Cholla cactus
Carnegiea-type	Saguaro cactus
Plantago	Plantain
Polemoniaceae	Phlox family
Solanaceae	Potato/tomato family
Sphaeralcea	Globe mallow
Tribulus	Puncture vine
Typha angustifolia	Cattail
Zea	Maize, corn
Indeterminate	
Spores	
Selaginella densa	Little clubmoss
Trilete	

either alone or with other food or may be packed around cholla buds when they are roasted. *Suaeda* greens were collected in April with cholla buds and were dried and stored for later use with the buds. The seeds were also ground into meal and frequently were mixed with cornmeal. The method of collection of these small seeds was to knock the seed heads from the plants into a basket. This technique also released pollen from the younger flowers on the same plant that had not yet matured and thus introduced Cheno-am pollen into the archaeological record. Various parts of Cheno-am plants have been gathered from early spring through the fall (Castetter and Bell 1942:61; Curtin 1984:47–71; Greenhouse, Gasser, and Gish 1981:238; Kearney and Peebles 1960:251, 255, 263, and 265).

The Liguliflorae include such genera of the Compositae as *Taraxacum* (dandelion) and *Chicorium* (chicory). The leaves of both plants may be eaten raw or used as a substitute for spinach. Dandelion roots, which are reputed to have medicinal properties, may also be eaten raw. The roots of chicory may be roasted to provide a substitute for coffee (Kearney and Peebles 1960:958, 964). The greens and roots may be harvested at any time during the growing season, although the greens would be more tender in the spring and early summer.

Several members of the Cruciferae family have been exploited for their greens, used as potherbs when the plant is young. In addition, seeds may have been parched, ground into flour, and used for making pinole, mush, or bread, or to thicken soup (Harrington 1967:308). *Descurainia* (tansy mustard) greens are eaten in the spring, and the seeds may also be parched and ground into meal or pinole (Colton 1974:309–310; Harrington 1967:307; Kearney and Peebles 1960:349; Stevenson 1915:60).

Euphorbia produces a milky, acrid juice and is frequently used medicinally. Certain *Euphorbia* species have been used to treat rattlesnake bites, and the Pima have used other species as an emetic. *Euphorbia* has also been used topically to treat warts, sores, and skin eruptions (Kearney and Peebles 1960:511; Krochmal and Krochmal 1973:101).

The seeds of several genera of the Gramineae family have been used as food by Southwestern peoples and are available during the summer and fall months. These include primarily *Sporobolus* (dropseed), *Oryzopsis* (Indian ricegrass), *Panicum* (panic grass), and *Sorghum* (Kearney and Peebles 1960:112–115, 119, 142).

Both *Allium* (wild onion) and *Calochortus* (sego lily) fall within the approximate size range of Liliaceae pollen recovered at this site. Many Native American groups have frequently exploited these plants. Wild onion may be utilized as a flavoring for stews or meats or may be boiled and eaten as a vegetable, or the juice may be used medicinally. Wild onions appear to have been extensively used in the past; processing methods include drying and storage for future use (Gilmore 1977:19; Harrington 1967:345–346; Hellson and Gadd 1974:100; Kearney and Peebles 1960:178; Smith 1974:271; Yanovsky 1936). Several parts of the sego lily are edible, including the greens, seeds, bulbs, and flowers. The bulbs constitute the most usable portion of the plant and were frequently boiled. They may also be stored for future use (Harrington 1967:159–161).

Cylindropuntia is an antiquated term for cholla cactus that palynologists have retained to distinguish cholla from prickly pear (*Opuntia*). Cholla buds have been collected during the spring and roasted. The cooked buds may also be dried and later ground. The buds are available in May, and the fruit ripens later in the summer. The fruits or tunas of both cholla and prickly pear cactus also may be collected for consumption, along with the joints of both cacti. The process of removing the spines from the cacti usually involves roasting or baking the plant in a pit, then rubbing off the spines (Greenhouse, Gasser, and Gish 1981; Kearney and Peebles 1960:581–586). Cholla buds were one of the three predominant ancient native foodstuffs, along with mesquite beans and saguaro fruits, for the Pima (Castetter and Bell 1942:63). The saguaro (*Carnegiea*) was important in Piman subsistence, providing both food and shelter. Fruits are produced annually, regardless of drought conditions, because of the saguaro's water storage capacity. The fruit matures in June and July and may be eaten fresh or stored as syrup or preserves. Juice from the fruit may be fermented to make an intoxicating beverage, and a type of butter may be made from the seeds (Castetter and Bell 1942:59–60; Kearney and Peebles 1960:569). *Carnegiea*-type pollen may include pollen from the saguaro, hedgehog (*Echinocereus*), barrel (*Ferocactus*), and fishhook (*Mammillaria*) cacti.

Members of the Solanaceae family, primarily *Solanum* and *Physalis*, have also been exploited for food. The berries of both plants are edible, as are the roots of *Solanum* (Robbins, Harrington, and Freire-Marreco 1916:59, 70–73; Stevenson 1915:70; Whiting 1939:90). Wild potato (*Solanum*) has been allowed to grow as a weed in otherwise carefully tended agricultural plots (Whiting 1939:16). *Solanum nigrum* presents a paradox for interpretation. It is reported as both an edible and a poisonous plant. Many people have consumed the fruits of nightshade without harmful effects. The poison may concentrate in immature berries or may vary with geography or soil differences. Sensitivity to this plant may also vary by individual. Harrington (1967:286) noted that the most likely explanation for the conflicting reports of both warnings about and consumption of this plant is that the toxin is more heavily concentrated in the green, unripe fruits, which are probably responsible for the poisonings that have occurred.

Typha is a rich source of nutrients. Steward (1938) and Chamberlin (1964) noted the utilization of cattail as food; Harrington (1967) described the use of both pollen and the seedlike fruits of cattail as food resources. The young, pollen-producing flowers may be stripped from the spikes, or the pollen may be removed by shaking the mature flowers. The resulting flowers and pollen may be mixed with flour. Flour made from cattail roots, which are best harvested in the fall, is similar with respect to quantities of fats, proteins, and carbohydrates to flour obtained from wheat, rice, and corn (Harrington 1967). The Lower Colorado River tribes have also eaten cattail roots (Castetter and Bell 1942).

Zea is an important cultivated food and can be prepared in a variety of ways. Historically Native American farmers widely used green corn and collected it from the regular fields. They ate mature ears following roasting, either wrapped in corn husks and boiled or else ground and cooked as cornmeal gruel. The Pimans harvested corn by pulling up the entire stalk after it was dry and making piles at the edges of the fields. Women and children removed unhusked ears from the stalks and then threw them into piles, ultimately carrying the ears to the dwelling in burden baskets. They frequently roasted unhusked ears by piling up corn and mesquite brush and setting this pile on fire. The fire burned much of the husk away, and the ears were pulled from the fire and dried on top of the house. The roasted, unhusked corn was then stored for later use. This corn was also sometimes shelled prior to storage. Shelled corn was stored in bins or granaries made of coiled arrowweed, which were set on a platform or the roof. The Tohono O'odham stored shelled corn in jars. Currently, the most popular method of storing corn is to keep unhusked ears on top of an ocotillo ramada for use as needed. Seed corn was stored in ollas with wood ashes or was braided in clusters that were dried outdoors, then suspended from the house ceiling (Castetter and Bell 1942:180–189).

Three pollen types noted as common agricultural weeds in portions of central Arizona are *Boerhaavia* (spiderling; a member of the Nyctaginaceae family), *Plantago* (plantain), and *Sphaeralcea* (globe mallow). Recovery of *Boerhaavia* pollen from various sites in the northern periphery by Bohrer (1984:258) has been interpreted as an indication of the utilization of a portion of this plant, possibly the green shoots. This pollen type was recovered in small quantities from the majority of samples from both the Agua Fria and New River drainages (Cummings 1989:850) and has also been observed in numerous noncultural or environmental samples collected throughout southern Arizona. Thus, its presence at Dutch Canal Ruin would not necessarily indicate utilization at the site. Fish (1984) used *Boerhaavia* as an indicator of cultivation, classifying it as an agricultural weed, and identified *Boerhaavia*, *Sphaeralcea*, and *Kallstroemia* (carpetweed) as common agricultural weeds near Tucson (Fish 1987:163); all may have been part of the population of agricultural weeds at Dutch Canal Ruin, although *Kallstroemia* was not positively identified at the site (Figure 16.1). *Plantago* was present at the site, possibly as another agricultural weed.

Research Areas

Area 1

Field crews collected two samples from the floors of pre-Classic period pit structures in Area 1 and one from a primary inhumation, presumably from the Classic period. Feature 1-1, an oval pit structure, was represented by a single sample (1-6) collected from the unprepared, compacted floor. Investigators recovered remnants of burned beams and organic material from this structure, but the lack of artifacts indicated that the inhabitants had intentionally abandoned the structure, which had later burned. This sample exhibited pollen frequencies generally typical of the record at this site. The Gramineae pollen frequency was slightly elevated, suggesting the possibility that the inhabitants had processed grass seeds in the structure. In a structure that has been purposefully abandoned, the pollen record is often a mixture of pollen that accumulated on the floor during the occupation and pollen that accumulated following abandonment but prior to destruction of the structure.

Feature 1-2, a roughly circular pit structure with an appended entryway, was also represented by a single sample collected from the floor (Sample 1-37), which had been compacted through use. Following abandonment, this feature was inundated, and a layer of coarse sand was deposited on the floor. The structure subsequently burned and collapsed. This sample yielded an elevated Cheno-am frequency compared to other frequencies within Area 1. *Prosopis* pollen (5% of the pollen in the sample) and Solanaceae pollen were present as well. The pollen record indicated that *Prosopis*, Cheno-ams, and a member of the Solanaceae family might have been processed in this structure, if the floor area sampled had been protected from the later flood deposits.

Feature 1-3, a primary inhumation probably buried in Feature 1-2 after this pit structure was abandoned, was represented by Sample 1-52 collected from the fill of the burial pit. The contents of this sample were similar to others from this area, except that the Gramineae frequency was elevated and a small aggregate was recovered. This suggested the possibility that grasses had been present in some form, perhaps as a grass mat or as seeds.

Area 2

Feature 2-2, a roughly circular pit structure, was represented by a single sample collected from floor contact (Sample 2-14). The structure burned, collapsed, and was finally covered by alluvial deposits. This sample displayed a small quantity of *Prosopis* pollen that might represent natural transport or possibly food processing activities. In addition, the level of Gramineae pollen was elevated, and both Cruciferae and *Zea mays* pollen were present, including an aggregate of *Zea mays* pollen, indicating that the area's inhabitants had processed *Zea mays* and probably members of the Cruciferae and Gramineae families in the structure.

Feature 2-2-2, the fire pit located in the northwest quadrant opposite the entry in the above-mentioned structure, exhibited a small quantity of Liliaceae pollen, suggesting that the occupants of the structure might have cooked this plant in the fire pit or processed Liliaceae near it. *Zea mays* pollen was absent from this sample, and analysts did not observe it in a scan of the slide. The presence of a small quantity of Liguliflorae-type Compositae pollen might reflect rodent activity or some other source of contamination, or possibly the presence of a member of this group as a prehistoric weed.

Area 3

Feature 3-1-1, a fire pit located in the southeast portion of an oval pit structure, exhibited very poorly preserved pollen in Sample 3-10. This sample contained a single very dark *Zea mays* pollen grain observed only upon a scan of the entire slide. The darkness of this pollen grain, compared to the color of pollen in other samples, suggested that it had been present when the fire pit was used and thus that the area's inhabitants had processed corn there.

Feature 3-2, a roughly square pit structure, was represented by Sample 3-33 from the compacted but unprepared floor. This sample exhibited small quantities of Cruciferae and *Zea mays* pollen, suggesting that both cultivated and native plants had been processed in this structure. The level of *Euphorbia* pollen was also slightly elevated, suggesting the possibility that the occupants had used or processed this plant.

Feature 3-2-1, a fire pit located near the center of this structure, was represented by Sample 3-57. This sample from the fire pit fill contained a small quantity of Cruciferae pollen, again suggesting that a member of this family had been processed in this area and possibly had been cooked in the fire pit. However, field crews recovered no *Zea mays* pollen. Analysts noted *Euphorbia* pollen in a larger quantity in the sample from the fire pit than in the sample from the floor of the structure, suggesting that processing activities had taken place there.

Feature 3-3, an unlined, oval, extramural pit that exhibited burning, yielded a small quantity of *Zea mays* pollen in Sample 3-41, suggesting that the area's inhabitants had processed corn there. The small quantity of Liguliflorae-type Compositae pollen noted in this sample might represent modern intrusion or possibly the presence of a member of this group prehistorically as a weedy plant. The level of *Euphorbia* pollen was again elevated in this pit, suggesting the possibility of processing. Alternatively, *Euphorbia* might have been present as a weed.

Feature 3-16, a basin-shaped pit structure oval in plan, was represented by a sample from floor contact (Sample 3-155). The depression of the abandoned and collapsed structure had apparently been used to discard refuse from a nearby roasting pit or horno. This sample yielded a very large quantity of Liguliflorae-type Compositae pollen (20%), as well as a small quantity of *Cylindropuntia* pollen. The large frequency of Liguliflorae-type pollen might have been directly caused by growth of these weedy plants in the depression. This structure was the only one to yield a large quantity of Liguliflorae pollen, and the casual mode of abandonment, collapse without burning, and subsequent use of the depression for trash disposal pointed to the probability that the Liguliflorae pollen recovered from this feature represented weedy growth rather than resource utilization. The presence of *Cylindropuntia* pollen suggested that inhabitants of the area might have processed cholla either in this pit or elsewhere and then discarded the remains in the pit depression.

Feature 3-18, an irregular, basin-shaped, extramural fire pit located in a probable extramural activity area, was represented by Sample 3-152. This sample produced pollen typical of the general pollen assemblage but not specifically indicative of subsistence activities. Pollen recovered from this sample might easily have been windblown. The recovery of a small quantity of *Zea mays* pollen accompanied by aggregates from this extramural fire pit indicated that the area's occupants had processed corn there.

Feature 3-19, a roughly oval pit structure exhibiting a shallow basin cross section, was represented by Sample 3-167 from the unprepared but compacted surface of the structure. The floor was sealed by a layer of burned roof material. Scattered deposits of coarse sand on the floor suggested that the structure had burned either during use or shortly after abandonment. The only element of this sample that appeared to represent subsistence activities was the *Zea mays* pollen.

Area 5

One small pit structure (Feature 5-1) was represented by two floor contact samples. This structure exhibited a poorly prepared, poorly preserved floor. It was located between two canals and had been built as a temporary shelter after both canals had been abandoned. Sample 5-17 was found beneath a metate fragment, while Sample 5-19 came from another area of floor contact. These samples were very similar to one another in pollen content. Both exhibited very large quantities of Cheno-am pollen, accompanied by aggregates, suggesting that the structure's occupants had processed Cheno-ams in both areas sampled. Analysts recorded a small quantity of *Zea mays* pollen during the scan of Sample 5-19, indicating that the residents had processed corn in this area of the structure as well.

Area 8

Analysts examined samples from one pit structure, one pit house, one hearth, and two pits from this area, all dating to the post-Classic period. Feature 8-2, a roughly rectangular pit house, was represented by Sample 8-191 from the unplastered, level, hard-packed floor. This house appeared to have been intentionally abandoned, with artifacts having been removed by the inhabitants. It also exhibited extensive disturbance by rodent activity. This sample yielded a very large quantity of Cheno-am pollen, as well as small quantities of *Carnegiea/Mammillaria* and *Typha* pollen, all of which suggest economic activities. Field crews collected Sample 8-108 from the fill of the fire pit (Feature 8-2-1), which was located just inside the structure and in front of the entry of this house. This sample exhibited very poorly preserved pollen, a much lower quantity of Cheno-am pollen, and no *Carnegiea/Mammillaria* or *Typha* pollen. *Euphorbia* pollen was slightly elevated in the fire pit, suggesting that the structure's occupants had processed that plant. The pollen record did not define definite economic activity for the fire pit.

Feature 8-3, an irregularly shaped pit structure, was represented by Sample 8-220, taken from the floor fill because of the difficulty in defining the floor, which was neither plastered nor hard-packed. The occupants of this structure had abandoned it in a casual manner. Sample 8-220 exhibited a small quantity and a small aggregate of *Typha* pollen, indicating that cattail had been processed in this structure; analysts noted no other evidence of economic activity. *Typha* pollen appeared in samples from Features 8-2 and 8-3 in this area but not in other samples examined from the 1990 field season.

Feature 8-12, a medium-sized, circular pit that exhibited signs of burning, was represented by Sample 8-210; Feature 8-14, a medium-sized, irregular pit that exhibited signs of burning, was represented by Sample 8-251 from the feature fill. Both Feature 8-12 and Feature 8-14 were basin-shaped in profile. Both contained *Cylindropuntia* pollen, although Feature 8-12 exhibited a larger quantity of Cheno-am pollen than Feature 8-14. The similarity in cholla pollen content within these samples, and the presence of either an elevated Cheno-am frequency or aggregates, suggested that residents of the area had used both pits to roast cholla buds and had also used Cheno-ams in the roasting process. *Cylindropuntia* pollen was absent from the other samples examined from Area 8.

Temporal Comparison

Pre-Classic Period Features

The pre-Classic period features at Dutch Canal Ruin yielded pollen from economic plants in eight samples from pit structures and five from pits (Table 16.1). All of the *Prosopis* pollen recovered from the pre-Classic period features came from three (37.5%) of the eight floor contact samples examined from pit structures (Table 16.3). Field crews recovered Cruciferae pollen from two of the eight pit structure samples (25.0%); Gramineae, *Cylindropuntia*, and Solanaceae pollen in single samples (12.5%); and *Zea mays* pollen in four of the eight (50.0%). Cheno-am pollen appeared in small to moderate quantities in most of the samples from the pre-Classic period pit structures and in large quantities in only two of the eight samples (25.0%). Generally, the frequencies were too similar to be interpreted as evidence of food processing activities.

Pits from the pre-Classic period exhibited Cruciferae pollen in one (20.0%) of the five samples examined. Liliaceae pollen was recovered from a single pit within a structure (20.0%). *Zea mays* pollen appeared in three (60.0%) of the five pit samples examined from the pre-Classic period.

Pollen samples collected from pit structures exhibited a much greater variety of types potentially indicative of food processing activities than did samples collected from pits. Seven pollen types recovered from pit structures appeared to be associated with economic activity, while only three pollen types recovered from pits appeared to represent economic activity. Pollen types associated with pit structures were *Prosopis,* Cruciferae, Gramineae, *Cylindropuntia*, Solanaceae, Cheno-ams, and *Zea mays*. Pollen types associated with pits included Cruciferae, Liliaceae, and *Zea mays*. Of this list, only Liliaceae pollen occurred in samples from pits but not pit structures.

The recovery of *Prosopis* pollen solely from pit structure floors suggested that the occupants of Dutch Canal Ruin had processed and possibly stored mesquite foodstuffs in these structures. The presence of pollen in Features 1-2, 2-2, and 3-2 implied the processing of mesquite pods. Mesquite pods mature during the summer (Gasser 1982b:226), indicating that occupation had occurred at that time. The recovery of *Zea mays* pollen from half of the pit structure samples similarly supported the notion of food processing or perhaps storage. The presence of *Zea mays* pollen suggested that the inhabitants had processed foodstuffs (and hence had occupied the site) following harvest. In a two-crop system this would have been during late June and October (Castetter and Bell 1942:179). Although cholla buds may be processed while fresh or may be dried for later use, they are available for collection and processing in May, suggesting that the area could have been occupied during this time as well.

Levels of Cheno-am pollen were elevated in both samples examined from Feature 5-1, a pit structure, suggesting that the occupants had processed Cheno-am seeds in this structure. One of the samples examined was a floor contact sample beneath a trough metate fragment, so possible postoccupational disturbance or the mode of abandonment could account for the introduction of this pollen type into the structure.

Table 16.3. Distribution of Pollen Types by Period and Feature Type

Period	Pollen Type	Pit Structures	Pits
Pre-Classic	*Prosopis*	37.5%	
Pre-Classic	Cruciferae	25.0%	20.0%
Pre-Classic	*Euphorbia*		40.0%
Pre-Classic	Gramineae	12.5%	
Pre-Classic	Liliaceae		20.0%
Pre-Classic	*Cylindropuntia*	12.5%	
Pre-Classic	Solanaceae	12.5%	
Pre-Classic	Cheno-am	25.0%	
Pre-Classic	*Zea*	50.0%	60.0%
Post-Classic	Cheno-am	50.0%	33.0%
Post-Classic	*Cylindropuntia*	66.0%	
Post-Classic	*Carnegiea*	50.0%	
Post-Classic	*Euphorbia*		33.0%
Post-Classic	*Typha*	100.0%	

Classic and Post-Classic Period Features

Analysts examined pollen samples from one pit house, one pit structure, two pits exhibiting burning, and a fire pit from post-Classic period contexts and a primary inhumation from a Classic period context. The majority of these features were in Area 8. The pit house (Feature 8-2) exhibited a high frequency of Cheno-am pollen, as well as small quantities of *Carnegiea*-type and *Typha angustifolia*-type pollens. The pit structure (Feature 8-3) yielded only a small quantity of *Typha angustifolia*-type pollen. This pattern of pollen distribution suggested that the occupants of Dutch Canal Ruin had processed, used, or stored cattail within pit houses and pit structures during the post-Classic period. They would also have processed, used, or stored Cheno-ams and *Carnegiea*-type cactus (including *Mammillaria*) within the pit house. The fire pit inside the pit house did not exhibit conclusive evidence of activities related to food processing.

Both pits that showed evidence of burning contained *Cylindropuntia* pollen. In addition, one of these pits (Feature 8-12) exhibited a large quantity of Cheno-am pollen. The other (Feature 8-14) exhibited Cheno-am aggregates. The pollen pattern revealed in samples recovered from these pits was typical of that of roasting pits that have been used to roast cholla buds. Macrofloral interpretation, however, suggested that remains from other roasting pits had been deposited in these pits as secondary refuse (Chapter 15).

Zea mays pollen was absent from all the post-Classic period samples. The primary inhumation (Feature 1-3) probably dated to the Soho phase of the Classic period. The only pollen suggestive of plant use or inclusion with the inhumation was the relatively large frequency of Gramineae pollen, accompanied by aggregates. Does the absence of *Zea mays* indicate that corn agriculture was not being practiced during the Polvorón phase and that the canal systems were no longer functional? If so, the wider variety of economic wild resources plus the appearance of cholla pollen may indicate that in this location such resources became dietary staples once corn agriculture was no longer possible.

IMPLICATIONS

Plant Use, Prehistoric Vegetation, and Site Settlement

The proximity of Dutch Canal Ruin to agricultural fields in which corn had been grown during the pre-Classic period was suggested by not only by the recovery of corn pollen from the pre-Classic period samples but also by the presence of *Boerhaavia* and *Sphaeralcea* pollen, which probably represented agricultural weeds, in many of the samples. *Plantago*, another possible agricultural weed, was less evident in the pollen record, and *Kallstroemia* was not noted. In samples from the Classic period components, although corn pollen was absent, both *Boerhaavia* and *Sphaeralcea* pollen types were present in frequencies similar to those from the pre-Classic period. Analysts also noted *Plantago* pollen in samples from a pit from the post-Classic period and the inhumation from the Classic period, suggesting that the Classic period inhabitants continued to practice some form of agriculture.

A reconstruction of the native vegetation for the Phoenix area (Turner 1974) indicates a desert saltbush (*Atriplex polycarpa*) series. This vegetation series is widely distributed and common in the area and offers a suitable environment for agriculture (Masse 1991:196; Turner 1974; Turner and Brown 1982:194). Turner (1974) also notes a transition to a creosotebush-bursage (*Larrea-Ambrosia*) community to the north of the project area. Analysts noted saltbush in the pollen record, which displayed overall Cheno-am frequencies of over 18%. Proximity to creosotebush and bursage communities was also evident from the *Larrea* pollen reported in a previous study at Dutch Canal Ruin (Cummings 1988), as well as from the recovery of small quantities of *Larrea* pollen from this study. Low-spine Compositae frequencies varying from 8% to 35% indicated that although bursage had grown in the general area, it had not been particularly abundant in the immediate vicinity of the site because this pollen, which is produced in great abundance, did not overwhelm other pollen types in this record.

The presence of *Prosopis* pollen in three of the pre-Classic period samples and its absence from post-Classic period samples suggested that mesquite may have been more abundant in the vicinity during the pre-Classic period than during the post-Classic period and, hence, would have been more available for exploitation earlier in the history of the site. *Prosopis* pollen is usually not abundant in archaeological sites, even when investigators know that the resource was utilized. In samples from the post-Classic period, cactus pollen (including both *Cylindropuntia* and *Carnegiea/Mammillaria*-type) was more abundant than in those from the pre-Classic period. *Typha* pollen also appeared in samples taken from post-Classic period features but not in those from the pre-Classic period. The pollen record suggested that mesquite (*Prosopis*) might have been more common along the canals and waterways during the pre-Classic period occupation, while cattail

(*Typha*) might have been more common along the canals and waterways during the post-Classic period occupation. Moreover, the occupants during the post-Classic period apparently used more cactus, including both cholla and saguaro, than did the people of the pre-Classic period occupation. Increases in the use of cactus suggested increasing exploitation of the bajada habitats.

Comparison with Previous Pollen Studies at Dutch Canal Ruin

Both the previous study (Cummings 1988) and the current study yielded greater pollen diversity for plants that might have been processed within pit structures rather than in fire pits. Better pollen preservation in the previous samples from Dutch Canal Ruin probably accounts for differences in the frequencies of *Prosopis* and *Larrea* pollen between the two studies. Both pollen types fall within an indistinguishable tricolpate/tricolporate category of indeterminate pollen when very poorly preserved. Therefore, differences in these pollen types are not significant in comparing the two studies.

Pit structures from the previous study (Cummings 1988) exhibited pollen evidence for the prehistoric processing of *Cylindropuntia, Opuntia, Carnegiea, Prosopis,* and *Zea mays.* Cheno-am frequencies fluctuated, and interpretation of food processing activities was problematic. Samples collected from occupation surfaces suggested that the inhabitants had probably processed *Opuntia* and *Zea mays.* Cheno-am frequencies from surfaces were generally lower than those from pit structures, suggesting that Cheno-ams were indeed among the foods processed within pit structures. Fire pits exhibited pollen evidence for processing *Prosopis* and possibly Cheno-ams, but analysts observed no *Zea mays* pollen in fire pit samples. All of these samples came from features dating to the pre-Classic period.

In comparison with Cummings's (1988) results, the pre-Classic period features examined in this study showed greater variety in the pollen record. The foods and possible medicinal plants that appeared to have been processed during the pre-Classic period included *Prosopis,* Cheno-ams, Cruciferae, *Euphorbia,* Gramineae, Liliaceae, *Cylindropuntia,* Solanaceae, and *Zea mays.* Fire pits exhibited considerably less variety in economic pollen than did pit structure samples. *Zea mays* pollen was, however, present in samples from both fire pits and pit structures. Post-Classic period features exhibited a smaller variety of pollen that might represent economic activity than did the pre-Classic period features, perhaps because the post-Classic period was represented by fewer samples than the pre-Classic period. Plants processed in the post-Classic period that left their mark in the pollen record included Cheno-ams, *Euphorbia, Cylindropuntia, Carnegiea,* and *Typha.*

Both studies of Dutch Canal Ruin indicated that the occupants had utilized resources such as mesquite and cattail that had been available along the canal or the Salt River. Cactus, which is commonly associated with the drier bajadas, appeared in samples from both the pre-Classic and the post-Classic period occupations at Dutch Canal Ruin. In addition, analysts recorded resources that should have been available in the local creosotebush-bursage communities outside the dwelling area or possibly the modified agricultural flats. That these people had cleared land for agriculture and planted at least corn was evident from the pollen record. Weeds that appeared to be associated with this agricultural activity included *Boerhaavia, Sphaeralcea,* and to a lesser extent *Plantago* and *Kallstroemia* (Cummings 1988) Plants that might have occurred as weeds and been encouraged or harvested in the agricultural plots included Cheno-ams, Cruciferae, and Solanaceae.

Pollen evidence for the exploitation of mesquite from the riparian habitat along the Salt River or the presumed mesquite bosque to the west of Pueblo Patricio and Dutch Canal Ruin, combined with evidence for utilization of cholla, prickly pear, and saguaro or *Mammillaria* cactus, indicated that the occupants of Dutch Canal Ruin had utilized resources from both the riparian and the drier bajada areas. The location of Dutch Canal Ruin was such that residents could have practiced corn agriculture while exploiting a wide variety of economic wild plants. Mitchell (1989b) postulated that resource utilization was directly related to local resource availability and that local resources were exploited as a matter of course. The recovery of both mesquite and cactus remains in the pollen record from Dutch Canal Ruin suggested that the occupants had positioned their site to take advantage of both riparian and bajada resources, as well as to pursue farming. With a shift in land-use and settlement strategies between the pre-Classic and post-Classic periods, changes in collection and use of economic resources may have taken place. These changes may have involved a greater emphasis on collecting wild resources and less variability in the types of resources utilized.

SUMMARY AND CONCLUSIONS

Investigators identified the pre-Classic period occupation at Dutch Canal Ruin as a field house locality oriented toward activities related to corn agriculture. The macrofloral record indicated that occupation had been seasonal or intermittent, possibly confined to the corn harvest seasons, with limited use of other plants having occurred (Chapter 15). The pollen record, however, indicated that the occupants of Dutch Canal Ruin had exploited a number of local resources, including those recovered along canals or possibly the Salt River such as mesquite, as well as cacti from the bajadas. Cheno-ams, Cruciferae, Gramineae, Liliaceae, Solanaceae, and possibly *Euphorbia* also appeared in the pollen record and might have been encouraged to grow near the site. Analysts recovered *Zea mays* pollen from both pit structures and fire pits, indicating that corn processing had been an important task during the pre-Classic period occupation.

By contrast, during the post-Classic period occupation of Dutch Canal Ruin, the variety of plants that appeared to have been processed declined in the pollen record. *Zea mays* pollen dropped from the record, as did *Prosopis* pollen. Cactus pollen, including both cholla and *Carnegiea/Mammillaria*-type, continued in the record, indicating probable exploitation of the bajadas. Analysts noted cattail pollen in samples from the post-Classic period occupation, indicating that cattails probably had grown along the canals and had been exploited. Cheno-am pollen represented a possible native or encouraged resource that had also been exploited during the post-Classic period. *Euphorbia* pollen occurred in elevated levels occasionally during both the pre-Classic and the post-Classic period occupations and might have been present as a weed or possibly exploited for its medicinal qualities.

The location of Dutch Canal Ruin in an area suitable for corn agriculture appeared to have been important. In addition to direct evidence of corn, agricultural weeds noted in the pollen record supported the interpretation of field house activities. Utilization of other local resources was reflected in the pollen record. Access to cacti appeared to have increased during the post-Classic period occupation. At no time, however, did cacti appear to have been an important or regular element of the subsistence base or processing activities at this site.

CHAPTER 17

FAUNAL REMAINS FROM DUTCH CANAL RUIN

Susan K. Stratton

Field crews recovered faunal remains totaling 374 bones and two pieces of eggshell from Areas 3, 7, 8, and 10 of Dutch Canal Ruin. Of those 374 remains, 162 came from screened contexts and 212 from unscreened units. The comparisons in this report will focus primarily on the screened materials.

Analysts examining the faunal assemblage addressed the research questions concerned primarily with subsistence and site function. Faunal remains were identified to the genus or species level for nine different genera and were placed in other, more generalized categories as necessary (i.e., family level and small, medium, and large mammal and bird groupings). Summary information breaking down taxonomic representation by provenience and element and further summary information on cultural modification (cutmarks, evidence of burning, tools, etc.) are presented in Appendix D.

This chapter includes a discussion of the methods and terminology employed in the analysis, a descriptive summary of the materials found (identifiable to genus or species level), and results and a summary of the analyses relating to the research questions concerning subsistence and site function. The descriptive summary includes a brief discussion regarding the natural habitat of each taxon.

ANALYTICAL METHODS AND TERMINOLOGY

The Museum of Southwest Biology at the University of New Mexico, Albuquerque, provided the comparative collections that analysts utilized to make taxonomic identifications. Researchers made no attempt to identify rodent phalanges, ribs, or vertebrae to the specific or generic level due to the difficulty of such a task (Klein and Cruz-Uribe 1984). When specific or generic identifications were not possible, analysts identified faunal remains to the family level. The categories of small, medium, and large mammal and bird provided some level of identification for fragmented specimens, vertebrae, ribs, and phalanges. Small mammals are those smaller than the genus *Sylvilagus* (cottontail rabbit). Medium mammals are those between *Sylvilagus* and *Canis* (coyote, dog) in size. Mammals larger than *Canis* fall into the large mammal category. Artiodactyls are specimens likely to be *Odocoileus* (deer), *Ovis* (sheep), or *Antilocapra* (pronghorn) that could not be identified more precisely. Small birds are sparrow-sized birds. Birds larger than sparrows and up to the size of Corvidae (crows and ravens) fall into the medium-sized bird category. Large birds are birds larger than Corvidae. These categories function as convenient groups for the purpose of assigning bone that could not be more precisely identified.

Analysts grouped identifications at the levels of species or genus with those identified as "compares favorably" (cf.) to make more meaningful comparisons. For example, counts of specimens identified as *Lepus* sp. (jackrabbit) were summed with specimens originally identified as cf. *Lepus* sp., resulting in larger sample sizes for analysis.

The unit of quantification employed was number of identified specimens (NISP). Analysts preferred NISP to minimum number of individuals (MNI) because of the problems inherent in the calculations of MNIs (Grayson 1984) and because MNI can be predicted from NISP. Both NISP and MNI provide ordinal measures of taxonomic abundance (Grayson 1984).

DESCRIPTIVE SUMMARY

Class Mammalia

Order Artiodactyla
Artiodactyls

Family Cervidae—Deer

Odocoileus hemionus—Mule Deer

   Screened: 1 cervical vertebra, 3 metapodials, 1 vertebra

   Unscreened: 1 thoracic vertebra

   Remarks: *Odocoileus hemionus* ranges throughout most of Arizona, except for the southwestern corner. Mule deer occupy a variety of habitats such as coniferous forests, chaparral, grassland with shrubs, and desert shrubs, with the presence of browse plants being a necessary element (Burt and Grossenheider 1976). In chaparral, deer usually need water within a distance of one mile or less and are especially dependent on free water during the summer dry periods (Hoffmeister 1986:543).

cf. *Odocoileus hemionus*

   Screened: 1 metacarpal, 1 tibia

Artiodactyl

   Screened: 2 innominates, 1 rib

   Unscreened: 3 humeri, 1 metatarsal, 1 tibia

cf. Artiodactyl

   Unscreened: 1 calcaneus, 1 humerus

Family Bovidae—Bison, Goats, Muskoxen, Sheep, Cows

Bos taurus—Cow

   Unscreened: 1 cervical vertebra, 2 innominates, 1 metapodial, 1 sacrum, 4 scapulae, 1 thoracic vertebra, 2 tibias, 4 vertebrae

   Remarks: *Bos taurus* represents an intrusive element.

cf. *Ovis canadensis*—Bighorn Sheep

Unscreened: 1 naviculo-cuboid

Remarks: *Ovis canadensis* is both a browser and a grazer, feeding on a great variety of plants (Burt and Grossenheider 1976). The preferred habitats are precipitous, rocky desert ranges or along washes, creek beds, and other natural water tanks, as water is as important as food (Hoffmeister 1986).

Order Lagomorpha
Rabbits, Hares, and Pikas

Family Leporidae—Rabbits and Hares

***Lepus* sp.—Jackrabbit**

Screened: 2 calcanei, 1 maxilla, 1 metacarpal, 3 metatarsals, 2 radii, 1 tibia

Unscreened: 1 humerus, 1 femur

Remarks: Two species of *Lepus* are found in Arizona: *L. alleni* (antelope jackrabbit), which is limited to the hot deserts and semiarid grasslands of south-central Arizona, and *L. californicus* (blacktail jackrabbit), which is abundant throughout the state at lower elevations and is the species most likely to occur in the research area (Hoffmeister 1986:139).

***Sylvilagus* sp.—Cottontail Rabbit**

Screened: 1 metacarpal, 1 metatarsal, 1 tibia, 1 ulna

Remarks: Three species of cottontail rabbit are found in Arizona: *S. floridanus* (eastern cottontail), which inhabits heavy brush, strips of forest with nearby open areas, weed patches, and edges of swamps; *S. nuttalli* (mountain cottontail), which occupies thickets, sagebrush, forests, mountains, and rocky cliff areas; and *S. audubonii* (desert cottontail), which occupies areas of grass, sagebrush, scattered piñons and junipers, open plains foothills, and low valleys (Burt and Grossenheider 1976; Hoffmeister 1986). Of the three species, *S. floridanus* and *S. audubonii* are the two species most likely to be located in the study area.

Order Rodentia
Rodents

Family Geomyidae—Pocket Gophers

***Thomomys bottae*—Botta's Pocket Gopher**

Screened: 1 tibia

Remarks: *Thomomys bottae* (also known as valley pocket gopher) lives in underground burrows and is found throughout the state in nearly every habitat. The habitat must have sufficient available tuberous roots and plant materials, and the soil must be suitable for digging (Hoffmeister 1986).

Family Heteromyidae—Pocket Mice, Kangaroo Mice, and Kangaroo Rats

Dipodomys sp.—Kangaroo Rat

Screened: 1 mandible

Remarks: Kangaroo rats are burrowing desert dwellers, five species of which are found in Arizona. Two species (*D. merriami* and *D. deserti*) commonly inhabit the research area (Hoffmeister 1986).

Order Carnivora
Carnivores

Family Canidae—Coyote, Wolves, Foxes, and Dog

Canis sp.—Coyote/Dog

Screened: 1 calcaneus, 2 humeri, 1 mandible

Unscreened: 1 cervical vertebra

Remarks: Materials designated coyote/dog (*Canis* sp.) may represent either coyote (*Canis latrans*) or domestic dog (*Canis familiaris*).

Class Aves
Birds

Order Galliformes
Grouse, Ptarmigan, Turkey, and Quail

Family Phasianidae—Quail, Partridges, and Pheasants

cf. Phasianidae

Unscreened: 2 humeri

Remarks: The phasianidae are small to large birds of field or open country. Quail, the smallest, are natives of the New and Old Worlds; partridges were introduced from Europe; and pheasants were introduced from Asia (Robbins, Bruun, and Zim 1966). Partridges and pheasants thus represent an intrusive element.

Order Passeriformes
Perching Birds

Family Corvidae—Jays, Magpies, and Crows

cf. *Aphelocoma* **sp.—Jays**

Unscreened: 1 humerus

Remarks: Two species of *Aphelocoma* are found in the study area, *A. coerulescens* (the scrub jay) and *A. ultramarina* (the Mexican jay). Both commonly inhabit oak and oak-pine forests (Robbins, Bruun, and Zim 1966).

DISCUSSION

Subsistence

Tables 17.1 and 17.2 list the total number of identified specimens in all taxonomic categories recovered from Dutch Canal Ruin. Analysts identified a total of nine taxa to the level of genus or better. Six mammalian taxa occurred in the screened contexts and five mammalian taxa and one bird taxon from the unscreened contexts. Table 17.3 lists in rank order the number of taxa that could be identified to species or genus (from screened contexts only). Rank-order abundance information is one method of assessing the relative importance of animals in the prehistoric economy (Wing and Brown 1979). Analysts did not consider the remains from unscreened contexts (n=212), 184 of which were the *Bos taurus* bones and a significant percentage of the unidentified large mammal bones, as representative of the prehistoric use of the site. Consequently, the remainder of the discussion focuses on those faunal remains recovered primarily from screened contexts.

The most abundant taxon identifiable to the genus or species level recovered from screened contexts at Dutch Canal Ruin was *Lepus* sp., which constituted 6.17% of the screened faunal remains (cf. designations were included in the totals). Following *Lepus* sp. in abundance were *Odocoileus hemionus* (4.32%) and *Canis* sp. and *Sylvilagus* sp. (2.47% each). These four genera constituted 15.43% of the screened assemblage. The medium mammal category represented the largest taxonomic category in the assemblage (from screened contexts only) with 14.20%, followed by the large mammal category with 11.73%.

A very small portion of the remains (from screened and unscreened contexts) exhibited cutmarks. Only seven (4.32%) pieces of faunal material from the screened contexts had cutmarks. The seven pieces were all large mammal fragments, so the cutmarks may have resulted from skinning or butchering activities. These seven large mammal fragments were recovered from Features 8-2 (a pit house), 8-15 (a secondary cremation), and 8-16 (a pit). Features 8-2 and 8-16 each contained one large mammal fragment with cutmarks. In Feature 8-15, two of five awl fragments from the bones of an unidentified large mammal exhibited cutmarks, as did three *Odocoileus hemionus* bone fragments. Only one bone (0.47%) recovered from the unscreened contexts had cutmarks; it was from *Bos taurus* and was intrusive. The paucity of cutmarks supported the idea that animals (particularly small and medium mammals such as *Lepus* and *Sylvilagus*) might have been flayed and drawn and roasted whole or stewed, methods of preparation that would have left no cutmarks (Stevenson 1904; Szuter 1991).

Table 17.1. Number of Identified Species Totals (screened contexts)

Taxon	Number	Percent
Artiodactyl	3	1.85
Canis sp.	4	2.47
Dipodomys sp.	1	0.62
Lepus sp.	10	6.17
Large Mammal	19	11.73
Large Ungulate	1	0.62
Medium Bird	2	1.23
Medium Mammal	23	14.20
Odocoileus hemionus	5	3.09
cf. *Odocoileus hemionus*	2	1.23
Small Mammal	11	6.79
Sylvilagus sp.	4	2.47
Thomomys bottae	1	0.62
Unidentified	76	46.91
Total	**162**	**100.00**

Table 17.2. Number of Identified Species Totals (unscreened contexts)

Taxon	Number	Percent
cf. *Aphelocoma* sp.	1	0.47
Artiodactyl	5	2.36
cf. Artiodactyl	2	0.94
Bos taurus	16	7.55
Canis sp.	1	0.47
Large Bird	1	0.47
Large Mammal	54	25.47
Lepus sp.	2	0.94
Medium Bird	2	0.94
Medium Mammal	6	2.84
Odocoileus hemionus	1	0.47
cf. *Ovis canadensis*	1	0.47
cf. Phasianidae	2	0.94
Unidentified	118	55.66
Total	**212**	**99.99**

Table 17.3. Number of Identified Species Totals: Genus or Better Identifications (screened contexts)

Taxon	Number	Percent	Rank*
Canis sp.	4	14.81	3
Dipodomys sp.	1	3.70	4
Lepus sp.	10	37.04	1
Odocoileus hemionus	7	25.93	2
Sylvilagus sp.	4	14.81	3
Thomomys bottae	1	3.70	4
Total	27	**99.99**	

Note: cf. collapsed with specific identifications
*Categories with same frequency given same ranking

A great portion of the bone recovered from the screened contexts (98 of the 162 bones, or 60.49%) was burned, while only five fragments (2.36%) from the unscreened contexts had evidence of burning. Many authors (e.g., Hamblin, Dirst, and Sparling 1978:226) interpret the presence of burning as evidence that the animal had been utilized for food and had been roasted. Of the 98 bones that exhibited evidence of burning, 78 (79.59%) were recovered from two features: Feature 3-12, a pit with evidence of burning, and Feature 8-15, the secondary cremation. Though these remains may have been utilized for food, they may also have been thrown into the pit as discarded trash or used as an offering with the secondary cremation.

Bone Tools

Field crews recovered five bone awl fragments (3.09% of screened faunal remains), all from large mammals, from Feature 8-15, the secondary cremation. All five fragments were burned, and they may have represented an offering. No bone tools appeared in the unscreened contexts.

Differential Inventory of Site Contents

Of the 162 bones from screened contexts, 159 (98.15%) were recovered from eight identifiable feature types (trash deposit, mixed artifact concentration, exterior fire pit, pit with burning, pit structure, pit house, secondary cremation, and pit not further specified [NFS]). The majority of the remains were contained within four types of features: the secondary cremation (31.45%), the pit structures (25.16%), pits with burning (19.50%), and exterior fire pits (15.72%). The other features contained less than 4% each of the remains. Tables 17.4 and 17.5 present a summary of the faunal remains in screened and unscreened contexts from the features.

Table 17.4. Summary of Faunal Remains by Feature Type (screened contexts)

Feature	Taxon	Number	Percent
Trash Deposit	artiodactyl	1	0.63
Mixed Artifact Concentration	large mammal	1	0.63
	cf. *Odocoileus hemionus*	2	1.26
Exterior Fire Pit	artiodactyl	1	0.63
	Canis sp.	1	0.63
	large mammal	7	4.40
	Lepus sp.	5	3.14
	medium mammal	2	1.26
	Sylvilagus sp.	2	1.26
	unidentifiable	7	4.40
Pit with Burning	*Lepus* sp.	1	0.63
	medium mammal	3	1.89
	Sylvilagus sp.	1	0.63
	Thomomys bottae	1	0.63
	unidentifiable	25	15.72
Pit Structure	*Dipodomys* sp.	1	0.63
	large mammal	1	0.63
	Lepus sp.	2	1.26
	medium mammal	2	1.26
	small mammal	11	6.92
	Sylvilagus sp.	1	0.63
	unidentifiable	22	13.84
Pit House	artiodactyl	1	0.63
	large mammal	1	0.63
	medium bird	1	0.63
	medium mammal	2	1.26
Secondary Cremation	*Canis* sp.	3	1.89
	large mammal	5	3.14
	Lepus sp.	1	0.63
	medium mammal	14	8.81
	Odocoileus hemionus	5	3.14
	unidentifiable	22	13.84
Pit NFS	large mammal	2	1.26
	Lepus sp.	1	0.63
	medium bird	1	0.63
Total		**159**	**100.00**

Table 17.5. Summary of Faunal Remains by Feature Type (unscreened contexts)

Feature	Taxon	Number	Percent
Canal Feature	unidentifiable	2	1.13
Farmstead Locale	artiodactyl	2	1.13
	Bos taurus	16	9.04
	large bird	1	0.56
	large mammal	41	23.16
Trash Deposit	*Canis* sp.	1	0.56
	Odocoileus hemionus	1	0.56
	large mammal	10	5.66
	medium bird	2	1.13
	medium mammal	4	2.27
	unidentifiable	97	54.80
Total		177	100.00

Areas 3 (37.04%) and 8 (51.23%) accounted for 88.27% of the recovered faunal remains from screened contexts. Area 3, the largest area investigated, contained features typical of field house and farmstead occupations. Area 3, which dated to the Colonial period, may have been occupied for longer periods (although seasonally) than other areas of Dutch Canal Ruin (Chapter 4). In contrast, Area 8 was a post-Classic period component representing a small hamlet (Chapter 9). Investigators recovered no large mammal remains from Area 3, but 27.91% (24 of 86) of the screened faunal material recovered from Area 8 consisted of large mammal remains (including artiodactyl and *Odocoileus hemionus*). This difference may be a result of sampling technique, differential preservation, a small sample size, or a difference in animal abundance, procurement processing locale, or utilization between the Colonial period (Area 3) and the post-Classic period (Area 8), or some combination of these factors.

Szuter (1989:324) discussed a variation in the exploitation of cottontails and jackrabbits according to settlement type. Her model integrates the size of settlement populations and their impact on the environment with the relative percentages of cottontails and jackrabbits in the immediate area. The smaller the population, the less impact on the environment; this condition could create a greater proportion of cottontails, which require relatively undisturbed groundcover. The larger, more aggregated village sites, which would have been more typical of the Classic period, would most likely alter the environment to a greater extent, which would lead to a reduced vegetative ground cover and create an environment more favorable for jackrabbits (Szuter 1989, 1991). Researchers have not examined post-Classic period settlements in this regard, and the sample sizes of *Lepus* and *Sylvilagus* at Areas 3 and 8 were too small for analysts to draw any conclusions regarding a difference in acquisition.

CONCLUSIONS

The analysis of the recovered faunal remains indicated that lagomorphs (*Lepus* and *Sylvilagus*) and, to a lesser extent, artiodactyls (*Odocoileus hemionus*) had formed the major protein base for the inhabitants at Dutch Canal Ruin. James (1989:322) also found this to be true of archaeofaunal remains from other Hohokam sites. The preponderance of rabbits and hares and the temporal differences in large faunal remains in the faunal assemblage may have been related to agriculture and an opportunistic hunting strategy (Anyon and LeBlanc 1984; Linares 1976; Semé 1984; Szuter 1989, 1991). The presence of artiodactyl remains may also have resulted from ceremonial activities and bone tool production as well as subsistence activities (Szuter 1989).

CHAPTER 18

CHRONOLOGY: RECONSTRUCTING TEMPORAL EVENTS AT DUTCH CANAL RUIN

Richard V. N. Ahlstrom
David H. Greenwald

Dutch Canal Ruin dated to the latter part of the Pioneer period through the post-Classic period, that is, between the seventh and fifteenth centuries. Evidence supporting these dates has already been discussed by area (Chapters 2 through 11) and as part of the ceramic analysis (Chapter 12) in this volume. The purpose of this chapter is to summarize and integrate these separate discussions and to place the dating of Dutch Canal Ruin in the broader context of Hohokam chronology.

CURRENT ISSUES IN HOHOKAM CHRONOLOGY

Several issues in Hohokam prehistory are important to the interpretation of chronological data from Dutch Canal Ruin. The first is the perennial question of the calendrical dating of the units of the Hohokam cultural sequence. These units (periods broken into phases) have been in existence since the 1930s, and although the sequence of units has been revised and questioned, it is still considered a valid and useful summary of Hohokam prehistory. Based primarily on associations of pottery types, the sequence is a relative dating tool consisting of units on an ordinal scale, that is, a scale that indicates the order of the units but not the amount of time that they represent. Archaeologists thus face the problem of calibrating this scale with reference to the scale of calendar years by assigning calendar dates to the beginning and ending of each period and phase in the cultural sequence. Two kinds of evidence are appropriate to this task. The first consists of trade pottery from the Anasazi and Mogollon, who lived to the north and northeast of the Hohokam. The pottery types in question have been dated in their areas of origin on the basis of tree-ring evidence. The second kind of evidence is chronometric: archaeomagnetic and radiocarbon dates.

Dean (1991) recently reviewed two approaches to the Hohokam chronology and, on the basis of a thorough analysis of the relevant data, produced a new calibration. His effort is but the most recent in a long series of attempts to date the Hohokam sequence. Archaeologists can consider all of these chronologies as hypotheses to be tested and revised as more and better data accumulate. From this perspective, Dean's chronology can be considered the "leading hypothesis" for the following reasons: first, he laid out the procedures for analyzing the data in considerable detail; second, he used a large body of data, including dates available as of 1988, in the analysis; and third, the resulting chronology takes into account the weaknesses in the data, in that it provides estimates of the uncertainty of the dates determined for the various period and phase boundaries.

A second issue is implicit in Dean's treatment of Hohokam chronology. Dean (1991:67–69) argued that only one viable procedure exists for applying chronometric dates to the time placement of periods and phases in the Hohokam cultural sequence. In this procedure, analysts examine aggregated data (i.e., the entire body of chronometric dates) without regard to the individual features and sites that yielded the dated samples. The alternative approach, what Dean (1991:67) calls "direct dating," includes two steps: the dating of individual features and sites, and then the use of these determinations to date units such as periods and phases. According to Dean (1991:67), "the low precision and resolving power of the chronometric techniques applicable in the desert coupled with the comparatively low number of dates vitiates this procedure as far as Hohokam chronology is concerned." Archaeologists find it difficult to be completely confident even of a "strong case," in which the dating of a feature or a site is based on the highest quality chronological evidence that one can reasonably hope for from a Hohokam site (archaeomagnetic or radiocarbon dates). In other words, there is a good chance that a dating inference based on archaeomagnetic or radiocarbon dates from a Hohokam feature or site will be in error.

To the extent that it is valid, Dean's negative assessment of chronometric data from Hohokam sites raises several serious questions for excavations such as those at Dutch Canal Ruin. That is, what can analysts do with the dates from individual sites? If Dean's chronology is a hypothesis to be tested with new data (Dean 1991:70, 81, 90), how many dates, and at what levels of resolution, will investigators require to perform a useful test? Few sites are likely to provide the requisite quantity and quality of data. If even a dating inference based on a strong case can be in error, how should archaeologists handle the typical, comparatively weak cases? Clearly, interpretations of chronometric data from Hohokam sites must be based on an understanding of the weaknesses in those data. In cases where any single temporal inference may be in error, the strength of a chronological argument must depend on either the redundancy or the reinforcement of chronological evidence. Redundancy can be defined as agreement in temporal information of one kind, such as the dating of several houses on the basis of archaeomagnetic dates; reinforcement is agreement in different kinds of temporal information, such as dating a structure by a combination of ceramic, archaeomagnetic, and radiocarbon dates.

A third issue in Hohokam chronology specifically relevant to the interpretation of Dutch Canal Ruin is the history of settlement in the Phoenix Basin. When and in what manner did the Hohokam utilize the portion of the floodplain containing the site, including the canals that crossed it? When did the Hohokam occupy the various areas of the site, and when did they build and use the canals that crossed the site?

STRATIGRAPHY

Using stratigraphy or stratigraphic association as a chronometric tool has its value as well as its shortcomings. Through archaeological and geomorphic observation, researchers can determine the relative order of events represented by cultural units such as sites, features, or deposits and by geomorphic units such as horizons or erosional and depositional events. In other words, stratigraphic association enables investigators to order events in the sequence in which they occurred. At Dutch Canal Ruin, soil formation processes have been restricted to the Holocene epoch, which roughly equates to the last 10,000 years. The geology and natural stratigraphy of the project area have been discussed previously as they relate to the Phoenix Basin and project sites (Volume 1:Chapter 3). Investigators had to understand the natural stratigraphy as it occurred at Dutch Canal Ruin before they could use the stratigraphic sequence as a chronometric tool.

Dutch Canal Ruin contained five basic stratigraphic units and a recently disturbed zone that postdated the prehistoric occupation. The recently disturbed zone illustrates only how prehistoric features and deposits have been affected by their position relative to the other strata. In Area 8, for instance, the base of the disturbed zone truncated many of the features as well as the greater extent of the Holocene soils at this location. The only consistent observation that archaeologists could make regarding the disturbed zone was that it routinely truncated Classic and post-Classic period remains.

One of the five stratigraphic units consisted of a deposit of coarse sand and gravel in the southern extremes of the site and in a former stream channel of the Salt River referenced here as Turney's Gully. Investigators did not find these coarse materials elsewhere at the site and inferred that they were deposits that had accumulated through streamflow action. The remaining four strata appeared almost consistently across the site area. The lowest stratum recognized was a weakly developed soil or paleosol that consisted of a thick deposit of alluvium composed of medium to coarse, massive sands. Carbonate concretions were occasionally present on these sands. This soil horizon probably developed during the early Holocene and may represent a preserved portion of the Blue Point terrace (Péwé 1978). Deposition of this unit predated all recognized cultural features, and only occasionally did prehistoric features intrude into the paleosol.

Poorly sorted sands generally occurred above the paleosol. Investigators could distinguish a textural boundary, although it was indistinct, between these two units. These sands were generally massive but appeared to have a fining upward sequence. Ash and charcoal lenses marked breaks in the depositional history of this unit, suggesting that it had accumulated as a result of sequential deposition episodes. The upper limits of this unit graded imperceptibly into the lower portion of the next unit, the clayey silts, possibly the result of a gradual change in the environmental conditions that affected the occurrence of this depositional sequence. Cultural materials and features surfaced in the poorly sorted sands, generally as intrusions from the upper levels.

The clayey silts consisted of several interbedded units. These units were often discontinuous from north to south across the site area, a result of low-velocity flood episodes demarcating the high-water level of flood events. This zone showed extensive bioturbation as well as prehistoric and historic disturbance: prehistoric features were excavated into this unit, and the clayey silts filled and covered early features. These data were useful in establishing the relative ages of the cultural deposits found within this stratum. As a general observation, Classic and post-Classic period features appeared to postdate these deposits or to co-occur within the upper levels of the clayey silts.

At Dutch Canal Ruin, the Holocene soils consisted of Avondale clay, Gilman loam, and Cashion clay. These are young soils that were formed in alluvium and have a high agricultural potential. Historic or recent use of the project area had disturbed the upper 30–40 cm of these soils. Clay enrichment also was apparent and was probably the result of both historic and prehistoric canal irrigation. These three soil types can be referred to as anthropic because of their alteration by human activities. The lower level of these soils sometimes extended as deep as the upper prehistoric deposits and occurred in relation to the Classic and post-Classic period occupations observed at Dutch Canal Ruin.

Canals at Dutch Canal Ruin frequently extended into the paleosol. The level of origin could occasionally be determined but often had been disturbed by later canal construction or historic activities. Later canals were stratigraphically higher than the earlier canals, a result of increased alluvial accumulations, and the disturbed zone truncated them. Investigators identified sequences of canals in several of the canal alignments that crossed Dutch Canal Ruin (Volume 1:Chapter 8). In certain cases, archaeologists recognized cut and fill episodes within the individual canals: "A fining upward sequence, characteristic of many canals, indicates that the efficiency of sediment transport decreases over time as fine sediment accumulates, necessitating dredging and maintenance" (Volume 1:125). Thus, an individual canal would have gone through a sequence of construction, deposition of sediment, cleaning, deposition of more sediment, and so on. When present, the fining-upward sequence suggested that the canal in question had been abandoned after an episode of sediment deposition rather than immediately after a cleaning. Furthermore, cleaning episodes were often represented by abrupt sediment changes, resulting in the channel being shallower, narrower, or slightly realigned in relation to the original cut. Channel migration, a result of erosional events within canals, indicated horizontal shifts and sequential events. Once the channel had been stabilized, sediment accumulation often indicated changes in velocity, which equated to changes in particle size. Canals often retained distinct characteristics that enabled them to be identified from trench to trench, sometimes over distances of several hundred meters. The composition of canal sediments also appeared to be temporally diagnostic, although on a relative scale. For example, early canals, those equated with the late Pioneer and early Colonial periods, consistently contained coarse basal sands grading to finer sediments with an increase in elevation. Canals that were dated to the late Colonial period contained fewer coarse sands in the lower portions, with alternating silts and clayey silts in the central portions. Dark, organic clays were common in the upper reaches. Classic period canals were filled with fine sediments that were lighter in color than those of earlier canals, the result of decreased amounts of organic matter. The recognition of distinct canal channels and their deposits made possible the sequential ordering of nearly all the canals at Dutch Canal Ruin.

CERAMIC IMPLICATIONS

Researchers can assign an archaeological unit, such as a feature or a site, to a period or phase in the Hohokam cultural sequence in two primary ways. First, they can date the unit to a period or phase per se. Archaeologists often base their assignment of Hohokam sites to periods and phases on assemblages of decorated ceramics, particularly red-on-buff and polychrome types (Chapter 12). The excavation of Dutch Canal Ruin and, in particular, of the numbered research areas within the site, yielded relatively few red-on-buff sherds, providing some limited help in this regard. Second, investigators can assign an archaeological unit to the time interval that corresponds to a period or phase. This kind of dating relies on chronometric techniques (specifically, archaeomagnetic and radiocarbon dating), which are discussed below.

Ceramic data (Tables 12.3 and 12.5; D. H. Greenwald 1988c:Table 10) showed that Dutch Canal Ruin was inhabited between the late Pioneer and post-Classic periods. The distribution of the various ceramic types and the features with which they were associated was somewhat spatially restricted. For example, Pioneer period remains were concentrated in the central portion of the site, whereas Colonial period remains were distributed over the majority of the site area. Sedentary period remains occurred in the southeastern and, possibly, the northeastern and extreme western portion of the site; Classic period remains were in the eastern portion; and post-Classic period remains were in the west-central portion. When analysts separated the temporally diagnostic ceramic types, they began to create temporal and spatial patterns (Figures 18.1 through 18.4). Researchers must use caution when using these data. Sherd counts from most areas were extremely low, a problem commonly encountered among field house loci, and these limited data are not considered statistically valid; nevertheless, archaeologists regard the ceramic associations as reliable.

CHRONOMETRICS: ARCHAEOMAGNETIC AND RADIOCARBON DATING

Researchers have used both archaeomagnetic and radiocarbon dates to calibrate the Hohokam cultural sequence and to assign individual features and sites to calendar intervals. Because an important theme in Hohokam chronology is the extent to which these two kinds of dates do or do not agree, they are best discussed together after some of their discrete features are considered.

Archaeomagnetic Dating

Archaeomagnetic dating can, for present purposes, be characterized as including three steps: collection of a sample, measurement of the sample's magnetism, and determination of a date from

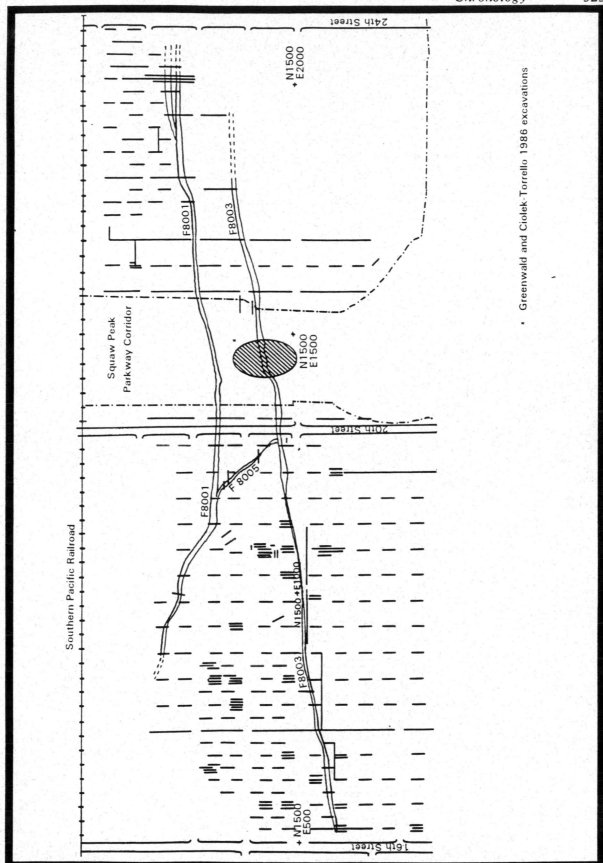

Figure 18.1. Pioneer period settlement, Dutch Canal Ruin. Alignment 8005 predates the eastern part of Alignment 8001 and may represent the earliest alignment of 8001 west to its junction with Alignment 8003.

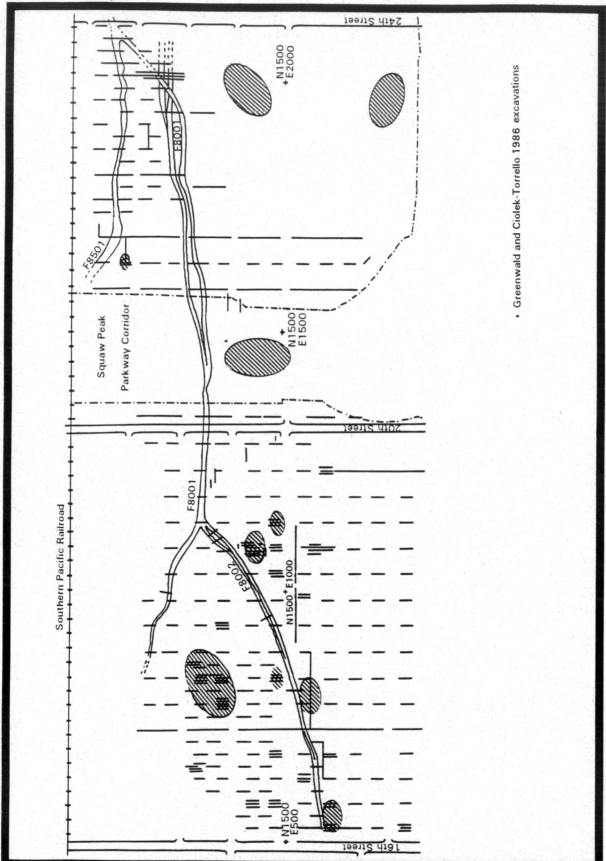

Figure 18.2. Colonial period settlement, Dutch Canal Ruin. The northwestern channel of Alignment 8001 probably was not in use after Alignment 8002 was constructed trending southwesterly.

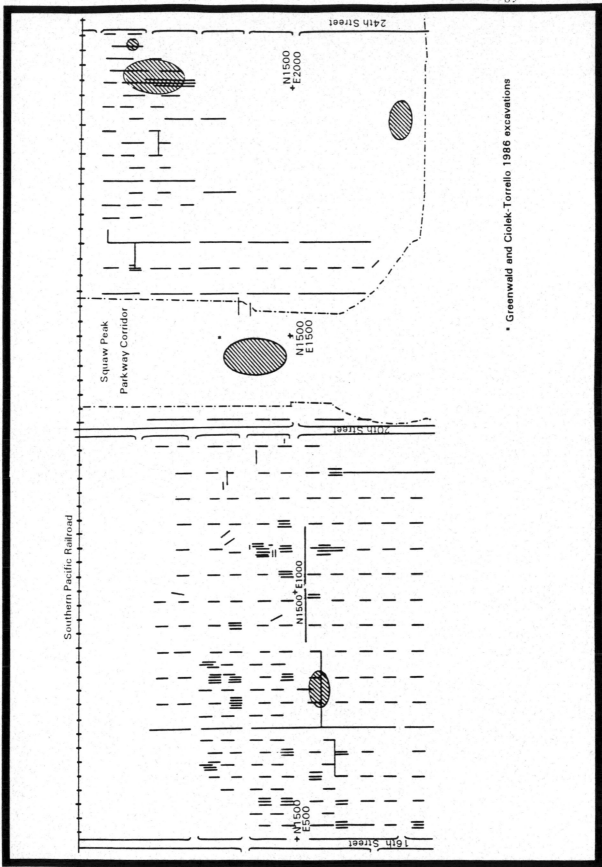

Figure 18.3. Sedentary period settlement, Dutch Canal Ruin.

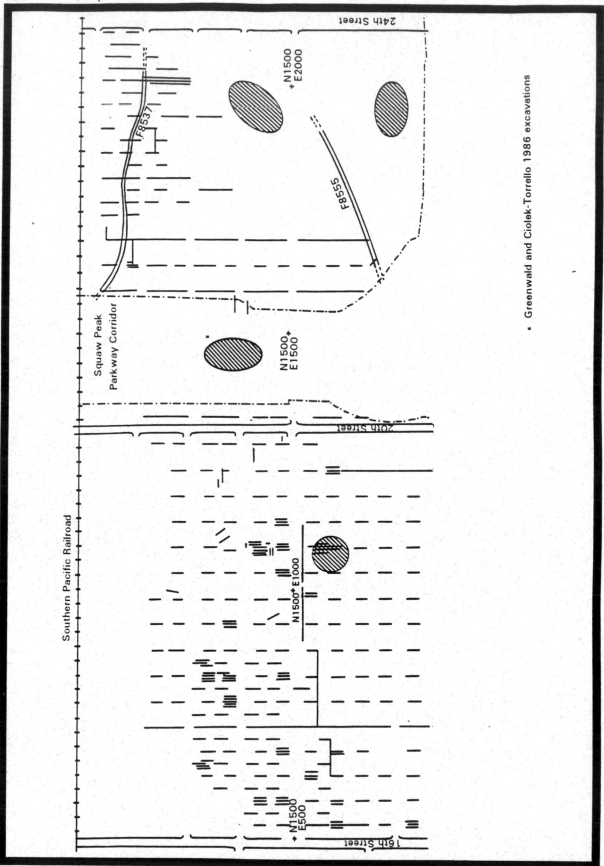

Figure 18.4. Classic and post-Classic period settlement, Dutch Canal Ruin.

that measurement. Sample collection, which has been described by Eighmy (1990), is not of concern here. The measurement of a sample's magnetism (technically, its remanent magnetization) is a "basic" datum, in that this measurement will not change even when procedures for dating the sample improve and change the date determined for that sample. The measurement is an "archaeomagnetic direction" pointing toward a virtual geomagnetic pole (VGP), which can be thought of as an approximation of the location of the north geomagnetic pole at the time the sample was magnetized. A sample consists of several specimens, which are measured separately. The amount of scatter of the individual specimen measurements (i.e., the error inherent in the VGP) is indicated by a statistic known as the alpha 95. The third step in archaeomagnetic dating, which is the actual dating of a sample in calendar intervals, consists of relating the sample VGP to a dated, or calibrated, VGP curve. Several procedures for determining this relationship have been developed, and more than one dated VGP curve exist.

A discussion of alternative dating procedures and methods for developing calibrated VGP curves is unnecessary here (Eighmy and Sternberg 1990). It should be noted, however, that archaeomagnetic dates are expressed as time intervals, that the intervals are estimates of the date of a sample's magnetization, and that the estimates cannot be characterized in terms of a probability distribution. In other words, one portion of the interval is as likely to include the true date of the event as another. Also, individual archaeomagnetic samples typically produce multiple date ranges, and analysts may or may not be able to reject one or more of these intervals on the basis of independent evidence.

Field crews took archaeomagnetic samples at Dutch Canal Ruin from two primary contexts: archaeological features (particularly burned hearths) and canal sediments. In dating the samples, the Archaeometric Laboratory at Colorado State University used a calibrated VGP curve known as SWCV590 (Appendix A). Researchers at the laboratory determined two sets of dates for the samples, one based on a visual technique and the other on a statistical technique. Following the laboratory's recommendations, SWCA analysts used the dates produced by the visual technique. Analysts also produced a third set of dates for sediment samples using a procedure known as declination dating (Appendix A). The inventory of dates included a number of determinations on samples that were collected during excavations by the Museum of Northern Arizona in the freeway corridor portion of the site in 1986. Investigators redated these samples using the current calibration curve, SWCV590.

Radiocarbon Dating

Radiocarbon dating is predicated on, first, a constant rate of production of radioactive carbon (^{14}C) in the upper atmosphere and, second, a stable concentration of radioactive carbon in the atmosphere. Research over the past several decades has demonstrated that neither of these assumptions is valid. Therefore, if radiocarbon dates are to be expressed in calendrical intervals, they need to be calibrated. Archaeologists have developed calibration procedures based on the measurement of radiocarbon in samples of known date, that is, pieces of wood that have been dated by dendrochronology.

This study uses Stuiver and Becker's (1986) calibration procedure as did Dean (1991). Use of this calibration has two advantages. First, it makes the dates from Dutch Canal Ruin directly comparable to those compiled by Dean. Second, Stuiver and Becker's calibration has been computerized (Stuiver and Reimer 1986), which simplifies the conversion of dates and produces more consistent results than alternative procedures based on the reading of graphs or the interpolating of data in tables (Stuiver and Becker 1986).

An uncalibrated radiocarbon date consists of a mean date and a standard error, which expresses the dispersion around that mean of a series of measurements taken on an individual sample. In other words, a radiocarbon date is a statement of probability. A standard error of one sigma (for example, in the date A.D. 950±60) means that the chances are about two out of three (67%) that the one sigma interval (in the example given, A.D. 890 to 1010) includes the true date of the event. The date is also a statement of central tendency. That is, the probability is highest for intervals near the mean and decreases with increasing distance from the mean. This is the first of two measurements of error that need to be considered. Because the calibration procedure is based on the radiocarbon dating of samples, it also includes a standard error. When an individual radiocarbon date is calibrated, the two standard errors (one applying to the date and the other to the calibration) are combined. Radiocarbon determinations are typically stated as ages B.P., that is, radiocarbon years before present, defined as A.D. 1950. The radiocarbon ages of the dated samples from Dutch Canal Ruin are given in Appendix B. For the sake of simplicity, the ages here are expressed as dates A.D.

When a radiocarbon date is converted to a calibrated date, the nature of the determination changes. For one thing, the calibrated date is expressed in terms of intercepts and ranges rather than as a mean and a standard error. The latter format is inappropriate because, although a calibrated date has a probability distribution, that distribution is not likely to be normal (Stuiver and Becker 1986:864). Nevertheless, the range does have a central tendency, in the sense that intervals near an intercept have a higher probability of including the true date of the event than do intervals away from the intercept. This is only helpful, however, when the calibration produces a single intercept. It is a curious result of the history of the production of radiocarbon in the earth's atmosphere that a single radiocarbon date can produce two or more intercepts and two or more date ranges. To return to the previous example, the radiocarbon date of A.D. 950±60 corresponds to the intercepts A.D. 1004, 1008, and 1019 and to the date ranges of A.D. 984-1038, 1102-1116, and 1141-1150 (Stuiver and Becker 1986:Table 3-K). Note that two of the date ranges do not include any intercepts. Unless one or more of the calibration ranges can be dismissed on the basis of independent evidence, which is rarely the case, the interpretation of multiple date ranges calls for a conservative approach: combining the individual ranges into a maximum of calibrated ranges, which would be, for the example given above, A.D. 984-1150 (Stuiver and Reimer 1986). This maximum range is not characterized by a central tendency, and investigators must assume that probabilities are equal throughout the range. Stuiver and Reimer (1986) present an alternative method of calibration that, for radiocarbon dates with multiple calibration ranges, specifies the relative amount of the total probability distribution that is accounted for by each of the date ranges. Analysts did not use this method in the Dutch Canal Ruin study because of its complexity and because it was not used by Dean (1991), with whose work the current study is correlated.

Archaeomagnetic and Radiocarbon Results

Habitation Features

Figure 18.5 summarizes the archaeomagnetic and radiocarbon dates from Dutch Canal Ruin. Several aspects of these data are noteworthy. For example, two groups of dates appear to be significant. Investigators identified two extramural fire pits in Area 3 as roughly contemporaneous on the basis of proximity and stratigraphic position. These fire pits (Features 3-10 and 3-11) produced similar archaeomagnetic dates, A.D. 830-925 and 850-940, respectively. In each case analysts rejected an earlier date range as incompatible with ceramic evidence. These features may have been absolutely contemporaneous, that is, in use during overlapping intervals, although archaeologists have difficulty proving contemporaneity (Ahlstrom 1985; Dean 1978).

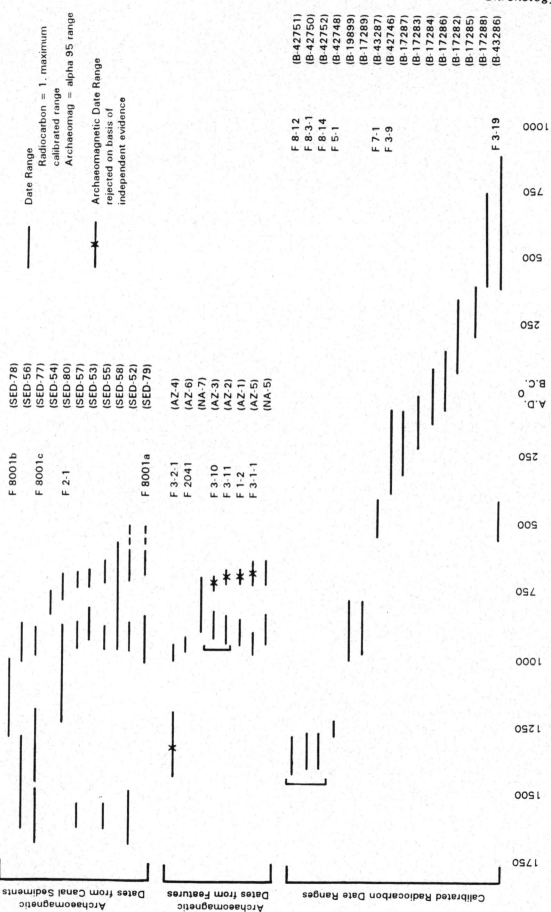

Figure 18.5. Chronometric dates (archaeomagnetic and radiocarbon), Dutch Canal Ruin.

The second group of dates came from a fire pit located in a structure (Feature 8-3-1) and from two extramural fire pits (Features 8-12 and 8-14) in Area 8. The three radiocarbon dates from these features produced maximum calibrated date ranges of A.D. 1280–1405, 1284–1423, and 1277–1405, respectively. The close agreement of these dates argued for their accuracy in placing the occupation of Area 8 between the late A.D. 1200s and early A.D. 1400s. This date range corresponded to Dean's (1991) dating of (1) the interval of certainty between the Soho and Civano phases and (2) the Civano phase. Ceramic data supported the assignment of Area 8 to the recently defined Polvorón phase of the post-Classic period, which Dean did not attempt to date but which apparently corresponds to what was previously considered the late Civano phase. If the assignment of Area 8 to the Polvorón phase is correct, the set of radiocarbon dates from this area suggests that the phase began by the early A.D. 1400s. Others have dated the Polvorón phase or late Classic period as early as A.D. 1350 (Ciolek-Torrello 1988; Sires 1984; see also Eighmy and McGuire 1988), and the results from Area 8 were consistent with those interpretations.

Canals

Canal features provided variable date intervals. In addition to the five archaeomagnetic samples submitted to Colorado State University during the current study, the samples collected by the Museum of Northern Arizona in 1986 were redated by Dr. Eighmy, who provided revised intervals based on the latest curve. Those dates did not change significantly, although the samples that had previously been dated to A.D. pre-700 were substantially adjusted. These included two samples from North Main Canal 1, one sample from North Main Canal 2, one sample from North Main Canal 3, and one sample from South Main Canal 1. With the exception of the A.D. pre-700 interval for North Main Canal 3, all of these adjusted dates can be accepted. Investigators rejected the early interval from North Main Canal 3 in favor of the second interval on the basis of stratigraphic association.

The sample collected from the basal oxides of the lower channel of the South Main Canal (CSUSED-55), which produced a date interval of A.D. 650–725, agreed with the sample collected from the upper channel (CSUSED-80), which produced a date interval of A.D. 700–800. However, the date interval does not necessarily indicate when the canal was constructed. Detrital samples such as these often provide date intervals reflective of later periods of sediment saturation, similar to the way hearths will generally reflect the date of the latest burning. Therefore, it is possible that the South Main Canal was constructed prior to A.D. 650 and that this date represented the last period of saturation of the basal oxide unit.

In the North Main Canal alignment, Sample 79 (CSUSED-79) produced a visual date range of A.D. 900–1000 for North Main Canal 1. This date range was inconsistent with previous results, ceramic associations, and stratigraphic associations. Samples 57 and 58, recovered in 1986, were recalculated and produced date intervals of A.D. 675–730 and A.D. 600–975, respectively. Further inspection of the VGP curve plot for Sample 77 could also include a date range of A.D. 650–725, given the wide alpha 95 for this sample as well as the declination curve (Appendix A). These corroborative intervals were only weakly supported; analysts must consider sample location within the canal. The date interval results did not necessarily support the provenience difference between the sample recovered from the basal oxides and those collected from the postabandonment clays. This difference may be partially explained by the apparently short use-life of this canal. Stratigraphic evidence suggested that North Main Canal 2 had replaced North Main Canal 1 shortly after construction of the earlier canal (Volume 4:Chapter 6).

The recalculated date interval for North Main Canal 2 (CSUSED-55) indicated that it may have been built during the early seventh century. However, North Main Canal 1 was probably not built until the middle or latter part of the seventh century, and North Main Canal 2 was stratigraphically later than North Main Canal 1. Sample 78, collected from Area 6 during the current study, provided no resolution. A metal pipe and other metal debris located within 3.0 m of this sample may have affected compass readings.

Reconstructing the temporal placement of North Main Canal 3 was a major issue during this study. North Main Canal 3 was dated by ceramic association and previous archaeomagnetic samples from the middle to the late Colonial period. Stratigraphically, it was placed in a time period later than both North Main Canals 1 and 2 and the South Main Canal. Sample 56 originally produced a date interval of A.D. 850–935 and a recalculated interval of A.D. 920–1020. Sample 79, collected from the upper clays, produced a date interval of A.D. 850–1015. The stratigraphic position and small ceramic assemblage indicated that use of North Main Canal 3 had probably occurred between A.D. 650 and A.D. 725 (Ciolek-Torrello and Greenwald 1988). Although the current archaeomagnetic results supported this range, archaeologists should consider the potentially devastating effects of annual discharge retrodictions calculated for the Salt River during A.D. 899; excessive streamflow during that year probably had severe impacts on prehistoric irrigation systems, including North Main Canal 3. The sediments sampled during the current study were the thick deposits of dark organic clays in the canal's upper fill.

The only other sample (CSUED-76) attempted during the Phoenix Sky Harbor Center investigations was from Canal Nuevo (Alignment 8537), the Classic period canal. The results when plotted on the VGP curve were located near the 80° N interval on the curve plot, and Canal Nuevo therefore was dated only through ceramic association and stratigraphic position.

RESEARCH DESIGN CONSIDERATIONS

Because of the importance of chronology for Hohokam prehistory, chronology building and refinement was a separate problem domain (Volume 1:Chapter 9). Investigators determined that four issues were relevant: (1) the absolute date of the transition from the Pioneer period to the Colonial period; (2) the absolute dates of the Soho, Civano, and Polvorón phases; (3) the absolute dates of the Polvorón phase and its relationship to the end of the Hohokam culture; and (4) the association of temporally diagnostic ceramics with each of the periods in the Hohokam chronology.

Although researchers identified cultural resources from the Pioneer period during the excavations conducted by the Museum of Northern Arizona for the Squaw Peak Parkway (Greenwald and Ciolek-Torrello 1988a), SWCA investigators discovered no additional features associated with this period as a result of the Phoenix Sky Harbor Center Project. The results of the earlier investigations and the current project indicated that the transition between the Pioneer and Colonial periods at Dutch Canal Ruin had occurred between the eighth and the middle of the ninth centuries A.D. This date corresponds with the conclusions drawn by Dean (1991:Table 3.1).

Chronometric dates for Area 8 were the only dates obtained for the Classic and post-Classic periods during these investigations at Dutch Canal Ruin. Calibrated radiocarbon dates produced intervals that included the last quarter of the thirteenth century and the first quarter of the fifteenth century A.D. The range of these dates was earlier than archaeologists expected from the ceramic evidence. Reasons for the discrepancy between the expected radiocarbon dates and those derived from the samples in the laboratory are discussed in Chapter 9 under "Chronology." For whatever reason, the chronometric data from Dutch Canal Ruin did little to elucidate the chronology of the Classic and post-Classic periods but supported the possibility that Area 8 had been occupied during the Polvorón phase.

Analysis of decorated ceramics and the association of diagnostic ceramics with Hohokam chronology have yielded no new associations. However, one important aspect of the ceramic analysis was the confirmation of a method using the ratios of Roosevelt Red Ware ceramics to identify Polvorón phase sites, areas, and features (Peterson and Abbott 1991). By applying this method, analysts supported the Polvorón phase association of Area 8 and assigned components of Areas 9 and 10 to the Civano phase (Chapter 12).

REVIEW AND CONCLUSIONS

Various lines of evidence have been presented in an attempt to reconstruct the chronological history of Dutch Canal Ruin. Archaeomagnetic dating and the limited ceramic assemblage indicated that canal construction and use had occurred as early as the late Pioneer period (the Snaketown phase), approximately A.D. 650. Although the Pioneer period occupation was limited to a single locus (Figure 18.1), Pioneer period settlements occurred in various locations along the route of the earliest canals (Figure 1.3) but at higher elevations.

Field houses and farmsteads at Dutch Canal Ruin increased dramatically in number during the Colonial period (Figure 18.2). Canal construction increased, and the area's occupants re-engineered some canals during this time. Near the end of the Colonial period, probably around A.D. 900, seasonal settlements declined in number.

Only four loci exhibited evidence of use or possible use during the subsequent Sedentary period (Figure 18.3). Howard notes, on the basis of testing data, an "apparent lack of Sedentary period occupation in the area of Dutch Canal Ruin" (Volume 1:Chapter 5). Dean (1991:Table 3.3) dated the Sacaton phase through independent dating methods to A.D. 950 to 1150. Although ceramic data recovered from Dutch Canal Ruin provided some support for a Sedentary period occupation, the affiliation was not well defined. Chronometric dating methods, principally archaeomagnetism, produced date intervals that extended into the early Sedentary period (Figure 18.5 and Appendix A).

Given the date ranges plotted in Figure 18.5, one could argue that a gap generally exists in the data between A.D. 1000 and 1200. This perception is based partly on the knowledge that Pueblo Salado, located a few hundred meters to the south but still, broadly speaking, in the "Dutch Canal Ruin area," was occupied in the Classic and post-Classic periods, that is, after A.D. 1200. Another nearby site, La Ciudad, was occupied during both the Sedentary and Classic periods, and the proximity of La Ciudad and other sites, such as Pueblo Salado, should be considered as a possible source of late pre-Classic and early Classic period ceramics at Dutch Canal Ruin.

Although researchers have discovered evidence indicating a hiatus in the occupation sequence of Dutch Canal Ruin, the Hohokam presumably never totally abandoned this portion of the geologic floodplain. In fact, it probably became a popular exploitation zone or secondary resource zone for gathering economically important plants and for hunting smaller mammals and waterfowl. The early canals may have enhanced certain forms of vegetation and encouraged others to grow. Mesquite and other useful plants may have increased in numbers along the abandoned canals, adding to the natural resource base of the former agricultural zone. Archaeologists in other research areas have defined many secondary resource zones and their features on the basis of limited surface remains (Doelle 1976; Goodyear 1975; Raab 1976). However, this luxury was not available in the Phoenix Sky Harbor Center due to historic and modern disturbance of the surface, a common phenomenon in urban settings.

Few recovered data suggested a Soho phase occupation. Only three radiocarbon intervals and four archaeomagnetic date ranges extended into the Soho phase. Analysts rejected three of the archaeomagnetic intervals because of the results of other dating methods or associations. The three radiocarbon intervals were confidently associated with the Civano and Polvorón phases. Therefore, few chronometric data exist to support a Soho phase occupation. However, four inhumations were discovered at Dutch Canal Ruin, including one discovered in the freeway corridor during the 1986 excavations, that had been interred in a more or less random pattern. The apparently random nature of the grave locations and the Classic period ceramics (possibly Soho phase) associated with them suggested that these individuals had not been related to any local settlements and that the burials reflected only a possible extended use of Dutch Canal Ruin during the Soho phase.

· Except for the few cases mentioned, the radiocarbon dates from Dutch Canal Ruin mostly predated A.D. 500 and were, by any reasonable interpretation of the site, too early by intervals of up to 500 years. Many of these early dates came from contexts within canal sediments, implying a bias, or directional error, in dates from this context. The apparent error observed in these samples may have been related to sample size, percent of organic matter present, or chemical alteration or organic contamination. The fact that the intervals were consistently early suggested that the soil matrix from which these samples were collected was rich in organic matter, which affects the accuracy of carbon samples. Alternatively, prolonged periods of saturation due to flood episodes may have caused chemical or organic contamination. Until researchers devise a regression analysis for this portion of the geologic floodplain, archaeologists should be cautious in using radiocarbon samples recovered from contexts that predate A.D. 1200, as their usefulness is currently limited.

Given the available data, analysts placed a relatively high level of confidence on the chronological reconstruction for Dutch Canal Ruin despite the poor results of the radiocarbon analysis, the extremely small ceramic frequencies from pre-Classic period habitation features, and the overall difficulty of dating canals. A major problem inherent in interpreting field house and farmstead sites is that samples of datable items are generally small. This was true of the pre-Classic period occupation at Dutch Canal Ruin, and analysis was hindered further by the lack of reliable radiocarbon dates and suitable samples for archaeomagnetic dating. The Classic and post-Classic period occupations, by contrast, were represented by high frequencies of diagnostic ceramics and produced consistent radiocarbon dates. Investigators dated this sequence of occupations with greater confidence than the preceding occupation because the relative permanence of this occupation resulted in a greater number of datable items.

In summary, Dutch Canal Ruin was first occupied during the late Pioneer period. Settlement, although seasonal or temporary, increased during the Colonial period, and occupation may have lasted into the early Sedentary period, with a dramatic decrease in settlement during the late Colonial period. A hiatus in occupation appeared to have occurred roughly during the latter half of the Sedentary period, with a continuation through the Soho phase. Sometime after A.D. 1250, Dutch Canal Ruin was resettled by small family groups on a permanent basis. Settlement continued into the Polvorón phase, with final abandonment around A.D. 1450. Efforts at chronology building and refinement were limited, since field crews recovered few chronometric samples or ceramic artifacts. Some datable samples collected previously were, however, in accord with those recovered during the current study. Corroborating evidence further supported the original date assignments for project canals, and analysts dated new canals by independent means when possible.

CHAPTER 19

DUTCH CANAL RUIN:
PERSPECTIVES ON THE SITE AND ITS REGIONAL ASSOCIATIONS

David H. Greenwald
Mark L. Chenault
M. Zyniecki

The data recovery project at Dutch Canal Ruin for the Phoenix Sky Harbor Center supported some of the findings of earlier investigations (Greenwald and Ciolek-Torrello 1988a) and also revealed new information. This project confirmed the Pioneer period affiliation of the earliest canals and demonstrated a hiatus in canal irrigation practices that began during the Sedentary period and extended through the Soho phase of the Classic period. Field house and farmstead locales were associated with the canals during the late Pioneer and Colonial periods. These locales were the focus of short-term occupation and seasonal activities associated with agriculture, wild plant and faunal resource exploitation, and canal construction and maintenance. Most likely, the inhabitants of Dutch Canal Ruin maintained permanent residences at nearby larger sites, such as Pueblo Patricio, Pueblo Grande, La Ciudad, Los Solares, and La Villa/Casa Chica. Sketchy evidence indicated that Dutch Canal Ruin may have been occupied during the late Classic period, or the Civano phase, and definite evidence existed for at least one Classic period canal alignment. The final occupation of Dutch Canal Ruin occurred during the Polvorón phase. During the Classic and post-Classic period occupations, settlement was permanent, with a full complement of activities taking place. One reason for the occupation of a high-risk area, the floodplain, during the Civano phase could have been that entrenchment of the river had reduced the risk of flooding. Another possibility is that environmental degradation accompanying a reduced moisture regime could have promulgated changes in the social structure of the Hohokam that prohibited construction and maintenance of large canal systems (Nials, Gregory, and Graybill 1989:69–70) and necessitated the occupation of previously abandoned high-risk areas.

Canal studies were a major focus of the investigation at Dutch Canal Ruin. The size of the Phoenix Sky Harbor Center provided an opportunity to examine extensive sections of canals in the Phoenix Basin. Investigators developed a sequence for most of the project canals and assigned dates to some that had not been previously found or dated. They also demonstrated that what Turney (1929) had considered a bifurcation was actually a re-engineering of the system, that is, it was not a system designed for water diversion to two channels at once but instead exhibited reuse of earlier channels to expedite canal construction.

One objective of the research at Dutch Canal Ruin was to determine its relationship to the remainder of Canal System 2 (Cable 1991; Cable and Mitchell 1991; Howard 1990; Howard and Huckleberry 1991). The excavations and analyses have shown that the canals within the Phoenix Sky Harbor Center represented the initial development of the canal system (Cable 1991; Greenwald and Ciolek-Torrello 1988a; Henderson 1989). The canals were on the floodplain, with fields situated nearby, but at a distance from the main river channel to reduce the risk of damage from annual flooding. Eventually the system included canals above the floodplain, and the Hohokam abandoned the canals on the first terrace (Howard and Wilcox 1988:930–933).

THE CANALS

Development of Canal Irrigation

Archaeologists generally agree that initial development of prehistoric canal irrigation in the Salt-Gila Basin was well underway by the late Pioneer period (Greenwald and Ciolek-Torrello 1988a; Howard 1991; Wilcox and Shenk 1977). Although a plethora of chronologies currently exist for the Hohokam, investigators for this project used Dean's (1991) chronological reconstruction (which is based on independent chronometry) to bring the canals into temporal perspective. During investigations at Snaketown, Haury (1976:140) identified an early canal, Pioneer Canal No. 1, that he assigned to the early Pioneer period. Using Dean's chronology, this would equate to A.D. 300–500. Some archaeologists have challenged Haury's interpretation of the canal's temporal assignment on the basis of mixed ceramic types (Wilcox and Shenk 1977), but most generally accept that the canal dated to at least the late Pioneer period. Many researchers have accepted the notion that the Hohokam culture began to evolve during the Vahki phase, but recently, some have presented evidence for a pre-Vahki Red Mountain phase (Cable and Doyel 1987; Doyel 1991:236–237). The Red Mountain and Vahki phases could well represent that period of time when cultural traits were developed that have since been designated "Hohokam." It is also reasonable to accept the premise of the evolution of indigenous desert peoples who, in adapting to changes in lifeways and subsistence strategies, assumed a collection of traits that have been recognized as Hohokam.

Incipient agriculture appears to have developed in the arid Southwest in locations that were favorable to early cultigens and, in some areas, that predated the earliest recognized Hohokam component (Eddy and Cooley 1983:46; Huckell 1987, 1988). Archaeologists have identified various farming techniques throughout the Southwest (e.g., Bryan 1929; Hack 1942), including dry farming, spring-fed fields, basin or runoff farming, and ak chin farming. In an attempt to reconstruct subsistence practices, researchers have described techniques associated with the Hohokam (e.g., Rankin and Katzer 1989; Rodgers 1985). Early efforts to farm the lower deserts probably focused on areas adjacent to springs or seeps, along water courses that had low potential for flooding, or in areas of increased direct precipitation and runoff. Evolution of irrigation systems in the alluvial valleys probably came about through small-scale ditch irrigation systems located either on the first terrace immediately adjacent to the stream channel or on alluvial bars within the channels themselves (Font 1775:48–52) or that extended a short distance onto the first terrace (Cable 1991; Henderson 1989). Evidence that would support these methods may have been removed from the archaeological record long since by flooding, erosion, prehistoric activities, and historic or modern disturbance of the stream channel banks and adjacent floodplain.

Even though the Hohokam had access to a large diversity of wild resources, they would have had difficulty in developing villages in the lower desert without some form of successful agriculture. The availability of resources undoubtedly influenced settlement location, especially of permanent sites. The establishment of villages such as Snaketown during the Pioneer period probably was linked directly to the agricultural potential of its hinterland. The Hohokam could have used various methods to farm the lower desert successfully. For example, by locating fields near water sources, early farmers could have carried water to their fields in ceramic containers, a time-consuming and tedious practice; moreover, field size would have been restricted by the farmer's ability to water the plants in the field. By bringing the water to the fields in greater volume or by locating fields closer to the water source, thereby reducing labor investment during the growing season, farmers could have planted larger areas. In most cases, however, such a strategy would have required that fields be located within the river channel or on adjacent floodplains.

The annual discharge of the Gila River is lower than that of the Salt and Verde rivers (Ackerly, Howard, and McGuire 1987:10). Because of this, floodplain or flood channel farming may have been more practical along the Gila River due to the lower probability of flooding during the growing season. Other rivers that compare to the Gila in this regard are the San Pedro and the Santa Cruz. As Gregory (1991:177) pointed out, periods of low flow probably coincided with the irrigation season. Through flood channel irrigation, farmers could have diverted water to fields through small ditches, using brush and earthen diversion barriers when needed, and could have built berms or dikes when water levels threatened fields. In many areas the Gila River is composed of braided channels or the channel itself is broad, which would allow rising water to spread shallowly across the flood channel until the entire channel is filled. Farmers could have managed low-velocity floods that did not fill the channel with some success using crude flood control devices. With a system similar to the model presented above, early Hohokam villages located in the broad alluvial valleys could have been established prior to the development of canal irrigation systems.

Archaeologists do not fully understand the evolution of the canal system and have constructed several different scenarios to explain how irrigation agriculture developed in alluvial basins. Nevertheless, at least three factors determined the success of canal irrigation in this area: (1) a reliable water source; (2) the means by which to transfer water from the source to fields; and (3) arable soils. If, for example, the stream channel had been cut too deeply, it might have been impractical to construct the required canal head, diversion dam, and channel. Therefore, the Hohokam would have preferred areas of low relief, which would have required less labor investment to build and maintain the system. Establishing field locations in areas less susceptible to flooding and linking them to canal systems would have reduced the annual effort required to re-establish fields in the flood channels destroyed during the period of high flow. The Hohokam could have constructed and maintained canals outside of the growing seasons, enabling the farmers to put more energy into production and thereby increasing their dependence on agricultural subsistence. Few streams provide the reliable water flow needed for successful irrigation agriculture, and those that do can also be destructive. Irrigation along Arizona streams may have evolved from small diversion systems located immediately adjacent to the stream channel, eventually expanding into areas of less risk at a distance from the channel.

Along the Salt River, the Hohokam apparently developed their canal system closest to the river initially, eventually expanding outward and onto parcels with less gradient (Howard and Wilcox 1988:930-936). Turney (1929) commented that early canals were located closest to the river, implying that construction of canals would have been easier in this location due to the gradient of the first terrace. Although Turney's interpretation apparently was correct, those canals adjacent to the river would have faced severe engineering problems. Washes and ravines would have posed serious problems for long canals, and erosional features that could not be crossed would have limited the growth potential of canal systems. Such limitations may not have been a serious problem for the early farmers. Since arable lands were readily available during the Pioneer period, the Hohokam could have redirected canal development to other areas or could have expanded existing systems by using distribution channels or branches.

Study of the canals that traversed Dutch Canal Ruin within the Phoenix Sky Harbor Center led investigators to several conclusions regarding early Hohokam subsistence patterns and land-use strategies. The occupants of Dutch Canal Ruin first constructed canals sometime after A.D. 600 but certainly prior to A.D. 650. After establishing the system, the farmers repeatedly built and, in some cases, re-engineered the canals. The Hohokam used the canals until sometime after A.D. 899, perhaps as late as A.D. 950, and then abandoned much of the geologic floodplain. The occupants of Dutch Canal Ruin did not use canals from the time they abandoned North Main Canal 3 until they established Canal Nuevo through the project area during the Classic period. The sequence of canal construction and use is presented below. The establishment of canal networks on the geologic floodplain through the project area would have made available an estimated 542 ha (1300 acres) for canal irrigation at any given time. The limit placed on the potential agricultural acreage associated with the canals that traversed Dutch Canal Ruin is artificial, because it is based on the length of the canals as indicated in archival records (Midvale 1934, 1968, n.d.; Patrick 1903; Turney 1924) and on the presence of physical features such as Turney's Gully. The Canal System 2 heads were located at various points along the river, extending as far as 3.2 km (2 miles) upstream of Pueblo Grande, near the bedrock mass associated with the Papago Buttes, and to a point nearly 2.4 km (1.5 miles) downstream of Pueblo Grande. Canals 4, 5, and 6 (Turney 1929) appeared to have been independent systems that served the lower reaches of Canal System 2. Arable lands associated with Canal System 2 occurred in the area immediately south of Pueblo Grande, extending westward for nearly 24.2 km (15 miles), the maximum extent of Canal System 2. The first terrace or geologic floodplain provided deep alluvial soils conducive to irrigation agriculture and was an important agricultural zone during the early pre-Classic periods.

The Dutch Canal Ruin Agricultural Zone

Given that the settlement of Dutch Canal Ruin during the pre-Classic periods was restricted to field house loci and farmsteads and given the presence of successive canals, investigators concluded that the first terrace was an area of agricultural specialization (Greenwald and Ciolek-Torrello 1988a:209). The data generated through investigations within the Phoenix Sky Harbor Center indicated that the Hohokam had established temporary agricultural settlements along the canals from the time of construction until abandonment. Investigators inferred that similar settlements lay along these canals beyond the limits of Dutch Canal Ruin. During the early use of the Dutch Canal Ruin canal network, these types of agricultural settlements probably occurred wherever farmers located their fields, except in the immediate vicinity of permanent settlements. Previous models using ethnographic data and analogy have discussed the distance that farmers would travel on a daily basis from their permanent homes to their fields (Boserup 1965; Chisholm 1962). Alternatively, or in addition to commuting, the Hohokam may have used a temporary habitation area for shelter from the elements, as storage for crops or tools, or as a device to mark one's property, a concept similar to the use of field boundary markers. Also, researchers have demonstrated the use of temporary structures for the exploitation of economically important wild resources (Cable and Doyel 1985), although the focus at Dutch Canal Ruin was on agriculture (Greenwald and Ciolek-Torrello 1988a).

Groups from neighboring villages developed and exploited the Dutch Canal Ruin agricultural zone, a term used here to denote the primary use of the general site area during the pre-Classic periods (Cable and Doyel 1984:269-275; Howard and Wilcox 1988:932). Although researchers can use models of central place to demonstrate the association of the agricultural potential and function of Dutch Canal Ruin with nearby sites, such as La Ciudad (Christaller 1933; Hodder and Orton 1976), they must also consider other factors, such as social interaction and geographical aspects. The linear pattern and length of the canals, when compared to the locations of permanent settlements, indicates that an organized system involving intersettlement structure was in place, or had been developed, to construct the canals and allocate water.

A model of intersite organization based on canal and field use illustrates this organizational structure and addresses possible strategies employed by early Hohokam farmers. The Hohokam could have built, operated, and maintained canals only through the organized efforts of all who would benefit directly from canal irrigation, resulting in a capital investment (Howard 1991:5.15; Neitzel 1991:180-181). Members of permanent settlements could have employed two possible strategies (or a combination of both). Given the linear pattern of canal networks, investigators concluded that the first strategy would have included access along the canal alignments to field areas and resource zones (Figure 19.1). Linear movement along the canals would have provided access to the Dutch Canal Ruin agricultural zone for residents of Pueblo Grande from the east and Pueblo Patricio and La Villa/Casa Chica from the west but would have excluded residents of the La Ciudad/Los Solares settlement approximately 1 km (.62 miles) to the north. The second possibility, a lateral strategy, would have cut across the canal alignments, traversing any physiographic features to access fields and other resources (Figure 19.2). Lateral movement would have provided settlements such as La Ciudad and Los Solares access to the first-terrace agricultural lands. This second strategy parallels central place theory in that land-use patterns associated with larger, permanent settlements would have transected canal systems. Although such strategies may be workable in theory, the practical application of linear land-use patterns has certain drawbacks. Linear features such as canals or rivers have caused social scientists considerable trouble when attempting to interpret resource exploitation. Canals, unlike rivers, are not self-sustaining and require a certain amount of labor investment to operate effectively. Furthermore, canals do not serve as natural barriers and are easily crossed. Because they are linear and, in the case of many of the canals in Canals System 2, traverse multiple settlement areas, successful operation undoubtedly required the implementation of some form of canal management and water allocation.

Following Chisholm's (1962) study of rural land use among peasant societies, in which he attempted to analyze catchment areas in terms of concentric circles, Cable and Doyel (1984:273) noted that at about 3–4 km (2–2.5 miles) from the village settlement, land-use strategies changed radically. Chisholm (1962) viewed this change as a direct reflection of the cost, in terms of labor, to effectively travel the required distance to complete the necessary tasks and return to the village in a single day. When the labor and travel costs became too high, temporary facilities, such as field houses, would be used. Cable and Doyel (1984), by superimposing hexagons around contemporaneous sites, illustrated that site spacing was such that La Ciudad's hexagons began to interface with those of neighboring sites at approximately 3 km (2 miles). The field houses at Pueblo Patricio were approximately this far from La Ciudad. Because of the linear nature of canals, considerable overlap in resource areas must have occurred. Therefore, the Hohokam may have used a combined strategy of linear and lateral movement, resulting in a higher density of field house and farmstead loci in certain areas. The spatial association of the above-mentioned sites indicated that both strategies had been used and that at times residents of different villages had interacted in constructing and maintaining canals as well as in farming and procuring resources. Investigators found support for the proposition that Pueblo Grande had served as an administrative settlement as early as the late Pioneer period. Its proximity to the canal heads would have fostered its sociopolitical development, as canal flow at the intake or head and flow to downcanal users would have had to be regulated. Sites such as La Ciudad would have been dependent on canal users upcanal for equitable allotments of water.

Early settlement of the Phoenix Basin was marked by the establishment of small villages on the first and second terraces, along the interface of the geologic floodplain and the lower bajada. Along the north side of the Salt River, downstream from the Papago Buttes, the interface is gradual and difficult to define. Establishment of settlements above the interface indicates that the Hohokam were aware of the flood potential that existed on the first terrace. Nevertheless, they constructed canals and farmed this zone of increased risk until at least the middle of the Colonial period. Although the deep, alluvial, well-drained soils found on the first terrace were conducive to irrigation agriculture, the canals located on this topographic feature were exceptionally susceptible to flood damage. Salt River floods could cause downcutting, which would erode the canals away, or infilling, requiring that entire lengths of canals be re-excavated before crops could be irrigated. Flood magnitudes of greater than 50 years would be required to cripple such a system, but the

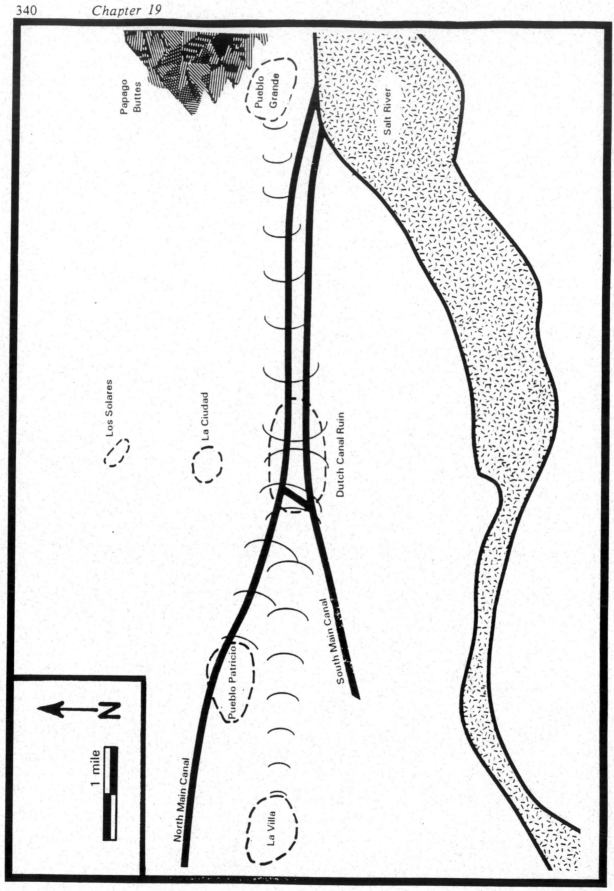

Figure 19.1. Stylized illustration of proposed land-use patterns in the Dutch Canal Ruin agricultural zone showing the linear strategy. (Compiled from Howard 1991; Midvale 1934; Turney 1929; and current investigations [T. Bostwick, personal communication 1992]. The extent of Dutch Canal Ruin is based on Midvale's 1934 map of the site.)

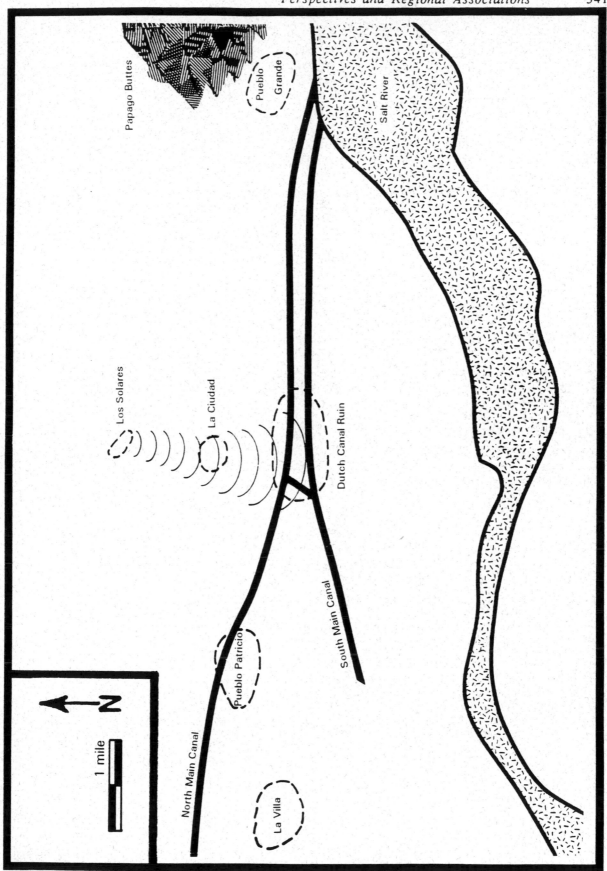

Figure 19.2. Stylized illustration of proposed land-use patterns in the Dutch Canal Ruin agricultural zone showing the lateral strategy. (Compiled from Howard 1991; Midvale 1934; Turney 1929; and current investigations [T. Bostwick, personal communication 1992]. The extent of Dutch Canal Ruin is based on Midvale's 1934 map of the site.)

effects on the system could have resulted in the loss of several irrigation seasons (Gregory 1991:184). During the pre-Classic periods, occupants of the project area constructed five canal alignments. The last canal appeared to have been abandoned at about the same time that a major flood occurred. The tree-ring record indicated that from the time of pre-Classic period abandonment until A.D. 1080 or 1100, the Salt River exhibited stability or decreased flow (Nials, Gregory, and Graybill 1989). Small-magnitude discharges occurred until about A.D. 1200, when streamflow decreased due to an apparent low-moisture regime that lasted for approximately 150 to 160 years.

The inhabitants of Dutch Canal Ruin abandoned their agricultural zone and its canal networks by the end of the Colonial period, with the possible exception of Areas 5 and 9. Researchers have posited flooding of the Salt River as the principal cause for abandonment, with the Hohokam agriculturists building new canals that followed a more northerly trajectory (Howard and Wilcox 1988). Because the first terrace was highly susceptible to flooding, the occupants would most certainly have recognized it as a zone of high risk. The current project and recent streamflow reconstructions for the Salt River (Graybill 1989) indicated that farmers had abandoned the first terrace as a zone of irrigation agriculture by A.D. 900. By this time, they had constructed new canals and new agricultural fields on the second terrace and were apparently no longer using Dutch Canal Ruin as an agricultural area, although they undoubtedly exploited the other economic resources there. Previous use of the first terrace may have resulted in microenvironments developing in abandoned fields, canals, and other land modifications that produced desired resources (Volume 4:Chapter 7). The Hohokam may have used this abandoned agricultural zone as a supplementary subsistence resource area.

Farmers did not use the northern portion of the project area as an agricultural zone again until the Classic period, when they established farmstead settlements, Areas 9 and 10, at Dutch Canal Ruin. A pattern of settlement and canal building similar to that in the previous period of use then evolved. The major difference between the two occupations was in the size of the settlements, their increased use, and their apparent permanency of occupation. There were fewer farmsteads during the Classic period than there were field house loci during the earlier occupation, possibly because larger parcels were being placed under cultivation by each irrigation settlement in the Classic period. These settlements were characterized by increased labor investment, indicated by more formalized structures and associated features, and a wider range of activities, represented by increased numbers of features and artifact disposal.

Although these farmsteads cannot be tied directly to nearby Classic period villages, an association similar to that posited for the pre-Classic periods can be argued. As in the pre-Classic periods, canals in the Classic period originated immediately downstream of Pueblo Grande. The Hohokam may have employed both linear and lateral land-use strategies, or members of the farmstead settlements and the larger neighboring villages may have established affinal or economic relations with each other. Other researchers have argued for interaction along canal systems, both upcanal and downcanal, to illustrate social and political relations between villages dependent on the same canal (Crown 1987; Doyel 1981; Gregory 1991; Gregory and Nials 1985; Howard 1991). Whatever the size of the habitation site, it would have depended on sites upcanal for a share of the water. Water allocation and dispersal, especially in drought years, would have necessitated some form of hierarchical sociopolitical system, although sociopolitical organization may have been linked most closely to sites within the same occupation zone (Howard and Wilcox 1988:938). All settlements within the irrigation community, regardless of size, would have had to comply with dispersal regulations, or downcanal settlements would have been affected by water shortages.

Canal System 2 reached its greatest extent during the Classic period (Howard 1991:5.21), and archaeologists have identified the extensive canal systems of this period as one of the hallmarks of the Hohokam culture (Turney 1929). Researchers have given various explanations for the growth of these irrigation systems and for the use of agriculturally marginal areas, including environmental factors (Doyel 1979; Plog 1980; Weaver 1972), population pressure (Grady 1976), and reorganization of Hohokam social and political structure (Doyel 1980; Gregory 1987; Gregory and Nials 1985; Wilcox 1979). Reoccupation of the Dutch Canal Ruin agricultural zone during the Classic period can be approached by any of the above three explanations.

From a regional perspective, the investigations of Dutch Canal Ruin within the Phoenix Sky Harbor Center provided an opportunity to examine segments of pre-Classic and Classic period canals nearly a mile long. These studies provided further documentation on agricultural land-use strategies in the Hohokam core area, including information on small settlements that have often been associated with resource specialization of outlying or peripheral areas. Perhaps because the first terrace can be considered a high-risk, less optimal area, it should be considered part of the inner periphery (Wilcox 1979) rather than included in the core. Dutch Canal Ruin contained not only some of the earliest Hohokam resources associated with canal irrigation, but also some of the latest settlements found in the Phoenix Basin. Post-Classic period occupation, recently named the Polvorón phase (Sires 1984), is just now being defined throughout the Hohokam area, and investigations at both Dutch Canal Ruin and Pueblo Salado have expanded our current understanding of late Classic and post-Classic period occupation in the Hohokam region.

Re-establishing Canal Irrigation on the First Terrace

With the reoccupation of the first terrace during the late Classic period, the canal irrigation system was re-established. Within the project area at Dutch Canal Ruin, investigators discovered two loci dating to the Classic period and one dating to the Polvorón phase of the post-Classic period. Settlement consisted of permanently occupied small habitations. Re-establishment of canal irrigation in the Dutch Canal agricultural zone occurred at the same time as or just prior to increased streamflow in the Salt River (Nials, Gregory, and Graybill 1989), sometime around A.D. 1300–1400. Archaeologists do not know the impetus for this shift in land-use strategies, but the shift indicates that the Hohokam were willing to utilize and live in this area of high risk. Occupation of the first terrace during the Classic period was not restricted to Dutch Canal Ruin. Pueblo Salado, occupied during the late Soho, Civano, and Polvorón phases, was also on the geologic floodplain (Volume 3).

Project investigators could not determine the duration of use of the Classic period canal at Dutch Canal Ruin, although two settlements (Areas 9 and 10) considered permanent habitations appeared to have been contemporaneous with one of the canals, Canal Nuevo (Alignment 8537). This canal was not well dated, but analysts assigned it to the late Classic period on the basis of its stratigraphic position and the occurrence of Gila Polychrome sherds in its basal fill. The presence of corn remains at the Classic period loci suggested that the inhabitants had practiced corn agriculture, supported by canal irrigation, during this time. The Polvorón phase remains at Pueblo Salado, located in the southern part of the project area, and at Area 8 of Dutch Canal Ruin appeared to be some of the latest occupations associated with canal irrigation along the north side of the Salt River.

Canal Sequence and Chronology

Prior to the Phoenix Sky Harbor Center data recovery phase, project investigators had several questions regarding the temporal associations and chronological ordering of canals in the project area. They were successful in addressing most of these questions, although diagnostic types were often underrepresented in ceramic assemblages. Although limited numbers of diagnostic ceramics were the mode for the entire pre-Classic period feature assemblage, ceramic assemblages were not skewed by the intrusion of later types. Therefore, analysts introduced their interpretations with relatively high confidence despite the limited numbers of diagnostic sherds. Stratigraphic evidence was fundamental to the following interpretations, as investigators had no doubt about the stratigraphic relationship of one event to another. Questions can always be raised about the temporal parameters placed on a given event from other dating methods, but stratigraphic relationships are absolute.

Prior to the data recovery phase, investigators had identified five issues concerning the function and temporal sequence of the canals within Dutch Canal Ruin. First, they needed to determine the relationship of the South Main Canal to the North Main Canal alignment. Second, they hoped to establish the function of Turney's "bifurcation," since they knew the sequence of use of the three canals involved in the area of the bifurcation. Third, they had not determined the function and age of Alignment 8005, although they knew this channel was associated with the South Main Canal and believed it to be a small crosscut channel. Fourth, in the eastern portion of the project area, investigators had observed another "bifurcation" and had targeted for further study the relationship and function of the two canals in question. The final issue involved assessing the mechanics of a suspected turnout, breach, or other anomalous canal-related event or feature in Area 7 and determining the age of the two canals associated with this feature. The answer to each of these questions is discussed below.

Analysts generated further support for the early age of the South Main Canal through archaeomagnetic dating and stratigraphic associations. Investigators considered the South Main Canal alignment to be one of the earliest canals in the project area (Greenwald and Ciolek-Torrello 1988a). However, they had never determined its temporal association with the North Main Canal alignment. Evidence gathered through additional sampling methods and archaeomagnetic and ceramic associations (and in the course of addressing the issue of the age and function of Alignment 8005) supported the inference that the South Main Canal was the earliest channel built through Dutch Canal Ruin. Alignment 8005 originated at the South Main Canal and followed a northwesterly trajectory until it intersected the North Main Canal alignment. By determining the stratigraphic relationship of Alignment 8005 to the channels of the North Main Canal alignment, investigators determined that Alignment 8005 predated the North Main Canal alignment and had not functioned as a crosscut canal. Since Alignment 8005 predated the North Main Canal alignment, so must the South Main Canal. The temporal difference between the South and North Main Canal alignments was not great, perhaps as little as 50 years.

Turney's bifurcation represented re-engineering events rather than a bifurcated system. Investigators found no evidence to support the contention that the division of the canal alignment had functioned in such a manner as to enable water to be diverted into two channels simultaneously. In fact, only North Main Canal 2 was split, which resulted from sequential use, as the profile of Trench 1144 clearly showed. This documentation indicated that the northwest channel of this alignment had been abandoned in favor of its southwest counterpart.

The study of the possible bifurcation in the eastern portion of the project area indicated that sequential events had created this feature. North Main Canal 3 was superimposed on Canal Viejo

(Alignment 8501), the earliest channel. North Main Canal 3 was often constructed within or adjacent to an earlier canal, as was apparently the situation in Area 5; however, investigators could not prove that supposition from the trench profile exposure alone. Stratigraphic associations dated Canal Viejo to a period earlier than approximately A.D. 830, the early range given for North Main Canal 3.

Analysis of the suspected turnout, breach, or bank modification recognized during testing in Area 7 determined that the deposits were normal canal deposits exposed by a backhoe trench that bisected the canal at an oblique angle. Exposure of Canal Viejo in plan view revealed that the alignment had turned sharply, so that investigators misidentified its deposits during the testing phase as being associated with some form of bank modification. Horizontal exposure also enabled the stratigraphic association of the two canals to be determined. Canal Nuevo was superimposed on Canal Viejo. The excavation of additional trenches led to the recovery of Gila Polychrome sherds from the lower depths of Canal Nuevo. An archaeomagnetic sample collected from this canal produced a date range that postdated the Hohokam sequence.

Archaeologists used these determinations to define the following canal sequence for Dutch Canal Ruin. South Main Canal was the earliest, dating between A.D. 600 and A.D. pre-700. North Main Canals 1 and 2 and probably Canal Viejo followed in sequence. North Main Canal 1 predated North Main Canal 2, according to stratigraphic association, and analysts dated these canals to A.D. 650–730. Canal Viejo predated North Main Canal 3, but analysis provided no further resolution. North Main Canal 3 had previously been dated to A.D. 850–950. Investigators during the present study revised this range, on the basis of new results, to A.D. 830–930. The North Main Canal alignment that has been equated with the northern branch of Canal Patricio was abandoned during the later use of North Main Canal 2. With the change in orientation to the southwest, the North Main Canal alignment no longer transferred water onto the second terrace. North Main Canal 3 continued the later alignment of North Main Canal 2, and both channels made use of the previously abandoned South Main Canal alignment until exiting the project area. Finally, Canal Nuevo represented a Classic period canal, a determination made on the basis of stratigraphic association and diagnostic ceramics. An assignment to the Civano phase or later appeared consistent with the ceramics and with the canal's proximity to late Classic period settlements. Canal Barranca, Alignment 8555, could not be confidently assigned to any time period, as investigators recovered no datable samples. It was dug into sands and gravels and filled with coarse sands. Field crews did not attempt to collect archaeomagnetic samples from the canal's sediments or base due to the coarseness of the materials. Neither diagnostic ceramics nor charcoal for radiocarbon sampling were present. Analysts made no attempt to assign Canal Barranca to any temporal period.

DUTCH CANAL RUIN SETTLEMENT PATTERNS AND LAND USE

The study of the prehistoric settlement patterns of a region involves the examination of complete systems and surrounding systems. Settlement pattern studies strive for a holistic view of ancient society and not for segmentation of a system or region (Vogt and Leventhal 1983). The prehistoric remains at Dutch Canal Ruin consisted of small field house loci and farmsteads, components of a larger settlement system. To examine settlement patterns in regard to Dutch Canal Ruin, researchers first surveyed the distribution of surrounding sites and information on regional settlement patterns (e.g., Gregory 1991). The field houses and farmsteads at Dutch Canal Ruin may have been associated with populations residing at sites along the same canals, such as Pueblo Patricio, La Villa/Casa Chica, and Pueblo Grande, or to other nearby sites such as La Ciudad/Los Solares.

Archaeologists can evaluate site structure, or "intra-site patterning" (Doyel 1987:14), for Dutch Canal Ruin from two perspectives: (1) by examining each locus of activity as an individual site; and (2) by considering the prehistoric features within the project area as components of a larger site (Figure 1.2). From the first perspective, each locus of activity represented habitation settlements at the simplest level. From the second perspective, the pre-Classic period field houses and farmsteads at Dutch Canal Ruin were functional extensions of other sites. The following statement made by Cable and Doyel regarding their findings for the Central Phoenix Redevelopment Project holds true for the situation at Dutch Canal Ruin:

> The dense occurrence of field house occupations indicates that the outlying zones around major Hohokam villages in the Salt River Valley were used intensively for agricultural activities and that a significant proportion of each year was spent in tending and harvesting crops [Cable and Doyel 1985:303].

Diachronic examination of the site structure of Dutch Canal Ruin revealed a pattern of small, isolated field houses and farmsteads situated on the floodplain of the Salt River. During any one temporal phase, only a few field house loci would have been in use, although such loci increased during the Colonial period, with the greatest numbers attributed to the Santa Cruz phase.

The exception to this pattern occurred during the Polvorón phase of the post-Classic period, when population levels throughout the entire Phoenix Basin decreased drastically and the Hohokam no longer established field houses and farmsteads associated with larger sites. Instead, the inhabitants located small habitation sites near the farming areas and probably near natural resource procurement areas, as they did with Area 8 of Dutch Canal Ruin. The settlement in this area represented an actual habitation site rather than an extension of a larger entity. Evidence for all of the activities usually associated with habitation sites appeared at the Polvorón phase occupation locus in Area 8 in the artifacts and the remains of domesticated crops and wild plant foods.

Investigators could not determine whether the Hohokam were still using any of the project area canals during the Polvorón phase. However, the occupants of the area may have maintained and used at least some portion of the canal system, perhaps on a more limited basis than during the pre-Classic and Classic periods. A variety of methods existed for the successful production of corn and other cultigens, and residents could have used any combination of these methods in lieu of canal irrigation.

The pre-Classic period features at Dutch Canal Ruin represented the utilization of potentially rich agricultural lands on the floodplain and were temporary use structures and limited activity loci related to farming and to procurement of natural resources. The Hohokam located their permanent habitation sites at higher elevations, in locations less susceptible to flooding. However, they could place field houses, requiring low labor investment for construction, near the canals and fields on the richer floodplain soils. The danger of losing crops and temporary structures to flooding was offset by the benefits of more fertile soil and greater ease in canal construction.

Investigators posited three main functions, alone or in combination, for the field houses and farmsteads at Dutch Canal Ruin: (1) agriculture; (2) resource procurement; and (3) canal construction and maintenance. Analysts found evidence for agricultural practices in the form of corn pollen at Areas 2, 3, and 5 (Chapter 16) and in carbonized corn remains at Areas 3 and 5 (Chapter 15). The location of the Dutch Canal Ruin habitation features along the prehistoric canals was also a strong argument for their function in agricultural pursuits.

The remains of numerous native plants recovered from Dutch Canal Ruin, such as mesquite, Cheno-ams, agave, and cholla (Chapters 15 and 16), provided evidence for resource procurement. The frequency of occurrence of these economic plants was lower, overall, than that of corn and agriculturally related plants (Chapters 15 and 16; Cummings 1988; Ruppé 1988). Locating field houses on the floodplain of the river would have allowed not only the use of rich alluvial soils but also the exploitation of wild plants growing within the narrow riverine environment and those common to the first terrace. This location would also have been one of the most productive for procurement of game, as the riverine zone would have served as habitat to a variety of species. The faunal assemblage showed remarkable consistency during the pre-Classic period occupation. Small mammals, dominated by jackrabbits, cottontail rabbits, and pocket gophers, constituted the assemblage. Analysts did not record evidence for the presence of large mammals before the Classic period occupation, when settlement occurred year-round.

Several areas at Dutch Canal Ruin showed evidence of canal maintenance activities. In Area 7, for example, a feature designed to reduce bank erosion had been constructed along the outer curve of the canal bank. Referenced as a baffle, this feature had effectively prevented lateral channel migration. Evidence of channel cleaning episodes also appeared. Siltation of channels would have been a common problem, but the Hohokam could have opened weirs and headgates, allowing water to flow at increased velocity and scour the channels. This process was probably similar to headward erosion, but residents may have facilitated it by loosening sediments manually, perhaps with digging sticks, and allowing the head of water to carry its bed load to the end of the canal. Researchers studied other forms of maintenance during the investigations of the Squaw Peak Parkway corridor when they found erosional features and canal breaches (D. H. Greenwald 1988a:65–70). Massive fill episodes coupled with an assortment of rocks and artifacts controlled bank breaches. The Hohokam could have managed canal breaches by opening downcanal gates and dams to reduce the volume of water in the canal at the location of the breach and making necessary repairs once the water level had dropped.

Canal maintenance may have taken place at any time of the year, but such activities would most logically have been associated with periods of noncultivation. Depending on the magnitude of the repairs, the Hohokam probably undertook canal maintenance on a yearly basis after the threat of winter, or high-season, flow regimes was past, in a cooperative effort involving members of all settlements who used the canals.

DUTCH CANAL RUIN AND ITS ASSOCIATION WITH CANAL SYSTEM 2

The association of Dutch Canal Ruin with Turney's Canal System 2 has been described only briefly in these volumes and elsewhere. Howard (Volume 1:Chapter 5) discusses the locational, environmental, and social relationships of Dutch Canal Ruin to Canal System 2. Although this section serves to some extent as a reiteration of Howard's earlier efforts, it also brings to light new interpretations that investigators advanced as a result of the data recovery phase in the Phoenix Sky Harbor Center.

The pre-Classic period canals in the Phoenix Sky Harbor Center represent initial stages of the growth and development of Canal System 2 as it is currently recognized. Earlier in this chapter, the evolution of canal irrigation in the arid southwest is discussed. The canals at Dutch Canal Ruin would constitute second-stage development, based on the earlier discussion, if such stages were strictly defined. First-stage development would involve short canal segments extending from the river channel to the adjacent floodplain. These early developmental canals would have had short use lives, being subjected to frequent episodes of infilling or scouring due to seasonal high flow regimes of the river. First-stage canal development would have required minimal sociopolitical organization, as construction and maintenance of these small canals and field systems could have been undertaken at the settlement level. Second-stage development, as conceived here, would include the construction of longer segments designed to provide water to larger tracts of arable lands somewhat removed from the threat of annual flooding. In the case of the Dutch Canal Ruin canals, this stage incorporated the upper reaches of the first terrace, eventually extending onto the second terrace. Placement of the late Pioneer and early Colonial canals in the project area may have been dictated by two physical features. The first defined the southern extent of the agricultural zone, based on the presence of a river channel that closely corresponds to the boundary between Sections 10 and 15 in T1N, R3E. The second defined the northern extent, based on topographic relief. Later canals indicated that the northern extent as defined by the associated irrigation zone of these canals was purely artificial, but expansion onto the second terrace did not occur until the Colonial period.

As Howard (1991:1.2) pointed out, Canal System 2 was a dynamic irrigation and settlement system. As a dynamic system, the pre-Classic period canals found in the Phoenix Sky Harbor Center represent an early part of the spectrum in the concept of the irrigation community. With the increased number of available dates and refinement of dating techniques (Chapter 18; see also Eighmy and Howard 1991; Waters, Howard, and Greenwald 1988), archaeologists can state that Hohokam irrigation communities developed as early as A.D. 600 in the Phoenix Basin. The prehistoric inhabitants used the area associated with the canals that traversed the project area in a very specific manner for agricultural production. This pattern of use required the development of a small canal system that included the establishment of fields along its course. Occupation of the agricultural zone was of a temporary or seasonal nature, with permanent habitations on the adjacent bajada slopes. The botanical record indicated that the Hohokam focused their subsistence strategies on agriculture during the pre-Classic periods at Dutch Canal Ruin, although they also exploited economically important wild plant resources and small game. Unlike early field house sites in other locations, Dutch Canal Ruin did not exhibit significant diversity within the resource base (Cable and Doyel 1985).

This land-use practice appears to have lasted until the mid Colonial period, after agriculturists established canals on the second terrace within Canal System 2. With the shift or expansion to the second terrace, the importance of the Dutch Canal Ruin agricultural zone diminished. The Hohokam abandoned their agricultural pursuits on the first terrace by A.D. 950, prior to the beginning of the Sedentary period (Greenwald and Ciolek-Torrello 1988a; Howard 1990, 1991). Although archaeologists have not determined the cause for the abandonment of the first terrace, environmental factors could have been a major contributing factor. If, as Gregory (1991:184) has suggested, a catastrophic flood of the magnitude projected through the streamflow reconstruction for the Salt River (Nials, Gregory, and Graybill 1989) occurred in A.D. 899, Hohokam farmers may have decided that the risks associated with farming the first terrace were greater than the benefits. The establishment of canals and fields in higher elevations on the second terrace also may have contributed to the decision to abandon the first terrace, as the severity of flood damage to canals would have been less on the second terrace than on the first terrace.

After abandoning the Dutch Canal Ruin agricultural zone, the Hohokam would have considered the first terrace to be of secondary importance, although the first terrace and its associated riverine zone provided a large diversity of resources, both plant and animal. Investigators recovered limited evidence of use during the Sedentary and early Classic periods during the current investigations; Area 9 and the extreme western portion of the project area appeared to have been used to some extent. Resettlement occurred during the latter part of the Soho phase at Pueblo Salado in the southern portion of the project area, and some activities identified at Dutch Canal Ruin may have been related to this new settlement.

The Hohokam reoccupied Dutch Canal Ruin during the Civano phase and constructed a canal through the northeastern portion of the project area. The trajectory of this new canal indicated that it had carried water onto the second terrace south of La Ciudad to the area near Pueblo Patricio. Although analysts projected that this canal would head downstream of the concentrated heads for Canal System 2 following the alignment of Canal Patricio, the area in question (approximately 2 miles between Park of the Four Waters near 44th Street on the east and 24th Street on the west, and between Turney's South Canal and Canal Patricio) has not been investigated. Given the presence of canals beyond the boundaries of this area, others almost certainly exist within it. During the Civano phase, small settlements, permanently occupied farmsteads, and hamlets were scattered along the canals on the first terrace. Dutch Canal Ruin's association with Canal System 2 at this time was again as an agricultural zone. Investigators recovered charred corn remains, including kernels and cobs, from Areas 9 and 10, indicating that the Hohokam had once again practiced corn agriculture in the project area. Investigators also recovered corn remains from Area 8, the Polvorón phase settlement in the western portion of Dutch Canal Ruin. However, corn was less common than other economic resources in this area, suggesting that inhabitants had begun to shift to a broader resource base. As suggested by Nials, Gregory, and Graybill (1989), this period co-occurred with the deepening and widening of the Salt River channel from high-magnitude floods. The floods probably destroyed canal heads and resulted in failure of the canal system. The Polvorón phase occupation at Dutch Canal Ruin may, like others throughout Canal System 2, represent relict populations that tried to make a living by implementing a diverse exploitation strategy.

APPENDIXES

APPENDIX A

ARCHAEOMAGNETIC DATA RESULTS AND GRAPHICS

Dr. Jeffrey L. Eighmy
Department of Anthropology
Colorado State University

ARCHAEOMAGNETIC LABORATORY REPORT

Archaeometric Laboratory
Department of Anthropology
Colorado State University
Fort Collins, Colorado 80523

(303) 491-7408 or 491-5784

Sample I.D. <u>AZ T:12:62(ASM)-1</u>   Feature I.D. <u>Feat. 1-2</u>

Site Latitude <u>33.44° N</u>   Site Longitude <u>247.96° E</u>

Site Declination <u>12.43° E</u>   Archaeological Guess Date <u>AD 750-850</u>

Collector <u>K.B/L.B.</u>   Date Collected <u>1990</u>

Laboratory Analysis

Demagnetization Steps (mT)	NRM	5.0	10.0
Alpha 95 (degrees)	13.77	10.37	6.44
Precision Parameter - k	10.90	18.49	46.37
Inclination (degrees dip)	51.26	51.42	51.59
Declination (degrees E)	7.83	8.13	6.17
Mean Sample Intensity (E-07 Tesla)	1.678	1.424	1.165
No. Specimens Collected/ No. Specimens Used	12/12	12/12	12/12
Specimen No. of Outlier(s)	none	none	none

Demagnetization Steps (mT)	15.0	20.0	25.0
Alpha 95 (degrees)	4.10	2.84	2.26
Precision Parameter - k	113.14	235.22	368.69
Inclination (degrees dip)	51.54	51.23	50.81
Declination (degrees E)	5.23	4.21	3.98
Mean Sample Intensity (E-07 Tesla)	.9429	.7624	.5207
No. Specimens Collected/ No. Specimens Used	12/12	12/12	12/12
Specimen No. of Outlier(s)	none	none	none

Sample AZ T:12:62(ASM)-1 (continued)

Demagnetization Steps (mT)	35.0
Alpha 95 (degrees)	2.47
Precision Parameter - k	309.76
Inclination (degrees dip)	50.86
Declination (degrees E)	4.19
Mean Sample Intensity (E-07 Tesla)	.4381
No. Specimens Collected/ No. Specimens Used	12/12
Specimen No. of Outlier(s)	none

Final Processing Results

Demagnetization Level Used	35.0
Paleolatitude (degrees N)	86.00
Paleolongitude (degrees E)	4.76
Error Along the Great Circle - EP (degrees)	2.25
Error Perpendicular to the Great Circle - EM (degrees)	3.33

Signed _____   Date _____

CURRENT SOUTHWEST VGP CURVE PLOTTED WITH SAMPLE AZT1262-1

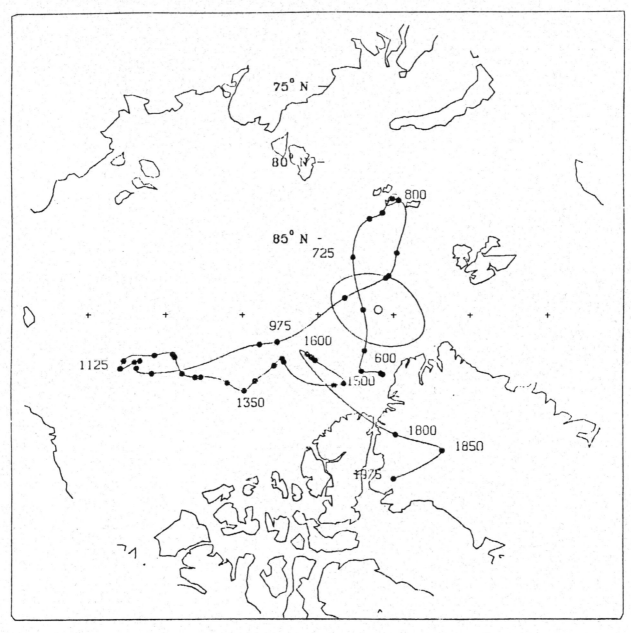

AD 675-725;
850-940

curve used: SWCV590

ARCHAEOMAGNETIC LABORATORY REPORT

Archaeometric Laboratory
Department of Anthropology
Colorado State University
Fort Collins, Colorado 80523

(303) 491-7408 or 491-5784

Sample I.D. __AZ T:12:62(ASM)-2__   Feature I.D. ___Feat. 3-11___

Site Latitude __33.44° N__   Site Longitude __247.96° E__

Site Declination __12.43° E__   Archaeological Guess Date__AD 700-900__

Collector __D. Block/T. Motsinger__   Date Collected ____1990____

Laboratory Analysis

Demagnetization Steps (mT)	NRM	5.0	10.0
Alpha 95 (degrees)	5.67	4.84	4.03
Precision Parameter - k	59.55	81.31	117.14
Inclination (degrees dip)	47.81	48.42	49.18
Declination (degrees E)	7.57	7.36	5.96
Mean Sample Intensity (E-07 Tesla)	1.351	1.185	0.948
No. Specimens Collected/ No. Specimens Used	12/12	12/12	12/12
Specimen No. of Outlier(s)	none	none	none

Demagnetization Steps (mT)	15.0	20.0	25.0
Alpha 95 (degrees)	3.41	3.13	3.01
Precision Parameter - k	163.02	192.72	209.16
Inclination (degrees dip)	49.63	50.17	50.51
Declination (degrees E)	5.36	4.91	4.89
Mean Sample Intensity (E-07 Tesla)	0.748	0.587	0.466
No. Specimens Collected/ No. Specimens Used	12/12	12/12	12/12
Specimen No. of Outlier(s)	none	none	none

Sample AZ T:12:62(ASM)-2 (continued)

Final Processing Results

Demagnetization Level Used   25.0

Paleolatitude (degrees N)   85.32

Paleolongitude (degrees E)   4.59

Error Along the Great Circle - EP (degrees)   2.72

Error Perpendicular to the Great Circle - EM   4.04
(degrees)

Signed _____ Date _____

CURRENT SOUTHWEST VGP CURVE PLOTTED WITH SAMPLE AZT1262–2

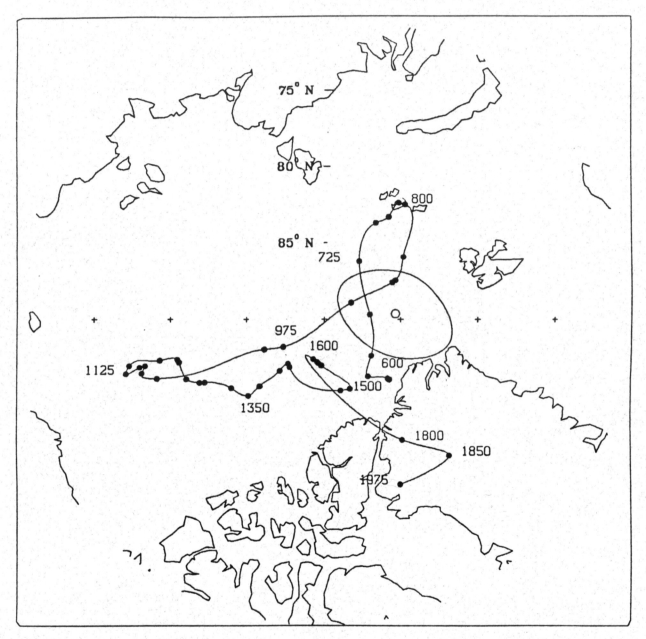

AD 675–725;
850–940

curve used: SWCV590

ARCHAEOMAGNETIC LABORATORY REPORT

Archaeometric Laboratory
Department of Anthropology
Colorado State University
Fort Collins, Colorado 80523

(303) 491-7408 or 491-5784

Sample I.D. __AZ T:12:62(ASM)-3__   Feature I.D. __Feat. 3-10__

Site Latitude __33.44__ N °   Site Longitude __247.96__ E °

Site Declination __12.43__ E °   Archaeological Guess Date __AD 700-900__

Collector __D. Block/T. Motsinger__   Date Collected _____1990_____

Laboratory Analysis

Demagnetization Steps (mT)	NRM	5.0	10.0
Alpha 95 (degrees)	3.49	2.27	2.29
Precision Parameter - k	192.94	453.12	446.79
Inclination (degrees dip)	48.68	48.13	48.34
Declination (degrees E)	5.03	3.91	3.39
Mean Sample Intensity (E-07 Tesla)	1.073	0.924	0.728
No. Specimens Collected/ No. Specimens Used	10/10	10/10	10/10
Specimen No. of Outlier(s)	none	none	none

Final Processing Results

Demagnetization Level Used	10.0
Paleolatitude (degrees N)	84.98
Paleolongitude (degrees E)	31.92
Error Along the Great Circle - EP (degrees)	1.97
Error Perpendicular to the Great Circle - EM (degrees)	3.00

Signed _____   Date _____

CURRENT SOUTHWEST VGP CURVE
PLOTTED WITH SAMPLE AZT1262—3

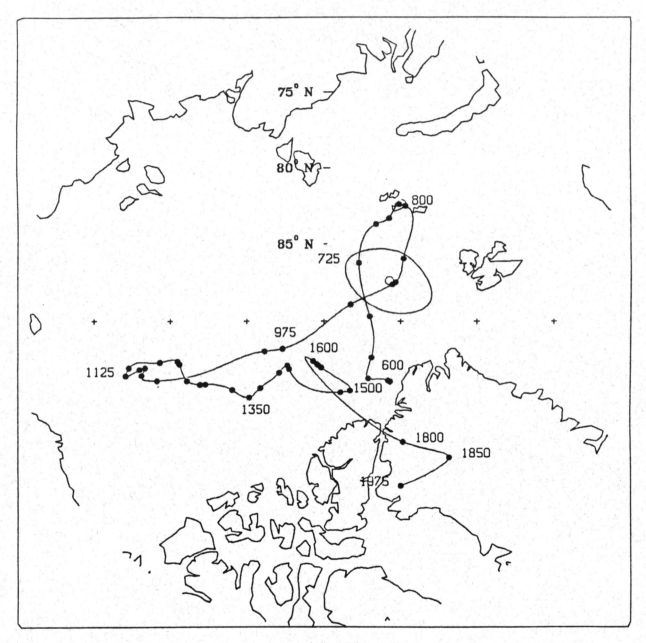

AD 700-750;
830-925

curve used: SWCV590

ARCHAEOMAGNETIC LABORATORY REPORT

Archaeometric Laboratory
Department of Anthropology
Colorado State University
Fort Collins, Colorado 80523

(303) 491-7408 or 491-5784

Sample I.D. __AZ T:12:62(ASM)-4__   Feature I.D. ___Feat. 3-2-1___

Site Latitude __33.44__ N   Site Longitude __247.96__ E

Site Declination __12.43__ E   Archaeological Guess Date __AD 700-850__

Collector ___TM/LB/JH___   Date Collected _____1990_____

Laboratory Analysis

Demagnetization Steps (mT)	NRM	5.0	15.0
Alpha 95 (degrees)	5.41	4.41	2.93
Precision Parameter – k	65.35	97.98	220.60
Inclination (degrees dip)	55.84	55.40	55.71
Declination (degrees E)	358.90	357.69	355.48
Mean Sample Intensity (E-07 Tesla)	3.731	3.023	1.580
No. Specimens Collected/ No. Specimens Used	12/12	12/12	12/12
Specimen No. of Outlier(s)	none	none	none

Demagnetization Steps (mT)	20.0	25.0
Alpha 95 (degrees)	2.47	2.45
Precision Parameter – k	308.93	313.96
Inclination (degrees dip)	55.96	55.81
Declination (degrees E)	354.85	354.74
Mean Sample Intensity (E-07 Tesla)	1.150	0.896
No. Specimens Collected/ No. Specimens Used	12/12	12/12
Specimen No. of Outlier(s)	none	none

Sample AZ T:12:62(ASM)-4 (continued)

Final Processing Results

Demagnetization Level Used	25.0
Paleolatitude (degrees N)	84.80
Paleolongitude (degrees E)	193.52
Error Along the Great Circle - EP (degrees)	2.52
Error Perpendicular to the Great Circle - EM (degrees)	3.52

Signed _____ Date _____

CURRENT SOUTHWEST VGP CURVE
PLOTTED WITH SAMPLE AZT1262-4

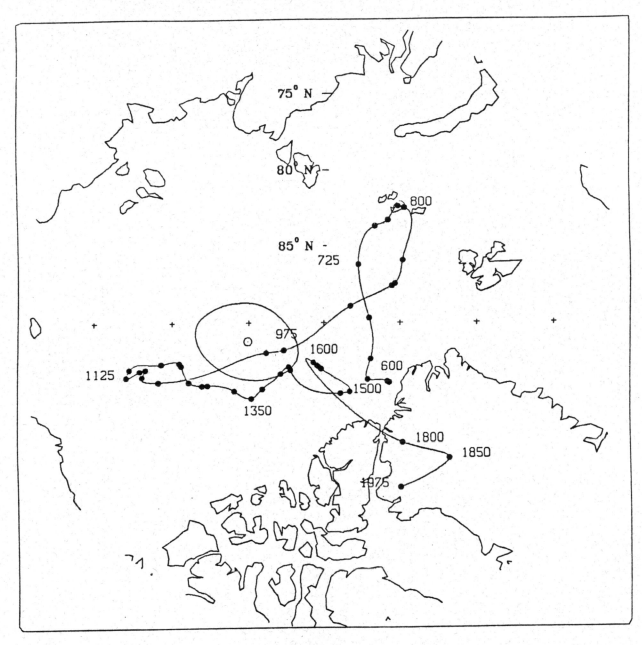

AD 940-1015;
1200-1450

curve used: SWCV590

ARCHAEOMAGNETIC LABORATORY REPORT

Archaeometric Laboratory
Department of Anthropology
Colorado State University
Fort Collins, Colorado 80523

(303) 491-7408 or 491-5784

Sample I.D. AZ T:12:62(ASM)-5   Feature I.D. ___Feat. 3-1-1___

Site Latitude 33.44 N   Site Longitude 247.96 E

Site Declination 12.43 E   Archaeological Guess Date AD 700-850

Collector ___TM/LB___   Date Collected ___1990___

Laboratory Analysis

Demagnetization Steps (mT)	NRM	5.0	10.0
Alpha 95 (degrees)	2.01	2.25	2.11
Precision Parameter - k	467.24	373.93	422.33
Inclination (degrees dip)	54.14	53.65	52.94
Declination (degrees E)	3.09	2.54	1.83
Mean Sample Intensity (E-07 Tesla)	1.141	0.984	0.779
No. Specimens Collected/ No. Specimens Used	12/12	12/12	12/12
Specimen No. of Outlier(s)	none	none	none

Demagnetization Steps (mT)	15.0	20.0
Alpha 95 (degrees)	2.03	2.48
Precision Parameter - k	457.39	306.12
Inclination (degrees dip)	52.53	52.55
Declination (degrees E)	1.07	1.01
Mean Sample Intensity (E-07 Tesla)	0.609	0.476
No. Specimens Collected/ No. Specimens Used	12/12	12/12
Specimen No. of Outlier(s)	none	none

Sample AZ T:12:62(ASM)-5 (continued)

Final Processing Results

Demagnetization Level Used   20.0

Paleolatitude (degrees N)   89.10

Paleolongitude (degrees E)   357.74

Error Along the Great Circle - EP (degrees)   2.36

Error Perpendicular to the Great Circle - EM   3.42
(degrees)

Signed _____   Date _____

CURRENT SOUTHWEST VGP CURVE PLOTTED WITH SAMPLE AZT1262-5

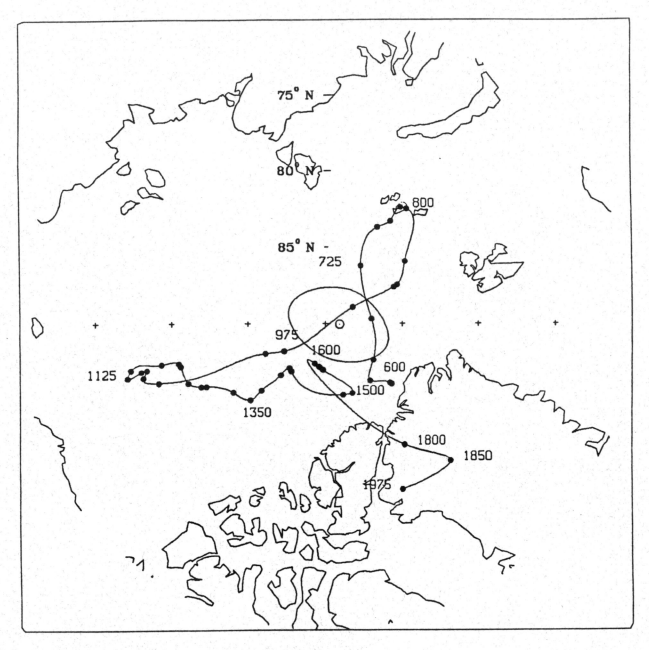

AD 650-725;
900-975

curve used: SWCV590

ARCHAEOMAGNETIC LABORATORY REPORT

Archaeometric Laboratory
Department of Anthropology
Colorado State University
Fort Collins, Colorado 80523

(303) 491-7408 or 491-5784

Sample I.D. __AZ T:12:62(ASM)-6__   Feature I.D. __Feat. 2041__

Site Latitude __33.44° N__   Site Longitude __247.97° E__

Site Declination __12.43° E__   Archaeological Guess Date _____

Collector __K. Schroeder__   Date Collected __4/7/89__

Laboratory Analysis

Demagnetization Steps (mT)	NRM	5.0	10.0	15.0
Alpha 95 (degrees)	1.43	1.62	1.59	1.67
Precision Parameter - k	916.02	718.34	748.69	679.66
Inclination (degrees dip)	53.94	53.52	52.94	52.64
Declination (degrees E)	359.96	359.86	359.50	359.24
Mean Sample Intensity (E-07 Tesla)	5.201	4.442	3.250	2.285
No. Specimens Collected/ No. Specimens Used	12/12	12/12	12/12	12/12
Specimen No. of Outlier(s)	none	none	none	none

Final Processing Results

Demagnetization Level Used	15.0
Paleolatitude (degrees N)	89.33
Paleolongitude (degrees E)	139.17
Error Along the Great Circle - EP (degrees)	1.58
Error Perpendicular to the Great Circle - EM (degrees)	2.30

Signed _____   Date __4/15/91__

CURRENT SOUTHWEST VGP CURVE
PLOTTED WITH SAMPLE AZT1262-6

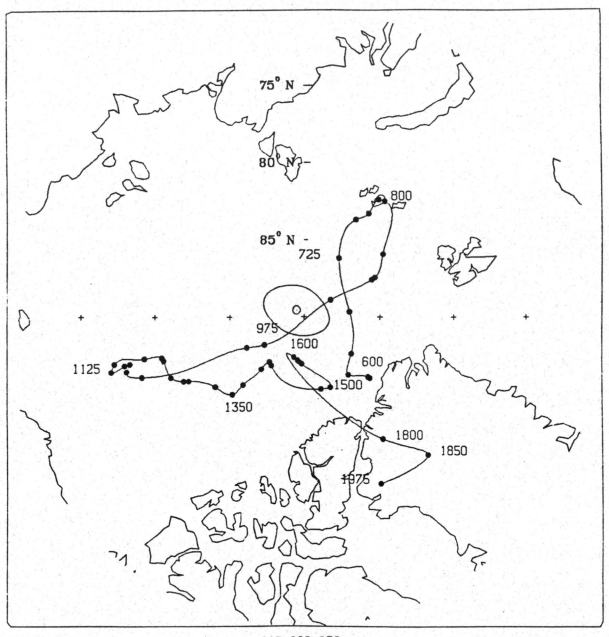

AD 925-975;
1550-1650

curve used: SWCV590

ARCHAEOMAGNETIC LABORATORY REPORT

Archaeometric Laboratory
Department of Anthropology
Colorado State University
Fort Collins, Colorado 80523

(303) 491-7408 or 491-5784

Sample I.D. __CSUSED-76__   Feature I.D. __Feat. 7-3__

Site Latitude __33.44 °N__   Site Longitude __247.97 °E__

Site Declination __12.43 °E__   Archaeological Guess Date __AD 1250-1400__

Collector ___T. Motsinger___   Date Collected __1990__

Laboratory Analysis

Demagnetization Steps (mT)	NRM	10.0	15.0	25.0
Alpha 95 (degrees)	2.93	3.06	3.10	3.65
Precision Parameter – k	219.98	202.03	196.90	142.02
Inclination (degrees dip)	37.95	34.45	33.34	33.38
Declination (degrees E)	358.94	358.36	357.82	357.71
Mean Sample Intensity (E-07 Tesla)	1.566	1.214	.9693	.6552
No. Specimens Collected/ No. Specimens Used	12/12	12/12	12/12	12/12
Specimen No. of Outlier(s)	none	none	none	none

Final Processing Results

Demagnetization Level Used	25.0
Paleolatitude (degrees N)	74.66
Paleolongitude (degrees E)	76.23
Error Along the Great Circle - EP (degrees)	2.36
Error Perpendicular to the Great Circle - EM (degrees)	4.16

Signed _____   Date _____

CURRENT SOUTHWEST VGP CURVE PLOTTED WITH SAMPLE CSUSED—76

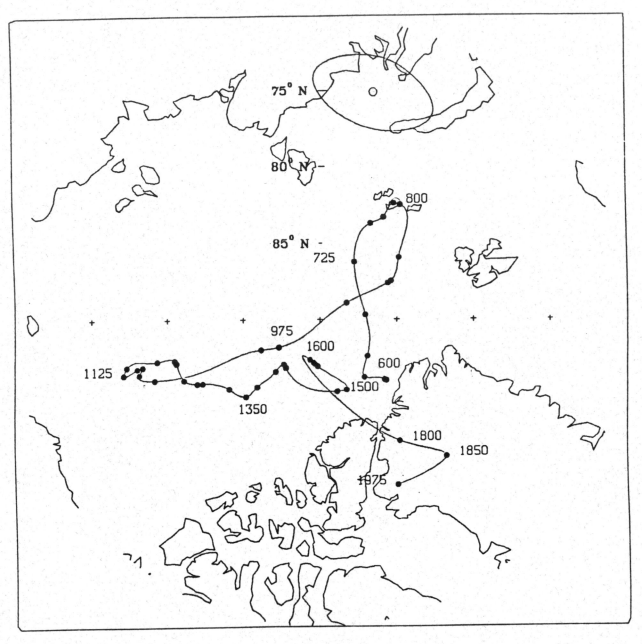

off curve

curve used: SWCV590

U.S. Southwest
Declination Curve: SWDC590

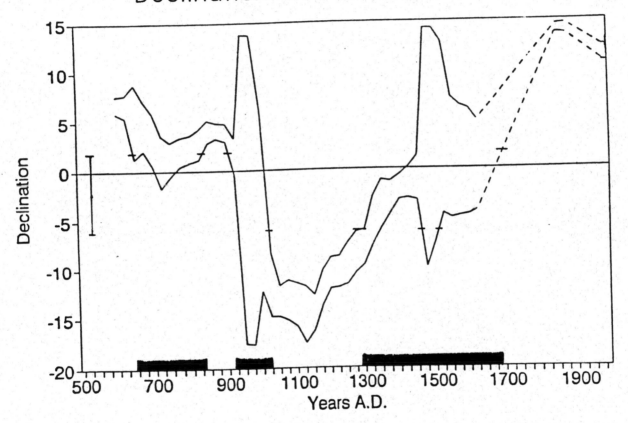

CSUSED-76
AD 650-850
925-1025
1275-1700

ARCHAEOMAGNETIC LABORATORY REPORT

Archaeometric Laboratory
Department of Anthropology
Colorado State University
Fort Collins, Colorado 80523

(303) 491-7408 or 491-5784

Sample I.D. ____CSUSED-77____   Feature I.D. __Feat. 8001-c__

Site Latitude _33.44° N_   Site Longitude _247.96° E_

Site Declination_12.43° E_   Archaeological Guess Date__AD 850-950__

Collector __D. Greenwald__   Date Collected ____1990____

Laboratory Analysis

Demagnetization Steps (mT)	NRM	10.0	15.0	20.0
Alpha 95 (degrees)	26.01	16.68	10.33	6.74
Precision Parameter - k	4.41	9.35	22.84	52.32
Inclination (degrees dip)	62.27	56.69	55.67	55.33
Declination (degrees E)	11.98	8.66	3.50	359.90
Mean Sample Intensity (E-07 Tesla)	1.752	1.106	0.863	0.689
No. Specimens Collected/ No. Specimens Used	10/10	10/10	10/10	10/10
Specimen No. of Outlier(s)	none	none	none	none

Demagnetization Steps (mT)	25.0	30.0	35.0	40.0
Alpha 95 (degrees)	5.07	4.13	3.66	3.54
Precision Parameter - k	91.82	137.76	175.55	187.66
Inclination (degrees dip)	54.70	54.68	54.32	54.22
Declination (degrees E)	357.04	356.10	355.53	355.46
Mean Sample Intensity (E-07 Tesla)	0.549	0.454	0.372	0.310
No. Specimens Collected/ No. Specimens Used	10/10	10/10	10/10	10/10
Specimen No. of Outlier(s)	none	none	none	none

Sample CSUSED-77 (continued)

Final Processing Results

Demagnetization Level Used	40.0
Paleolatitude (degrees N)	86.02
Paleolongitude (degrees E)	178.39
Error Along the Great Circle - EP (degrees)	3.49
Error Perpendicular to the Great Circle - EM (degrees)	4.97

Signed _____   Date _____

CURRENT SOUTHWEST VGP CURVE
PLOTTED WITH SAMPLE CSUSED-77

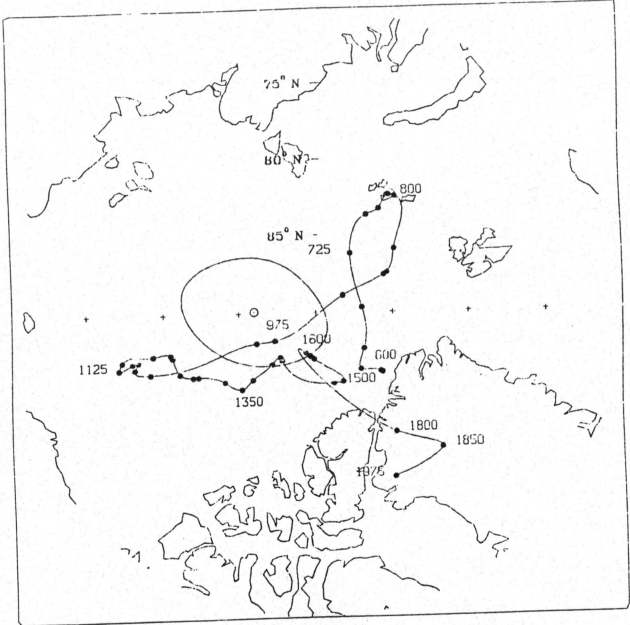

AD 900–1000;
1200–1475;
1500–1700

curve used: SWCV590

U.S. Southwest
Declination Curve: SWDC590

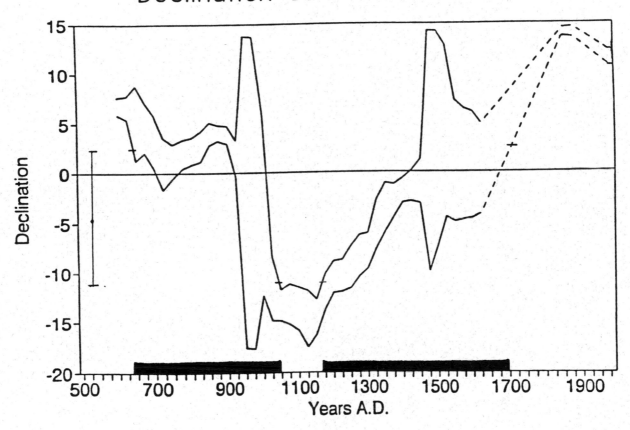

CSUSED-77

AD 625-1050
1175-1700

ARCHAEOMAGNETIC LABORATORY REPORT

Archaeometric Laboratory
Department of Anthropology
Colorado State University
Fort Collins, Colorado 80523

(303) 491-7408 or 491-5784

Sample I.D. __CSUSED-78__   Feature I.D. __Feat. 8001b__

Site Latitude __33.44__ °N   Site Longitude __247.96__ °E

Site Declination __12.43__ °E   Archaeological Guess Date __AD 700-800__

Collector __D. Greenwald__   Date Collected __1990__

Laboratory Analysis

Demagnetization Steps (mT)	NRM	5.0	10.0
Alpha 95 (degrees)	4.86	5.05	5.26
Precision Parameter - k	130.91	121.48	112.06
Inclination (degrees dip)	58.64	57.76	57.36
Declination (degrees E)	343.01	341.95	342.06
Mean Sample Intensity (E-07 Tesla)	1.318	1.142	.9296
No. Specimens Collected/ No. Specimens Used	9/8	9/8	9/8
Specimen No. of Outlier(s)	# 6	# 6	# 6

Final Processing Results

Demagnetization Level Used	10.0
Paleolatitude (degrees N)	74.77
Paleolongitude (degrees E)	180.39
Error Along the Great Circle - EP (degrees)	5.61
Error Perpendicular to the Great Circle - EM (degrees)	7.68

Signed _____   Date _____

CURRENT SOUTHWEST VGP CURVE
PLOTTED WITH SAMPLE CSUSED–78

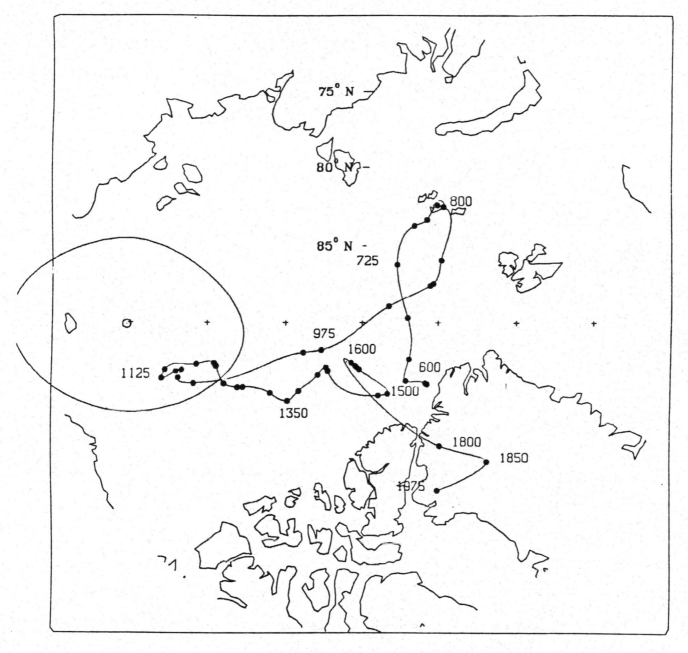

AD 1015–1300

curve used: SWCV590

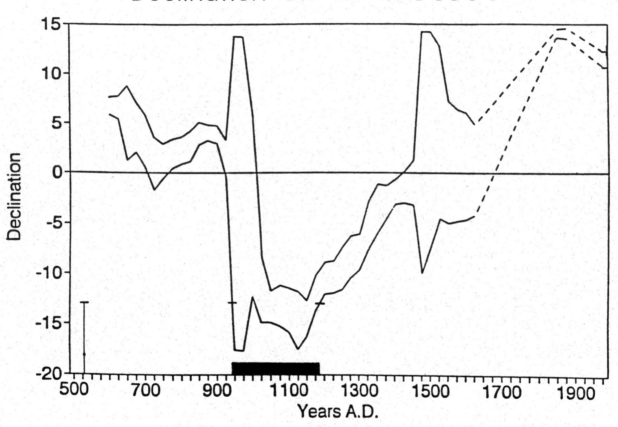

U.S. Southwest
Declination Curve: SWDC590

CSUSED-78

AD 950-1200

ARCHAEOMAGNETIC LABORATORY REPORT

Archaeometric Laboratory
Department of Anthropology
Colorado State University
Fort Collins, Colorado 80523

(303) 491-7408 or 491-5784

Sample I.D. _CSUSED-79_   Feature I.D. ___Feat. 8001a___

Site Latitude _33.44° N_   Site Longitude _247.96° E_

Site Declination _12.43° E_   Archaeological Guess Date _AD 700-800_

Collector _D. Greenwald_   Date Collected ___1990___

Laboratory Analysis

Demagnetization Steps (mT)	NRM	5.0	15.0
Alpha 95 (degrees)	7.87	5.86	4.57
Precision Parameter - k	50.54	90.24	148.01
Inclination (degrees dip)	57.73	56.79	56.95
Declination (degrees E)	2.22	4.41	5.24
Mean Sample Intensity (E-07 Tesla)	.6823	.6206	.5254
No. Specimens Collected/ No. Specimens Used	9/8	9/8	9/8
Specimen No. of Outlier(s)	# 7	# 7	# 7

Demagnetization Steps (mT)	15.0	20.0
Alpha 95 (degrees)	4.52	4.49
Precision Parameter - k	150.82	153.35
Inclination (degrees dip)	57.10	56.92
Declination (degrees E)	6.66	6.09
Mean Sample Intensity (E-07 Tesla)	.4223	.3369
No. Specimens Collected/ No. Specimens Used	9/8	9/8
Specimen No. of Outlier(s)	# 7	# 7

Sample CSUSED-79 (continued)

Final Processing Results

Demagnetization Level Used	20.0
Paleolatitude (degrees N)	83.59
Paleolongitude (degrees E)	296.89
Error Along the Great Circle - EP (degrees)	4.74
Error Perpendicular to the Great Circle - EM (degrees)	6.52

Signed _____ Date _____

CURRENT SOUTHWEST VGP CURVE
PLOTTED WITH SAMPLE CSUSED–79

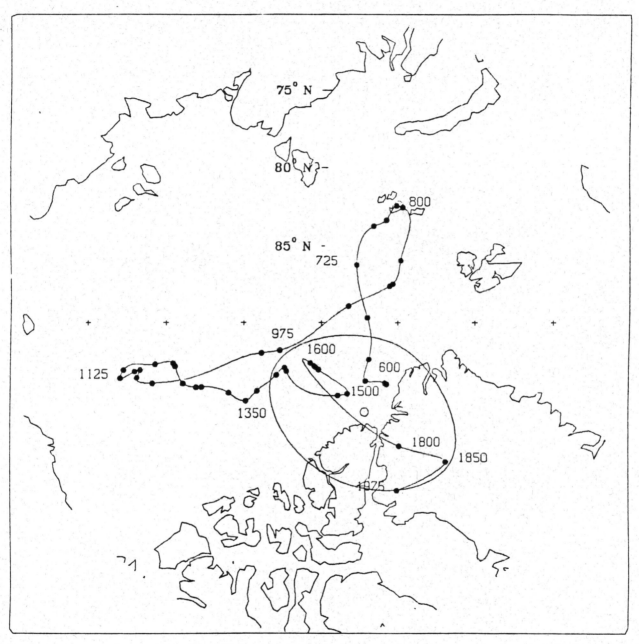

curve used: SWCV590

pre 600-700;
850-1015
1375-modern

U.S. Southwest
Declination Curve: SWDC590

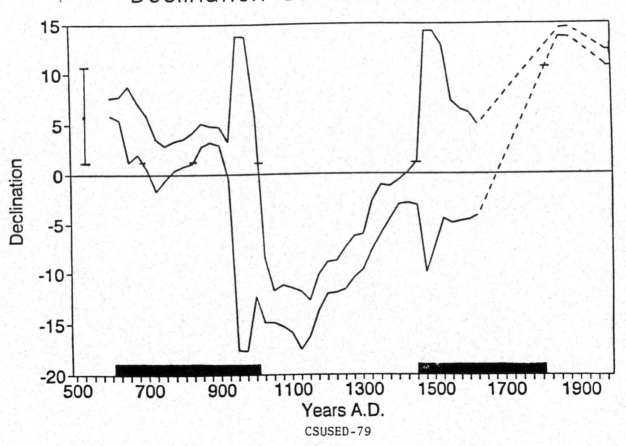

Years A.D.

CSUSED-79

AD 600-1025
1450-1800

ARCHAEOMAGNETIC LABORATORY REPORT

Archaeometric Laboratory
Department of Anthropology
Colorado State University
Fort Collins, Colorado 80523

(303) 491-7408 or 491-5784

Sample I.D. ___CSUSED-80___   Feature I.D. __Feat. 2-1__

Site Latitude _33.44° N_   Site Longitude _247.96° E_

Site Declination _12.43° E_   Archaeological Guess Date__AD 600-700__

Collector __TM/KB__   Date Collected ___1990___

Laboratory Analysis

Demagnetization Steps (mT)	NRM	5.0	10.0	15.0
Alpha 95 (degrees)	18.14	15.62	11.89	8.88
Precision Parameter - k	6.69	8.68	14.28	24.86
Inclination (degrees dip)	57.91	56.54	55.38	54.00
Declination (degrees E)	342.52	343.28	344.86	347.32
Mean Sample Intensity (E-07 Tesla)	0.900	0.735	0.550	0.414
No. Specimens Collected/ No. Specimens Used	12/12	12/12	12/12	12/12
Specimen No. of Outlier(s)	none	none	none	none

Demagnetization Steps (mT)	20.0	25.0	30.0	35.0
Alpha 95 (degrees)	7.63	5.86	5.31	6.56
Precision Parameter - k	33.29	55.86	67.74	44.73
Inclination (degrees dip)	52.91	51.29	49.81	49.82
Declination (degrees E)	348.49	350.53	351.65	350.28
Mean Sample Intensity (E-07 Tesla)	0.316	0.252	0.207	0.161
No. Specimens Collected/ No. Specimens Used	12/12	12/12	12/12	12/12
Specimen No. of Outlier(s)	none	none	none	none

Sample CSUSED-80 (continued)

Final Processing Results

Demagnetization Level Used	35.0
Paleolatitude (degrees N)	81.30
Paleolongitude (degrees E)	141.77
Error Along the Great Circle - EP (degrees)	5.84
Error Perpendicular to the Great Circle - EM (degrees)	8.75

Signed _____   Date _____

CURRENT SOUTHWEST VGP CURVE
PLOTTED WITH SAMPLE CSUSED–80

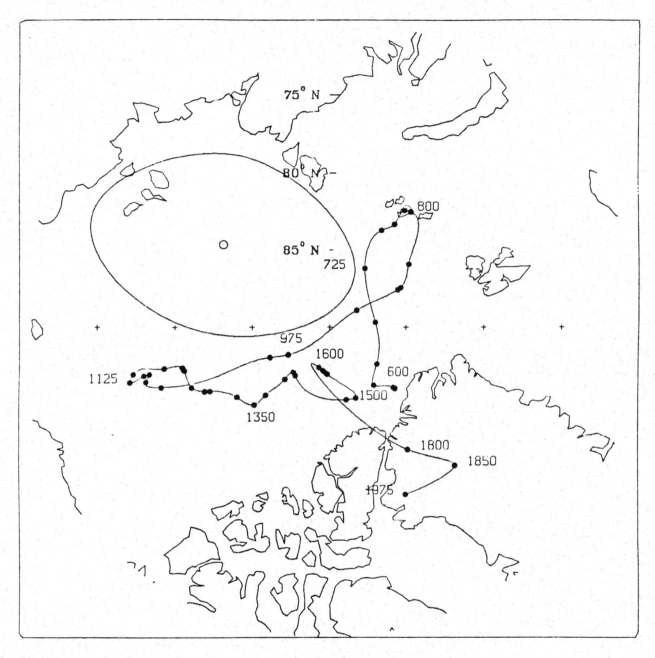

AD 700-800;
875-1250

curve used: SWCV590

U.S. Southwest
Declination Curve: SWDC590

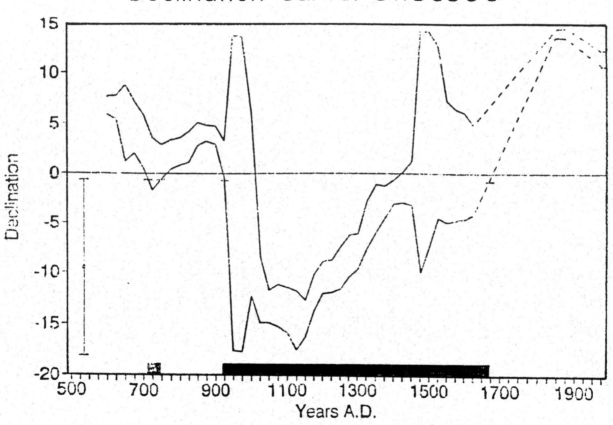

CSUSED-80

AD 925-1675

CURRENT SOUTHWEST VGP CURVE AND SAMPLE NA19324-5

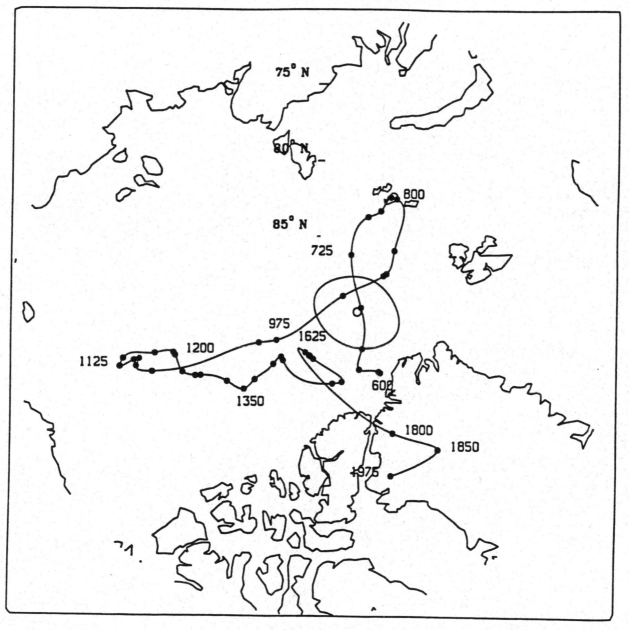

AD 625-720;
850-940;
1550-1650

curve used: SWCV590

CURRENT SOUTHWEST VGP CURVE AND SAMPLE NA19324−7

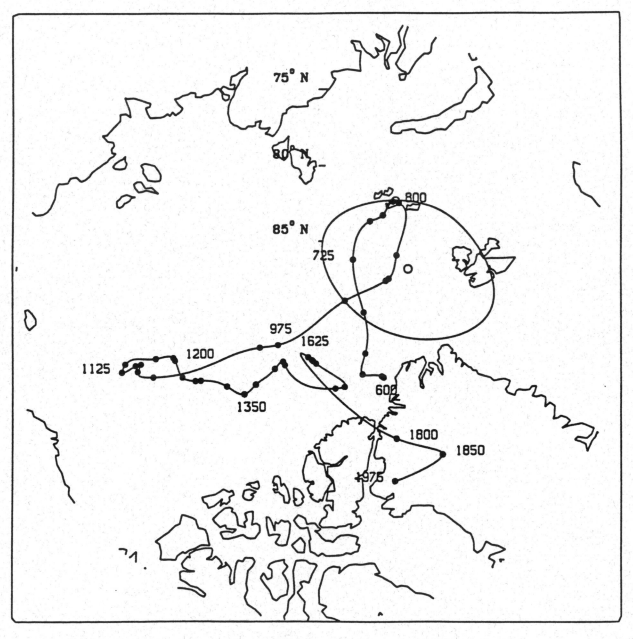

AD 700-950

curve used: SWCV590

PLOT I   11.43.38 MON 5 AUG, 1991   JOB=eighey , COLORADO STATE UNIVERSITY DISSPLA 11.0

CURRENT SOUTHWEST VGP CURVE AND SAMPLE CSUSED–52

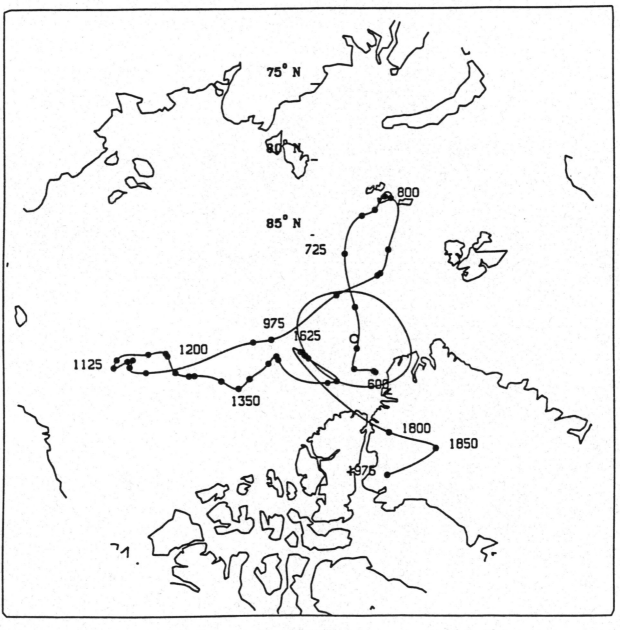

75° N

80° N

85° N

800

725

975

1625

1200

1125

1350

600

1800

1850

1975

pre AD 600-720;
900-975;
1500-1700

curve used: SWCV590

PLOT I   10.33.34   MON   5 AUG, 1991   JDB-eigbey , COLORADO STATE UNIVERSITY   DISSPLA 11.0

U.S. Southwest
Declination Curve: SWDC590

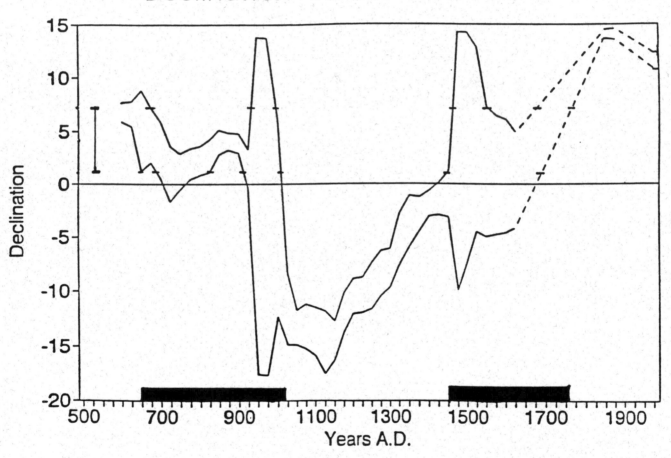

CSUSED-52
AD 650-1025;
1450-1750

CURRENT SOUTHWEST VGP CURVE AND SAMPLE CSUSED–53

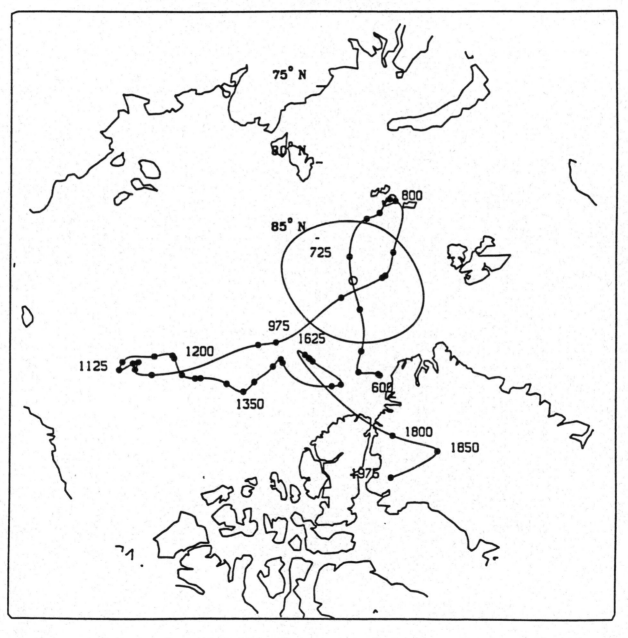

AD 675-750;
830-940

curve used: SWCV590

U.S. Southwest
Declination Curve: SWDC590

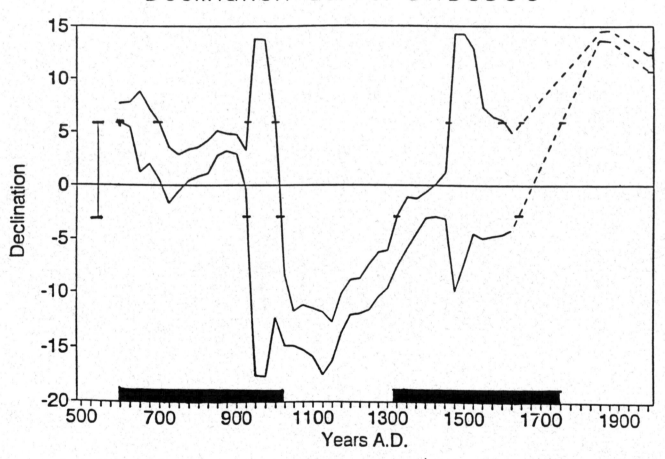

CSUSED-53
AD 600-1025;
1325-1750

CURRENT SOUTHWEST VGP CURVE AND SAMPLE CSUSED–54

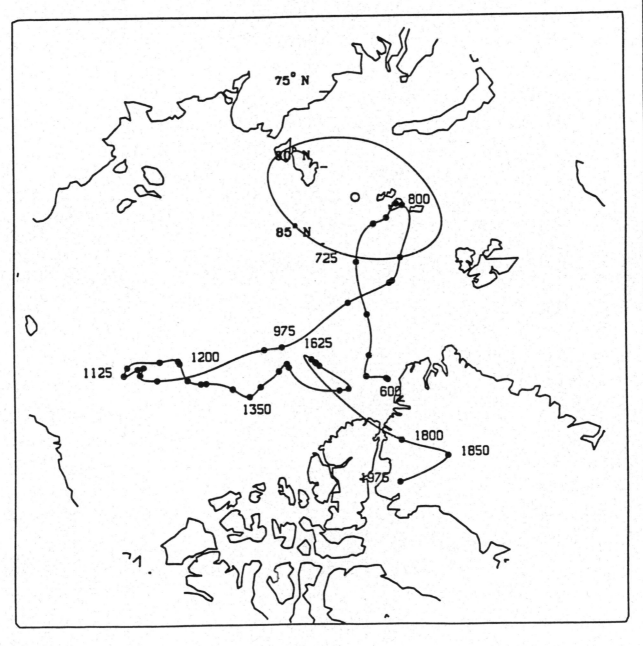

AD 725-850

curve used: SWCV590

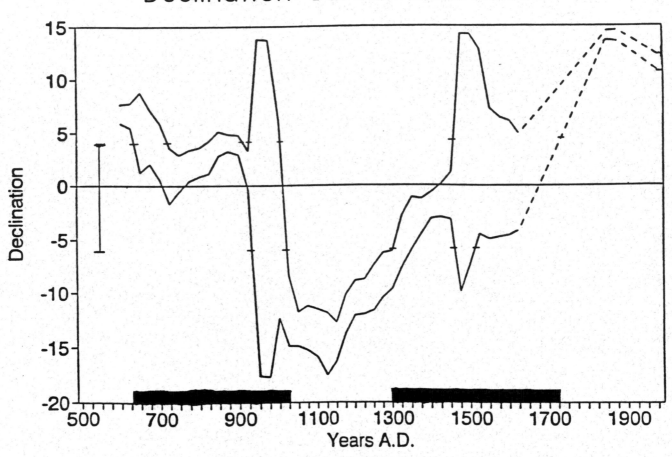

U.S. Southwest
Declination Curve: SWDC590

CSUSED-54
AD 625-1025;
1300-1725

CURRENT SOUTHWEST VGP CURVE AND SAMPLE CSUSED–55

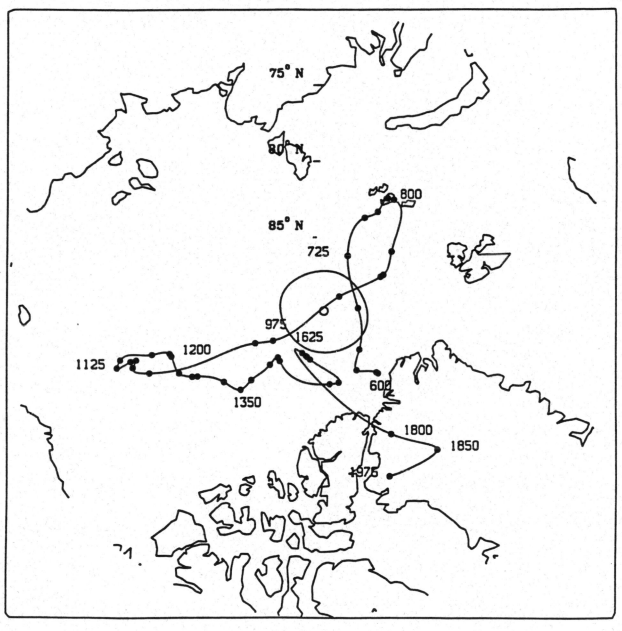

75° N

80° N

85° N

800

725

975

1625

1200

1125

1350

600

1800

1850

975

AD 650-725;
900-975;
1550-1650

curve used: SWCV590

PLOT 1   10.10.15 MON 5 AUG, 1991   JOB=sigbuy , COLORADO STATE UNIVERSITY DISSPLA 11.0

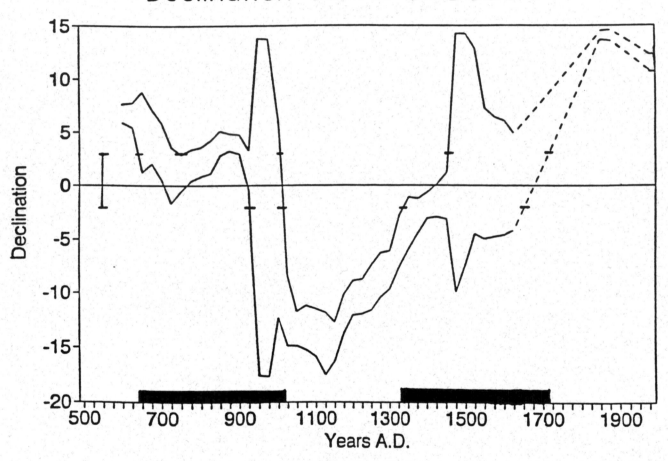

U.S. Southwest
Declination Curve: SWDC590

CSUSED-55
AD 650-1025;
1325-1725

CURRENT SOUTHWEST VGP CURVE AND SAMPLE CSUSED−56

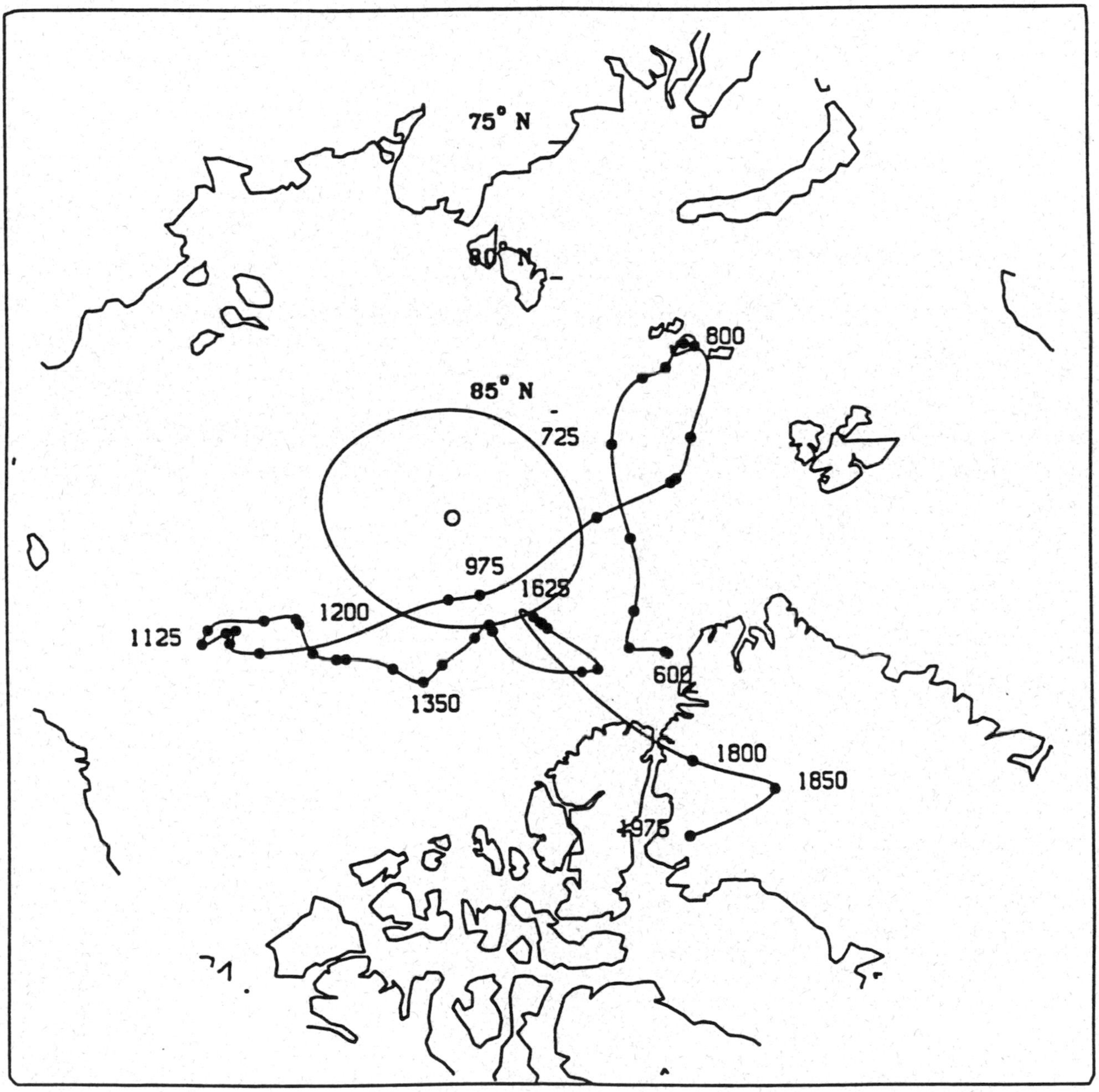

75° N

80° N

85° N

800

725

975

1625

1200

1125

1350

600

1800

1850

+975

AD 920-1020;
1350-1650

curve used: SWCV590

PLOT 1   10.43.03 MON 5 AUG, 1991   JOB-eighmy , COLORADO STATE UNIVERSITY DISSPLA 11.0

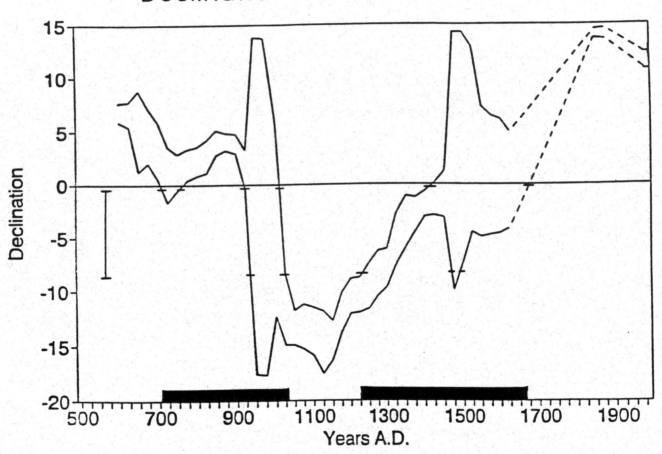

U.S. Southwest
Declination Curve: SWDC590

CSUSED-56
AD 700-1025;
1225-1675

CURRENT SOUTHWEST VGP CURVE AND SAMPLE CSUSED–57

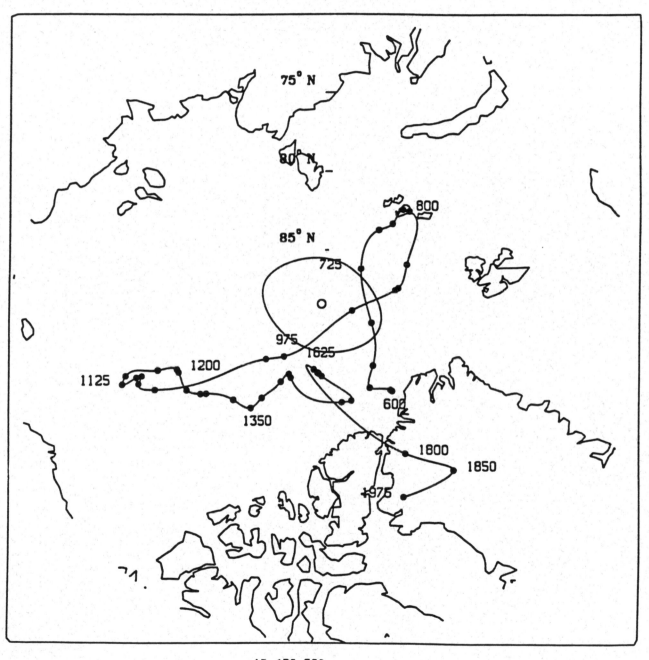

AD 675-730;
900-975;
1550-1650

curve used: SWCV590

PLOT 1   10.44.15 MON   5 AUG, 1991   JOB=sighmy , COLORADO STATE UNIVERSITY DISSPLA 11.0

U.S. Southwest
Declination Curve: SWDC590

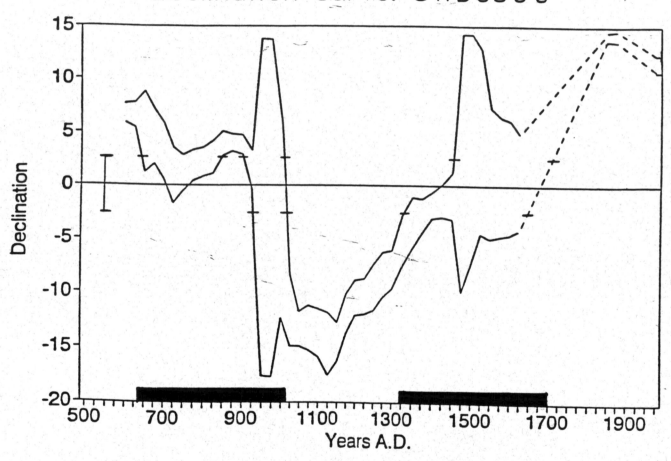

CSUSED-57
AD 650-1025;
1325-1700

CURRENT SOUTHWEST VGP CURVE AND SAMPLE CSUSED–58

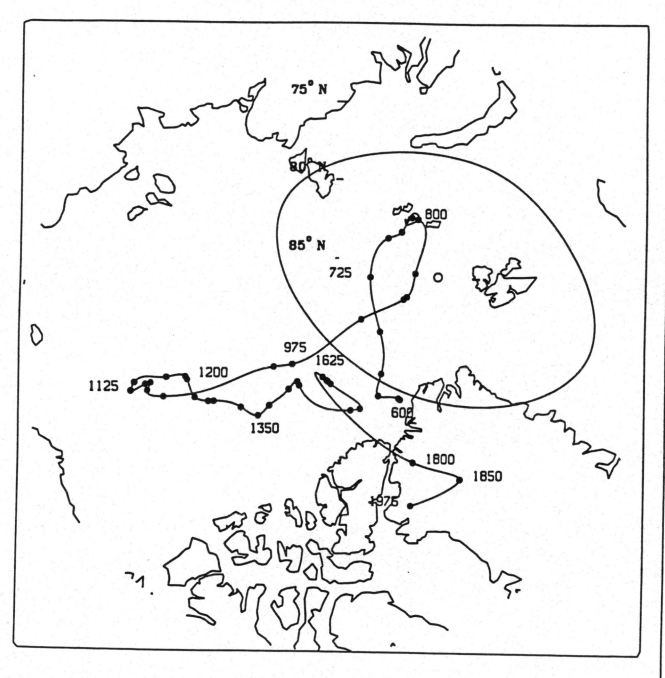

AD 600-975

curve used: SWCV590

U.S. Southwest
Declination Curve: SWDC590

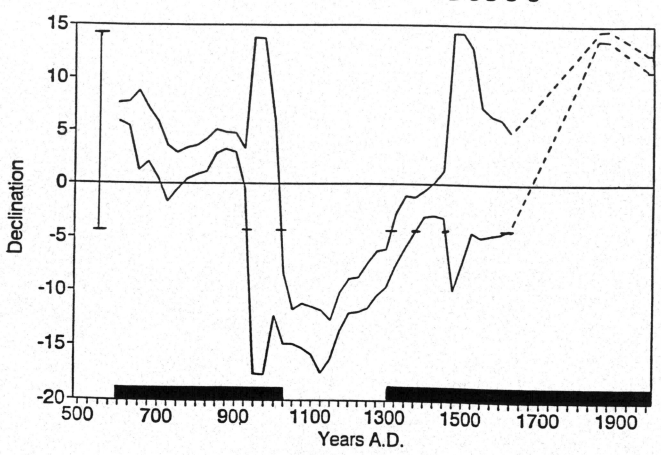

CSUSED- 58
AD 600-1025;
1300-modern

APPENDIX B

CALIBRATED RADIOCARBON DATES
(including dates from the Squaw Peak Parkway corridor investigations)

APPENDIX B
CALIBRATED RADIOCARBON DATES

DUTCH CANAL RUIN

Area 3

Feature 3-9

Beta Sample Number: 42746

Radiocarbon Age: B.P. 1810.0±120.0 **Calibrated Age(s):**   cal A.D. 180, 194, 225
                    cal B.P. 1770, 1756, 1725

      cal A.D./B.C. (cal B.P.) age ranges obtained from intercepts (Method A):

            one sigma*   cal A.D. 60–360(1890–1590); 371–381(1579–1569)
            two sigma**   cal B.C. 92–77(2041–2026); 70–A.D. 430(2019–1520)
                             453–466(1497–1484); 502–514(1448–1436)
                             515–531(1435–1419)

      summary of above:
      minimum of cal age ranges (cal ages) maximum of cal age ranges

      one sigma   cal A.D. 60 (180, 194, 225) 381
                     cal B.P. 1890 (1770, 1756, 1725) 1569

      two sigma   cal B.C. 92 (cal A.D. 180, 194, 225) cal A.D. 531
                     cal B.P. 2041 (1770, 1756, 1725) 1419

cal A.D./B.C. age ranges (cal ages as above) from probability distribution (Method B):

% area enclosed	cal B.C. (cal B.P.) age ranges	relative area under probability distribution
68.3 (one sigma)	cal A.D. 70–260(1880–1690)	.73
	cal A.D. 265–343(1685–1607)	.26
	cal A.D. 373–377(1577–1573)	.01
95.4 (two sigma)	cal B.C. 92–77(2041–2026)	.01
	cal B.C. 70–cal A.D. 430(2019–1520)	.97
	cal B.C. 501–531(1449–1419)	.01

Feature 3-19

Beta Sample Number: 43286

Radiocarbon Age: B.P. 2530.0±220.0 **Calibrated Age(s):**   cal B.C. 769, 671, 663
                    cal B.P. 2718, 2620, 2612

cal A.D./B.C. (cal B.P.) age ranges obtained from intercepts (Method A):

one sigma*   cal B.C. 900–390(2849–2339)
two sigma**   cal B.C. 1257–1251(3206–3200); 1245–1232(3194–3181)
               cal B.C. 1215–1199(3164–3148); 1194–1139(3143–3088)
               cal B.C. 1130–160(3079–2109); 150–111(2099–2060)

summary of above:
minimum of cal age ranges (cal ages) maximum of cal age ranges

one sigma   cal B.C. 900 (769, 671, 663) 390
               cal B.P. 2849 (2718, 2620, 2612) 2339

two sigma   cal B.C. 1257 (769, 671, 663) 111
               cal B.P. 3206 (2718, 2620, 2612) 2060

Area 5

Feature 5-1

Beta Sample Number: 42748; eth-7672 (processed using AMS technique)

Radiocarbon Age: B.P. 735.00±60.0   **Calibrated Age(s):**   cal A.D. 1277
                                                                    cal B.P. 673

cal A.D./B.C. (cal B.P.) age ranges obtained from intercepts (Method A):

one sigma*   cal A.D. 1231–1241(719–709); 1256–1283(694–667)
two sigma**   cal A.D. 1190–1320(760–630); 1367–1388(583–562)

summary of above:
minimum of cal age ranges (cal ages) maximum of cal age ranges

one sigma   cal A.D. 1231 (1277) 1283
               cal B.P. 719 (673) 667

two sigma   cal A.D. 1190 (1277) 1388
               cal B.P. 760 (673) 562

cal A.D./B.C. age ranges (cal ages as above) from probability distribution Method B):

% area enclosed	cal A.D. (cal B.P.) age ranges	relative area under probability distribution
68.3 (one sigma)	cal A.D. 1218–1297(732–653)	.98
	cal A.D. 1374–1376(576–574)	.02
95.4 (two sigma)	cal A.D. 1161–1183(789–767)	.03
	cal A.D. 1186–1327(764–623)	.86
	cal A.D. 1351–1390(599–560)	.10

Area 7

Feature 7-1

Beta Sample Number: 43287; eth-7730 (Processed using AMS technique)

Radiocarbon Age: B.P. 1575.0±55.0   **Calibrated Age(s):**   cal A.D. 432, 519, 529
cal B.P. 1518, 1431, 1421

cal A.D./B.C. (cal B.P.) age ranges obtained from intercepts (Method A):

one sigma*   cal A.D. 413–542(1537–1408)
two sigma**   cal A.D. 344–372(1606–1578); 380–600(1570–1350)

summary of above:
minimum of cal age ranges (cal ages) maximum of cal age ranges

one sigma   cal A.D. 413 (432, 519, 529) 542
cal B.P. 1537 (1518, 1431, 1421) 1408

two sigma   cal A.D. 344 (432, 519, 529) 600
cal B.P. 1606 (1518, 1431, 1421) 1350

cal A.D./B.C. age ranges (cal ages as above) from probability distribution (Method B):

% area enclosed	cal A.D. (cal B.P.) age ranges	relative area under probability distribution
68.3 (one sigma)	cal A.D. 420–540(1530–1410)	1.00
95.4 (two sigma)	cal A.D. 343–372(1607–1578)	.04
	cal A.D. 380–602(1570–1348)	.96

Area 8

Feature 8-3-1

Beta Sample Number: 42750

Radiocarbon Age: B.P. 630.0±70.0   **Calibrated Age(s):**   cal A.D. 1304, 1371, 1384
cal B.P. 646, 579, 566

cal A.D./B.C. (cal B.P.) age ranges obtained from intercepts (Method A):

one sigma*   cal A.D. 1280–1334(670–616); 1337–1405(613–545)
two sigma**   cal A.D. 1260–1430(690–520)

summary of above:
minimum of cal age ranges (cal ages) maximum of cal age ranges

one sigma cal A.D. 1280 (1304, 1371, 1384) 1405
      cal B.P. 670 (646, 579, 566) 545

two sigma cal A.D. 1260 (1304, 1371, 1384) 1430
      cal B.P. 690 (646, 579, 566) 520

cal A.D./B.C. age ranges (cal ages as above) from probability distribution (Method B):

% area enclosed	*cal A.D. (cal B.P.) age ranges*	*relative area under probability distribution*
68.3 (one sigma)	cal A.D. 1283–1392(667–621)	.52
	cal A.D. 1348–1329(602–558)	.48
95.4 (two sigma)	cal A.D. 1263–1422(687–528)	1.00

Feature 8-12

Beta Sample Number: 42751

Radiocarbon Age: B.P. 580.0±80.0 **Calibrated Age(s):** cal A.D. 1330, 1347, 1393
                      cal B.P. 620, 603, 557

cal A.D./B.C. (cal B.P.) age ranges obtained from intercepts (Method A):

one sigma* cal A.D. 1284–1423(666–527)
two sigma** cal A.D. 1280–1440(670–510)

summary of above:
minimum of cal age ranges (cal ages) maximum of cal age ranges

one sigma cal A.D. 1284 (1330, 1347, 1393) 1423
      cal B.P. 666 (620, 603, 557) 527

two sigma cal A.D. 1280 (1330, 1347, 1393) 1440
      cal B.P. 670 (620, 603, 557) 510

cal A.D./B.C. age ranges (cal ages as above) from probability distribution (Method B):

% area enclosed	*cal A.D. (cal B.P.) age ranges*	*relative area under probability distribution*
68.3 (one sigma)	cal A.D. 1302–1372(648–578)	.69
	cal A.D. 1382–1414(568–536)	.31
95.4 (two sigma)	cal A.D. 1276–1446(674–504)	.99

Feature 8-14

Beta Sample Number: 42752

Radiocarbon Age: B.P. 650.0±90.0   **Calibrated Age(s):**   cal A.D. 1296, 1375
                                                                        cal B.P. 654, 575

> **cal A.D./B.C. (cal B.P.) age ranges obtained from intercepts (Method A):**
>
>> one sigma*   cal A.D. 1277–1334(673–616); 1338–1405(612–545)
>> two sigma**   cal A.D. 1210–1430(740–520)
>
>> *summary of above:*
>> *minimum of cal age ranges (cal ages) maximum of cal age ranges*
>
>>> one sigma   cal A.D. 1277 (1296, 1375) 1405
>>>   cal B.P. 673 (654, 575) 545
>
>>> two sigma   cal A.D. 1210 (1296, 1375) 1430
>>>   cal B.P. 740 (654, 575) 520
>
> **cal A.D./B.C. age ranges (cal ages as above) from probability distribution (Method B):**

% area enclosed	*cal A.D. (cal B.P.) age ranges*	*relative area under probability distribution*
68.3 (one sigma)	cal A.D. 1277–1333(673–617)	.51
	cal A.D. 1340–1398(610–522)	.49
95.4 (two sigma)	cal A.D. 1209–1436(741–514)	1.00

SQUAW PEAK PARKWAY CORRIDOR
(Greenwald and Ciolek-Torrello 1988a)

North Main Canal 2

Beta Sample Number: 17282

Radiocarbon Age: B.P. 2150.0±90.0 **Calibrated Age(s):**   cal B.C. 335, 202
                                                                  cal B.P. 2284, 2151

cal A.D./B.C. (cal B.P.) age ranges obtained from intercepts (Method A):

    one sigma*   cal B.C. 363–268(2312–2217); 264–93(2213–2042)
                   cal B.C. 77–73(2026–2022)
    two sigma**   cal B.C. 400–cal A.D. 30(2349–1920); 41–49(1909–1901)

summary of above:
minimum of cal age ranges (cal ages) maximum of cal age ranges

    one sigma   cal B.C. 363 (335, 202) 73
                   cal B.P. 2312 (2284, 2151) 2022

    two sigma   cal B.C. 400 (335, 202) cal A.D. 49
                   cal B.P. 2349 (2284, 2151) 1901

cal A.D./B.C. age ranges (cal ages as above) from probability distribution (Method B):

% area enclosed	*cal B.C. (cal B.P.) age ranges*	*relative area under probability distribution*
68.3 (one sigma)	cal B.C. 357–289(2306–2238)	.31
	cal B.C. 280–273(2229–2222)	.03
	cal B.C. 259–247(2208–2196)	.05
	cal B.C. 233–95(2182–2044)	.61
95.4 (two sigma)	cal B.C. 395–cal A.D. 2(2344–1948)	1.00

North Main Canal 3

Beta Sample Number: 17283

Radiocarbon Age: B.P. 1900.0±80.0 **Calibrated Age(s):**   cal A.D. 77
                                                                  cal B.P. 1873

cal A.D./B.C. (cal B.P.) age ranges obtained from intercepts (Method A):

    one sigma*   cal A.D. 4–9(1946–1941); 21–176(1929–1774)
                   cal A.D. 197–219(1753–1731)
    two sigma**   cal B.C. 94–76(2043–2025); 70–cal A.D. 260(2019–1690)
                   cal B.C. 281–290(1669–1660); 299–323(1651–1627)

summary of above:
minimum of cal age ranges (cal ages) maximum of cal age ranges

one sigma   cal A.D. 4 (77) 219
                   cal B.P. 1946 (1873) 1731

two sigma   cal B.C. 94 (cal A.D. 77) cal A.D. 323
                   cal B.P. 2043 (1873) 1627

cal A.D./B.C. age ranges (cal ages as above) from probability distribution (Method B):

% area enclosed	*cal A.D. (cal B.P.) age ranges*	*relative area under probability distribution*
68.3 (one sigma)	cal A.D. 3–11(1947–1939)	.03
	cal A.D. 19–179(1931–1771)	.83
	cal A.D. 194–225(1756–1725)	.13
95.4 (two sigma)	cal B.C. 94–cal A.D. 260(2043–1690)	.96
	cal B.C. 297–330(1653–1620)	.03

Lateral Canal 2

Beta Sample Number: 17284

Radiocarbon Age: B.P. 1970.0±80.0 **Calibrated Age(s):**   cal A.D. 26, 42, 48
                                                                   cal B.P. 1924, 1908, 1902

cal A.D./B.C. (cal B.P.) age ranges obtained from intercepts (Method A):

one sigma*   cal B.C. 92–78(2041–2027); 73–cal A.D. 81(2022–1869)
               cal B.C. 111–126(1839–1824)
two sigma**   cal B.C. 197–193(2146–2142); 170–cal A.D. 180(2119–1770)
               cal B.C. 192–226(1758–1724)

summary of above:
minimum of cal age ranges (cal ages) maximum of cal age ranges

one sigma   cal B.C. 92 (cal A.D. 26, 42, 48) cal A.D. 126
                   cal B.P. 2041 (1924, 1908, 1902) 1824

two sigma   cal B.C. 197 (cal A.D. 26, 42, 48) cal A.D. 226
                   cal B.P. 2146 (1924, 1908, 1902) 1724

cal A.D./B.C. age ranges (cal ages as above) from probability distribution (Method B):

% area enclosed	cal B.C. (cal B.P.) age ranges	relative area under probability distribution
68.3 (one sigma)	cal B.C. 92–78(2041–2027)	.07
	cal B.C. 73–cal A.D. 81(2022–1869)	.84
	cal B.C. 95–97(1855–1853)	.01
	cal B.C. 107–126(1843–1824)	.08
95.4 (two sigma)	cal B.C. 174–cal A.D. 184(2123–1766)	.96
	cal B.C. 187–228(1763–1722)	.04

Feature 103

Beta Sample Number: 17285

Radiocarbon Age: B.P. 2230.0±60.0 **Calibrated Age(s):** cal B.C. 359, 286, 284, 271, 261, 241, 235
cal B.P. 2308, 2235, 2233, 2220, 2210, 2190, 2184

cal A.D./B.C. (cal B.P.) age ranges obtained from intercepts (Method A):

one sigma*   cal B.C. 391–345(2340–2294); 323–206(2272–2155)
two sigma**   cal B.C. 400–170(2349–2119); 142–122(2091–2071)

summary of above:
minimum of cal age ranges (cal ages) maximum of cal age ranges

one sigma   cal B.C. 391 (359, 286, 284, 271, 261, 241, 235) 206
cal B.P. 2340 (2308, 2235, 2233, 2220, 2210, 2190, 2184) 2155

two sigma   cal B.C. 400 (359, 286, 284, 271, 261, 241, 235) 122
cal B.P. 2349 (2308, 2235, 2233, 2220, 2210, 2190, 2184) 2071

cal A.D./B.C. age ranges (cal ages as above) from probability distribution (Method B):

% area enclosed	cal B.C. (cal B.P.) age ranges	relative area under probability distribution
68.3 (one sigma)	cal B.C. 388–349(2337–2298)	.28
	cal B.C. 319–308(2268–2257)	.07
	cal B.C. 303–228(2252–2177)	.55
	cal B.C. 222–209(2171–2158)	.09
95.4 (two sigma)	cal B.C. 401–169(2350–2118)	.98
	cal B.C. 139–127(2088–2076)	.02

Lateral Canal 3

Beta Sample Number: 17286

Radiocarbon Age: B.P. 2020.0±90.0 **Calibrated Age(s):** cal B.C. 89, 82, 68, 60, 41, 9, 3
cal B.P. 2038, 2031, 2017, 2009, 1990, 1958, 1952

cal A.D./B.C. (cal B.P.) age ranges obtained from intercepts (Method A):

one sigma*   cal B.C. 169–140(2118–2089); 125–cal A.D. 65(2074–1885)
two sigma** cal B.C. 352–294(2301–2243); 231–219(2180–2168)
             cal B.C. 210–cal A.D. 130(2159–1820); 203–208(1747–1742)

summary of above:
minimum of cal age ranges (cal ages) maximum of cal age ranges

one sigma   cal B.C. 169 (89, 82, 68, 60, 41, 9, 3) cal A.D. 65
             cal B.P. 2118 (2038, 2031, 2017, 2009, 1990, 1958, 1952) 1885

two sigma   cal B.C. 352 (89, 82, 68, 60, 41, 9, 3) cal A.D. 208
             cal B.P. 2301 (2038, 2031, 2017, 2009, 1990, 1958, 1952) 1742

cal A.D./B.C. age ranges (cal ages as above) from probability distribution (Method B):

% area enclosed	cal B.C. (cal B.P.) age ranges	relative area under probability distribution
68.3 (one sigma)	cal B.C. 169–140(2118–2089)	.11
	cal B.C. 125–cal A.D. 65(2074–1885)	.89
95.4 (two sigma)	cal B.C. 354–291(2303–2240)	.06
	cal B.C. 232–217(2181–2166)	.01
	cal B.C. 214–cal A.D. 134(2163–1816)	.91

Feature 110

Beta Sample Number: 17287

Radiocarbon Age: B.P. 1840.0±80.0 **Calibrated Age(s):** cal A.D. 133, 203, 207
cal B.P. 1817, 1747, 1743

cal A.D./B.C. (cal B.P.) age ranges obtained from intercepts (Method A):

one sigma*   cal A.D. 68–256(1882–1694); 304–317(1646–1633)
two sigma** cal B.C. 87–83(2036–2032); 66–64(2015–2013)
             cal B.C. 37–32(1986–1981); 21–12(1970–1961)
             cal A.D. 1–360(1949–1590); 367–385(1583–1565)

summary of above:
minimum of cal age ranges (cal ages) maximum of cal age ranges

one sigma   cal A.D. 68 (133, 203, 207) 317
                 cal B.P. 1882 (1817, 1747, 1743) 1633

two sigma   cal B.C. 87 (cal A.D. 133, 203, 207) cal A.D. 385
                 cal B.P. 2036 (1817, 1747, 1743) 1565

cal A.D./B.C. age ranges (cal ages as above) from probability distribution (Method B):

% area enclosed	cal B.C. (cal B.P.) age ranges	relative area under probability distribution
68.3 (one sigma)	cal A.D. 70–253(1880–1697)	.97
	cal A.D. 308–316(1642–1634)	.03
95.4 (two sigma)	cal B.C.	.00
	cal B.C. 20–13(1969–1962)	.01
	cal A.D. 1–362(1949–1588)	.98
	cal A.D. 368–384(1582–1566)	.01

North Main Canal 1

Beta Sample Number: 17288

Radiocarbon Age: B.P. 2400.0±90.0 **Calibrated Age(s):**   cal B.C. 411
                                                                   cal B.P. 2360

cal A.D./B.C. (cal B.P.) age ranges obtained from intercepts (Method A):

one sigma*   cal B.C. 761–682(2710–2631); 659–632(2608–2581)
              cal B.C. 620–611(2569–2560); 594–394(2543–2343)
two sigma**   cal B.C. 790–360(2739–2309); 290–276(2239–2225)
              cal B.C. 275–258(2224–2207); 250–233(2199–2182)

summary of above:
minimum of cal age ranges (cal ages) maximum of cal age ranges

one sigma   cal B.C. 761 (411) 394
                 cal B.P. 2710 (2360) 2343

two sigma   cal B.C. 790 (411) 233
                 cal B.P. 2739 (2360) 2182

cal A.D./B.C. **age ranges (cal ages as above) from probability distribution (Method B):**

% area enclosed	cal B.C. (cal B.P.) age ranges	relative area under probability distribution
68.3 (one sigma)	cal B.C. 758–687(2707–2636)	.27
	cal B.C. 657–640(2606–2589)	.06
	cal B.C. 591–573(2540–2522)	.06
	cal B.C. 552–457(2501–2406)	.37
	cal B.C. 455–398(2404–2347)	.23
95.4 (two sigma)	cal B.C. 792–359(2741–2308)	.98

Feature 100

Beta Sample Number: 17289

Radiocarbon Age: B.P. 1150.0±60.0   **Calibrated Age(s):**   cal A.D. 889
cal B.P. 1061

cal A.D./B.C. **(cal B.P.) age ranges obtained from intercepts (Method A):**

one sigma*   cal A.D. 778–792(1172–1158); 799–980(1151–970)
two sigma**   cal A.D. 689–702(1261–1248); 708–751(1242–1199)
cal A.D. 760–1000(1190–950)

summary of above:
minimum of cal age ranges (cal ages) maximum of cal age ranges

one sigma   cal A.D. 778 (889) 980
cal B.P. 1172 (1061) 970

two sigma   cal A.D. 689 (889) 1000
cal B.P. 1261 (1061) 950

cal A.D./B.C. **age ranges (cal ages as above) from probability distribution (Method B):**

% area enclosed	cal A.D. (cal B.P.) age ranges	relative area under probability distribution
68.3 (one sigma)	cal A.D. 781–789(1169–1161)	.06
	cal A.D. 804–823(1146–1127)	.13
	cal A.D. 827–841(1123–1109)	.10
	cal A.D. 858–904(1092–1046)	.32
	cal A.D. 916–976(1034–974)	.40
95.4 (two sigma)	cal A.D. 714–724(1236–1226)	.01
	cal A.D. 725–746(1225–1204)	.03
	cal A.D. 768–997(1182–953)	.96

Backhoe Trench 30

Beta Sample Number: 19899

Radiocarbon Age: B.P. 1100.0±70.0   **Calibrated Age(s):**   cal A.D. 910, 915, 977
                                                                     cal B.P. 1040, 1035, 973

cal A.D./B.C. (cal B.P.) age ranges obtained from intercepts (Method A):

one sigma*   cal A.D. 783–787(1167–1163); 872–998(1078–952)
two sigma**   cal A.D. 770–1030(1180–920); 1144–1147(806–803)

summary of above:
minimum of cal age ranges (cal ages) maximum of cal age ranges

one sigma   cal A.D. 783 (910, 915, 977) 998
               cal B.P. 1167 (1040, 1035, 973) 952

two sigma   cal A.D. 770 (910, 915, 977) 1147
               cal B.P. 1180 (1040, 1035, 973) 803

cal A.D./B.C. age ranges (cal ages as above) from probability distribution (Method B):

% area enclosed	cal A.D. (cal B.P.) age ranges	relative area under probability distribution
68.3 (one sigma)	cal A.D. 783–787(1167–1163)	.02
	cal A.D. 812–816(1138–1134)	.02
	cal A.D. 833–837(1117–1113)	.02
	cal A.D. 868–999(1082–951)	.93
95.4 (two sigma)	cal A.D. 728–744(1222–1206)	.01

References for computer program: Stuiver and Reimer 1986
References for datasets and intervals used: Stuiver and Becker 1986

Note: 0* represents a "negative" age B.P.

*one sigma = square root of (sample std. dev.^{2} + curve std. dev.^{2})
**two sigma = 2 × square root of (sample std. dev.^{2} + curve std. dev.^{2})

APPENDIX C

RAW FLOTATION AND MACROBOTANICAL DATA

Scott Kwiatkowski

Context/ Provenience	Bag No.	Volumes: Sample Size (lt. fraction) [wood charcoal]	Carbonized Plant Taxa	Uncarbonized Plant Taxa	Other
Pit structure Feature 1-1 floor fill, south half	1-1	2.0 l (14.0 ml) [0.5 ml]	(UNPRODUCTIVE)		9 Fecal pellets 9 Insect exoskeleton fragments 330 Macrospores[a] 2 Ostracod valves 9 Snail shells
Pit structure Feature 1-2 floor fill[b]	1-23	4.0 l (5.5 ml) [0.5 ml]	(UNPRODUCTIVE) 2 Miscellaneous spiral twists		Numerous fecal pellets[c] 19 Insect exoskeleton fragments 1000 Macrospores[a] 6 Snail shells 1 Unburned bone fragment
Primary inhumation Feature 1-3 fill	1-51	1.5 l (3.0 ml) [<0.5 ml]	(UNPRODUCTIVE)	1 cf. Cheno-am seed fragment 1 Compositae achene 1 Magnoliopsida leaf 1 *Physalis* seed 2 cf. *Schismus* grain fragments	14 Fecal pellets 8 Insect exoskeleton fragments <200 Macrospores[a] Numerous termite pellets[c]
Pit structure Feature 2-2, floor fill, west half[b]	2-13	2.75 l (7.5 ml) [<0.5 ml]	(UNPRODUCTIVE) 2 Miscellaneous round fibers (1 CaO)[d]		Numerous fecal pellets[c] 30 Fecal pellet aggregates 8 Insect exoskeleton fragments 1000 Macrospores[a]
Pit structure Feature 3-1 floor fill[b]	3-12	2.0 l (2.5 ml) [0.5 ml]	1 Cheno-am seed 1 Indeterminate seed fragment 3 *Leptochloa* type grains 4 cf. *Leptochloa* type grain fragments 1 Miscellaneous endosperm fragment		25 Fecal pellets 34 Insect exoskeleton fragments 21 Macrospores 8 Snail shells

Pit structure Feature 3-2 floor fill[b]	3-32	4.0 l (12.0 ml) [1.0 ml]	4 Indeterminate seed fragments 3 Miscellaneous endosperm fragments 2 *Zea mays* cupule fragments 2 cf. *Zea mays* cupule fragments WOOD CHARCOAL: 1 Arboreal legume cf. *Prosopis*	4 Cheno-am seeds 6 Cheno-am seed fragments 1 Indeterminate seed fragment	Numerous fecal pellets[c] Numerous insect exoskeleton fragments[c] 1 Fly 2000 Macrospores[a] 1 Ostracod 2 Ostracod valves 13 Snail shells
Pit structure fire pit Feature 3-2-1 fill	3-56	4.0 l (73.0 ml) [2.0 ml]	(UNPRODUCTIVE) 1 cf. Cheno-am seed fragment 138 Miscellaneous endosperm fragments 2 cf. *Zea mays* cupule fragments 3 cf. *Zea mays* kernel fragments	1 Cheno-am seed 2 Cheno-am seed fragments 2 Indeterminate fruit fragments	4 Burned bone fragments 3 Charred termite pellets 25 Fecal pellets 1 Insect case 121 Insect exoskeleton fragments 535 Macrospores[a] 5 Snail shells 3 Snail shell fragments 1 Unburned bone 10 Unburned bone fragments
Pit Feature 3-3 fill	3-38	macro-botanical sample	5 *Zea mays* cupules 8 *Zea mays* cupule fragments 1 *Zea mays* glume fragment		
Pit Feature 3-3 fill	3-42	2.0 l (3.0 ml) [<0.5 ml]	2 Indeterminate seed fragments 2 *Zea mays* cupules 60 *Zea mays* cupule fragments 124 cf. *Zea mays* cupule fragments 5 *Zea mays* glume fragments 2 cf. *Zea mays* glume fragments		1 Charred termite pellet 19 Fecal pellets 92 Insect exoskeleton fragments 343 Macrospores[a] 1 Macrospore cluster 2 Ostracod valves 17 Snail shells
Pit structure Feature 3-9 floor fill[b]	3-100	4.0 l (29.0 ml) [2.0 ml]	1 Indeterminate spine fragment 2 Miscellaneous fragments (CaO)[d] 2 Miscellaneous round fibers 1 *Zea mays* cupule fragment		1 Burned bone fragment 12 Fecal pellets 7 Insect exoskeleton fragments 600 Macrospores[a] 1 Shell fragment 1 Snail shell
Pit structure Feature 3-16 floor fill	3-156	4.0 l (41.0 ml) [0.5ml]	(UNPRODUCTIVE)	1 *Baccharis* achene	Numerous fecal pellets[c] 27 Insect exoskeleton fragments 500 Macrospores[a] 1 Ostracod 1 Shell fragment 5 Snail shells

Extramural fire pit Feature 3-18 fill	3-149	4.0 l (70.0 ml) [1.0 ml]	(UNPRODUCTIVE) 12 Miscellaneous endosperm fragments 2 Miscellaneous fragments (CaO)[d]	1 *Baccharis* achene 1 Cheno-am seed 1 Cheno-am seed fragment 1 cf. Cheno-am seed fragment 1 *Physalis* seed	1 Fecal pellet 8 Insect exoskeleton fragments 220 Macrospores[a] 2 Ostracods 1 Ostracod valve 15 Snail shells 37 Snail shell fragments 1 Unburned bone fragment
Pit structure Feature 3-19, floor fill, west half	3-168	4.0 l (2.0 ml) [0.5 ml]	(UNPRODUCTIVE) 2 Indeterminate seed fragments 1 Miscellaneous endosperm fragment 1 Miscellaneous round fiber 1 cf. *Zea mays* cupule fragment WOOD CHARCOAL: 1 cf. *Larrea tridentata*		1 Charred termite pellet fragment 60 Fecal pellets 2 Insect exoskeleton fragments 300 Macrospores[a] 3 Snail shells 1 Termite pellet
Pit Feature 4-1 fill, north half	4-20	4.0 l (1.0 ml) [1.0 ml]	1 cf. Cheno-am seed coat fragment 1 Indeterminate seed fragment 4 Miscellaneous endosperm fragments 1 *Trianthema portulacastrum* seed coat fragment		1 *Chara* oogonium 1 Charophyceae oogonium 1 Charophyceae oogonium fragment 2 Charred fecal pellets 1 Charred termite pellet 1 Charred termite pellet fragment 1 Insect exoskeleton fragment 900 Macrospores[a] 3 Ostracods 3 Ostracod valves 11 Snail shells

Pit structure Feature 5-1 floor contact	5-20	4.0 l (44.5 ml) [3.0 ml]	4 *Agave* fibers (CaO)[d] 2 Cheno-am seed fragments 6 Indeterminate seed fragments 1 Miscellaneous endosperm fragments 1 Miscellaneous fragment (CaO)[d] 6 Miscellaneous round fibers (4 CaO)[d] 1 *Zea mays* cupule 2 cf. *Zea mays* cupule fragment 1 cf. *Zea mays* kernel fragment **WOOD CHARCOAL:** 6 Arboreal legume cf. *Prosopis*	1 cf. Cheno-am seed fragment 2 Indeterminate seed fragments 1 *Physalis* seed	5 Charophyceae oogonia 1 Charophyceae oogonium fragment 1 Charred termite pellet 16 Insect exoskeleton fragments 51 Macrospores[a] 4 Ostracod valves 40 Snail shells 15 Snail shell fragments 1 Unburned bone fragment
Pit house Feature 8-2 floor fill[b]	8-110	2.5 l (12.0 ml) [5.0 ml]	99 *Agave* fibers (CaO)[a,d] 2 Cheno-am seeds 5 Cheno-am seed fragments 9 Indeterminate seed fragments 1 *Leptochloa* type grain 14 Miscellaneous endosperm fragments 1 Miscellaneous flat fiber (CaO)[d] 133 Miscellaneous fragments (CaO)[a,d] 66 Miscellaneous round fibers (63 CaO)[a,d] 1 cf. *Phalaris* grain fragment **WOOD CHARCOAL:** 2 Arboreal legume cf. *Cercidium* 1 Arboreal legume cf. *Prosopis*	2 Cheno-am seeds 2 Compositae achenes 3 Indeterminate seed fragments 1 *Physalis* seed 1 *Physalis* seed fragment 1 *Schismus arabicus* floret 1 cf. *Schismus* grain fragment 1 *Trianthema portulacastrum* seed	2 Charophyceae oogonia 26 Fecal pellets[a] Numerous insect exoskeleton fragments[c] 4 Ostracods 4 Snail shells 3 Unburned bones

| Pit house hearth Feature 8-2-1 fill[b] | 8-109 | 2.0 l (25.0 ml) [6.0 ml] | 141 *Agave* fibers (140 CaO)[a,d]
4 Cheno-am seeds
8 Cheno-am seed fragments
1 *Euphorbia* seed
1 *Euphorbia* seed fragment
1 Gramineae culm fragment
1 Gramineae grain
3 *Hordeum* grain fragments
69 Indeterminate seed fragments[a]
50 Miscellaneous endosperm fragments[a]
211 Miscellaneous fragments (CaO)[a,d]
273 Miscellaneous round fibers (244 CaO)[a,d]
2 Miscellaneous round fiber bundles (CaO)[d]
2 *Phalaris* grains
4 cf. *Phalaris* grain fragments
2 *Prosopis* seed fragments
2 cf. *Prosopis* pod fragments
1 *Suaeda* seed
1 *Zea mays* cupule fragment

WOOD CHARCOAL:
6 Arboreal legume cf. *Prosopis* | 1 *Euphorbia* seed
1 *Euphorbia* seed fragment
2 Indeterminate seed fragments
5 *Physalis* seeds
1 *Portulaca* seed fragment
2 *Trianthema portulacastrum* seed fragments | 2 Charophyceae oogonia
10 Fecal pellets
Numerous insect exoskeleton fragments[c]
200 Macrospores[a]
1 Ostracod
1 Ostracod valve
8 Snail shells
4 Unburned bone fragments |

Pit structure fire pit Feature 8-3-1 fill[b]	8-56	2.0 l (13.0 ml) [13.0 ml]	463 *Agave* fibers (462 CaO)[a,d] 1 Cheno-am seed fragment 1 *Gossypium hirsutum* var. *punctatum* seed fragment 2 cf. *Gossypium hirsutum* var. *punctatum* seed fragments 165 Indeterminate seed fragments[a] 1 Indeterminate spine fragment 1 Miscellaneous D-shaped fiber (CaO)[d] 58 Miscellaneous endosperm fragments[a] 2 Miscellaneous flat fibers (CaO)[d] 259 Miscellaneous fragments (CaO)[a,d] 552 Miscellaneous round fibers (534 CaO)[a,d] 6 Miscellaneous round fiber bundles (CaO)[d] 1 *Phalaris* grain 2 *Phalaris* grain fragments 3 cf. *Phalaris* grain fragments 1 *Prosopis* seed fragment 3 *Suaeda* seeds 3 *Zea mays* cupule fragments 1 *Zea mays* glume fragment 1 cf. *Zea mays* kernel fragment WOOD CHARCOAL: 3 Arboreal legume cf. *Prosopis*	5 *Euphorbia* seeds	2 Charophyceae oogonia 5 Fecal pellets 7 Insect exoskeleton fragments 267 Macrospores[a] 3 Ostracods 4 Snail shells

Pit Feature 8-12 fill	8-211	4.0 l (36.0 ml) [28.0 ml]	322 *Agave* fibers (321 CaO)[a,d]	8 *Euphorbia* seeds	7 Charophyceae oogonia
			2 *Agave* fragments of round and trough-shaped fibers and parenchyma (CaO)[d]	1 *Euphorbia* seed fragment	2 Charred termite pellets
			6 *Bouteloua* type grains	4 cf. *Euphorbia* seed fragments	45 Fecal pellets[a]
			1 *Bromus-Elymus* type grain fragment	1 Indeterminate seed fragment	16 Insect exoskeleton fragments[a]
			8 Cheno-am seeds	4 *Physalis* seeds	9 Macrospores[a]
			16 Cheno-am seed fragments		11 Ostracods
			20 cf. Cheno-am seed fragments		10 Ostracod valves
			3 *Descurainia* seeds		1 Shell fragment
			8 cf. *Descurainia* seed fragments		15 Snail shells
			2 Gramineae culm fragments		2 Unburned bones
			6 cf. Gramineae grain fragments		10 Unburned bone fragments
			99 Indeterminate seed fragments[a]		
			1 Leguminosae cf. *Phaseolus* bean (2.4 mm long)		
			9 *Leptochloa* type grains		
			4 cf. *Leptochloa* type grain fragments		
			1 Miscellaneous D-shaped fiber (CaO)[d]		
			Numerous miscellaneous endosperm fragments[c]		
			1 Miscellaneous flat fiber (CaO)[d]		
			244 Miscellaneous fragments (CaO)[a,d]		
			3 Miscellaneous fragments of round fibers scattered in parenchyma (CaO)[d]		
			327 Miscellaneous round fibers (326 CaO)[a,d]		
			7 Miscellaneous round fiber bundles (CaO)[d]		
			1 *Nicotiana* cf. *trigonophylla* seed		
			21 *Phalaris* grains		
			12 *Phalaris* grain fragments		
			6 cf. *Phalaris* grain fragments		
			1 *Prosopis* seed		
			7 *Prosopis* seed fragments		

Pit Feature 8-12 fill, continued			9 cf. *Prosopis* seed fragments 7 cf. *Prosopis* pod fragments 21 *Sporobolus* type grains 3 *Sporobolus* type grain fragments 14 cf. *Sporobolus* type grain fragments 1 *Suaeda* seed 20 *Zea mays* cupule fragments 11 cf. *Zea mays* cupule fragments 1 cf. *Zea mays* embryo fragment 2 *Zea mays* glume fragments 1 cf. *Zea mays* glume fragment 5 cf. *Zea mays* kernel fragments WOOD CHARCOAL: 7 Arboreal legume cf. *Prosopis*		

| Pit Feature 8-14 fill | 8-249 | 4.0 l (18.0 ml) [13.5 ml] | 150 *Agave* fibers (CaO)[a,d]
7 Cheno-am seeds
11 Cheno-am seed fragments
4 cf. Cheno-am seed fragments
1 *Descurainia* seed
1 cf. *Gossypium hirsutum* var. *punctatum* seed fragment
1 cf. Gramineae culm fragment
50 Indeterminate seed fragments[a]
6 *Leptochloa* type grains
2 cf. *Leptochloa* type grain fragments
1 Miscellaneous D-shaped fiber (CaO)[d]
92 Miscellaneous endosperm fragments[a]
161 Miscellaneous fragments (CaO)[a,d]
2 Miscellaneous fragments of one round fiber and parenchyma (CaO)[d]
212 Miscellaneous round fibers (211 CaO)[a,d]
4 Miscellaneous round fiber bundles (CaO)[d]
2 *Phalaris* grains
1 *Phalaris* grain fragment
3 *Prosopis* seed fragments
6 cf. *Prosopis* seed fragments
1 *Suaeda* seed
2 *Zea mays* cupule fragments
1 cf. *Zea mays* cupule fragment
1 cf. *Zea mays* glume fragment

WOOD CHARCOAL:
6 Arboreal legume cf. *Prosopis* | 2 *Euphorbia* seeds
1 *Euphorbia* seed fragment
2 cf. *Portulaca* seed fragments | 2 *Chara* oogonia
1 Charophyceae oogonium fragment
10 Fecal pellets
56 Insect exoskeleton fragments[a]
1268 Macrospores[a]
7 Ostracods
4 Ostracod valves
2 Red resin globules
13 Snail shells
3 Snail shell fragments
1 Unburned bone
4 Unburned bone fragments |

Note: All are flotation samples unless otherwise indicated. Analysts examined material <0.25 mm mesh in the first eight samples analyzed. When this fine residue produced no identifiable charred plant remains except for 9 *Agave* fibers in Feature 8-2, Bag 8-110, they did not analyze it subsequently.

[a]Estimated number.

[b]Material <0.25 mm mesh was examined, but results are not reported (see text).

[c]*Numerous* is defined as more than 50 parts per liter.

[d]CaO = White styloid crystals present.

430

APPENDIX D

FAUNAL ANALYSIS, SCREENED AND UNSCREENED DATA

Susan K. Stratton

Table D.1. Faunal Analysis: Screened Data

#	Bag #	Side	Element	Taxon	Landmark or Part	Screen	Bone Count	Epiphyses	Bone Tool	Cut Mark	Burning	Carnivore	Rodent	FEA #
1	10-25	L	FEMUR	LG. UNGULATE	HEAD FRAGMENT	1/4	1	N						0
2	10-25	U	UNID	LG. MAMMAL	FRAGMENT	1/4	1	N						0
3	10-4	U	INNOMINATE	ARTIODACTYL	ACETABULUM FRAGME...	1/4	1	N						10
4	10-70	U	TIBIA	cf. ODOCOILEUS HEMIONUS	DISTAL FRAGMENT	1/4	1	N						10-1
5	10-70	U	METACARPAL	cf. ODOCOILEUS HEMIONUS	DISTAL FRAGMENT	1/4	1	N			X			10-9
6	10-70	U	UNID	LG. MAMMAL	FRAGMENT	1/4	1	N			X			10-9
7	3-109	U	METATARSAL	SYLVILAGUS SP.	DISTAL	1/4	1	N			X			10-9
8	3-109	U	UNID	MD. MAMMAL	SHAFT FRAGMENTS	1/4	3	N			X			10-9
9	3-109	R	CALCANEUM	LEPUS SP.	PROXIMAL	1/4	1	N			X (ALL)			3-12
10	3-109	L	TIBIA	THOMOMYS BOTTAE	PROXIMAL	1/4	1	N			X			3-12
11	3-109	U	UNID	UNID	FRAGMENTS	1/4	3	N			X			3-12
12	3-109	U	UNID	UNID	FRAGMENTS	1/4	3	N						3-12
13	3-126	U	UNID	SM. MAMMAL	FRAGMENTS	1/4	2	N			X (ALL)			3-12
14	3-126	R	ULNA	SYLVILAGUS SP.	PROX. END & SHAFT	1/4	1	N			X (ALL)			3-14-1
15	3-140	U	UNID	UNID	FRAGMENTS	1/4	1	N						3-14-1
16	3-151	U	UNID	MD. MAMMAL	SHAFT FRAGMENT	1/4	2	N						3-9
17	3-26	U	UNID	MD. MAMMAL	SHAFT FRAGMENTS	1/4	1	N						3-18
18	3-26	R	RADIUS	LEPUS SP.	PROXIMAL	1/4	2	N						3-2
19	3-26	L	RADIUS	LEPUS SP.	PROXIMAL	1/4	1	N						3-2
20	3-31	U	UNID	UNID	FRAGMENT	1/4	1	N						3-2
21	3-31	L	MANDIBLE	DIPODOMYS SP.	HORIZONTAL RAMUS	1/4	1	N						3-2
22	3-49	U	UNID	UNID	FRAGMENTS	1/4	6	N						3-2
23	3-76	U	UNID	UNID	FRAGMENT	1/4	1	N						3-2
24	3-93	U	TIBIA	LEPUS SP.	SHAFT FRAGMENT	1/4	1	N						3-2
25	7-35	U	UNID	UNID	FRAGMENTS	1/4	12	N			X			3-11
26	8-107	U	UNID	LG. MAMMAL	SHAFT FRAGMENT	1/4	1	N		X	X			7-1
27	8-121	U	UNID	LG. MAMMAL	SHAFT FRAGMENT	1/4	1	N			X			8-2
28	8-164	U	UNID	MD. BIRD	SHAFT FRAGMENT	1/4	1	N						8-3
29	8-164	U	UNID	MD. MAMMAL	SHAFT FRAGMENTS	1/4	2	N						8-2
30	8-198	L	TIBIA	SYLVILAGUS SP.	DISTAL	1/4	1	N						8-2
31	8-198	L	METATARSAL	LEPUS SP.	PROXIMAL	1/4	1	Y						8-10
32	8-198	U	CALCANEUM	LEPUS SP.	FRAGMENT	1/4	1	N						8-10
33	8-201	L	MANDIBLE	CANIS SP.	HORIZONTAL RAMUS	1/4	1	N		X	X			8-10
34	8-201	U	UNID	UNID	FRAGMENT	1/4	1	N						8-10
35	8-204	U	UNID	LG. MAMMAL	SHAFT FRAGMENTS	1/4	7	N						8-10
36	8-239	U	INNOMINATE	ARTIODACTYL	FRAGMENT	1/4	1	N						8-12
37	8-239	U	UNID	UNID	FRAGMENTS	1/4	6	N						8-12
38	8-239	L	METACARPAL	SYLVILAGUS SP.	COMPLETE	1/4	1	Y						8-12
39	8-266	U	METACARPAL	LEPUS SP.	COMPLETE	1/4	1	Y						8-14
40	8-282	U	UNID	LG. MAMMAL	SHAFT FRAGMENT	1/4	1	N						8-16
41	8-282	U	RIB	MD. BIRD	FRAGMENT	1/4	1	N						8-16
42	8-282	L	UNID	LG. MAMMAL	SHAFT FRAGMENT	1/4	1	N		X	X			8-16
43	8-282	L	METATARSAL	LEPUS SP.	PROXIMAL	1/8	1	N						8-21
44	8-295	L	METATARSAL	LEPUS SP.	PROXIMAL	1/8	1	N		X				8-21
45	8-295	U	RIB	MD. MAMMAL	FRAGMENT	1/8	1	N						8-15
46	8-330	A	SKULL	MD. MAMMAL	FRAGMENTS	1/8	14	N			X (ALL)			8-15
47	8-330	U	UNID	LG. MAMMAL	BONE AWL FRAGMENTS	1/8	3	N	Y		X (ALL)			8-15
48	8-330	U	UNID	LG. MAMMAL	BONE AWL FRAGMENTS	1/8	2	N	Y	X	X (ALL)			8-15
49	8-330	A	AXIS (CERV. VERT)	ODOCOILEUS HEMIONUS	TRANSVERSE PROCESS	1/8	1	N			X (ALL)			8-15
50	8-330	L	CALCANEUM	CANIS SP.	FRAGMENT	1/4	1	N			X			8-15
51	8-330	R	HUMERUS	CANIS SP.	PROXIMAL FRAGMENT	1/4	1	Y			X			8-15
52	8-330	U	UNID	UNID	FRAGMENTS	1/8	2	N			X			8-15
53	8-330	U	METAPODIAL	ODOCOILEUS HEMIONUS	DISTAL FRAGMENTS	1/8	3	Y		X	X (ALL)			8-15
54	8-330	L	HUMERUS	CANIS SP.	PROXIMAL FRAGMENT	1/8	1	Y			X (ALL)			8-15
55	8-330	R	MAXILLA	LEPUS SP.	FRAGMENT	1/8	1	N			X			8-15
56	8-360	A	VERTEBRA	ODOCOILEUS HEMIONUS	FRAGMENTS	1/4	1	N			X			8-15
57	8-5	A	EGGSHELL	AVIAN	FRAGMENTS		2	N						8-15
58	8-69	U	UNID	LG. MAMMAL	SHAFT FRAGMENT	1/4	1	N			X			8-15
59	8-92	U	RIB	ARTIODACTYL	FRAGMENT	1/4	1	N						8-2

Table D.2. Faunal Analysis: Unscreened Data

#	Bag #	Side	Element	Taxon	Landmark or Part	Screened	Bone Count	Epiphyses	Bone Tool	Cut Mark	Burning	Carnivore	Rodent	FEA. #
1	10-33	L	TIBIA	BOS TAURUS	PROX. EPIPHYSIS	N	1	Y	-	-	-	-	-	8
2	10-33	U	METAPODIAL	BOS TAURUS	FRAGMENT	N	1	N	-	-	-	-	-	8
3	10-33	A	THORACIC VERTEBRA	BOS TAURUS	SPINOUS PROCESS	N	1	N	-	X	-	-	-	8
4	10-33	U	SCAPULA	BOS TAURUS	FRAGMENTS	N	3	N	-	-	-	-	-	8
5	10-33	R	SCAPULA	BOS TAURUS	FRAGMENT	N	1	N	-	-	-	-	-	8
6	10-33	U	UNID	LG. MAMMAL	FRAGMENTS	N	35	N	-	-	-	-	-	8
7	10-33	U	RIB	LG. MAMMAL	FRAGMENTS	N	6	N	-	-	-	-	-	8
8	10-33	R	TIBIA	BOS TAURUS	PROX. SHAFT FRAG.	N	1	N	-	-	-	-	-	8
9	10-33	U	UNID	LG. BIRD	FRAGMENT	N	1	N	-	-	-	-	-	8
10	10-33	U	INNOMINATE	BOS TAURUS	AURICULAR SURFACE	N	1	N	-	-	-	-	-	8
11	10-33	U	METATARSAL	ARTIODACTYL	SHAFT FRAGMENT	N	1	N	-	-	-	-	-	8
12	10-33	L	HUMERUS	ARTIODACTYL	PROX. EPIPHYSIS	N	1	Y	-	-	-	-	-	8
13	10-33	A	CERVICAL VERTEBRA	BOS TAURUS	ATLAS	N	1	N	-	-	-	-	-	8
14	10-33	R	INNOMINATE	BOS TAURUS	PUBIS	N	1	N	-	-	-	-	-	8
15	10-33	A	SACRUM	BOS TAURUS	FRAGMENT	N	1	N	-	-	-	-	-	8
16	10-33	A	VERTEBRA	BOS TAURUS	FRAGMENT	N	4	N	-	-	-	-	-	8
17	10-46	U	RIBS	LG. MAMMAL	PROXIMAL END	N	3	N	-	-	-	-	-	10-1
18	10-46	A	VERTEBRAE	MD. BIRD	FRAGMENTS	N	2	N	-	-	-	-	-	10-1
19	10-46	A	VERTEBRAE	LG. MAMMAL	FRAGMENTS	N	7	N	-	-	-	-	-	10-1
20	10-46	A	VERTEBRA	MD. MAMMAL	FRAGMENT	N	1	N	-	-	-	-	-	10-1
21	10-46	U	RIB	MD. MAMMAL	FRAGMENTS	N	2	N	-	-	-	-	-	10-1
22	10-46	A	CERVICAL VERTEBRA	CANIS SP.	FRAGMENT	N	1	N	-	-	-	-	-	10-1
23	10-46	A	THORACIC VERTEBRA	ODOCOILEUS HEMIONUS	SPINE	N	1	N	-	-	-	-	-	10-1
24	10-46	A	MANDIBLE	MD. MAMMAL	FRAGMENT	N	1	N	-	-	-	-	-	10-1
25	10-46	U	UNID	UNID	FRAGMENTS	N	97	N	-	-	-	-	-	10-1
26	7-46	U	UNID	UNID	FRAGMENTS	N	2	N	-	-	-	-	-	7-3
27	8-163	U	UNID	MD. MAMMAL	SHAFT FRAGMENT	N	1	N	-	-	-	-	-	-
28	8-185	U	UNID	UNID	FRAGMENT (JOINED)	N	2	N	-	-	-	-	-	-
29	8-185	U	UNID	UNID	FRAGMENTS	N	8	N	-	-	-	-	-	-
30	8-185	U	HUMERUS	cf. ARTIODACTYL	WORKED BONE RING	N	1	N	-	-	-	-	-	-
31	8-20	U	UNID	UNID	FRAGMENTS	N	2	N	-	-	-	-	-	-
32	8-20	R	HUMERUS	LEPUS SP.	PROXIMAL & SHAFT	N	1	N	-	-	-	-	-	-
33	8-20	R	NAVICULO-CUBOID	cf. OVIS CANADENSIS		N	1	N	-	-	-	-	-	-
34	8-20	U	UNID	LG. MAMMAL	SHAFT FRAGMENTS	N	2	N	-	-	X (ALL)	-	-	-
35	8-231	R	HUMERUS	ARTIODACTYL	PROXIMAL	N	1	N	-	-	-	-	-	-
36	8-274	U	HUMERUS	LG. MAMMAL	SHAFT FRAGMENT	N	1	N	-	-	-	-	-	-
37	8-30	L	HUMERUS	ARTIODACTYL	SHAFT FRAGMENT	N	1	N	-	-	-	-	-	-
38	8-30	R	TIBIA	cf. ARTIODACTYL	PROXIMAL	N	1	Y	-	-	-	-	-	-
39	8-30	L	CALCANEUM	cf. ARTIODACTYL	PROXIMAL	N	1	N	-	-	-	-	-	-
40	8-35	U	UNID	UNID	FRAGMENTS	N	4	N	-	-	-	-	-	-
41	8-35	R	HUMERUS	cf. PHASIANIDAE	DISTAL & SHAFT	N	2	Y	-	-	-	-	-	-
42	8-35	L	HUMERUS	cf. APHELOCOMA SP.	PROXIMAL	N	1	Y	-	-	-	-	-	-
43	8-35	A	VERTEBRAE	MD. MAMMAL	JUVENILE	N	1	Y	-	-	-	-	-	-
44	8-35	L	FEMUR	LEPUS SP.	MID SHAFT FRAGMENT	N	1	N	-	-	-	-	-	-
45	8-4	U	UNID	UNID	FRAGMENTS	N	3	N	-	-	X (ALL)	-	-	-

APPENDIX E

HUMAN REMAINS EXCAVATED DURING TESTING

Kathryn L. Wullstein
David H. Greenwald

During the testing phase, investigators excavated two extended inhumations, Features 1021 and 1029. These burials were exposed in backhoe trenches in areas that were not selected for further investigation and where no associated features were found. Feature 1021 was within the channel of North Main Canal 1 near the northern project boundary. Feature 1029 was along the south side of the South Main Canal along the extreme western project boundary.

Feature 1021

Horizontal Provenience: N 1765.85–1766.50; E 885.65
Vertical Provenience: 4.76–5.28 meters below datum (mbd)

The burial pit, within the channel of North Main Canal 1, was oriented in an east-west direction, in alignment with the canal. The pit appeared to contain a bench-like feature on each side, narrowing and extending to its maximum depth in the center. The body had been placed within the deepest portion of the pit. The pit measured 2.40 m east-west by 0.85 m north-south and was 1.72 m wide from bench wall to bench wall. The body had been placed in a supine position with the head to the east. Five vessels, all Hohokam Red Ware, were located in the eastern end of the pit, and a large sherd covered with a layer of charcoal and ash was found at the extreme western end. The legs were slightly flexed, perhaps to accommodate placement of the grave goods. The burial was dug into the canal after the canal had been abandoned, as investigators found no water-laid deposits in the burial pit. This burial was similar to another found in North Main Canal 3 during excavations in the Squaw Peak Parkway corridor. Both were in abandoned canals, were oriented to the east, and were extended.

Analysts could not determine the sex of the interred, although an earlier examination indicated that this individual was a female, approximately 35 to 45 years of age at the time of death (Volume 1:Appendix C). The skeletal remains were very fragmented at the time of analysis, although they were reported to be in good condition when discovered. Portions of the left and right scapula, clavicles, and humeri were present, as well as the left radius and ulna. The right innominate was extremely fragmented. The left femur, right and left patella, right and left tibia, and both fibulae were present. There was a minimum of eleven ribs. It was possible to identify the fifth lumbar and first two sacral vertebrae. Two cervical, three thoracic, and two lumbar vertebrae were too fragmented for determination of their exact location. A few hand bones were recovered. The metatarsals and tarsals were present for both feet.

Based on this second examination of the remains, the following determinations were made: the individual was an adult of unknown age or sex. No pathologies could be identified. Due to the poor preservation of the bone, metric data could not be gathered. However, the difference between the initial and subsequent analyses indicates the importance of conducting analytical documentation of human remains as soon as possible after their exposure, especially when the application of a preservative is not an option.

Feature 1029

Horizontal Provenience: N 1408.88–1409.90; E 535.65
Vertical Provenience: 4.57–5.00 meters below datum

This burial was placed in a rectangular pit, oriented northeast-southwest, that measured 2.00 × 0.52 m. The burial was in a supine position with the head to the northeast. The only grave accompaniment recovered was a Hohokam Red Ware scoop, found in the cranial region. Temporally, Feature 1029 compared with Feature 1021, as indicated by the presence of the redware vessels.

This inhumation was fairly complete, lacking only the feet, which were removed during excavation of the backhoe trench. Although fragmented, all cranial bones were present. The right humerus and both radii and ulnae were recovered. Portions of the left innominate were present. Both femora and tibias and the left patella were also recovered. Two thoracic, four lumbar, and five sacral vertebrae were present. The left hand was represented by the capitate, hamate, trapezium, scaphoid, trapezoid, and second metacarpal. The left calcaneus and cuboid were all that remained of the foot bones.

Analysts determined that this individual was a female, approximately 30 to 40 years of age at the time of death. They could not calculate stature or height from the remains. The maxilla contained all teeth except both third molars. The right maxillary second premolar exhibited an occlusal caries. Alveolar resorption was evident on the right mandible, where the first and second molars were located. There was postmortem loss of many mandibular teeth. Only the right incisors, left canine, left premolars, and left first molar were present. The left central incisor also exhibited an occlusal caries.

This adult female had lost several teeth before her death. Due to the poor preservation of the skeletal remains investigators could not obtain any metric data. Analysts observed no pathological conditions.

APPENDIX F

GLOSSARY

APPENDIX F
GLOSSARY

ash pit
A pit containing a high concentration of ash, usually found spatially associated with hearths or fire pits, implying the ash was secondarily deposited in the pit, after originating in a nearby hearth or fire pit.

baffle
An embankment constructed on the downslope side of a canal, consisting of canal clays reinforced by rocks and artifacts, sometimes post-reinforced in the same manner as modern rip-rap. These canal maintenance features functioned to slow erosive water currents and change the course of channel flow.

buccal-lingual wear
Striations created on the occlusal surfaces of the teeth next to the cheek and the tongue.

cache
A concentration or collection of cultural items intentionally buried in a pit.

ceramic concentration
A large number of ceramic items in a well-defined, usually small area, such as in a pit, in a structure, or on an extramural surface. Multiple pot breaks or discarded sherds may be included.

eluvium
Fine soil or sand moved and deposited by the wind.

entryway/doorway
Access to a structure consisting of a simple wall opening or slightly appended walls. Remnants should be recognized by clay pads or "threshold" stones lying to either side of the entryway.

fire pit
A pit that exhibits in situ burning, i.e., has oxidized edges but lacks a clay lining.

fire pit with rock
A fire pit containing rocks that exhibit heat or fire alteration.

hearth
A formally prepared pit for cooking or heating with a clay lining, generally found inside a structure.

horno
An earth oven with sides that are oxidized and coated with organic slag from burning; the oven may contain slag accumulations, fire-cracked rock, charcoal, other rocks, and other burned materials. These features are typically large.

lateral canal
The terminal segment of an irrigation system, which conveys water directly onto field areas.

main canal
The segment of a canal system that extends from the canal intake to the first major junction at which the channel size is significantly reduced, such as at a distribution canal.

manuport
An unmodified stone item not naturally occurring in excavated context; presumed to have been intentionally transported.

mixed artifact concentration
A well-defined concentration of at least two types of artifacts (ceramics, flaked stone, or ground stone) or one of those artifact classes plus one of any other category of artifacts such as nonhuman bone, shell, mineral, or vegetal material.

pit house
A structure with a symmetrical shape usually possessing a centered hearth, an entry, and a floor, which may or may not be plastered. Floor features should have recognizable patterning.

pit NFS
Any pit for which a function cannot be assigned or for which other morphology does not exist.

pit structure
A feature that lacks the formal characteristics (hearth and a plastered floor) of a pit house, often is irregular in shape, and displays no patterning of floor features or entryway orientation.

pit structure NFS
A structure in which one or more of the features characteristic of a pit structure have not been preserved, but whose basic morphology is that of a pit structure.

pit with burning
A pit that does not exhibit evidence of pit edge burning or oxidation but has charcoal or ash in its fill. It differs from a fire pit in that it lacks direct evidence of in situ burning on the pit edges.

pit with rock
A pit feature that contains rock. No function can be inferred, and there is no evidence of burning.

possible structure
A feature displaying inconclusive evidence for a structure, usually very disturbed or almost destroyed. Walls and floor are indistinct or not present. Distribution of ash, charcoal, and the presence of an oxidized or use-compacted surface may suggest some type of structure.

primary inhumation
The original location of a burial; the human remains have not been moved since interment.

rock cluster/rock concentration
An unnatural grouping of rocks. The cluster of stones may represent construction materials or rock that was used for heating or burning but that was not found in association with another feature, such as a pit.

secondary cremation
The calcined bone and teeth of a cremation along with charcoal from the fire. No evidence of in situ burning is present. Cremations in pottery vessels are usually included in this category.

trash deposit, other   Usually a large area of concentrated refuse. This feature type includes all trash deposits other than trash in pits, such as middens and sheet trash deposits.

trash pit   Any pit filled with trash. The feature must show evidence of intentional trash deposition or contain high quantities of trash.

true pit house   A pit house in which the sides of the pit form the walls of the structure.

wall segment   A portion of an adobe wall with no contiguous or adjoining walls.

444

REFERENCES

Abbott, David R.
   1991   Rock-type Descriptions as They Appear in Hohokam Pottery Using a Binocular Microscope. Ms. on file, Soil Systems, Inc., Phoenix.

   1993   Rough-Sort Analysis of the Pottery from the Hohokam Expressway Project. In The Pueblo Grande Project: Ceramics and Production and Exchange of Pottery in the Central Phoenix Basin, vol. 3. Soil Systems Publications in Archaeology, Phoenix. Draft.

Abbott, David R., Kim E. Beckwith, Patricia L. Crown, R. Thomas Euler, David A. Gregory, J. Ronald London, Marilyn B. Saul, Larry A. Schwalbe, Mary Bernard-Shaw, Christine R. Szuter, and Arthur W. Vokes
   1988   *The 1982–1984 Excavations at Las Colinas: Material Culture.* Arizona State Museum Archaeological Series No. 162, Vol. 4. The University of Arizona, Tucson.

Abbott, David R., and David A. Gregory
   1988   Hohokam Ceramic Wares and Types. In *The 1982–1984 Excavations at Las Colinas: Material Culture,* by David R. Abbott, Kim E. Beckwith, Patricia L. Crown, R. Thomas Euler, David A. Gregory, J. Ronald London, Marilyn B. Saul, Larry A. Schwalbe, Mary Bernard-Shaw, Christine R. Szuter, and Arthur W. Vokes, pp. 5–28. Arizona State Museum Archaeological Series No. 162, Vol. 4. The University of Arizona, Tucson.

Abbott, David R., and David M. Schaller
   1990   Electron Microprobe and Petrographic Analyses of Prehistoric Hohokam Pottery to Determine Ceramic Exchange within the Salt River Valley, Arizona. Paper presented at the Spring Meeting of the Material Research Society, San Francisco.

Abbott, David R., David M. Schaller, and Robert I. Birnie
   1991   Compositional Analysis of Hohokam Pottery from the Salt River Valley, Arizona. Paper presented at the 62nd Annual Meeting of the Southwestern Anthropological Association, Tucson.

Ackerly, Neal W.
   1979   The Southern Desert Study Area. In *An Archaeological Survey of the Cholla-Saguaro Transmission Line Corridor,* edited by Lynn Teague and Linda Mayro, pp. 267–406. Arizona State Museum Archaeological Series No. 135. The University of Arizona, Tucson.

Ackerly, Neal W., Jerry B. Howard, and Randall H. McGuire
   1987   *La Ciudad Canals: A Study of Hohokam Irrigation Systems at the Community Level.* Office of Cultural Resource Management Anthropological Field Studies No. 17. Arizona State University, Tempe.

Ackerly, Neal W., JoAnn E. Kisselburg, and Richard J. Martynec
   1989   Canal Junctions and Water Control Features. In *Prehistoric Agricultural Activities on the Lehi-Mesa Terrace: Perspectives on Hohokam Irrigation Cycles,* edited by Neal W. Ackerly and T. Kathleen Henderson, pp. 146–183. Northland Research, Inc., Flagstaff.

Adams, Jenny L.
   1991   The Mechanics of Manos and Metates. Paper presented at the 62nd Annual Meeting of the Southwestern Anthropological Association, Tucson.

Adams, Karen R.
   1986   A Model to Assess New World *Hordeum pusillum* (Barley) Characteristics in Natural and Human Habitats. Paper presented at the 9th Annual Ethnobiology Conference, Albuquerque.

   1987   Little Barley (*Hordeum pusillum* Nutt.) as a Possible New World Domesticate. In *La Ciudad: Specialized Studies in the Economy, Environment, and Culture of La Ciudad,* part 3, edited by JoAnn E. Kisselburg, Glen E. Rice, and Brenda L. Shears, pp. 203–237. Arizona State University Anthropological Field Studies No. 20. Tempe.

Ahlstrom, Richard V. N.
   1985   *The Interpretation of Archaeological Tree-Ring Dates.* Ph.D. dissertation, Department of Anthropology, The University of Arizona. University Microfilms, Ann Arbor.

Antieau, John M.
   1981   *The Palo Verde Archaeological Investigations, Hohokam Settlement at the Confluence: Excavations along the Palo Verde Pipeline.* Museum of Northern Arizona Research Paper No. 20. Flagstaff.

Anyon, Roger, and Steven A. LeBlanc
   1984   *The Galaz Ruin, a Prehistoric Mimbres Village in Southwestern New Mexico.* University of New Mexico Press, Albuquerque.

Barker, Graeme
   1975   Early Neolithic Land Use in Yugoslavia. *Proceedings of the Prehistoric Society* 41:85–104.

Beckwith, Kim E.
   1988   Intrusive Ceramic Wares and Types. In *The 1982–1984 Excavations at Las Colinas: Material Culture,* by David R. Abbott, Kim E. Beckwith, Patricia L. Crown, R. Thomas Euler, David A. Gregory, J. Ronald London, Marilyn B. Saul, Larry A. Schwalbe, Mary Bernard-Shaw, Christine R. Szuter, and Arthur W. Vokes, pp. 199–256. Arizona State Museum Archaeological Series No. 162, Vol. 4. The University of Arizona, Tucson.

Benitez, Alex
   1991   A Preliminary Study on the Shaping of Groundstone Artifacts from the Homol'ovi III Site. Paper presented at the 62nd Annual Meeting of the Southwestern Anthropological Association, Tucson.

Bernard-Shaw, Mary
   1983   The Stone Tool Assemblage of the Salt-Gila Aqueduct Project Sites. In *Hohokam Archaeology along the Salt-Gila Aqueduct, Central Arizona Project: Material Culture,* edited by Lynn S. Teague and Patricia L. Crown, pp. 381–443. Arizona State Museum Archaeological Series No. 150, Vol. 8. The University of Arizona, Tucson.

   1984   A Systematic and Comparative Study of Hohokam Tabular Tool Use. Paper presented at the 49th Annual Meeting of the Society for American Archaeology, Portland, Oregon.

1988　Chipped Stone Artifacts. In *The 1982–1984 Excavations at Las Colinas: Material Culture*, by David R. Abbott, Kim E. Beckwith, Patricia L. Crown, R. Thomas Euler, David A. Gregory, J. Ronald London, Marilyn B. Saul, Larry A. Schwalbe, Mary Bernard-Shaw, Christine Szuter, and Arthur W. Vokes, pp. 273–298. Arizona State Museum Archaeological Series No. 162, Vol. 4. The University of Arizona, Tucson.

Binford, L., and G. Quimby
1963　Indian Sites and Chipped Stone Materials in the Northern Lake Michigan Area. *Fieldiana, Anthropology* 36(12).

Birnie, Robert I., and Mary-Ellen Walsh-Anduze
1991　*Archaeological Test Excavations at Pueblo Grande Cultural Park, AZ U:9:7 (ASM), Phoenix, Arizona: The Museum Expansion Project.* Soil Systems Technical Report No. 91-1. Phoenix.

Bohrer, Vorsila L.
1975　The Prehistoric and Historic Role of the Cool-Season Grasses in the Southwest. *Economic Botany* 29:199–207.

1984　Domesticated and Wild Crops in the CAEP Study Area. In *Prehistoric Cultural Development in Central Arizona: Archaeology of the Upper New River Region*, edited by Patricia M. Spoerl and George J. Gumerman, pp. 183–259. Southern Illinois University Center for Archaeological Investigations Occasional Paper No. 5. Carbondale.

1986　Guideposts in Ethnobotany. *Journal of Ethnobiology* 6:27–43.

1987　The Plant Remains from La Ciudad, a Hohokam Site in Phoenix. In *La Ciudad: Specialized Studies in the Economy, Environment, and Culture of La Ciudad*, part 3, edited by JoAnn E. Kisselburg, Glen E. Rice, and Brenda L. Shears, pp. 67–202. Arizona State University Anthropological Field Studies No. 20. Tempe.

Boserup, Ester
1965　*The Conditions of Agricultural Growth: The Economics of Agrarian Change under Population Pressure.* Aldine, Chicago.

Bostwick, Todd W.
1988　Centennial Wash Ground Stone Assemblages. In *An Investigation of Archaic Subsistence and Settlement in the Harquahala Valley, Maricopa County, Arizona*, assembled by Todd W. Bostwick, pp. 136–166. Northland Research, Inc., Flagstaff.

Bostwick, Todd W., and James H. Burton
1989　An Initial Study of Basalt Sources in Central Arizona Used for Hohokam Grinding Implements. Paper presented at the 1989 Spring Conference of the Arizona Archaeological Council, Flagstaff.

Breternitz, David A.
1966　*An Appraisal of Tree-Ring Dated Pottery in the Southwest.* Anthropological Papers of The University of Arizona No. 2. Tucson.

Bruder, J. Simon
    1983   *Archaeological Investigations in the Adobe Dam Project Area.* Museum of Northern Arizona Research Paper No. 27. Flagstaff.

Brumfiel, Elizabeth
    1976   Regional Growth in the Eastern Valley of Mexico: A Test of the "Population Pressure" Hypothesis. In *The Early Mesoamerican Village,* edited by Kent V. Flannery, pp. 234–249. Academic Press, New York.

Brunson, Judy L.
    1989   *The Social Organization of the Los Muertos Hohokam: A Reanalysis of Cushing's Hemenway Expedition Data.* Unpublished Ph.D. dissertation, Department of Anthropology, Arizona State University, Tempe.

BRW (Bennett, Ringrose, Wolsfeld, Jarvis, and Gardner, Inc.)
    1986   *Sky Harbor Center Pedestrian Survey and Cultural Resources Program.* Prepared for Phoenix Sky Harbor Center, Community and Economic Development Department, City of Phoenix.

Bryan, Kirk
    1929   Flood-water Farming. *Geographical Review* 19:444–456.

Burt, William H., and Richard P. Grossenheider
    1976   *A Field Guide to the Mammals of America North of Mexico.* Houghton-Mifflin, Boston.

Burton, Robert J.
    1976   An Archaeological Survey of Bureau of Land Management Exchange Parcels near Phoenix, Kingman, and Tucson, Arizona. Ms. on file, Bureau of Land Management. Phoenix.

Cable, John S.
    1983a   Part I: Summary and Concluding Remarks. In *City of Phoenix, Archaeology of the Original Townsite: Block 28-North,* edited by John S. Cable, Susan L. Henry, and David E. Doyel, pp. 201–203. Soil Systems Publications in Archaeology No. 2. Phoenix.

    1983b   Site Structure and Function of Block 28-North: A Seasonally Used Hohokam Field House Site. In *City of Phoenix, Archaeology of the Original Townsite: Block 28-North,* edited by John S. Cable, Susan L. Henry, and David E. Doyel, pp. 153–182. Soil Systems Publications in Archaeology No. 2. Phoenix.

    1985   Constructing Hypothetical Occupations within an Absolute Chronological Framework. In *City of Phoenix, Archaeology of the Original Townsite, Block 24-East,* edited by John S. Cable, Kathleen S. Hoffman, David E. Doyel, and Frank Ritz, pp. 85–126. Soil Systems Publications in Archaeology No. 8. Phoenix.

    1991   The Role of Irrigation Agriculture in the Formation and Sociopolitical Development of Early Hohokam Villages in the Lowlands of the Phoenix Basin, Arizona. In *Prehistoric Irrigation in Arizona: Symposium 1988,* edited by Cory Dale Breternitz, pp. 107–137. Soil Systems Publications in Archaeology No. 17. Phoenix.

Cable, John S., and David E. Doyel
1983 The Organization and Scheduling of Subsistence Strategies at Colonial and Sedentary Period Field Houses in the Hohokam Core Area. In *City of Phoenix, Archaeology of the Original Townsite: Block 28-North*, edited by John S. Cable, Susan L. Henry, and David E. Doyel, pp. 183–199. Soil Systems Publications in Archaeology No. 2. Phoenix.

1984 The Implications for Field Houses for Modeling Hohokam Agricultural Systems. In *City of Phoenix Archaeology of the Original Townsite: The Murphy's Addition*, edited by John S. Cable, Susan L. Henry, and David E. Doyel, pp. 259–279. Soil Systems Publications in Archaeology No. 3. Phoenix.

1985 Hohokam Land-Use Patterns along the Terraces of the Lower Salt River Valley: The Central Phoenix Project. In *Proceedings of the 1983 Hohokam Symposium, Part I*, edited by Alfred E. Dittert, Jr., and Donald E. Dove, pp. 263–310. Arizona Archaeological Society Occasional Paper No. 2. Phoenix.

1987 Pioneer Period Village Structure and Settlement Pattern in the Phoenix Basin. In *The Hohokam Village: Site Structure and Organization*, edited by David E. Doyel, pp. 21–70. Southwestern and Rocky Mountain Division of the American Association for the Advancement of Science, Glenwood Springs, Colorado.

Cable, John S., and Ronald R. Gould
1988 The Casa Buena Ceramic Assemblage: A Study of Typological Systematics and Ceramic Change in Classic Period Assemblages. In *Excavations at Casa Buena: Changing Hohokam Land Use along the Squaw Peak Parkway*, edited by Jerry B. Howard, pp. 271–357. Soil Systems Publications in Archaeology No. 11, Vol. 1. Phoenix.

Cable, John S., Kathleen S. Hoffman, David E. Doyel, and Frank Ritz (editors)
1985 *City of Phoenix, Archaeology of the Original Townsite: Block 24-East*. Soil Systems Publications in Archaeology No. 8. Phoenix.

Cable, John S., and Douglas R. Mitchell
1988 La Lomita Pequeña in Regional Perspective. In *Excavations at La Lomita Pequeña: A Santa Cruz/Sacaton Phase Hamlet in the Salt River Valley*, edited by Douglas R. Mitchell, pp. 95–177. Soil Systems Publications in Archaeology No. 10. Phoenix.

1991 Settlement Growth and Integration in a Large Hohokam Canal System. In *Prehistoric Irrigation in Arizona: Symposium 1988*, edited by Cory Dale Breternitz, pp. 89–106. Soil Systems Publications in Archaeology No. 17. Phoenix.

Castetter, Edward F., and Willis H. Bell
1942 *Pima and Papago Indian Agriculture*. Inter-American Studies No. 1. University of New Mexico Press, Albuquerque.

Castetter, Edward F., Willis H. Bell, and Alvin R. Grove
1938 *Ethnobiological Studies in the American Southwest: The Early Utilization and the Distribution of Agave in the American Southwest*. University of New Mexico Bulletin No. 6. Albuquerque.

Castetter, Edward F., and Ruth Underhill
    1935   *Ethnobiological Studies in the American Southwest 2: The Ethnobiology of the Papago Indians.* University of New Mexico Bulletin No. 275, Biological Series No. 4, Vol. 8. Albuquerque.

Chamberlin, Ralph V.
    1964   The Ethnobotany of the Gosiute Indians of Utah. Reprinted. *American Anthropological Association Memoirs* 2:329–405. Kraus Reprint Corp., New York. Originally printed 1911.

Chisholm, M. D. I.
    1962   *Rural Settlement and Land Use: An Essay in Location.* Hutchinson, London.

Christaller, W.
    1933   *Die Zentralen Orte in Suddeutschland.* Jena.

Ciolek-Torrello, Richard
    1988   Chronology. In *Hohokam Settlement along the Slopes of the Picacho Mountains: Synthesis and Conclusions, Tucson Aqueduct Project,* edited by Richard Ciolek-Torrello and David R. Wilcox, pp. 42–120. Museum of Northern Arizona Research Paper No. 35, Vol. 6. Flagstaff.

Ciolek-Torrello, Richard, and David H. Greenwald
    1988   Chronology. In *Archaeological Investigations at the Dutch Canal Ruin, Phoenix, Arizona: Archaeology and History along the Papago Freeway Corridor,* edited by David H. Greenwald and Richard Ciolek-Torrello, pp. 153–167. Museum of Northern Arizona Research Paper No. 38. Flagstaff.

Collins, M. B.
    1975   Lithic Technology as a Means of Processual Inference. In *Lithic Technology,* edited by E. H. Swanson, pp. 15–34. Aldine, Chicago.

Colton, Harold S.
    1939   *Prehistoric Culture Units and Their Relationships in Northern Arizona.* Museum of Northern Arizona Bulletin No. 17. Flagstaff.

    1974   Hopi History and Ethnobotany. In *Hopi Indians,* by Harold S. Colton. Garland, New York.

Cowan, C. Wesley
    1978   The Prehistoric Use and Distribution of Maygrass in Eastern North America: Cultural and Phytogeographical Implications. In *The Nature and Status of Ethnobotany,* edited by Richard I. Ford, pp. 263–288. University of Michigan Anthropological Papers No. 67. Ann Arbor.

    1985   Understanding the Evolution of Plant Husbandry in Eastern North America: Lessons from Botany, Ethnography, and Archaeology. In *Prehistoric Food Production in North America,* edited by Richard I. Ford, pp. 205–243. University of Michigan Anthropological Papers No. 75. Ann Arbor.

Crabtree, Donald E.
1972 *An Introduction to Flintworking*. Occasional Papers of the Idaho State University Museum No. 28. Pocatello.

Crosswhite, Frank S.
1981 Desert Plants, Habitat and Agriculture in Relation to the Major Pattern of Cultural Differentiation in the O'odham People of the Sonoran Desert. *Desert Plants* 2(3):47–76.

Crown, Patricia L.
1981 Analysis of the Las Colinas Ceramics. In *The 1968 Excavations at Mound 8, Las Colinas Ruins Group, Phoenix, Arizona*, edited by Laurens C. Hammack and Alan P. Sullivan, pp. 87–170. Arizona State Museum Archaeological Series No. 154. The University of Arizona, Tucson.

1983 Introduction: Field Houses and Farmsteads in South-Central Arizona. In *Hohokam Archaeology along the Salt-Gila Aqueduct, Central Arizona Project: Material Culture*, edited by Lynn S. Teague and Patricia L. Crown, pp. 205–248. Arizona State Museum Archaeological Series No. 150, Vol. 8. The University of Arizona, Tucson.

1984 The Morphology and Function of Hohokam Small Structures. Paper presented at the 49th Annual Meeting of the Society for American Archaeology, Portland, Oregon.

1985a Morphology and Function of Hohokam Small Structures. *The Kiva* 50:75–94.

1985b Intrusive Ceramics and the Identification of Hohokam Exchange Networks. In *Proceedings of the 1983 Hohokam Conference, Part II*, edited by Alfred E. Dittert, Jr., and Donald E. Dove, pp. 439–458. Arizona Archaeological Society Occasional Paper No. 2. Phoenix.

1987 Classic Period Hohokam Settlement and Land Use in the Casa Grande Ruins Area, Arizona. *Journal of Field Archaeology* 14:147–162.

1991 The Hohokam: Current Views of Prehistory and the Regional System. In *Chaco and Hohokam: Prehistoric Regional Systems in the American Southwest*, edited by Patricia L. Crown and W. James Judge, pp. 135–157. School of American Research Press, Santa Fe.

Crown, Patricia L., and Ronald L. Bishop
1987 Convergence in Ceramic Manufacturing Traditions in the Late Prehistoric Southwest. Paper presented at the 52nd Annual Meeting of the Society for American Archaeology, Toronto.

Crown, Patricia L., Larry A. Schwalbe, and J. Ronald London
1988 X-ray Fluorescence Analysis of Materials Variability in Las Colinas Ceramics. In *The 1982–1984 Excavations at Las Colinas: Material Culture*, by David R. Abbott, Kim E. Beckwith, Patricia L. Crown, R. Thomas Euler, David A. Gregory, J. Ronald London, Marilyn B. Saul, Larry A. Schwalbe, Mary Bernard-Shaw, Christine Szuter, and Arthur W. Vokes, pp. 29–71. Arizona State Museum Archaeological Series No. 162, Vol. 4. The University of Arizona, Tucson.

Crown, Patricia L., and Earl W. Sires, Jr.
   1984   The Hohokam Chronology and Salt-Gila Aqueduct Research. In *Hohokam Archaeology along the Salt-Gila Aqueduct, Central Arizona Project: Synthesis and Conclusions*, edited by Lynn S. Teague and Patricia L. Crown, pp. 73–86. Arizona State Museum Archaeological Series No. 150, Vol. 9. The University of Arizona, Tucson.

Cummings, Linda Scott
   1988   Pollen Analysis. In *Archaeological Investigations at the Dutch Canal Ruin, Phoenix, Arizona: Archaeology and History along the Papago Freeway Corridor*, edited by David H. Greenwald and Richard Ciolek-Torrello, pp. 180–196. Museum of Northern Arizona Research Paper No. 38. Flagstaff.

   1989   Pollen Analysis for the Waddell Data Recovery Project: An Examination of the Subsistence Base from the Agua Fria and New River Drainages. In *Settlement, Subsistence, and Specialization in the Northern Periphery: The Waddell Project*, edited by Margerie Green, pp. 801–850. Archaeological Consulting Services, Ltd., Cultural Resources Report No. 65. Tempe.

Curtin, Leonora Scott Muse
   1984   *By the Prophet of the Earth, Ethnobotany of the Pima*. Reprinted. The University of Arizona Press, Tucson. Originally published 1949, San Vicente Foundation, Santa Fe.

Dean, Jeffrey S.
   1978   Independent Dating in Archaeological Analysis. In *Advances in Archaeological Method and Theory*, vol. 1, edited by Michael B. Schiffer, pp. 223–255. Academic Press, New York.

   1991   Thoughts on Hohokam Chronology. In *Exploring the Hohokam: Prehistoric Desert Peoples of the American Southwest*, edited by George J. Gumerman, pp. 61–149. An Amerind Foundation Publication, Dragoon, Arizona. University of New Mexico Press, Albuquerque.

Di Peso, Charles C.
   1956   *The Upper Pima of San Cayetano del Tumacacori: An Archaeo-historical Reconstruction of the Ootam of Pimeria Alta*. The Amerind Foundation No. 7. Dragoon, Arizona.

Dodd, Walter A., Jr.
   1979   The Wear and Use of Battered Tools at Armijo Rockshelter. In *Lithic Use-Wear Analysis*, edited by Brian Hayden, pp. 231–242. Academic Press, New York.

Doelle, William H.
   1976   *Desert Resources and Hohokam Subsistence: The CONOCO Florence Project*. Arizona State Museum Archaeological Series No. 103. The University of Arizona, Tucson.

Doyel, David E.
   1974   *Excavations in the Escalante Ruin Group, Southern Arizona*. Arizona State Museum Archaeological Series No. 37. The University of Arizona, Tucson.

   1979   The Prehistoric Hohokam of the Arizona Desert. *American Scientist* 67:544–554.

1980   Hohokam Social Organization and the Sedentary to Classic Transition. In *Current Issues in Hohokam Prehistory*, edited by David E. Doyel and Fred Plog, pp. 23–40. Arizona State University Anthropological Research Papers No. 23. Tempe.

1981   *Late Hohokam Prehistory in Southern Arizona*. Contributions to Archaeology No. 2. Gila Press, Scottsdale, Arizona.

1985   Exchange and Interaction. In *Hohokam Settlement and Economic Systems in the Central New River Drainage, Arizona*, edited by David E. Doyel and Mark D. Elson, pp. 715–725. Soil Systems Publications in Archaeology No. 4. Phoenix.

1987   The Hohokam Village. In *The Hohokam Village: Site Structure and Organization*, edited by David E. Doyel, pp. 1–20. Southwest and Rocky Mountain Division of the American Association for the Advancement of Science, Glenwood Springs, Colorado.

1991   Hohokam Exchange and Interaction. In *Chaco and Hohokam: Prehistoric Regional Systems in the American Southwest*, edited by Patricia L. Crown and W. James Judge, pp. 225–252. School of American Research Press, Santa Fe.

Eddy, Frank W., and Maurice E. Cooley
  1983   *Cultural and Environmental History of Cienega Valley, Southeastern Arizona*. Anthropological Papers No. 43. The University of Arizona Press, Tucson.

Eighmy, Jeffrey L.
  1990   Archaeomagnetic Dating: Practical Problems for the Archaeologist. In *Archaeomagnetic Dating*, edited by Jeffrey L. Eighmy and Robert S. Sternberg, pp. 33–64. The University of Arizona Press, Tucson.

Eighmy, Jeffrey L., and Kathleen A. Baker
  1991   Archaeomagnetic Laboratory Results for Sites AZ T:12:47 (ASM) with an Updated Analysis of Samples from NA19324. Ms. on file, SWCA, Inc., Environmental Consultants, Flagstaff.

Eighmy, Jeffrey L., and Jerry B. Howard
  1991   Direct Dating of Prehistoric Canal Sediments Using Archaeomagnetism. *American Antiquity* 56:88–102.

Eighmy, Jeffrey L., and Pamela Y. Klein
  1991   Archaeomagnetic Laboratory Results for the Sky Harbor Project. Ms. on file, SWCA, Inc., Environmental Consultants, Flagstaff.

Eighmy, Jeffrey L., and Randall H. McGuire
  1988   *Archaeomagnetic Dates and the Hohokam Phase Sequence*. Colorado State University Archaeomagnetic Technical Series No. 3. Fort Collins, Colorado.

Eighmy, Jeffrey L., and Robert S. Sternberg (editors)
  1990   *Archaeomagnetic Dating*. The University of Arizona Press, Tucson.

Ellis, Florence H.
1978   Small Structures Used by Historic Pueblo Peoples and Their Immediate Ancestors. In *Limited Activity and Occupation Sites: A Collection of Conference Papers*, edited by Albert E. Ward, pp. 59–68. Center for Anthropological Studies Contributions to Anthropological Studies No. 1. Albuquerque.

Elstien, Elizabeth
1989   Chipped Stone Artifact Analysis. In *Settlement, Subsistence, and Specialization in the Northern Periphery: The Waddell Project*, edited by Margerie Green, pp. 643–735. Archaeological Consulting Services, Ltd., Cultural Resources Report No. 65. Tempe.

Ezell, Paul H.
1961   *The Hispanic Acculturation of the Gila River Pimas*. Memoirs of the American Anthropological Association No. 90. Menasha, Wisconsin.

Felger, Richard S., and Mary Beck Moser
1985   *People of the Desert and Sea, Ethnobotany of the Seri Indians*. The University of Arizona Press, Tucson.

Findlow, Frank J., and Marisa Bolognese
1982   A Preliminary Analysis of Prehistoric Obsidian Use within the Mogollon Area. In *Mogollon Archaeology: Proceedings of the 1980 Mogollon Conference*, edited by Patrick H. Beckett, pp. 297–315. Acomar Books, Ramona, California.

Fish, Suzanne K.
1984   Agriculture and Subsistence Implications of the Salt-Gila Aqueduct Project Pollen Analysis. In *Hohokam Archaeology along the Salt-Gila Aqueduct, Central Arizona Project: Environment and Subsistence*, edited by Lynn S. Teague and Patricia L. Crown, pp. 111–139. Arizona State Museum Archaeological Series No. 150, Vol. 7. The University of Arizona, Tucson.

1987   Marana Sites Pollen Analysis. In *Studies in the Hohokam Community of Marana*, edited by Glen E. Rice, pp. 161–170. Office of Cultural Resource Management Anthropological Field Studies No. 15. Arizona State University, Tempe.

Fish, Suzanne K., Paul R. Fish, Charles Miksicek, and John Madsen
1985   Prehistoric Agave Cultivation in Southern Arizona. *Desert Plants* 7:107–112.

Fish, Suzanne K., Robert E. Gasser, and Jeanne K. Swarthout
1985   Site Function and Subsistence Patterns. In *Studies in the Hohokam and Salado of the Tonto Basin*, edited by Glen Rice. Office of Cultural Resource Management Report No. 63. Arizona State University, Tempe.

Flannery, Kent V.
1976   Empirical Determination of Site Catchments in Oaxaca and Tehuacan. In *The Early Mesoamerican Village*, edited by Kent V. Flannery, pp. 103–117. Academic Press, New York.

Flenniken, J. Jeffery
1981   *Replicative Systems Analysis: A Model Applied to the Vein Quartz Artifacts from the Hoko River Site*. Washington State University Report No. 59. Pullman.

Font, Pedro
    1775   Diario que forme el P. Fr. Pedro Font. Colegio de la Santa Cruz de Querétaro, Sacado del Borrodor que Escribo en el Camino, del Viage que Hizo a Monterey y Puerte de San Francisco. Copies in the Archivo General y Pública de la Nación, Mexico City, and the Bancroft Library, University of California, Berkeley.

Fratt, Lee, and Maggie Biancaniello
    1991   Looking Beyond the Norm: A Study of Raw Material Used for Ground Stone Artifacts from Homol'ovi III. Paper presented at the 62nd Annual Meeting of the Southwestern Anthropological Association, Tucson.

Gasser, Robert E.
    1982a   Flotation and Macroplant Analysis, Blocks 1 and 2, Phoenix City Project. In *City of Phoenix, Archaeology of the Original Townsite: Blocks 1 and 2*, edited by John S. Cable, Susan L. Henry, and David E. Doyel, pp. 417–430. Soil Systems Publications in Archaeology No. 1. Phoenix.

    1982b   Hohokam Use of Desert Plant Foods. *Desert Plants* 3:6–235.

    1983   Analysis of Prehistoric Flotation Samples, Block 28-North, Phoenix City Project. In *City of Phoenix, Archaeology of the Original Townsite: Block 28-North*, edited by John S. Cable, Susan L. Henry, and David E. Doyel, pp. 441–446. Soil Systems Publications in Archaeology No. 2. Phoenix.

    1984   Analysis of Prehistoric Flotation Samples from the Murphy's Addition. In *City of Phoenix, Archaeology of the Original Townsite: The Murphy's Addition*, edited by John S. Cable, Susan L. Henry, and David E. Doyel, pp. 181–194. Soil Systems Publications in Archaeology No. 3. Phoenix.

    1985   Appendix C: Macrobotanical Analysis. In *City of Phoenix, Archaeology of the Original Townsite: Block 24-East*, edited by John S. Cable, Kathleen S. Hoffman, David E. Doyel, and Frank Ritz, pp. 391–394. Soil Systems Publications in Archaeology No. 8. Phoenix.

    1988   The Casa Buena Flotation Analysis. In *Excavations at Casa Buena: Changing Hohokam Land Use along the Squaw Peak Parkway*, edited by Jerry B. Howard, pp. 561–586. Soil Systems Publications in Archaeology No. 11. Phoenix.

Gasser, Robert E., and Scott M. Kwiatkowski
    1991   Food for Thought: Recognizing Patterns in Hohokam Subsistence. In *Exploring the Hohokam: Prehistoric Desert Peoples of the American Southwest*, edited by George J. Gumerman, pp. 417–460. University of New Mexico Press, Albuquerque.

Geib, Phil R.
    1986   *Grinding Implement Production near Riviera, Arizona: Archaeological Investigations at the Riviera Substation*. Northern Arizona University Archaeological Report No. 760. Flagstaff.

Gilmore, Melvin R.
    1977   *Uses of Plants by the Indians of the Missouri River Region*. Reprinted. University of Nebraska Press, Lincoln. Originally published 1919, Bureau of American Ethnology, Washington, D.C.

Gladwin, Harold S.
 1948 *Excavations at Snaketown IV: Review and Conclusions.* Medallion Papers No. 38. Gila
       Pueblo, Globe, Arizona.

Gladwin, Harold S., Emil W. Haury, E. B. Sayles, and Nora Gladwin
 1937 *Excavations at Snaketown: Material Culture.* Medallion Papers No. 25. Gila Pueblo,
       Globe, Arizona.

Goodyear, Albert C., III
 1975 *Hecla II and III: An Interpretive Study of Archaeological Remains from the Lakeshore
       Project, Papago Reservation, South Central Arizona.* Arizona State University
       Anthropological Research Paper No. 9. Tempe.

Grady, Mark A.
 1976 *Aboriginal Agrarian Adaptation to the Sonoran Desert: A Regional Synthesis and Research
       Design.* Unpublished Ph.D. dissertation, Department of Anthropology, The University of
       Arizona, Tucson.

Graybill, Donald A.
 1989 The Reconstruction of Prehistoric Salt River Streamflow. In *The 1982–1984 Excavations
       at Las Colinas: Environment and Subsistence,* by Donald A. Graybill, David A. Gregory,
       Fred L. Nials, Suzanne K. Fish, Robert E. Gasser, Charles H. Miksicek, and Christine
       R. Szuter, pp. 25–38. Arizona State Museum Archaeological Series No. 162, Vol. 5. The
       University of Arizona, Tucson.

Graybill, Donald A., and Fred L. Nials
 1989 Aspects of Climate, Streamflow, and Geomorphology Affecting Irrigation Patterns in the
       Salt River Valley. In *The 1982–1984 Excavations at Las Colinas: Environment and
       Subsistence,* by Donald A. Graybill, David A. Gregory, Fred L. Nials, Suzanne K. Fish,
       Robert E. Gasser, Charles H. Miksicek, and Christine R. Szuter, pp. 5–23. Arizona State
       Museum Archaeological Series No. 162, Vol. 5. The University of Arizona, Tucson.

Grayson, Donald K.
 1984 *Quantitative Zooarchaeology, Topics in the Analysis of Archaeological Faunas.* Academic
       Press, New York.

Green, Margerie
 1989 Turquoise Source Analysis. In *Settlement, Subsistence, and Specialization in the Northern
       Periphery: The Waddell Project,* edited by Margerie Green, pp. 776–779. Archaeological
       Consulting Services, Ltd., Cultural Resources Report No. 65. Tempe.

Greenhouse, Ruth, Robert E. Gasser, and Jannifer W. Gish
 1981 Cholla Bud Roasting Pits: An Ethnoarchaeological Example. *The Kiva* 46:227–242.

Greenwald, David H.
 1988a Water Conveyance Features. In *Archaeological Investigations at the Dutch Canal Ruin,
       Phoenix, Arizona: Archaeology and History along the Papago Freeway Corridor,* edited
       by David H. Greenwald and Richard Ciolek-Torrello, pp. 54–89. Museum of Northern
       Arizona Research Paper No. 38. Flagstaff.

1988b Introduction. In *Archaeological Investigations at the Dutch Canal Ruin, Phoenix, Arizona: Archaeology and History along the Papago Freeway Corridor*, edited by David H. Greenwald and Richard Ciolek-Torrello, pp. 1–23. Museum of Northern Arizona Research Paper No. 38. Flagstaff.

1988c Material Culture. In *Archaeological Investigations at the Dutch Canal Ruin, Phoenix, Arizona: Archaeology and History along the Papago Freeway Corridor*, edited by David H. Greenwald and Richard Ciolek-Torrello, pp. 122–152. Museum of Northern Arizona Research Paper No. 38. Flagstaff.

Greenwald, David H., and Mark L. Chenault
1991 Differentiating the Effects of Urbanization, Alluviation, and Time on Hohokam Sites in the Phoenix Basin. Paper presented at the 56th Annual Meeting of the Society for American Archaeology, New Orleans. Ms. on file, SWCA, Inc., Environmental Consultants, Tucson.

Greenwald, David H., and Richard Ciolek-Torrello
1987 An Early Hohokam Canal System on the First Terrace of the Salt River. Paper presented at the 1987 Hohokam Symposium, Arizona State University, Tempe.

1988a *Archaeological Investigations at the Dutch Canal Ruin, Phoenix, Arizona: Archaeology and History along the Papago Freeway Corridor*. Museum of Northern Arizona Research Paper No. 38. Flagstaff.

1988b Locus E. In *Hohokam Settlement along the Slopes of the Picacho Mountains: The Brady Wash Sites, Tucson Aqueduct Project*, edited by Richard Ciolek-Torrello, Martha M. Callahan, and David H. Greenwald, pp. 254–365. Museum of Northern Arizona Research Paper No. 35, Vol. 2. Flagstaff.

Greenwald, Dawn M.
1988a Ground Stone. In *Hohokam Settlement along the Slopes of the Picacho Mountains: Material Culture, Tucson Aqueduct Project*, edited by Martha M. Callahan, pp. 127–220. Museum of Northern Arizona Research Paper No. 35, Vol. 4. Flagstaff.

1988b Lithic Studies. In *Investigations of the Baccharis Site and Extension Arizona Canal: Historic and Prehistoric Land Use Patterns in the Northern Salt River Valley*, by David H. Greenwald, pp. 103–138. Museum of Northern Arizona Research Paper No. 40. Flagstaff.

1988c Flaked Stone. In *Hohokam Settlements along the Slopes of the Picacho Mountains: Material Culture, Tucson Aqueduct Project*, edited by Martha M. Callahan, pp. 221–282. Museum of Northern Arizona Research Paper No. 35, Vol. 4. Flagstaff.

1990 *A Functional Evaluation of Hohokam Food Grinding Systems*. Unpublished master's thesis, Department of Anthropology, Northern Arizona University, Flagstaff.

1991 Prehistoric Food Grinding Systems in the Hohokam Region. Paper presented at the 62nd Annual Meeting of the Southwestern Anthropological Association, Tucson.

Gregory, David A.
1987   The Morphology of Platform Mounds and the Structure of Classic Period Hohokam Sites. In *The Hohokam Village: Site Structure and Organization*, edited by D. E. Doyel, pp. 183–210. Southwestern and Rocky Mountain Division of the American Association for the Advancement of Science, Glenwood Springs, Colorado.

1988   The Changing Spatial Structure of the Mound 8 Precinct. In *The 1982–1984 Excavations at Los Colinas: The Mound 8 Precinct*, edited by David A. Gregory, pp. 25–49. Arizona State Museum Archaeological Series No. 162, Vol. 3. The University of Arizona, Tucson.

1991   Form and Variation in Hohokam Settlement Patterns. In *Chaco and Hohokam: Prehistoric Regional Systems in the American Southwest*, edited by Patricia L. Crown and W. James Judge, pp. 159–193. School of American Research Press, Santa Fe.

Gregory, David A., and Fred L. Nials
1985   Observations Concerning the Distribution of Classic Period Hohokam Platform Mounds. In *Proceedings of the 1983 Hohokam Symposium, Part I*, edited by Alfred E. Dittert, Jr., and Donald E. Dove, pp. 373–388. Arizona Archaeological Society Occasional Paper No. 2. Phoenix.

Hack, John T.
1942   *The Changing Physical Environment of the Hopi Indians of Arizona*. Papers of the Peabody Museum of American Archaeology and Ethnology No. 35, Vol. 1. Harvard University, Cambridge.

Halbirt, Carl D.
1985   *Pollen Analysis of Metate Wash Samples: Evaluating Technique for Determining Metate Function*. Unpublished master's thesis, Department of Anthropology, Northern Arizona University, Flagstaff.

1989   Worked Stone Implements from La Cuenca del Sedimento. In *Prehistoric Agricultural Activities on the Lehi-Mesa Terrace: Excavations at La Cuenca del Sedimento*, edited by T. Kathleen Henderson, pp. 190–204. Northland Research, Inc., Flagstaff.

Hamblin, Kenneth W.
1985   *The Earth's Dynamic Systems: A Textbook in Physical Geology*. Burgess Publishing, Minneapolis, Minnesota. 4th ed.

Hamblin, Nancy L., Victoria Dirst, and John B. Sparling
1978   An Analysis of the Antelope House Faunal Collection, Canyon de Chelly National Monument, Arizona. *The Kiva* 43:201-230.

Harrington, H. D.
1967   *Edible Native Plants of the Rocky Mountains*. University of New Mexico Press, Albuquerque.

Haury, Emil W.
1945   *The Excavation of Los Muertos and Neighboring Ruins in the Salt River Valley, Southern Arizona*. Papers of the Peabody Museum of American Archaeology and Ethnology, Vol. 24, No. 1. Harvard University, Cambridge.

1950 *The Stratigraphy and Archaeology of Ventana Cave*. The University of Arizona Press, Tucson.

1976 *The Hohokam, Desert Farmers and Craftsmen: Excavations at Snaketown, 1964–1965*. The University of Arizona Press, Tucson.

Hayden, Brian
1980 Confusion in the Bipolar Works: Bashed Pebbles and Splintered Pieces. *Lithic Technology* 9(1):2–7.

Heacock, Laura A.
1988 Ceramic Studies. In *Investigations of the Baccharis Site and Extension Arizona Canal: Historic and Prehistoric Land Use Patterns in the Northern Salt River Valley*, edited by David H. Greenwald, pp. 80–103. Museum of Northern Arizona Research Paper No. 40. Flagstaff.

Hellson, John C., and Morgan Gadd
1974 *Ethnobotany of the Blackfoot Indians*. National Museums of Canada, Ottawa.

Henderson, T. Kathleen (editor)
1989 *Prehistoric Agricultural Activities on the Lehi-Mesa Terrace: Excavations at La Cuenca del Sedimento*. Northland Research, Inc., Flagstaff.

Henderson, T. Kathleen, and Vera Morgan
1989 Archaeological Features within the Tempe Outer Loop Corridor. In *Prehistoric Agricultural Activities on the Lehi-Mesa Terrace: Excavations at La Cuenca del Sedimento*, edited by T. Kathleen Henderson, pp. 37–73. Northland Research, Inc., Flagstaff.

Herskovitz, Robert M.
1982 AZ U:9:46: A Dual Component Hohokam Site in Tempe, Arizona. *The Kiva* 47(1–2).

Hodder, Ian, and Clive Orton
1976 *Spatial Analysis in Archaeology*. Cambridge University Press, Cambridge.

Hoffman, C. Marshall
1988 Lithic Technology and Tool Production at Casa Buena. In *Excavations at Casa Buena: Changing Hohokam Land Use along the Squaw Peak Parkway*, edited by Jerry B. Howard, pp. 359–457. Soil Systems Publications in Archaeology No. 11. Phoenix.

Hoffman, Theresa L.
1985 Pecked and Ground Stone Artifacts. In *Hohokam Settlement and Economic Systems in the Central New River Drainage, Arizona*, edited by David E. Doyel and Mark D. Elson, pp. 565–592. Soil Systems Publications in Archaeology No. 4. Phoenix.

Hoffman, Theresa L., and David E. Doyel
1985 Ground Stone Tool Production in the New River Basin. In *Hohokam Settlement and Economic Systems in the Central New River Drainage, Arizona*, edited by David E. Doyel and Mark D. Elson, pp. 521–564. Soil Systems Publications in Archaeology No. 4. Phoenix.

Hoffmeister, Donald F.
 1986   *Mammals of Arizona.* The University of Arizona Press and the Arizona Game and Fish Department, Tucson.

Howard, Ann Valdo, Richard W. Effland, Jr., Mark R. Hackbarth, Mary-Ellen Walsh, and Donald Irwin
 1990   Study Methodology. In *Archaeological Investigations at La Ciudad de Los Hornos: Lassen Substation Parcel,* compiled by Richard W. Effland, Jr., pp. 11–24. The Arizona Archaeologist No. 24. Arizona Archaeological Society, Phoenix.

Howard, Jerry B.
 1987   The Lehi Canal System: Organization of a Classic Period Community. In *The Hohokam Village: Site Structure and Organization,* edited by David E. Doyel, pp. 211–222. Southwestern and Rocky Mountain Division of the American Association for the Advancement of Science, Glenwood Springs, Colorado.

 1988   Casa Buena Architectural Types and Descriptions. In *Excavations at Casa Buena: Changing Hohokam Land Use along the Squaw Peak Parkway,* edited by Jerry B. Howard, pp. 67–144. Soil Systems Publications in Archaeology No. 11. Phoenix.

 1990   *Paleohydraulics: Techniques for Modeling the Operation and Growth of Prehistoric Canal Systems.* Unpublished master's thesis, Department of Anthropology, Arizona State University, Tempe.

 1991   System Reconstruction: The Evolution of an Irrigation System. In *The Operation and Evolution of an Irrigation System: The East Papago Canal Study,* by Jerry B. Howard and Gary Huckleberry, pp. 5.1–5.33. Soil Systems Publications in Archaeology No. 18. Phoenix.

Howard, Jerry B., and Gary Huckleberry
 1991   *The Operation and Evolution of an Irrigation System: The East Papago Canal Study.* Soil Systems Publications in Archaeology No. 18. Phoenix.

Howard, Jerry B., and David R. Wilcox
 1988   The Place of Casa Buena and Locus 2 in the Evolution of Canal System 2. In *Excavations at Casa Buena: Changing Hohokam Land Use along the Squaw Peak Parkway,* edited by Jerry B. Howard, pp. 903–939. Soil Systems Publications in Archaeology No. 11. Phoenix.

Hubbard, R. N. L. B.
 1980   Development of Agriculture in Europe and the Near East: Evidence from Quantitative Studies. *Economic Botany* 34:51–67.

Huckell, Bruce B.
 1981   The Las Colinas Flaked Stone Assemblage. In *The 1968 Excavations at Mound 8, Las Colinas Ruins Group, Phoenix, Arizona,* edited by Laurens C. Hammack and Alan P. Sullivan, pp. 171–200. Arizona State Museum Archaeological Series No. 154. The University of Arizona, Tucson.

 1987   The Milagro Site. Paper presented at the 1987 Hohokam Symposium, Arizona State University, Tempe.

1988 Late Archaic Archaeology of the Tucson Basin: A Status Report. In *Recent Research on Tucson Basin Prehistory: Proceedings of the Second Tucson Basin Conference*, edited by William H. Doelle and Paul R. Fish, pp. 57–80. Institute for American Research Anthropological Papers No. 10. Tucson.

James, Steven R.
1989 Archaeofaunal Remains from the Tempe Outer Loop Corridor. In *Prehistoric Agricultural Activities on the Lehi-Mesa Terrace: Excavations at La Cuenca del Sedimento*, edited by T. Kathleen Henderson, pp. 303–333. Northland Research, Inc., Flagstaff.

Jarman, M. R., and D. Webley
1975 Settlement and Land Use in Capitanata, Italy. In *Paleoeconomy*, edited by Eric S. Higgs, pp. 177–221. Cambridge University Press, Cambridge.

Kearney, Thomas H., and Robert H. Peebles
1960 *Arizona Flora*. University of California Press, Berkeley.

Keepax, Carole I.
1977 Contamination of Archaeological Deposits by Seeds of Modern Origin with Particular Reference to the Use of Flotation. *Journal of Archaeological Science* 4:221–229.

Klein, Richard G., and Kathryn Cruz-Uribe
1984 *The Analysis of Animal Bones from Archaeological Sites*. University of Chicago Press, Chicago.

Knight, Paul J.
1978 *The Role of Seed Morphology in Identification of Archaeological Remains*. Unpublished master's thesis, Department of Biology, University of New Mexico, Albuquerque.

Krochmal, Arnold, and Connie Krochmal
1973 *A Guide to the Medicinal Plants of the United States*. Quadrangle, The New York Times Book Co., New York.

Kwiatkowski, Scott
1988a *The Effects of Postoccupational Disturbance on Archaeobotanical Data from AZ U:9:24 (ASU)*. Unpublished master's thesis, Department of Anthropology, Arizona State University, Tempe.

1988b Flotation, Macrobotanical and Charcoal Analyses. In *Excavations at La Lomita Pequeña: A Santa Cruz/Sacaton Phase Hamlet in the Salt River Valley*, edited by Douglas R. Mitchell, pp. 231–289. Soil Systems Publications in Archaeology No. 10. Phoenix.

1988c The Macrobotanical and Flotation Analyses of Small Sites in the East Papago Freeway Corridor, Including an Archaeobotanical Study of Hohokam Fieldhouses. In *Arizona Department of Transportation Archaeological Testing Program: Part 2, East Papago Freeway*, edited by Daniel G. Landis, pp. 207–231. Soil Systems Publications in Archaeology No. 13. Phoenix.

1989a El Caserío Flotation and Wood Charcoal Studies. In *El Caserío: Colonial Period Settlement along the East Papago Freeway*, edited by Douglas R. Mitchell, pp. 143–178. Soil Systems Publications in Archaeology No. 14. Phoenix.

1989b   The Paleoethnobotany of the Grand Canal Ruins: Results from Flotation, Macrobotanical, and Wood Charcoal Analyses. In *Archaeological Investigations at the Grand Canal Ruins: A Classic Period Site in Phoenix, Arizona,* edited by Douglas R. Mitchell, pp. 497–558. Soil Systems Publications in Archaeology No. 12. Phoenix.

1990   La Lomita Flotation, Macrobotanical, and Wood Charcoal Results. In *The La Lomita Excavations: 10th Century Hohokam Occupation in South-Central Arizona,* edited by Douglas R. Mitchell, pp. 169–186. Soil Systems Publications in Archaeology No. 15. Phoenix.

1991   *Flotation and Macrobotanical Results from the 1990 Excavations at the Dutch Canal Ruin, Phoenix, Arizona.* Soil Systems Technical Report No. 91-10. Phoenix.

Landis, Daniel G.
1989   Chipped Stone Assemblages of the Grand Canal Ruins. In *Archaeological Investigations at the Grand Canal Ruins: A Classic Period Site in Phoenix, Arizona,* edited by Douglas R. Mitchell, pp. 385–442. Soil Systems Publications in Archaeology No. 12. Phoenix.

Lane, Anne Marie
1986   Hohokam Use of Stone at Casa Buena. Ms. on file, Soil Systems, Inc., Phoenix.

1989   The Grand Canal Ceramic Assemblage: Ceramic Type Descriptions. In *Archaeological Investigations at the Grand Canal Ruins: A Classic Period Site in Phoenix, Arizona,* edited by Douglas R. Mitchell, pp. 249–267. Soil Systems Publications in Archaeology No. 12. Phoenix.

Lewenstein, Suzanne
1986   Hohokam Use of Stone at Casa Buena. Ms. on file, Soil Systems, Inc., Phoenix.

Linares, Olga F.
1976   Garden Hunting in the American Tropics. *Human Ecology* 4:331-350.

Logan, Erik
1991   A Preliminary Analysis of the Occurrence of Pigment on the Groundstone from Homol'ovi III. Paper presented at the 62nd Annual Meeting of the Southwestern Anthropological Association, Tucson.

McQuestion, Kathleen S., and William R. Gibson
1987   West-Wing-Sunrise Mountain Archaeological Investigations. Ms. on file, U.S. Department of Interior, Bureau of Land Management, Phoenix District Office.

Martin, Alexander C., and William D. Barkley
1961   *Seed Identification Manual.* University of California Press, Berkeley.

Masse, W. Bruce (editor)
1980   *Excavations at Gu Achi: A Reappraisal of Hohokam Settlement and Subsistence in the Arizona Papagueria.* Western Archaeological Center Publications in Anthropology No. 12. Tucson.

1981   Prehistoric Irrigation Systems in the Salt River Valley, Arizona. *Science* 214:408–415.

1987   *Archaeological Investigations of Portions of the Las Acequias–Los Muertos Irrigation System*. Arizona State Museum Archaeological Series No. 176. The University of Arizona, Tucson.

1991   The Quest for Subsistence, Sufficiency, and Civilization in the Sonoran Desert. In *Chaco and Hohokam: Prehistoric Regional Systems in the American Southwest*, edited by Patricia L. Crown and W. James Judge, pp. 195–223. School of American Research Press, Santa Fe.

Midvale, Frank
1934   The Frank Midvale Collections. Ms. on file, Mesa Southwest Museum, Mesa, Arizona.

1945   Map of Prehistoric Canals and Sites of the Salt River Valley. On file, Department of Anthropology, Arizona State University, Tempe.

1968   Prehistoric Irrigation in the Salt River Valley. *The Kiva* 34:28–32.

n.d.   The Frank Midvale Collections. Ms. on file, Site Record Collections, Department of Anthropology, Arizona State University, Tempe.

Miksicek, Charles H.
1987   Formation Processes of the Archaeobotanical Record. In *Advances in Archaeological Method and Theory*, vol. 10, edited by Michael B. Schiffer, pp. 211–247. Academic Press, New York.

1988   Rethinking Hohokam Paleoethnobotanical Assemblages: A Progress Report for the Tucson Basin. In *Recent Research on Tucson Basin Prehistory: Proceedings of the Second Tucson Basin Conference*, edited by William H. Doelle and Paul R. Fish, pp. 47–56. Institute for American Research Anthropological Papers No. 10. Tucson.

1989   Snails, Seeds, and Charcoal: Biological Remains from La Cuenca del Sedimento. In *Prehistoric Agricultural Activities on the Lehi-Mesa Terrace: Excavations at La Cuenca del Sedimento*, edited by T. Kathleen Henderson, pp. 222–242. Northland Research, Inc., Flagstaff.

Miksicek, Charles H., and Robert E. Gasser
1989   Hohokam Plant Use at Las Colinas: The Flotation Evidence. In *The 1982–1984 Excavations at Las Colinas: Environment and Subsistence*, by Donald A. Graybill, David A. Gregory, Fred L. Nials, Suzanne K. Fish, Robert E. Gasser, Charles H. Miksicek, and Christine R. Szuter, pp. 95–115. Arizona State Museum Archaeological Series No. 162, Vol. 5. The University of Arizona, Tucson.

Minnis, Paul E.
1978   Paleoethnobotanical Indicators of Prehistoric Environmental Disturbance: A Case Study. In *The Nature and Status of Ethnobotany*, edited by Richard I. Ford, pp. 347–366. University of Michigan Anthropology Papers No. 67. Ann Arbor.

1981   Seeds in Archaeological Sites: Sources and Some Interpretive Problems. *American Antiquity* 46:143–152.

Mitchell, Douglas R.

1988   *La Lomita Pequeña: A Santa Cruz/Sacaton Phase Hamlet in the Salt River Valley.* Soil Systems Publications in Archaeology No. 10. Phoenix.

1989a   *Archaeological Investigations at the Grand Canal Ruins: A Classic Period Site in Phoenix, Arizona.* Soil Systems Publications in Archaeology No. 12. Phoenix.

1989b   La Lomita Pequeña: Relationships between Plant Resource Variability and Settlement Patterns in the Phoenix Basin. *Kiva* 54:127–146.

Muto, Guy Roger

1971   *A Technological Analysis of the Early Stages in the Manufacture of Lithic Artifacts.* Unpublished master's thesis, Department of Anthropology, Idaho State University, Pocatello.

Neitzel, Jill

1991   Hohokam Material Culture and Behavior: The Dimensions of Organizational Change. In *Exploring the Hohokam: Prehistoric Desert Peoples of the American Southwest,* edited by George J. Gumerman, pp. 177–230. University of New Mexico Press, Albuquerque.

Neusius, Phillip D.

1988   Functional Analysis of Selected Flakes Lithic Assemblages from the Dolores River Valley: A Low-Power Microwear Approach. In *Dolores Archaeological Program Supporting Studies: Additive and Reductive Technologies,* edited by Eric Blinman, Carl J. Phagan, and Richard H. Wilshusen, pp. 209–282. U.S. Department of the Interior, Washington, D.C.

Nials, Fred L., and David A. Gregory

1989   Irrigation Systems in the Lower Salt River Valley. In *The 1982–1984 Excavations at Las Colinas: Environment and Subsistence,* edited by Donald A. Graybill, David A. Gregory, Fred L. Nials, Suzanne K. Fish, Robert E. Gasser, Charles H. Miksicek, and Christine R. Szuter, pp. 39–58. Arizona State Museum Archaeological Series No. 162, Vol. 5. The University of Arizona, Tucson.

Nials, Fred L., David A. Gregory, and Donald A. Graybill

1989   Salt River Streamflow and Hohokam Irrigation Systems. In *The 1982–1984 Excavations at Las Colinas: Environment and Subsistence,* edited by Donald A. Graybill, David A. Gregory, Fred L. Nials, Suzanne K. Fish, Robert E. Gasser, Charles H. Miksicek, and Christine R. Szuter, pp. 59–76. Arizona State Museum Archaeological Series No. 162, Vol. 5. The University of Arizona, Tucson.

Odell, George Hamley, and Frieda Odell-Vereecken

1980   Verifying the Reliability of Lithic Use-Wear Assessments by "Blind Tests": The Low-Power Approach. *Journal of Field Archaeology* 7:87–120.

Osborne, Carolyn M.

1965   The Preparation of Yucca Fiber: An Experimental Study. In *Contributions of the Wetherill Mesa Archaeological Project,* assembled by Douglas Osborne, pp. 45–50. Memoirs of the Society for American Archaeology No. 19. Washington, D.C.

Parry, William J., and Robert L. Kelly
1987   Expedient Core Technology and Sedentism. In *The Organization of Core Technology*, edited by Jay K. Johnson and Carol A. Morrow, pp. 285–304. Westview Special Studies in Archaeological Research. Westview Press, Boulder, Colorado.

Patrick, H. R.
1903   *The Ancient Canal Systems and Pueblos of the Salt River Valley*. Phoenix Free Museum Bulletin No. 1. Phoenix.

Patterson, L. W.
1982   The Importance of Flake Size Distribution. *Contract Abstracts and Cultural Resource Management Archaeology* 3(1):70–72.

Peterson, Jane D., and David R. Abbott
1991   Indices of Salado Polychrome Ceramic Production, the Evidence from Pueblo Grande, Arizona. Paper presented at the 56th Annual Meeting of the Society for American Archaeology, New Orleans.

Péwé, Troy L.
1978   Terraces of the Lower Salt River Valley in Relation to the Late Cenozoic History of the Phoenix Basin, Arizona. In *Guidebook to the Geology of Central Arizona*, edited by Donald M. Burt and Troy L. Péwé, pp. 1–45. Arizona Bureau of Geology and Mineral Technology Special Paper No. 2. Tucson.

Phagan, Carl
1976   Technology: Flake Analysis. In *Prehistory of the Ayacucho Basin, Peru*, vol. 3, edited by Richard S. MacNeish, Robert K. Vierra, Antoinette Nelkin-Terner, and Carl J. Phagan, pp. 233–281. University of Michigan Press, Ann Arbor.

Plog, Fred
1980   Explaining Culture Change in the Hohokam Preclassic. In *Current Issues in Hohokam Prehistory*, edited by David E. Doyel and Fred Plog, pp. 4–22. Arizona State University Anthropological Research Papers No. 23. Tempe.

Proper, Earl
1990   The Chipped Stone Assemblage. In *Hohokam Utilization of the Intermontane Mazatzal Region: The State Route 87 Pine Creek Project*, edited by Margerie Green, pp. 289–340. Archaeological Consulting Services, Ltd., Cultural Resources Report No. 66. Tempe.

Raab, Mark L.
1976   *The Structure of Prehistoric Community Organization at Santa Rosa Wash, Southern Arizona*. Unpublished Ph.D. dissertation, Arizona State University, Tempe.

Rankin, Adrianne G., and Keith L. Katzer
1989   Agricultural Systems in the ACS Waddell Project Area. In *Settlement, Subsistence, and Specialization in the Northern Periphery: The Waddell Project*, edited by Margerie Green, pp. 981–1020. Archaeological Consulting Services, Ltd., Cultural Resources Report No. 65. Tempe.

Robbins, Chandler S., Bertel Bruun, and Herbert S. Zim
1966   *Birds of North America*. Western Publishing Company, Racine, Wisconsin.

Robbins, W. W., J. P. Harrington, and Barbara Freire-Marreco
1916   *Ethnobotany of the Tewa Indians*. Bureau of American Ethnology Bulletin No. 55. Washington, D.C.

Rodgers, James B.
1985   Prehistoric Agricultural Variability in the Hohokam Northern Periphery. In *Hohokam Settlement and Economic Systems in the Central New River Drainage, Arizona*, edited by David E. Doyel and Mark D. Elson, pp. 249–296. Soil Systems Publications in Archaeology No. 4. Phoenix.

Rodgers, James B., and David H. Greenwald
1988   Historic Resources. In *Archaeological Investigations at the Dutch Canal Ruin, Phoenix, Arizona: Archaeology and History along the Papago Freeway Corridor*, edited by David H. Greenwald and Richard Ciolek-Torrello, pp. 24–53. Museum of Northern Arizona Research Paper No. 38. Flagstaff.

Rogers, Malcolm J.
1945   An Outline of Yuman Prehistory. *Southwestern Journal of Archaeology* 1(2):167–198.

Rossman, David L.
1976   A Site Catchment Analysis of San Lorenzo, Veracruz. In *The Early Mesoamerican Village*, edited by Kent V. Flannery, pp. 95–103. Academic Press, New York.

Ruppé, Patricia A.
1988   Macrobotanical Analysis. In *Archaeological Investigations at the Dutch Canal Ruin, Phoenix, Arizona*, edited by David H. Greenwald and Richard Ciolek-Torrello, pp. 168–180. Museum of Northern Arizona Research Paper No. 38. Flagstaff.

Russell, Frank
1908   *The Pima Indians*. Twenty-sixth Annual Report of the Bureau of American Ethnology, 1904–1905. Smithsonian Institution, Washington, D.C.

1975   *The Pima Indians*. Reprinted. The University of Arizona Press, Tucson. Originally published 1908, Bureau of American Ethnology, Washington, D.C.

Russell, Scott C.
1978   The Agricultural Fieldhouse: A Navajo Limited Occupation and Special Use Site. In *Limited Activity and Occupation Sites: A Collection of Conference Papers*, edited by Albert E. Ward, pp. 35–40. Center for Anthropological Studies Contributions to Anthropological Studies No. 1. Albuquerque.

Schaller, David M.
1985   Petrographic Analysis of Ground Stone Artifacts. In *Hohokam Settlement and Economic Systems in the Central New River Drainage, Arizona*, edited by David E. Doyel and Mark D. Elson, pp. 779–781. Soil Systems Publications in Archaeology No. 4, Vol. 2. Phoenix.

1987   Petrographic Analysis of Lithic Artifacts. In *Excavations at La Lomita Pequeña: A Santa Cruz/Sacaton Phase Hamlet in the Salt River Valley*, edited by Douglas R. Mitchell, pp. C.1–C.4. Soil Systems Publications in Archaeology No. 10. Phoenix.

1988 Mineralogy and Petrology of Ground Stone Artifacts. In *Settlement, Subsistence, and Specialization in the Northern Periphery: The Waddell Project*, edited by Margerie Green, pp. 754–759. Archaeological Consulting Services, Ltd., Cultural Resources Report No. 65. Tempe.

Schiffer, M.
1982 Hohokam Chronology: An Essay on History and Method. In *Hohokam and Patayan: Prehistory of Southwestern Arizona*, edited by Randall M. McGuire and Michael B. Schiffer, pp. 299–344. Academic Press, New York.

1987 Formation Processes and Archaeological Inference: Hohokam Chronology. In *Formation Processes of the Archaeological Record*, by Michael B. Schiffer, pp. 305–322. University of New Mexico Press, Albuquerque.

Semé, Michelle
1984 The Effects of Agricultural Fields on Faunal Assemblage Variation. In *Papers on the Archaeology of Black Mesa, Arizona*, vol. 2, edited by Stephen Plog and Shirley Powell, pp. 139–157. Southern Illinois University Press, Carbondale and Edwardsville.

Shackley, M. Steven
1988 Sources of Archaeological Obsidian in the Southwest: An Archaeological, Petrological, and Geochemical Study. *American Antiquity* 53:752–772.

Shantz, H. L., and R. L. Piemeisel
1924 Indicator Significance of the Natural Vegetation of the Southwestern Desert Regions. *Journal of Agricultural Research* 28:721–801.

Sheehy, James J.
1990 Agave as a Fuel Resource for Middle Horizon Teotihuacán. Paper presented at the 89th Annual Meeting of the American Anthropological Association, New Orleans.

Sires, Earl W., Jr.
1984 Excavations at El Polvorón. In *Hohokam Archaeology along the Salt-Gila Aqueduct Central Arizona Project: Prehistoric Occupation of the Queen Creek Delta*, edited by Lynn S. Teague and Patricia L. Crown, pp. 221–354. Arizona State Museum Archaeological Series No. 150, Vol. 4. The University of Arizona, Tucson.

Smith, Anne M.
1974 *Ethnography of the Northern Utes*. Papers in Anthropology No. 17. Museum of New Mexico Press, Albuquerque.

Soil Systems, Inc.
1993 The Pueblo Grande Project. Soil Systems Publication in Archaeology, Phoenix. Draft.

Spears, Carol S.
1975 Hammers, Nuts and Jolts, Cobbles, Cobbles, Cobbles: Experiments in Cobble Technologies in Search of Correlates. In *The Arkansas Eastman Archaeological Project*, edited by Charles M. Baker, pp. 83–108. Arkansas Archaeological Survey Research Report No. 6. Little Rock.

Stevenson, Matilda Coxe
1904   *The Zuni Indians.* Twenty-third Annual Report of the Bureau of American Ethnology, 1901–1902. Washington, D.C.

1915   *Ethnobotany of the Zuni Indians.* Thirtieth Annual Report of the Bureau of American Ethnology. Washington, D.C.

Steward, Julian H.
1938   *Basin-Plateau Aboriginal Sociopolitical Groups.* Bureau of American Ethnology Bulletin No. 120. Washington, D.C.

Stuiver, Minze, and Bernd Becker
1986   High-Precision Decadal Calibration of the Radiocarbon Time Scale, AD 1950–2500 BC. *Radiocarbon* 28(2B):863–910.

Stuiver, Minze, and Paula J. Reimer
1986   A Computer Program for Radiocarbon Age Calibration. *Radiocarbon* 28(2B):1022–1030.

Sullivan, Alan P., and Kenneth C. Rozen
1985   Debitage Analysis and Archaeological Interpretation. *American Antiquity* 50:755–779.

Swidler, Nina B.
1989   Variability in the Chipped Stone Assemblage from La Cuenca del Sedimento. In *Prehistoric Agricultural Activities on the Lehi-Mesa Terrace: Excavations at La Cuenca del Sedimento,* edited by T. Kathleen Henderson, pp. 154–189. Northland Research, Inc., Flagstaff.

Szuter, Christine Rose
1989   *Hunting by Prehistoric Horticulturalists in the American Southwest.* Ph.D. dissertation, The University of Arizona. University Microfilms International, Ann Arbor, Michigan.

1991   Hunting by Hohokam Desert Farmers. *Kiva* 56:277–291.

Teague, George A.
1981   The Nonflaked Stone Artifacts from Las Colinas. In *The 1968 Excavations at Mound 8, Las Colinas Ruins Group, Phoenix, Arizona,* edited by Laurens C. Hammack and Alan P. Sullivan, pp. 201–247. Arizona State Museum Archaeological Series No. 154. The University of Arizona, Tucson.

Turner, Raymond M.
1974   Map Showing Vegetation in the Phoenix Area, Arizona. In *Folio of the Phoenix Area,* Map I-845-I. U.S. Geological Survey, Denver.

Turner, Raymond M., and David E. Brown
1982   Sonoran Desertscrub. *Desert Plants* 4:181–222.

Turney, Omar
1924   *Land of the Stone Hoe.* Arizona Republican Print Shop, Phoenix.

1929   *Prehistoric Irrigation.* Arizona Historical Review No. 2, Vol. 5. Phoenix.

Upham, Steadman, and Glen E. Rice
  1980   Up the Canal without a Pattern: Modeling Hohokam Interaction and Exchange. In *Current Issues in Hohokam Prehistory*, edited by David E. Doyel and Fred Plog, pp. 78–105. Arizona State University Anthropological Research Papers No. 23. Tempe.

U.S. Department of Agriculture, Soil Conservation Service
  1934   Aerial Survey Maps. On file in the Cartographic Division, Salt River Project, Phoenix.

U.S. General Land Office
  1870a   *Township No. 1 North, Range No. 2 East, Gila and Salt River Meridian*. Survey plat on file, U.S. Bureau of Land Management, Phoenix.

  1870b   *Township No. 1 North, Range No. 3 East, Gila and Salt River Meridian*. Survey plat on file, U.S. Bureau of Land Management, Phoenix.

Vita-Finzi, Claudio, and Eric S. Higgs
  1970   Prehistoric Economy in the Mount Carmel Area of Palestine: Site Catchment Analysis. *Proceedings of the Prehistoric Society* 36:1–37.

Vogt, Evon Z., and Richard M. Leventhal (editors)
  1983   *Prehistoric Settlement Patterns: Essays in Honor of Gordon R. Willey*. University of New Mexico Press, Albuquerque.

Vokes, Arthur W.
  1984   The Shell Assemblage of the Salt-Gila Aqueduct Project Sites. In *Hohokam Archaeology along the Salt-Gila Aqueduct, Central Arizona Project: Material Culture*, edited by Lynn S. Teague and Patricia L. Crown, pp. 465–574. Arizona State Museum Archaeological Series No. 150, Vol. 8. The University of Arizona, Tucson.

  1988   Shell Artifacts. In *The 1982–1984 Excavations at Las Colinas: Material Culture*, by David R. Abbott, Kim E. Beckwith, Patricia L. Crown, R. Thomas Euler, David A. Gregory, J. Ronald London, Marilyn B. Saul, Larry A. Schwalbe, Mary Bernard-Shaw, Christine R. Szuter, and Arthur W. Vokes, pp. 319–384. Arizona State Museum Archaeological Series No. 162, Vol. 4. The University of Arizona, Tucson.

Walker, William
  1991   Under the Weather: A Systematic Study of Weathering on Groundstone Artifacts Recovered from Survey. Paper presented at the 62nd Annual Meeting of the Southwestern Anthropological Association, Tucson.

Ward, Albert E. (editor)
  1978   *Limited Activity and Occupation Sites: A Collection of Conference Papers*. Center for Anthropological Studies Contributions to Anthropological Studies No. 1. Albuquerque.

Waters, Michael R.
  1982a   Hohokam Chronology: An Essay on History and Method. In *Hohokam and Patayan: Prehistory of Southwestern Arizona*, edited by Randall H. McGuire and Michael B. Schiffer, pp. 299–344. Academic Press, New York.

1982b The Lowland Patayan Ceramic Typology. In *Hohokam and Patayan: Prehistory of Southwestern Arizona*, edited by Randall H. McGuire and Michael B. Schiffer, pp. 275–298. Academic Press, New York.

Waters, Michael R., Jerry Howard, and David H. Greenwald
1988   Radiometric Dating of Prehistoric Canal Features: Progress and Problems. Hohokam Canal Symposium, Fall Meeting of the Arizona Archaeological Council, Pueblo Grande, Phoenix.

Weaver, Donald E., Jr.
1972   A Cultural-Ecological Model for the Classic Hohokam Period in the Lower Salt River Valley, Arizona. *The Kiva* 38:57–94.

1977   Investigations Concerning the Hohokam Classic Period in the Lower Salt River Valley, Arizona. *The Arizona Archaeologist* 9:69–70.

West, M.
1980   Prehistoric Resource Exploitation in the Viru Valley, Peru. In *Catchment Analysis–Essays on Prehistoric Resource Space*, edited by Frank Findlow and J. Ericson, pp. 137–156. Department of Anthropology, University of California, Los Angeles.

Whiting, Alfred F.
1939   *Ethnobotany of the Hopi*. Museum of Northern Arizona Bulletin No. 15. Flagstaff.

Wilcox, David R.
1978   The Theoretical Significance of Field Houses. In *Limited Activity and Occupation Sites*, edited by Albert E. Ward, pp. 25–34. Center for Anthropological Studies Contributions to Anthropological Studies No. 1. Albuquerque.

1979   The Hohokam Regional System. In *An Archaeological Test of Sites in the Gila Butte–Santan Region, South-Central Arizona*, edited by Glen Rice, David R. Wilcox, Kevin Rafferty, and James Schoenwetter, pp. 77–116. Arizona State University Anthropological Research Papers No. 18. Tempe.

1987   *The Frank Midvale Investigations of the Site of La Ciudad*. Office of Cultural Resource Management Anthropological Field Studies No. 19. Arizona State University, Tempe.

1991   Hohokam Social Complexity. In *Chaco and Hohokam: Prehistoric Regional Systems in the American Southwest*, edited by Patricia L. Crown and W. James Judge, pp. 253–275. School of American Research Press, Santa Fe.

Wilcox, David R., and Lynette O. Shenk
1977   *The Architecture of the Casa Grande and Its Interpretation*. Arizona State Museum Archaeological Series No. 115. The University of Arizona, Tucson.

Wing, E. S., and A. B. Brown
1979   *Paleonutrition: Methods and Theory in Prehistoric Foodways*. Academic Press, New York.

Wood, J. Scott
1987   *Checklist of Pottery Types for the Tonto National Forest*. The Arizona Archaeologist No. 21. Arizona Archaeological Society, Phoenix.

Yanovsky, E.
    1936   *Food Plants of the North American Indians*. U.S. Department of Agriculture
            Miscellaneous Publication No. 237:1–83. Washington, D.C.

Zarbin, Earl
    1979   *The Swilling Legacy*. Salt River Project, Phoenix.

    1980   Salt River Valley Canals: 1867–1875. Ms. on file, Salt River Project, Phoenix.